Methods in Enzymology

Volume 313
ANTISENSE TECHNOLOGY
Part A
General Methods, Methods of Delivery,
and RNA Studies

METHODS IN ENZYMOLOGY

EDITORS-IN-CHIEF

John N. Abelson   Melvin I. Simon

DIVISION OF BIOLOGY
CALIFORNIA INSTITUTE OF TECHNOLOGY
PASADENA, CALIFORNIA

FOUNDING EDITORS

Sidney P. Colowick and Nathan O. Kaplan

Methods in Enzymology

Volume 313

Antisense Technology

Part A
General Methods, Methods of Delivery,
and RNA Studies

EDITED BY

M. Ian Phillips

UNIVERSITY OF FLORIDA COLLEGE OF MEDICINE
GAINESVILLE, FLORIDA

ACADEMIC PRESS

San Diego London Boston New York Sydney Tokyo Toronto

Academic Press
A Harcourt Science and Technology Company
525 B Street, Suite 1900, San Diego, California 92101-4495, USA
http://www.academicpress.com

Academic Press Limited
24-28 Oval Road, London NW1 7DX, UK
http://www.hbuk.co.uk/ap/

International Standard Book Number: 0-12-182214-1

PRINTED IN THE UNITED STATES OF AMERICA
99 00 01 02 03 04 MM 9 8 7 6 5 4 3 2 1

Table of Contents

Section I. General Methods

v

Section II. Methods of Delivery

Section III. RNA Studies

Contributors to Volume 313

Article numbers are in parentheses following the names of contributors.
Affiliations listed are current.

SURESH ALAHARI (19), *Department of Pharmacology, School of Medicine, University of North Carolina, Chapel Hill, North Carolina 27599*

SIDNEY ALTMAN (26), *Department of Molecular, Cellular and Developmental Biology, Yale University, New Haven, Connecticut 06520*

ANNA ASTRIAB (19), *Department of Pharmacology, School of Medicine, University of North Carolina, Chapel Hill, North Carolina 27599*

DAVID BELLIDO (14), *Unitat de Biologia Cellular, Departament de Bioquímica i Fisiologia, Universitat de Barcelona, E-08028 Barcelona, Spain*

LYUBA BENIMETSKAYA (16), *Columbia University, New York, New York 10032*

ECKHART BUDDECKE (15), *Division of Molecular Cardiology, Institute for Arteriosclerosis Research, University of Münster, D-48149 Münster, Germany*

JEFFREY S. BUZBY (22), *Hematology Research Laboratory, Children's Hospital of Orange County, Orange, California 92868*

ALAN CARLETON (7), *Institut Alfred Fessard, CNRS, 91198 Gif-sur-Yvette Cedex, France*

DANIELA CASTANOTTO (23), *Department of Molecular Biology, Beckman Research Institute of the City of Hope, Duarte, California 91010*

DOUGLAS L. COLE (12), *Manufacturing Process Department, ISIS Pharmaceuticals, Inc., Carlsbad, California 92008*

STANLEY T. CROOKE (1), *ISIS Pharmaceuticals, Inc., Carlsbad, California 92008*

JOHN M. DAGLE (24), *Department of Pediatrics, University of Iowa, Iowa City, Iowa 52242*

CHARLOTTE DARRAH (29), *Department of Human Anatomy and Genetics, Oxford University, Oxford O51 3QX, United Kingdom*

SCOTT F. DEAMOND (17), *Department of Biochemistry, The Johns Hopkins School of Hygiene and Public Health, Baltimore, Maryland 21225*

RANJIT R. DESHMUKH (12), *Manufacturing Process Department, ISIS Pharmaceuticals, Inc., Carlsbad, California 92008*

DAVID DESMAISONS (7), *Institut Alfred Fessard, CNRS, 91198 Gif-sur-Yvette Cedex, France*

SONIA DHEUR (3), *Laboratoire de Biophysique, INSERM U201, CNRS URA481, Muséum National d'Histoire Naturelle, 75005 Paris, France*

BEIHUA DONG (31), *Department of Cancer, Lerner Research Institute, The Cleveland Clinic Foundation, Cleveland, Ohio 44195*

ROBERT J. DUFF (17), *Department of Biochemistry, The Johns Hopkins School of Hygiene and Public Health, Baltimore, Maryland 21225*

GEORGE L. ELICEIRI (25), *Department of Pathology, Saint Louis University School of Medicine, St. Louis, Missouri 63104-1028*

RAMON ERITJA (14), *European Molecular Biology Laboratory, D-69012 Heidelberg, Germany*

LOUISE EVERATT (29), *Department of Human Anatomy and Genetics, Oxford University, Oxford OS1 3QX, United Kingdom*

JEAN-CHRISTOPHE FRANÇOIS (4), *Laboratoire de Biophysique, INSERM U201, CNRS UMR8646, Muséum National d'Histoire Naturelle, 75005 Paris, France*

CHANDRAMALLIKA GHOSH (6), *AVI BioPharma, Inc., Corvallis, Oregon 97333*

RICHARD V. GILES (5), *Department of Haematology, The University of Liverpool, Royal Liverpool University Hospital, Liverpool L7 8XP, United Kingdom*

LINDA GORMAN (30), *Lineberger Cancer Center, University of North Carolina, Chapel Hill, North Carolina 27599*

VLADIMIR V. GORN (9), *Epoch Pharmaceuticals, Inc., Redmond, Washington 98052*

CECILIA GUERRIER-TAKADA (26), *Department of Molecular, Cellular and Developmental Biology, Yale University, New Haven, Connecticut 06520*

TROY O. HARASYM (18), *Inex Pharmaceuticals Corporation, Burnaby, British Columbia, Canada V5J 5J8*

KAIZHANG HE (13), *Department of Chemistry, Duke University, Durham, North Carolina 27708-0346*

MICHAEL J. HOPE (18), *Inex Pharmaceuticals Corporation, Burnaby, British Columbia, Canada V5J 5J8*

JEFF HUGHES (19), *Department of Pharmaceutics, University of Florida, Gainesville, Florida 32610*

MASAYORI INOUYE (28), *Department of Biochemistry, Robert Wood Johnson Medical School, Piscataway, New Jersey 08854*

PATRICK IVERSEN (6), *AVI BioPharma, Inc., Corvallis, Oregon 97333*

EMMA R. JAKOI (27), *Department of Physiology, Medical College of Virginia/Virginia Commonwealth University, Richmond, Virginia 23298*

R. L. JULIANO (19), *Department of Pharmacology, School of Medicine, University of North Carolina, Chapel Hill, North Carolina 27599*

SHIN-HONG KANG (30), *Lineberger Cancer Center, University of North Carolina, Chapel Hill, North Carolina 27599*

HARUKO KATAYAMA (20), *Department of Neurosurgery, Teikyo University Ichihara Hospital, Ichihara City, Chiba 299-0111, Japan*

MICHAEL W. KILPATRICK (29), *Department of Pediatrics, University of Connecticut Health Center, Farmington, Connecticut 06030*

SANDRA K. KLIMUK (18), *Department of Biochemistry and Molecular Biology, The University of British Columbia, Vancouver, British Columbia, Canada V6T 1Z3*

RYSZARD KOLE (30), *Lineberger Cancer Center and Department of Pharmacology, University of North Carolina, Chapel Hill, North Carolina 27599*

IGOR KUTYAVIN (9), *Epoch Pharmaceuticals, Inc., Redmond, Washington 98052*

JÉROME LACOSTE (4), *Plasticité et expression des génomes microbiens, CNRS EP2029, CEA LRC12, CERMO, Université Joseph Fourier, 38041 Grenoble, France*

LAURENT LACROIX (4), *Laboratoire de Biophysique, INSERM U201, CNRS UMR8646, Muséum National d'Histoire Naturelle, 75005 Paris, France*

BERNARD LEBLEU (11), *Institut de Génétique Moléculaire de Montpellier, UMR5535, CNRS, F-34293 Montpellier, France*

EARVIN LIANG (19), *Department of Pharmaceutics, University of Florida, Gainesville, Florida 32610*

PIERRE-MARIE LLEDO (7), *Institut Alfred Fessard, CNRS, 91198 Gif-sur-Yvette Cedex, France*

EUGENE A. LUKHTANOV (9), *Epoch Pharmaceuticals, Inc., Redmond, Washington 98052*

RATAN K. MAITRA (31), *HIV Core Facility, Lerner Research Institute, The Cleveland Clinic Foundation, Cleveland, Ohio 44195*

DESPINA MANIOTIS (29), *Department of Human Anatomy and Genetics, Oxford University, Oxford OS1 3QX, United Kingdom*

AKIRA MATSUNO (20), *Department of Neurosurgery, Teikyo University Ichihara Hospital, Ichihara City, Chiba 299-0111, Japan*

JEAN-LOUIS MERGNY (4), *Laboratoire de Biophysique, INSERM U201, CNRS UMR8646, Muséum National d'Histoire Naturelle, 75005 Paris, France*

DAVID MILESI (9), *Epoch Pharmaceuticals, Inc., Redmond, Washington 98052*

OLEG MIROCHNITCHENKO (28), *Department of Biochemistry, Robert Wood Johnson Medical School, Piscataway, New Jersey 08854*

PAUL A. MORCOS (10), *Gene Tools, LLC, Corvallis, Oregon 97333*

TADASHI NAGASHIMA (20), *Department of Neurosurgery, Teikyo University Ichihara Hospital, Ichihara City, Chiba 299-0111, Japan*

PETER E. NIELSEN (8), *Department of Medical Biochemistry and Genetics, The Panum Institute, University of Copenhagen, DK-2200 Copenhagen N, Denmark*

ANUSCH PEYMAN (15), *Chemical Research G 838, Hoechst Marion Roussel Deutschland GmbH, D-65926 Frankfurt am Main, Germany*

M. IAN PHILLIPS (2), *Department of Physiology, University of Florida College of Medicine, Gainesville, Florida 32610*

LEONIDAS A. PHYLACTOU (29), *Cyprus Institute of Neurology and Genetics, 1683 Nicosia, Cyprus*

JAUME PIULATS (14), *Laboratorio de Bioinvestigación, Merck Farma y Química, S.A., E-08010 Barcelona, Spain*

MARK R. PLAYER (31), *3-Dimensional Pharmaceutical, Inc., Exton, Pennsylvania 19341*

KEN PORTER (13), *Department of Chemistry, Duke University, Durham, North Carolina 27708-0346*

VLADIMIR RAIT (13), *Department of Chemistry, Duke University, Durham, North Carolina 27708-0346*

MICHAEL W. REED (9), *Epoch Pharmaceuticals, Inc., Redmond, Washington 98052*

IAN ROBBINS (11), *Institut de Génétique Moléculaire de Montpellier, UMR5535, CNRS, F-34293 Montpellier, France*

CLINTON ROBY (17), *Department of Biochemistry, The Johns Hopkins School of Hygiene and Public Health, Baltimore, Maryland 21225*

JOHN J. ROSSI (23), *Department of Molecular Biology, Beckman Research Institute of the City of Hope, Duarte, California 91010*

ANTONINA RYTE (15), *Lombardi Cancer Center, Georgetown University Medical Center, Washington, DC 20007-2197*

E. TULA SAISON-BEHMOARAS (3), *Laboratoire de Biophysique, INSERM U201, CNRS URA481, Muséum National d'Histoire Naturelle, 75005 Paris, France*

YOGESH S. SANGHVI (12), *Manufacturing Process Department, ISIS Pharmaceuticals, Inc., Carlsbad, California 92008*

MICHAELA SCHERR (23), *Abteilung Haematologie und Onkologie, Medizinische Hochschule Hannover, D-30625 Hannover, Germany*

ANNETTE SCHMIDT (15), *Division of Molecular Cardiology, Institute for Arteriosclerosis Research, University of Münster, D-48149 Münster, Germany*

SEAN C. SEMPLE (18), *Inex Pharmaceuticals Corporation, Burnaby, British Columbia, Canada V5J 5J8*

DMITRI SERGUEEV (13, 19), *Department of Chemistry, Duke University, Durham, North Carolina 27708-0346*

ZINAIDA SERGUEEVA (13), *Department of Chemistry, Duke University, Durham, North Carolina 27708-0346*

W. L. SEVERT (27), *Department of Physiology, Medical College of Virginia/Virginia Commonwealth University, Richmond, Virginia 23298*

BARBARA RAMSAY SHAW (13, 19), *Department of Chemistry, Duke University, Durham, North Carolina 27708-0346*

HALINA SIERAKOWSKA (30), *Lineberger Cancer Center, University of North Carolina, Chapel Hill, North Carolina 27599*

ROBERT H. SILVERMAN (31), *Department of Cancer, Lerner Research Institute, The Cleveland Clinic Foundation, Cleveland, Ohio 44195*

DAVID G. SPILLER (5), *School of Biological Sciences, The University of Liverpool, Liverpool L69 7ZB, United Kingdom*

C. A. STEIN (16), *Columbia University, New York, New York 10032*

DAVID STEIN (6), *AVI BioPharma, Inc., Corvallis, Oregon 97333*

JACK SUMMERS (13), *Department of Chemistry, Duke University, Durham, North Carolina 27708-0346*

AKIRA TAMURA (20), *Department of Neurosurgery, Tokyo University Hospital, Itabashi-ku, Tokyo 173-0003, Japan*

ANA M. TARI (21), *Department of Bioimmunotherapy, University of Texas MD Anderson Cancer Center, Houston, Texas 77030*

GEMMA TARRASÓN (14), *Laboratorio de Bioinvestigación, Merck Farma y Química, S.A., E-08010 Barcelona, Spain*

DAVID M. TIDD (5), *School of Biological Sciences, The University of Liverpool, Liverpool L69 7ZB, United Kingdom*

JOHN TONKINSON (16), *Columbia University, New York, New York 10032*

PAUL F. TORRENCE (31), *Section on Biomedical Chemistry, Laboratory of Medicinal Chemistry, National Institute of Diabetes and Digestive and Kidney Diseases, National Institutes of Health, Bethesda, Maryland 20892-0805*

PAUL O. P. TS'O (17), *Department of Biochemistry, The Johns Hopkins School of Hygiene and Public Health, Baltimore, Maryland 21225*

EUGEN UHLMANN (15), *Chemical Research G 838, Hoechst Marion Roussel Deutschland GmbH, D-65926 Frankfurt am Main, Germany*

SENÉN VILARÓ (14), *Unitat de Biologia Cellular, Departament de Bioquímica i Fisiologia, Universitat de Barcelona, E-08028 Barcelona, Spain*

JEAN-DIDIER VINCENT (7), *Institut Alfred Fessard, CNRS, 91198 Gif-sur-Yvette Cedex, France*

DANIEL L. WEEKS (24), *Department of Biochemistry, University of Iowa, Iowa City, Iowa 52242*

DWIGHT WELLER (6), *AVI BioPharma, Inc., Corvallis, Oregon 97333*

SHIRLEY A. WILLIAMS (22), *Hematology Research Laboratory, Children's Hospital of Orange County, Orange, California 92868*

HOON YOO (19), *Department of Pharmacology, School of Medicine, University of North Carolina, Chapel Hill, North Carolina 27599*

Y. CLARE ZHANG (2), *Department of Physiology, University of Florida College of Medicine, Gainesville, Florida 32610*

YUANZHONG ZHOU (17), *Cell Works, Inc., Baltimore, Maryland 21227*

Preface

Antisense technology reached a watershed year in 1998 with the FDA approval of the antisense-based therapy, Vitravene, developed by ISIS. This is the first drug based on antisense technology to enter the marketplace and makes antisense technology a reality for therapeutic applications. However, antisense technology still needs further development, and new applications need to be explored.

Contained in this Volume 313 (Part A) of *Methods in Enzymology* and its companion Volume 314 (Part B) are a wide range of methods and applications of antisense technology in current use. We set out to put together a single volume, but it became obvious that the variations in methods and the numerous applications required at least two volumes, and even these do not, by any means, cover the entire field. Nevertheless, the articles included represent the work of active research groups in industry and academia who have developed their own methods and techniques. This volume, Part A: General Methods, Methods of Delivery, and RNA Studies, includes several methods of antisense design and construction, general methods of delivery, and antisense used in RNA studies. In Part B: Applications, chapters cover methods in which antisense is designed to target membrane receptors and antisense application in the neurosciences, as well as in nonneuronal tissues. The therapeutic applications of antisense technology, the latest area of new interest, complete the volume.

Although *Methods in Enzymology* is designed to emphasize methods, rather than achievements, I congratulate all the authors on their achievements that have led them to make their methods available. In compiling and editing these two volumes I could not have made much progress without the excellent secretarial services of Ms. Gayle Butters of the University of Florida, Department of Physiology.

M. IAN PHILLIPS

METHODS IN ENZYMOLOGY

VOLUME XVII. Metabolism of Amino Acids and Amines (Parts A and B)
Edited by HERBERT TABOR AND CELIA WHITE TABOR

VOLUME XVIII. Vitamins and Coenzymes (Parts A, B, and C)
Edited by DONALD B. MCCORMICK AND LEMUEL D. WRIGHT

VOLUME XIX. Proteolytic Enzymes
Edited by GERTRUDE E. PERLMANN AND LASZLO LORAND

VOLUME XX. Nucleic Acids and Protein Synthesis (Part C)
Edited by KIVIE MOLDAVE AND LAWRENCE GROSSMAN

VOLUME XXI. Nucleic Acids (Part D)
Edited by LAWRENCE GROSSMAN AND KIVIE MOLDAVE

VOLUME XXII. Enzyme Purification and Related Techniques
Edited by WILLIAM B. JAKOBY

VOLUME XXIII. Photosynthesis (Part A)
Edited by ANTHONY SAN PIETRO

VOLUME XXIV. Photosynthesis and Nitrogen Fixation (Part B)
Edited by ANTHONY SAN PIETRO

VOLUME XXV. Enzyme Structure (Part B)
Edited by C. H. W. HIRS AND SERGE N. TIMASHEFF

VOLUME XXVI. Enzyme Structure (Part C)
Edited by C. H. W. HIRS AND SERGE N. TIMASHEFF

VOLUME XXVII. Enzyme Structure (Part D)
Edited by C. H. W. HIRS AND SERGE N. TIMASHEFF

VOLUME XXVIII. Complex Carbohydrates (Part B)
Edited by VICTOR GINSBURG

VOLUME XXIX. Nucleic Acids and Protein Synthesis (Part E)
Edited by LAWRENCE GROSSMAN AND KIVIE MOLDAVE

VOLUME XXX. Nucleic Acids and Protein Synthesis (Part F)
Edited by KIVIE MOLDAVE AND LAWRENCE GROSSMAN

VOLUME XXXI. Biomembranes (Part A)
Edited by SIDNEY FLEISCHER AND LESTER PACKER

VOLUME XXXII. Biomembranes (Part B)
Edited by SIDNEY FLEISCHER AND LESTER PACKER

VOLUME XXXIII. Cumulative Subject Index Volumes I–XXX
Edited by MARTHA G. DENNIS AND EDWARD A. DENNIS

VOLUME XXXIV. Affinity Techniques (Enzyme Purification: Part B)
Edited by WILLIAM B. JAKOBY AND MEIR WILCHEK

VOLUME XXXV. Lipids (Part B)
Edited by JOHN M. LOWENSTEIN

Section I

General Methods

[1] Progress in Antisense Technology: The End of the Beginning

By STANLEY T. CROOKE

Introduction

During the past decade, intense efforts to develop and exploit antisense technology have been mounted. With the recent FDA approval of Vitravene, the first drug based on antisense technology to be commercialized, the technology has achieved an important milestone. Nevertheless, the technology is still in its infancy. Although the basic questions have been answered, there are still many more unanswered than answered questions.

The objectives of this article are to provide an overview of the progress in converting the antisense concept into broad therapeutic reality and to provide advice about appropriate experimental design and interpretation of data with regard to the therapeutic potential of the technology.

Proof of Mechanism

Factors That May Influence Experimental Interpretations

Clearly, the ultimate biological effect of an oligonucleotide will be influenced by the local concentration of the oligonucleotide at the target RNA, the concentration of the RNA, the rates of synthesis and degradation of the RNA, the type of terminating mechanism, and the rates of the events that result in termination of the activity of RNA. At present, we understand essentially nothing about the interplay of these factors.

Oligonucleotide Purity. Currently, phosphorothioate oligonucleotides can be prepared consistently and with excellent purity.[1] However, this has only been the case since the mid-1990s. Prior to that time, synthetic methods were evolving and analytical methods were inadequate. In fact, our laboratory reported that different synthetic and purification procedures resulted in oligonucleotides that varied in cellular toxicity[2] and that potency varied from batch to batch. Although there are no longer synthetic problems with phosphorothioates, they undoubtedly complicated earlier studies. More

---

[1] S. T. Crooke and C. K. Mirabelli, "Antisense Resarch and Applications." CRC Press, Boca Raton, FL, 1993.

[2] R. M. Crooke, *Anti-Cancer Drug Design* **6,** 609 (1991).

importantly, with each new analog class, new synthetic, purification, and analytical challenges are encountered.

Oligonucleotide Structure. Antisense oligonucleotides are designed to be single stranded. We now understand that certain sequences, e.g., stretches of guanosine residues, are prone to adopt more complex structures.[3] The potential to form secondary and tertiary structures also varies as a function of the chemical class. For example, higher affinity 2′-modified oligonucleotides have a greater tendency to self-hybridize, resulting in more stable oligonucleotide duplexes than would be expected based on rules derived from oligodeoxynucleotides.[3a]

RNA Structure. RNA is structured. The structure of the RNA has a profound influence on the affinity of the oligonucleotide and on the rate of binding of the oligonucleotide to its RNA target.[4,5] Moreover, the RNA structure produces asymmetrical binding sites that then result in very divergent affinity constants depending on the position of oligonucleotide in that structure.[5-7] This in turn influences the optimal length of an oligonucleotide needed to achieve maximal affinity. We understand very little about how RNA structure and RNA protein interactions influence antisense drug action.

Variations in in Vitro Cellular Uptake and Distribution. Studies in several laboratories have clearly demonstrated that cells in tissue culture may take up phosphorothioate oligonucleotides via an active process and that the uptake of these oligonucleotides is highly variable depending on many conditions.[2,8] Cell type has a dramatic effect on total uptake, kinetics of uptake, and pattern of subcellular distribution. At present, there is no unifying hypothesis to explain these differences. Tissue culture conditions, such as the type of medium, degree of confluence, and the presence of serum, can all have enormous effects on uptake.[8] The oligonucleotide chemical class obviously influences the characteristics of uptake as well as the mechanism of uptake. Within the phosphorothioate class of oligonucleo-

---

[3] J. R. Wyatt, T. A. Vickers, J. L. Roberson, R. W. Buckheit, Jr., T. Klimkait, E. DeBaets, P. W. Davis, B. Rayner, J. L. Imbach, and D. J. Ecker, *Proc. Natl. Acad. Sci. U.S.A.* **91**, 1356 (1994).

[3a] S. M. Freier, unpublished results.

[4] S. M. Freier, in "Antisense Research and Applications" (S. T. Crooke and B. Lebleu, eds.), p. 67. CRC Press, Boca Raton, FL, 1993.

[5] D. J. Ecker, in "Antisense Research and Applications" (S. T. Crooke and R. Lebleu, eds.) p. 387. CRC Press, Boca Raton, FL, 1993.

[6] W. F. Lima, B. P. Monia, D. J. Ecker, and S. M. Freier, *Biochemistry* **31**, 12055 (1992).

[7] D. J. Ecker, T. A. Vickers, T. W. Bruice, S. M. Freier, R. D. Jenison, M. Manoharan, and M. Zounes, *Science* **257**, 958 (1992).

[8] S. T. Crooke, L. R. Grillone, A. Tendolkar, A. Garrett, M. J. Fratkin, J. Leeds, and W. H. Barr, *Clin. Pharmacol. Ther.* **56**, 641 (1994).

tides, uptake varies as a function of length, but not linearly. Uptake varies as a function of sequence and stability in cells is also influenced by the sequence.[8,9]

Given the foregoing, it is obvious that conclusions about *in vitro* uptake must be made very carefully and generalizations are virtually impossible. Thus, before an oligonucleotide could be said to be inactive *in vitro,* it should be studied in several cell lines. Furthermore, while it may be absolutely correct that receptor-mediated endocytosis is a mechanism of uptake of phosphorothioate oligonucleotides,[10] it is obvious that a generalization that all phosphorothioates are taken up by all cells *in vitro* primarily by receptor-mediated endocytosis is simply unwarranted.

Finally, extrapolations from *in vitro* uptake studies to predictions about *in vivo* pharmacokinetic behavior are entirely inappropriate and, in fact, there are now several lines of evidence in animals and humans that demonstrate that even after careful consideration of all *in vitro* uptake data, one cannot predict *in vivo* pharmacokinetics of the compounds.[8,11–13]

Binding to and Effects of Binding to Nonnucleic Acid Targets. Phosphorothioate oligonucleotides tend to bind to many proteins and those interactions are influenced by many factors. The effects of binding can influence cell uptake, distribution, metabolism, and excretion. They may induce nonantisense effects that can be mistakenly interpreted as antisense or complicate the identification of an antisense mechanism. By inhibiting RNase H, protein binding may inhibit the antisense activity of some oligonucleotides. Finally, binding to proteins can certainly have toxicological consequences.

In addition to proteins, oligonucleotides may interact with other biological molecules, such as lipids or carbohydrates, and such interactions, like those with proteins, will be influenced by the chemical class of oligonucleotide studied. Unfortunately, essentially no data bearing on such interactions are currently available.

An especially complicated experimental situation is encountered in many *in vitro* antiviral assays. In these assays, high concentrations of drugs, viruses, and cells are often coincubated. The sensitivity of each virus to

[9] S. T. Crooke, *in* "Burger's Medicinal Chemistry and Drug Discovery" (M. E. Wolff, ed.), vol. 1, p. 863. Wiley, New York, 1995.

[10] S. L. Loke, C. A. Stein, X. H. Zhang, K. Mori, M. Nakanishi, C. Subasinghe, J. S. Cohen, and L. M. Neckers, *Proc. Natl. Acad. Sci. U.S.A.* **86,** 3474 (1989).

[11] P. A. Cossum, H. Sasmor, D. Dellinger, L. Truong, L. Cummins, S. R. Owens, P. M. Markham, J. P. Shea, and S. Crooke, *J. Pharmacol. Exp. Ther.* **267,** 1181 (1993).

[12] P. A. Cossum, L. Truong, S. R. Owens, P. M. Markham, J. P. Shea, and S. T. Crooke, *J. Pharmacol. Exp. Ther.* **269,** 89 (1994).

[13] H. Sands, L. J. Gorey-Feret, S. P. Ho, Y. Bao, A. J. Cocuzza, D. Chidester, and F. W. Hobbs, *Mol. Pharmacol.* **47,** 636 (1995).

nonantisense effects of oligonucleotides varies depending on the nature of the virion proteins and the characteristics of the oligonucleotides.[14,15] This has resulted in considerable confusion. In particular for human immune deficiency virus (HIV), herpes simplex viruses, cytomegaloviruses, and influenza virus, the nonantisense effects have been so dominant that identifying oligonucleotides that work via an antisense mechanism has been difficult. Given the artificial character of such assays, it is difficult to know whether nonantisense mechanisms would be as dominant *in vivo* or result in antiviral activity.

Terminating Mechanisms. It has been amply demonstrated that oligonucleotides may employ several terminating mechanisms. The dominant terminating mechanism is influenced by RNA receptor site, oligonucleotide chemical class, cell type, and probably many other factors.[16] Obviously, as variations in terminating mechanism may result in significant changes in antisense potency and studies have shown significant variations from cell type to cell type *in vitro,* it is essential that the terminating mechanism be well understood. Unfortunately, at present, our understanding of terminating mechanisms remains rudimentary.

Effects of "Control Oligonucleotides." A number of types of control oligonucleotides have been used, including randomized oligonucleotides. Unfortunately, we know little to nothing about the potential biological effects of such "controls," and the more complicated a biological system and test the more likely that "control" oligonucleotides may have activities that complicate interpretations. Thus, when a control oligonucleotide displays a surprising activity, the mechanism of that activity should be explored carefully before concluding that the effects of the "control oligonucleotide" prove that the activity of the putative antisense oligonucleotide is not due to an antisense mechanism.

Kinetics of Effects. Many rate constants may affect the activities of antisense oligonucleotides, e.g., the rate of synthesis and degradation of the target RNA and its protein, the rates of uptake into cells, the rates of distribution, extrusion, and metabolism of an oligonucleotide in cells, and similar pharmacokinetic considerations in animals. Despite this, relatively few time courses have been reported and *in vitro* studies have been reported that range from a few hours to several days. In animals, we have a growing body of information on pharmacokinetics, but in most studies reported to

[14] L. M. Cowsert, *in* "Antisense Research and Applications" (S. T. Crooke and B. Lebleu, eds.), p. 521. CRC Press, Boca Raton, FL, 1993.
[15] R. F. Azad, V. B. Driver, K. Tanaka, R. M. Crooke, and K. P. Anderson, *Antimicrob. Agents Chemother.* **37,** 1945 (1993).
[16] S. T. Crooke, "Therapeutic Applications of Oligonucleotides." R. G. Landes Company, Austin, TX, 1995.

date, the doses and schedules were chosen arbitrarily and, again, little information on duration of effect and onset of action has been presented.

Clearly, more careful kinetic studies are required and rational *in vitro* and *in vivo* dose schedules must be developed.

Recommendations

Positive Demonstration of Antisense Mechanism and Specificity. Until more is understood about how antisense drugs work, it is essential to positively demonstrate effects consistent with an antisense mechanism. For RNase H-activating oligonucleotides, Northern blot analysis showing selective loss of the target RNA is the best choice and many laboratories are publishing reports *in vitro* and *in vivo* of such activities.[17–20] Ideally, a demonstration that closely related isotypes are unaffected should be included.

More recently, in our laboratories we have used RNA protection assays and DNA chip arrays.[20a] These assays provide a great deal of information about the levels of various RNA species. Coupled to careful kinetic analysis, such approaches can help assure that the primary mechanism of action of the drug is antisense and can identify events that are secondary to antisense inhibition of a specific target. This can then support the assignment of a target to a particular pathway, the analysis of the roles of a particular target, and the factors that regulate its activity. We have adapted all these methods for use in animals and will be determining their utility in clinical trials.

In brief, then, for proof of mechanism, the following steps are recommended.

Perform careful dose–response curves *in vitro* using several cell lines and methods of *in vitro* delivery.

Correlate the rank order potency *in vivo* with that observed *in vitro* after thorough dose–response curves are generated *in vivo*.

Perform careful "gene walks" for all RNA species and oligonucleotide chemical classes.

Perform careful time courses before drawing conclusions about potency.

[17] M. Y. Chiang, H. Chan, M. A. Zounes, S. M. Freier, W. F. Lima, and C. F. Bennett, *J. Biol. Chem.* **266,** 18162 (1991).

[18] N. M. Dean and R. McKay, *Proc. Natl. Acad. Sci. U.S.A.* **91,** 11762 (1994).

[19] T. Skorski, M. Nieborowska-Skorska, N. C. Nicolaides, C. Szczylik, P. Iversen, R. V. Iozzo, G. Zon, and B. Calabretta, *Proc. Natl. Acad. Sci. U.S.A.* **91,** 4504 (1994).

[20] N. Hijiya, J. Zhang, M. Z. Ratajczak, J. A. Kant, K. DeRiel, M. Herlyn, G. Zon, and A. M. Gewirtz, *Proc. Natl. Acad. Sci. U.S.A.* **91,** 4499 (1994).

[20a] J. F. Taylor, Q. Q. Zhang, B. P. Monia, E. G. Marcusson, and N. M. Dean, *Oncogene,* in press.

Directly demonstrate the proposed mechanism of action by measuring
the target RNA and/or protein.

Evaluate specificity and therapeutic indices via studies on closely related isotypes and with appropriate toxicological studies.

Perform sufficient pharmacokinetics to define rational dosing schedules
for pharmacological studies.

When control oligonucleotides display surprising activities, determine
the mechanisms involved.

Molecular Mechanisms of Antisense Drugs

Occupancy-Only Mediated Mechanisms

Classic competitive antagonists are thought to alter biological activities
because they bind to receptors preventing natural agonists from binding
the inducing normal biological processes. Binding of oligonucleotides to
specific sequences may inhibit the interaction of the RNA with proteins,
other nucleic acids, or other factors required for essential steps in the
intermediary metabolism of the RNA or its utilization by the cell.

Inhibition of Splicing. A key step in the intermediary metabolism of
most mRNA molecules is the excision of introns. These "splicing" reactions
are sequence specific and require the concerted action of spliceosomes.
Consequently, oligonucleotides that bind to sequences required for splicing
may prevent the binding of necessary factors or physically prevent the
required cleavage reactions. This would then result in inhibition of the
production of the mature mRNA. Although there are several examples of
oligonucleotides directed to splice junctions, none of the studies present
data showing inhibition of RNA processing, accumulation of splicing intermediates, or a reduction in mature mRNA. Nor are there published data
in which the structure of the RNA at the splice junction was probed and
the oligonucleotides demonstrated to hybridize to the sequences for which
they were designed.[21-24] Activities have been reported for anti-c-*myc* and
antiviral oligonucleotides with phosphodiester, methylphosphonate, and

---

[21] M. E. McManaway, L. M. Neckers, S. L. Loke, A. A. Al-Nasser, R. L. Redner, B. T.
Shiramizu, W. L. Goldschmidts, B. E. Huber, K. Bhatia, and I. T. Magrath, *Lancet* **335,**
808 (1990).

[22] M. Kulka, C. C. Smith, L. Aurelian, R. Fishelevich, K. Meade, P. Miller, and P. O. P. Ts'o,
Proc. Natl. Acad. Sci. U.S.A. **86,** 6868 (1989).

[23] P. C. Zamecnik, J. Goodchild, Y. Taguchi, and P. S. Sarin, *Proc. Natl. Acad. Sci. U.S.A.*
83, 4143 (1986).

[24] C. C. Smith, L. Aurelian, M. P. Reddy, P. S. Miller, and P. O. P. Ts'o, *Proc. Natl. Acad.
Sci. U.S.A.* **83,** 2787 (1986).

phosphorothioate backbones. An oligonucleotide has been reported to induce alternative splicing in a cell-free splicing system and, in that system, RNA analyses confirmed the putative mechanism.[25]

In our laboratory, we have attempted to characterize the factors that determine whether splicing inhibition is effected by an antisense drug.[26] To this end, a number of luciferase-reporter plasmids containing various introns were constructed and transfected into HeLa cells. The effects of antisense drugs designed to bind to various sites were then characterized. The effects of RNase H-competent oligonucleotides were compared to those of oligonucleotides that do not serve as RNase H substrates. The major conclusions from this study were, first, that most of the earlier studies in which splicing inhibition was reported were probably due to nonspecific effects. Second, less effectively spliced introns are better targets than those with strong consensus splicing signals. Third, the 3'-splice site and branchpoint are usually the best sites to which to target to the oligonucleotide to inhibit splicing. Fourth, RNase H-competent oligonucleotides are usually more potent than even higher affinity oligonucleotides that inhibit by occupancy only.

Translational Arrest. A mechanism for which the many oligonucleotides have been designed is to arrest the translation of targeted protein by binding to the translation initiation codon. The positioning of the initiation codon within the area of complementarily of the oligonucleotide and the length of oligonucleotide used have varied considerably. Again, unfortunately, only in relatively few studies have the oligonucleotides, in fact, been shown to bind to the sites for which they were designed, and data that directly support translation arrest as the mechanism have been lacking.

Target RNA species that have been reported to be inhibited by a translational arrest mechanism include HIV, vesicular stomatitis virus (VSV), n-*myc*, and a number of normal cellular genes.[27–33] In our laboratories, we have shown that a significant number of targets may be inhibited by binding to translation initiation codons. For example, ISIS 1082 hybrid-

[25] Z. Dominski and R. Kole, *Proc. Natl. Acad. Sci. U.S.A.* **90**, 8673 (1993).

[26] D. Hodges and S. T. Crooke, *Mol. Pharmacol.* **48**, 905 (1995).

[27] S. Agrawal, J. Goodchild, M. P. Civeira, A. H. Thornton, P. S. Sarin, and P. C. Zamecnik, *Proc. Natl. Acad. Sci. U.S.A.* **85**, 7079 (1988).

[28] M. Lemaitre, B. Bayard, and B. Lebleu, *Proc. Natl. Acad. Sci. U.S.A.* **84**, 648 (1987).

[29] A. Rosolen, L. Whitesell, N. Ikegaki, R. H. Kennett, and L. M. Neckers, *Cancer Res.* **50**, 6316 (1990).

[30] G. Vasanthakumar and N. K. Ahmed, *Cancer Commun.* **1**, 225 (1989).

[31] Sburlati, A. R., R. E. Manrow, and S. L. Berger, *Proc. Natl. Acad. Sci. U.S.A.* **88**, 253 (1991).

[32] H. Zheng, B. M. Sahai, P. Kilgannon, A. Fotedar, and D. R. Green, *Proc. Natl. Acad. Sci. U.S.A.* **86**, 3758 (1989).

[33] J. A. Maier, P. Voulalas, D. Roeder, and T. Maciag, *Science* **249**, 1570 (1990).

izes to the AUG codon for the UL13 gene of herpes virus types 1 and 2. RNase H studies confirmed that it binds selectively in this area. *In vitro* protein synthesis studies confirmed that it inhibited the synthesis of the UL13 protein, and studies in HeLa cells showed that it inhibited the growth of herpes type 1 and type 2 with IC_{50} of 200–400 nM by translation arrest.[34] Similarly, ISIS 1753, a 30-mer phosphorothioate complementary to the translation initiation codon and surrounding sequences of the E2 gene of bovine papilloma virus, was highly effective and its activity was shown to be due to translation arrest. ISIS 2105, a 20-mer phosphorothioate complementary to the same region in human papilloma virus, was shown to be a very potent inhibitor. Compounds complementary to the translation initiation codon of the E2 gene were the most potent of the more than 50 compounds studied complementary to various other regions in the RNA.[35] We have shown inhibition of translation of a number of other mRNA species by compounds designed to bind to the translation codon as well.

In conclusion, translation arrest represents an important mechanism of action for antisense drugs. A number of examples purporting to employ this mechanism have been reported, and studies on several compounds have provided data that unambiguously demonstrate that this mechanism can result in potent antisense drugs. However, very little is understood about the precise events that lead to translation arrest.

Disruption of Necessary RNA Structure. RNA adopts a variety of three-dimensional structures induced by intramolecular hybridization, the most common of which is the stem loop. These structures play crucial roles in a variety of functions. They are used to provide additional stability for RNA and as recognition motifs for a number of proteins, nucleic acids, and ribonucleoproteins that participate in the intermediary metabolism and activities of RNA species. Thus, given the potential general activity of the mechanism, it is surprising that occupancy-based disruption RNA has not been exploited more extensively.

As an example, we designed a series of oligonucleotides that bind to the important stem–loop present in all RNA species in HIV, the TAR element. We synthesized a number of oligonucleotides designed to disrupt TAR and showed that several indeed did bind to TAR, disrupt the structure, and inhibit TAR-mediated production of a reporter gene.[36] Furthermore,

[34] C. K. Mirabelli, C. F. Bennett, K. Anderson, and S. T. Crooke *Anti-Cancer Drug Design* **6,** 647 (1991).

[35] L. M. Cowsert, M. C. Fox, G. Zon, and C. K. Mirabelli, *Antimicrob. Agents Chemother.* **37,** 171 (1993).

[36] T. Vickers, B. F. Baker, P. D. Cook, M. Zounes, R. W. Buckheit, Jr., J. Germany, and D. J. Ecker, *Nucleic Acids Res.* **19,** 3359 (1991).

general rules useful in disrupting stem–loop structures were developed as well.[7]

Although designed to induce relatively nonspecific cytotoxic effects, two other examples are noteworthy. Oligonucleotides designed to bind to a 17 nucleotide loop in *Xenopus* 28 S RNA required for ribosome stability and protein synthesis inhibited protein synthesis when injected into *Xenopus* oocytes.[37] Similarly, oligonucleotides designed to bind to highly conserved sequences in 5.8 S RNA inhibited protein synthesis in rabbit reticulocyte and wheat germ systems.[38]

Occupancy-Activated Destabilization

RNA molecules regulate their own metabolism. A number of structural features of RNA are known to influence stability, various processing events, subcellular distribution, and transport. It is likely that as RNA intermediary metabolism is better understood, many other regulatory features and mechanisms will be identified.

5'-Capping. A key early step in RNA processing is 5'-capping (Fig. 1). This stabilizes pre-mRNA and is important for the stability of mature mRNA. It also is important in binding to the nuclear matrix and transport of mRNA out of the nucleus. As the structure of the cap is unique and understood, it presents an interesting target.

Several oligonucleotides that bind near the cap site have been shown to be active, presumably by inhibiting the binding of proteins required to cap the RNA. For example, the synthesis of SV40 T-antigen was reported to be most sensitive to an oligonucleotide linked to polylysine and targeted to the 5'-cap site of RNA.[39] However, again, in no published study has this putative mechanism been demonstrated rigorously. In fact, in no published study have the oligonucleotides been shown to bind to the sequences for which they were designed.

In our laboratory, we have designed oligonucleotides to bind to 5'-cap structures and reagents to specifically cleave the unique 5'-cap structure.[40] These studies demonstrate that 5'-cap-targeted oligonucleotides were capable of inhibiting the binding of the translation initiation factor eIF-4a.[41]

Inhibition of 3'-Polyadenylation. In the 3'-untranslated region of pre-mRNA molecules are sequences that result in the posttranscriptional

[37] S. K. Saxena and E. J. Ackerman, *J. Biol. Chem.* **265**, 3263 (1990).
[38] K. Walker, S. A. Elela, and R. N. Nazar, *J. Biol. Chem.* **265**, 2428 (1990).
[39] P. Westermann, B. Gross, and G. Hoinkis, *Biomed. Biochim Acta* **48**, 289 (1989).
[40] B. F. Baker, *J. Am. Chem. Soc.* **115**, 3378 (1993).
[41] B. F. Baker, L. Miraglia, and C. H. Hagedorn, *J. Biol. Chem.* **267**, 11495 (1992).

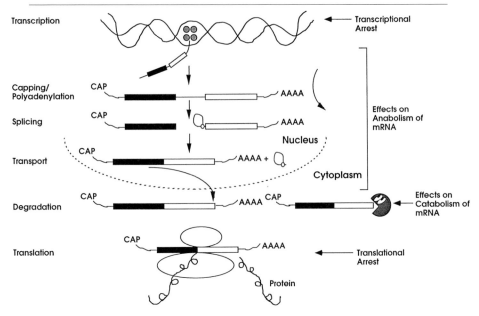

Transcription — Transcriptional Arrest

Capping/ Polyadenylation — CAP — AAAA

Splicing — CAP — AAAA — Nucleus

Effects on Anabolism of mRNA

Transport — CAP — AAAA + — Cytoplasm

Degradation — CAP — AAAA — CAP — Effects on Catabolism of mRNA

Translation — CAP — AAAA — Translational Arrest — Protein

FIG. 1. RNA processing.

addition of long (hundreds of nucleotides) tracts of polyadenylate. Polyadenylation stabilizes mRNA and may play other roles in the intermediary metabolism of RNA species. Theoretically, interactions in the 3′-terminal region of pre-mRNA could inhibit polyadenylation and destabilize the RNA species. Although there are a number of oligonucleotides that interact in the 3′-untranslated region and display antisense activities, to date, no study has reported evidence for alterations in polyadenylation.[17]

Other Mechanisms

In addition to 5′-capping and 3′-adenylation, there are clearly other sequences in the 5′- and 3′-untranslated regions of mRNA that affect the stability of the molecules. Again, there are a number of antisense drugs that may work by these mechanisms.

Zamecnik and Stephenson[42] reported that 13-mer targeted to untranslated 3′- and 5′-terminal sequences in Rous sarcoma viruses was active. Oligonucleotides conjugated to an acridine derivative and targeted to a 3′-terminal sequence in type A influenza viruses were reported to be active. Against several RNA targets, studies in our laboratories have shown that

---

[42] P. C. Zamecnik and M. L. Stephenson, *Proc. Natl. Acad. Sci. U.S.A.* **75,** 289 (1978).

sequences in the 3'-untranslated region of RNA molecules are often the most sensitive.[43-45] For example, ISIS 1939, a 20-mer phosphorothioate that binds to and appears to disrupt a predicted stem–loop structure in the 3'-untranslated region of the mRNA for the intracellular adhesion molecule (ICAM), is a potent antisense inhibitor. However, inasmuch a 2'-methoxy analog of ISIS 1939 was much less active, it is likely that, in addition to destabilization to cellular nucleolytic activity, the activation of RNase H (see later) is also involved in the activity of ISIS 1939.[17]

Activation of RNase H

RNase H is an ubiquitous enzyme that degrades the RNA strand of an RNA–DNA duplex. It has been identified in organisms as diverse as viruses and human cells.[46] At least two classes of RNase H have been identified in eukaryotic cells. Multiple enzymes with RNase H activity have been observed in prokaryotes.[46]

Although RNase H is involved in DNA replication, it may play other roles in the cell and is found in the cytoplasm as well as the nucleus.[47] However, the concentration of the enzyme in the nucleus is thought to be greater, and some of the enzyme found in cytoplasmic preparations may be due to nuclear leakage.

RNase H activity is quite variable in cells. It is absent or minimal in rabbit reticulocytes but is present in wheat germ extracts.[46,48] In HL-60 cells, for example, the level of activity in undifferentiated cells is greatest, relatively high in dimethyl sulfoxide- and vitamin D-differentiated cells, and much lower in phorbol ester-differentiated cells (Hoke, unpublished data).

The precise recognition elements for RNase H are not known. However, it has been shown that oligonucleotides with DNA-like properties as short as tetramers can activate RNase H.[49] Changes in the sugar influence RNase H activation as sugar modifications that result in RNA-like oligonucleotides, e.g., 2'-fluoro or 2'-methoxy do not appear to serve as substrates for

[43] A. Zerial, N. T. Thuong, and C. Helene, *Nucleic Acids Res.* **15,** 9909 (1987).
[44] N. T. Thuong, U. Asseline, and T. Monteney-Garestier, *in* "Oligodeoxynucleotides: Antisense Inhibitors of Gene Expression," p. 25. CRC Press, Baca Raton, FL, 1989.
[45] C. Helene and J.-J. Toulme, *in* "Oligonucleotides: Antisense Inhibitors of Gene Expression" (J. S. Cohen, ed.), p. 137. CRC Press, Boca Raton, FL, 1989.
[46] R. J. Crouch and M.-L. Dirksen, *in* "Nucleases" (S. M. Linn and R. J. Roberts, eds.), p. 211. Cold Spring Harbor Laboratory Press, Cold Spring Harbor, NY, 1985.
[47] C. Crum, J. D. Johnson, A. Nelson, and D. Roth, *Nucleic Acids Res.* **16,** 4569 (1988).
[48] M. T. Haeuptle, R. Frank, and B. Dobberstein, *Nucleic Acids Res.* **14,** 1427 (1986).
[49] H. Donis-Keller, *Nucleic Acids Res.* 7 (1979).

RNase H.[50,51] Alterations in the orientation of the sugar to the base can also affect RNase H activation as α-oligonucleotides are unable to induce RNase H or may require parallel annealing.[52,53] Additionally, backbone modifications influence the ability of oligonucleotides to activate RNase H. Methylphosphonates do not activate RNase H.[54,55] In contrast, phosphorothioates are excellent substrates.[34,56,57] In addition, chimeric molecules have been studied as oligonucleotides that bind to RNA and activate RNase H.[58,59] For example, oligonucleotides composed of wings of 2'-methoxyphosphonates and a five-base gap of deoxyoligonucleotides bind to their target RNA and activate RNase H.[58,59] Furthermore, a single ribonucleotide in a sequence of deoxyribonucleotides was shown to be sufficient to serve as a substrate for RNase H when bound to its complementary deoxyoligonucleotide.[60]

That it is possible to take advantage of chimeric oligonucleotides designed to activate RNase H and have greater affinity for their RNA receptors and to enhance specificity has also been demonstrated.[61,62] In a recent study, RNase H-mediated cleavage of target transcript was much more selective when deoxyoligonucleotides composed of methylphosphonate deoxyoligonucleotide wings and phosphodiester gaps were compared to full phosphodiester oligonucleotides.[62]

Despite the information about RNase H and the demonstration that many oligonucleotides may activate RNase H in lysate and purified enzyme assays, relatively little is known yet about the role of structural features in RNA targets in activating RNase H.[63-65] In fact, direct proof that RNase

[50] A. M. Kawasaki, M. D. Casper, S. M. Freier, E. A. Lesnik, M. C. Zounes, L. L. Cummins, C. Gonzalez, and P. D. Cook, *J. Med. Chem.* **36**, 831 (1993).
[51] B. S. Sproat, A. I. Lamond, B. Beijer, P. Neuner, and U. Ryder, *Nucleic Acids Res.* **17**, 3373 (1989).
[52] F. Morvan, B. Rayner, and J. L. Imbach, *Anticancer Drug Des.* **6**, 521 (1991).
[53] C. Gagnor, B. Rayner, J. P. Leonetti, J. L. Imbach, and B. Lebleu, *Nucleic Acids Res.* **17**, 5107 (1989).
[54] L. J. Maher, III, B. Wold, and P. B. Dervan, *Science* **245**, 725 (1989).
[55] P. S. Miller, *in* "Oligodeoxynucleotides: Antisense Inhibitors of Gene Expression" (J. S. Cohen, ed.), p. 79. CRC Press, Boca Raton, FL, 1989.
[56] C. A. Stein and Y.-C. Cheng, *Science* **261**, 1004 (1993).
[57] C. Cazenave, C. A. Stein, N. Loreau, N. T. Thuong, L. M. Neckers, C. Subasinghe, C. Heléne, J. S. Cohen, and J.-J. Toulm, *Nucleic Acids Res.* **17**, 4255 (1989).
[58] R. S. Quartin, C. L. Brakel, and J. G. Wetmur, *Nucleic Acids Res.* **17**, 7253 (1989).
[59] P. J. Furdon, Z. Dominski, and R. Kole, *Nucleic Acids Res.* **17**, 9193 (1989).
[60] P. S. Eder and J. A. Walder, *J. Biol. Chem.* **266**, 6472 (1991).
[61] B. P. Monia, E. A. Lesnik, C. Gonzalez, W. F. Lima, D. McGee, C. J. Guinosso, A. M. Kawasaki, P. D. Cook, and S. M. Freier, *J. Biol. Chem.* **268**, 14514 (1993).
[62] R. V. Giles and D. M. Tidd, *Nucleic Acids Res.* **20**, 763 (1992).
[63] R. Y. Walder and J. A. Walder, *Proc. Natl. Acad. Sci. U.S.A.* **85**, 5011 (1988).

H activation is, in fact, the mechanism of action of oligonucleotides in cells is, to a large extent, lacking.

Studies in our laboratories provide additional, albeit indirect, insights into these questions. ISIS 1939 is a 20-mer phosphorothioate complementary to a sequence in the 3'-untranslated region of ICAM-1 RNA.[17] It inhibits ICAM production in human umbilical vein endothelial cells, and Northern blots demonstrate that ICAM-1 mRNA is degraded rapidly. A 2'-methoxy analog of ISIS 1939 displays higher affinity for the RNA than the phosphorothioate, is stable in cells, but inhibits ICAM-1 protein production much less potently than ISIS 1939. It is likely that ISIS 1939 destabilizes the RNA and activates RNase H. In contrast, ISIS 1570, an 18-mer phosphorothioate that is complementary to the translation initiation codon of the ICAM-1 message, inhibited production of the protein, but caused no degradation of the RNA. Thus, two oligonucleotides that are capable of activating RNase H had different effects depending on the site in the mRNA at which they bound.[17]

A more direct demonstration that RNase H is likely a key factor in the activity of many antisense oligonucleotides was provided by studies in which reverse-ligation polymerase chain reaction (RT-PCR) was used to identify cleavage products from bcr-abl mRNA in cells treated with phosphorothioate oligonucleotides.[66]

Given the emerging role of chimeric oligonucleotides with modifications in the 3'- and 5'-wings designed to enhance affinity for the target RNA and nuclease stability and a DNA-type gap to serve as a substrate for RNase H, studies focused on understanding the effects of various modifications on the efficiency of the enzyme(s) are also of considerable importance. In one such study on *Escherichia coli* RNase H, we have reported that the enzyme displays minimal sequence specificity and is processive. When a chimeric oligonucleotide with 2'-modified sugars in the wings was hybridized to the RNA, the initial site of cleavage was the nucleotide adjacent to the methoxy–deoxy junction closest to the 3' end of the RNA substrate. The initial rate of cleavage increased as the size of the DNA gap increased, and the efficiency of the enzyme was considerably less against an RNA target duplexed with a chimeric antisense oligonucleotide than a full DNA-type oligonucleotide.[67]

[64] J. Minshull and T. Hunt, *Nucleic Acids Res.* **14,** 6433 (1986).

[65] C. Gagnor, J. R. Bertrand, S. Thenet, M. Lemaitre, F. Morvan, B. Rayner, C. Malvy, B. Lebleu, J. L. Imbach, and C. Paoletti, *Nucleic Acids Res.* **15,** 10419 (1987).

[66] R. V. Giles, D. G. Spiller, and D. M. Tidd, *Antisense Res. Dev.* **5,** 23 (1995).

[67] S. T. Crooke, K. M. Lemonidis, L. Neilson, R. Griffey, E. A. Lesnik, and B. P. Monia, *Biochem. J.* **312,** 599 (1995).

In subsequent studies, we have evaluated the interactions of antisense oligonucleotides with structured and unstructured targets and the impacts of these interactions on RNase H in more detail.[68] Using a series of non-cleavable substrates and Michaelis–Menten analyses, we were able to evaluate both binding and cleavage. We showed that, in fact, *E. coli* RNase H1 is a double-strand RNA-binding protein. The K_d for RNA duplex was 1.6 μM; the K_d for a DNA duplex was 176 μM; and the K_d for single-strand DNA was 942 μM. In contrast, the enzyme could only cleave RNA in an RNA–DNA duplex. Any 2′ modification in the antisense drug at the cleavage site inhibited cleavage, but a significant charge reduction and 2′ modifications were tolerated at the binding site. Finally, placing a positive charge (e.g., 2′-propoxyamine) in the antisense drug reduced affinity and cleavage.

We have also examined the effects of antisense oligonucleotide-induced RNA structures on the activity of *E. coli* RNase H1.[69] Any structure in the duplex substrate was found to have a significant negative effect on the cleavage rate. Further, cleavage of selected sites was inhibited entirely, and this was explained by steric hindrance imposed by the RNA loop traversing either the minor or the major grooves or the heteroduplex.

We have succeeded in cloning, expressing, and characterizing a human RNase H that is homologous to *E. coli* RNase H1 and has properties comparable to the type 2 enzyme.[70] Additionally, we have cloned and expressed a second RNase H homologous to *E. coli* RNase H2.[70a] Given these steps, we are now in position to evaluate the roles of each of these enzymes in cellular activities and antisense pharmacology. We are also characterizing these proteins and their enzymological properties.

Activation of Double-Strand RNase

By using phosphorothioate oligonucleotides with 2′-modified wings and a ribonucleotide center, we have shown that mammalian cells contain enzymes that can cleave double-stranded RNAs.[70] This is an important step forward because it adds to the repertoire of intracellular enzymes that may be used to cleave target RNAs and because chimeric oligonucleotide 2′-modified wings and oligoribonucleotide gaps have a higher affinity for RNA targets than chimeras with oligodeoxynucleotide gaps.

[68] W. F. Lima and S. T. Crooke, *Biochemistry* **36**, 390 (1997).
[69] W. F. Lima, M. Venkatraman, and S. T. Crooke, *J. Biol. Chem.* **272**, 18191 (1997).
[70] H. Wu, W. F. Lima, and S. T. Crooke, *Antisense Nucleic Acid Drug Dev.* **8**, 53 (1998).
[70a] H. Wu and S. T. Crooke, unpublished observations.

Selection of Optimal RNA-Binding Site

It has been amply demonstrated that a significant fraction of every RNA species is not accessible to phosphorothioate oligodeoxynucleotides in a fashion that permits antisense effects (for a review, see Crooke[71]). Thus, substantial efforts have been directed to the development of methods that might predict optimal sites for binding within RNA species. Although a number of screening methods have been proposed,[72–75] in our experience the correlation between these antisense effects in cells is insufficient to warrant their use.[75a]

Consequently, we have developed rapid throughput systems that use a 96-well format and a 96-channel oligonucleotide synthesizer coupled to an automated RT-PCR instrument. This provides rapid screening of up to 80 sites in an RNA species for two chemistries under consistent highly controlled experimental conditions. It is hoped that based on such a system (we are currently evaluating two genes per week), we will be able to develop improved methods that predict optimal sites.

Characteristics of Phosphorothioate Oligodeoxynucleotides

Introduction

Of the first-generation oligonucleotide analogs, the class that has resulted in the broadest range of activities and about which the most is known is the phosphorothioate class. Phosphorothioate oligonucleotides were first synthesized in 1969 when a poly(rIrC)phosphorothioate was synthesized.[76] This modification clearly achieves the objective of increased nuclease stability. In this class of oligonucleotides, one of the oxygen atoms in the phosphate group is replaced with a sulfur. The resulting compound is negatively charged, is chiral at each phosphorothioate phosphodiester, and is much more resistant to nucleases than the parent phosphorothioate.[77]

[71] S. T. Crooke, *FASEB J.* **7,** 533 (1993).
[72] T. W. Bruice and W. F. Lima, *Biochemistry* **36,** 5004 (1997).
[73] O. Matveeva, B. Felden, S. Audlin, R. F. Gesteland, and J. F. Atkins, *Nucleic Acids Res.* **25,** 5010 (1997).
[74] E. M. Southern, Ciba Foundation, Wiley, London, 1977.
[75] S. P. Ho, Y. Bao, T. Lesher, R. Malhotra, L. Y. Ma, S. J. Fluharty, and R. R. Sakai, *Nat. Biotechnol.* **16,** 59 (1998).
[75a] Wyatt, unpublished results.
[76] E. De Clercq, F. Eckstein, and T. C. Merigan, *Science* **165,** 1137 (1969).
[77] J. S. Cohen, *in* "Antisense Research and Applications" (S. T. Crooke and B. Lebleu, eds.), p. 205. CRC Press, Boca Raton, FL, 1993.

Hybridization

The hybridization of phosphorothioate oligonucleotides to DNA and RNA has been characterized thoroughly.[1,78–80] The T_m of a phosphorothioate oligodeoxynucleotide for RNA is approximately 0.5° less per nucleotide than for a corresponding phosphodiester oligodeoxynucleotide. This reduction in T_m per nucleotide is virtually independent of the number of phosphorothioate units substituted for phosphodiesters. However, the sequence context has some influence as the ΔT_m can vary from -0.3 to 1.0°, depending on the sequence. Compared to RNA and RNA duplex formation, a phosphorothioate oligodeoxynucleotide has a T_m approximately $-2.2°$ lower per unit.[4] This means that to be effective *in vitro,* phosphorothioate oligodeoxynucleotides must typically be 17- to 20-mer in length and that invasion of double-stranded regions in RNA is difficult.[6,36,61,81]

Association rates of phosphorothioate oligodeoxynucleotide to unstructured RNA targets are typically 10^6–10^7 M^{-1} sec^{-1} independent of oligonucleotide length or sequence.[4,6] Association rates to structured RNA targets can vary from 10^2 to 10^8 M^{-1} sec^{-1}, depending on the structure of the RNA, site of binding in the structure, and other factors.[4] Said another way, association rates for oligonucleotides that display acceptable affinity constants are sufficient to support biological activity at therapeutically achievable concentrations. Interestingly, in a study using phosphodiester oligonucleotides coupled to fluorescein, hybridization was detectable within 15 min after microinjection into K562 cells.[82]

The specificity of hybridization of phosphorothioate oligonucleotides is, in general, slightly greater than phosphodiester analogs. For example, a T–C mismatch results in a 7.7 or 12.8° reduction in T_m, respectively, for a phosphodiester or phosphorothioate oligodeoxynucleotide 18 nucleotides in length with the mismatch centered.[4] Thus, from this perspective, the phosphorothioate modification is quite attractive.

Interactions with Proteins

Phosphorothioate oligonucleotides bind to proteins. Interactions with proteins can be divided into nonspecific, sequence specific, and structure-

[78] S. T. Crooke, *Bio/Technology* **10**, 882 (1992).

[79] R. M. Crooke, *in* "Antisense Research and Applications" (S. T. Crooke and B. Lebleu, eds.), p. 427. CRC Press, Boca Raton, FL, 1993.

[80] S. T. Crooke, *Annu. Rev. Pharmacol. Toxicol.* **32**, 329 (1992).

[81] B. P. Monia, J. F. Johnston, D. J. Ecker, M. A. Zounes, W. F. Lima, and S. M. Freier, *J. Biol. Chem.* **267**, 19954 (1992).

[82] D. L. Sokol, X. Zhang, P. Lu, and A. M. Gewirtz, *Proc. Natl. Acad. Sci. U.S.A.* **95**, 11538 (1998).

specific binding events, each of which may have different characteristics and effects. Nonspecific binding to a wide variety of proteins has been demonstrated. Exemplary of this type of binding is the interaction of phosphorothioate oligonucleotides with serum albumin. The affinity of such interactions is low. The K_d for albumin is approximately 200 μM, thus, in a similar range with aspirin or penicillin.[83,84] Furthermore, in this study, no competition between phosphorothioate oligonucleotides and several drugs that bind to bovine serum albumin was observed. In this study, binding and competition were determined in an assay in which electrospray mass spectrometry was used. In contrast, in a study in which an equilibrium dissociation constant was derived from an assay using albumin loaded on a CH-Sephadex column, the K_m ranged from 1–5 $\times$ 10^{-5} M for bovine serum albumin to 2–3 $\times$ 10^{-4} M for human serum albumin. Moreover, warfarin and indomethacin were reported to compete for binding to serum albumin.[85] Clearly, much more work is required before definitive conclusions can be drawn.

Phosphorothioate oligonucleotides can interact with nucleic acid-binding proteins such as transcription factors and single-strand nucleic acid-binding proteins. However, very little is known about these binding events. Additionally, it has been reported that phosphorothioates bind to an 80-kDa membrane protein that was suggested to be involved in cellular uptake processes.[10] However, again, little is known about the affinities, sequence, or structure specificities of these putative interactions. More recently, interactions with 30- and 46-kDa surface proteins in T15 mouse fibroblasts were reported.[86]

Phosphorothioates interact with nucleases and DNA polymerases. These compounds are metabolized slowly by both endo- and exonucleases and inhibit these enzymes.[78,87] The inhibition of these enzymes appears to be competitive, which may account for some early data suggesting that phosphorothioates are almost infinitely stable to nucleases. In these studies, the oligonucleotide to enzyme ratio was very high and, thus, the enzyme was inhibited. Phosphorothioates also bind to RNase H when in an RNA–DNA duplex and the duplex serves as a substrate for RNase H.[88] At higher

[83] S. T. Crooke, M. J. Graham, J. E. Zuckerman, D. Brooks, B. S. Conklin, L. L. Cummins, M. J. Greig, C. J. Guinosso, D. Kornburst, M. Manoharan, H. M. Sasmor, T. Schleich, K. L. Tivel, R. H. Griffey *et al., J. Pharmacol. Exp. Ther.* **277,** 923 (1996).
[84] R. W. Joos and W. H. Hall, *J. Pharmacol. Exp. Ther.* **166,** 113 (1969).
[85] S. K. Srinivasan, H. K. Tewary, and P. L. Iversen, *Antisense Res. Dev.* **5,** 131 (1995).
[86] P. Hawley and I. Gibson, *Antisense Nucleic Drug Dev.* **6,** 185 (1996).
[87] R. M. Crooke, M. J. Graham, M. E. Cooke, and S. T. Crooke, *J. Pharmacol. Exp. Ther.* **275,** 462 (1995).
[88] W.-Y. Gao, F.-S. Han, C. Storm, W. Egan, and Y.-C. Cheng, *Mol. Pharmacol.* **41,** 223

concentrations, presumably by binding as a single strand to RNase H, phosphorothioates inhibit the enzyme.[67,78] Again, the oligonucleotides appear to be competitive antagonists for the DNA–RNA substrate.

Phosphorothioates have been shown to be competitive inhibitors of DNA polymerase α and β with respect to the DNA template and noncompetitive inhibitors of DNA polymerases γ and δ.[88] Despite this inhibition, several studies have suggested that phosphorothioates might serve as primers for polymerases and be extended.[9,56,89] In our laboratories, we have shown extensions 2-3 nucleotides only. At present, a full explanation as to why longer extensions are not observed is not available.

Phosphorothioate oligonucleotides have been reported to be competitive inhibitors for HIV-reverse transcriptase and inhibit RT-associated RNase H activity.[90,91] They have been reported to bind to the cell surface protein, CD4, and to protein kinase C (PKC).[92] Various viral polymerases have also been shown to be inhibited by phosphorothioates.[56] Additionally, we have shown potent, nonsequence-specific inhibition of RNA splicing by phosphorothioates.[26]

Like other oligonucleotides, phosphorothioates can adopt a variety of secondary structures. As a general rule, self-complementary oligonucleotides are avoided, if possible, to avoid duplex formation between oligonucleotides. However, other structures that are less well understood can also form. For example, oligonucleotides containing runs of guanosines can form tetrameric structures called G-quartets, which appear to interact with a number of proteins with relatively greater affinity than unstructured oligonucleotides.[3]

In conclusion, phosphorothioate oligonucleotides may interact with a wide range of proteins via several types of mechanisms. These interactions may influence the pharmacokinetic, pharmacologic, and toxicologic properties of these molecules. They may also complicate studies on the mechanism of action of these drugs and may, in fact, obscure an antisense activity. For example, phosphorothioate oligonucleotides were reported to enhance lipopolysaccharide-stimulated synthesis or tumor necrosis factor.[93] This would obviously obscure antisense effects on this target.

[89] S. Agrawal, J. Temsamani, and J. Y. Tang, *Proc. Natl. Acad. Sci. U.S.A.* **88**, 7595 (1991).

[90] C. Majumdar, C. A. Stein, J. S. Cohen, S. Broder, and S. H. Wilson, *Biochemistry,* **28**, 1340 (1989).

[91] Y. Cheng, W. Gao, and F. Han, *Nucleosides Nucleotides,* **10**, 155 (1991).

[92] C. A. Stein, M. Neckers, B. C. Nair, S. Mumbauer, G. Hoke, and R. Pal, *J. Acq. Immune Deficiency Syndr.* **4**, 686 (1991).

[93] G. Hartmann, A. Krug, K. Waller-Fontaine, and S. Endres, *Mol. Med.* **2**, 429 (1996).

Pharmacokinetic Properties

To study the pharmacokinetics of phosphorothioate oligonucleotides, a variety of labeling techniques have been used. In some cases, 3'- or 5'- [32]P end-labeled or fluorescently labeled oligonucleotides have been used in *in vitro* or *in vivo* studies. These are probably less satisfactory than internally labeled compounds because terminal phosphates are removed rapidly by phosphatases and fluorescently labeled oligonucleotides have physicochemical properties that differ from unmodified oligonucleotides. Consequently, either uniformly ^{35}S-labeled or base-labeled phosphorothioates are preferable for pharmacokinetic studies. In our laboratories, a tritium exchange method that labels a slowly exchanging proton at the C-8 position in purines was developed and proved to be quite useful.[94] A method that added radioactive methyl groups via *S*-adenosylmethionine has also been used successfully.[95] Finally, advances in extraction, separation, and detection methods have resulted in methods that provide excellent pharmacokinetic analyses without radiolabeling.[83]

Nuclease Stability. The principle metabolic pathway for oligonucleotides is cleavage via endo- and exonucleases. While quite stable to various nucleases, phosphorothioate oligonucleotides are competitive inhibitors of nucleases.[16,88,96–98] Consequently, the stability of phosphorothioate oligonucleotides to nucleases is probably a bit less than initially thought, as high concentrations (that inhibited nucleases) of oligonucleotides were employed in early studies. Similarly, phosphorothioate oligonucleotides are degraded slowly by cells in tissue culture with a half-life of 12–24 hr and are metabolized slowly in animals.[11,16,96] The pattern of metabolites suggests primarily exonuclease activity with perhaps modest contributions by endonucleases. However, a number of lines of evidence suggest that, in many cells and tissues, endonucleases play an important role in the metabolism of oligonucleotides. For example, 3'- and 5'-modified oligonucleotides with phosphodiester backbones have been shown to be degraded relatively rapidly in cells and after administration to animals.[13,99] Thus,

[94] M. J. Graham, S. M. Freier, R. M. Crooke, D. J. Ecker, R. N. Maslova, and E. A. Lesnik, *Nucleic Acid. Res.* **21**, 3737 (1993).
[95] H. Sands, L. J. Gorey-Feret, A. J. Cocuzza, F. W. Hobbs, D. Chidester, and G. L. Trainor, *Mol. Pharmacol.* **45**, 932 (1994).
[96] G. D. Hoke, K. Draper, S. M. Freier, C. Gonzalez, V. B. Driver, M. C. Zounes, and D. J. Ecker, *Nucleic Acids Res.* **19**, 5743 (1991).
[97] E. Wickstrom, *J. Biochem. Biophys. Methods* **13**, 97 (1986).
[98] J. M. Campbell, T. A. Bacon, and E. Wickstrom, *J. Biochem. Biophys. Methods* **20**, 259 (1990).
[99] T. Miyao, Y. Takakura, T. Akiyama, F. Yoneda, H. Sezaki, and M. Hashida, *Antisense Res Dev.* **5**, 115 (1995).

strategies in which oligonucleotides are modified at only the 3′ and 5′ terminus as a means of enhancing stability have not proven to be successful.

In Vitro Cellular Uptake. Phosphorothioate oligonucleotides are taken up by a wide range of cells *in vitro.*[2,16,88,100,101] In fact, the uptake of phosphorothioate oligonucleotides into a prokaryote, *Vibrio parahaemolyticus,* has been reported, as has uptake into *Schistosoma mansoni.*[102,103] Uptake is time and temperature dependent. It is also influenced by cell type, cell culture conditions, media and sequence, and length of the oligonucleotide.[16] No obvious correlation between the lineage of cells, whether the cells are transformed or whether the cells are infected virally, and uptake has been identified.[16] Nor are the factors that result in differences in the uptake of different sequences of oligonucleotide understood. Although several studies have suggested that receptor-mediated endocytosis may be a significant mechanism of cellular uptake, data are not yet compelling enough to conclude that receptor-mediated endocytosis accounts for a significant portion of the uptake in most cells.[10]

Numerous studies have shown that phosphorothioate oligonucleotides distribute broadly in most cells once taken up.[16,79] Again, however, significant differences in the subcellular distribution between various types of cells have been noted.

Cationic lipids and other approaches have been used to enhance the uptake of phosphorothioate oligonucleotides in cells that take up little oligonucleotide *in vitro.*[104–106] Again, however, there are substantial variations from cell type to cell type. Other approaches to enhanced intracellular uptake *in vitro* have included streptolysin D treatment of cells and the use of dextran sulfate and other liposome formulations as well as physical means such as microinjections.[16,66,107]

[100] R. M. Crooke, *in* "Antisense Research and Applications" (S. T. Crooke and B. Lebleu, eds.), p. 471. CRC Press, Boca Raton, FL, 1993.

[101] L. M. Neckers, *in* "Antisense Research and Applications" (S. T. Crooke and B. Lebleu, eds.), p. 451. CRC Press, Boca Raton, FL, 1993.

[102] L. A. Chrisey, S. E. Walz, M. Pazirandeh, and J. R. Campbell, *Antisense Res. Dev.* **3,** 367 (1993).

[103] L. F. Tao, K. A. Marx, W. Wongwit, Z. Jiang, S. Agrawal, and R. M. Coleman, *Antisense Res. Dev.* **5,** 123 (1995).

[104] C. F. Bennett, M. Y. Chiang, H. Chan, and S. Grimm, *J. Liposome Res.* **3,** 85 (1993).

[105] C. F. Bennett, M. Y. Chiang, H. Chan, J. E. E. Shoemaker, and C. K. Mirabelli, *Mol. Pharmacol.* **41,** 1023 (1992).

[106] A. Quattrone, L. Papucci, N. Schiavone, E. Mini, and S. Capaccioli, *Anti-Cancer Drug Design* **9,** 549 (1994).

[107] S. Wang, R. J. Lee, G. Cauchon, D. G. Gorenstein, and P. S. Low, *Proc. Natl. Acad. Sci. U.S.A.* **92,** 3318 (1995).

In Vivo Pharmacokinetics. Phosphorothioate oligonucleotides bind to serum albumin and α_2-macroglobulin. The apparent affinity for albumin is quite low (200–400 μM) and comparable to the low-affinity binding observed for a number of drugs, e.g., aspirin and penicillin.[83–85] Serum protein binding, therefore, provides a repository for these drugs and prevents rapid renal excretion. As serum protein binding is saturable, at higher doses, intact oligomer may be found in urine.[89,108] Studies in our laboratory suggest that in rats, oligonucleotides administered intravenously at doses of 15–20 mg/kg saturate the serum protein-binding capacity.[108a]

Phosphorothioate oligonucleotides are absorbed rapidly and extensively after parenteral administration. For example, in rats, after an intradermal dose of 3.6 mg/kg of [^{14}C]ISIS 2105, a 20-mer phosphorothioate, approximately 70% of the dose was absorbed within 4 hr and total systemic bioavailability was in excess of 90%.[12] After intradermal injection in humans, absorption of ISIS 2105 was similar to that observed in rats.[8] Subcutaneous administration to rats and monkeys results in somewhat lower bioavailability and greater distribution to lymph as would be expected.[108a]

Distribution of phosphorothioate oligonucleotides from blood after absorption or intravenous administration is extremely rapid. We have reported distribution half-lives of less than 1 hr, and similar data have been reported by others.[11,12,89,108] Blood and plasma clerance is multiexponential with a terminal elimination half-life from 40 to 60 hr in all species except humans. In humans the terminal elimination half-life may be somewhat longer.[8]

Phosphorothioates distribute broadly to all peripheral tissues. Liver, kidney, bone marrow, skeletal muscle, and skin accumulate the highest percentage of a dose, but other tissues display small quantities of drug.[11,12] No evidence of significant penetration of the blood–brain barrier has been reported. Rates of incorporation and clearance from tissues vary as a function of the organ studied, with liver accumulating drug most rapidly (20% of a dose within 1–2 hr) and other tissues accumulating drug more slowly. Similarly, elimination of drug is more rapid from liver than any other tissue, e.g., terminal half-life from liver: 62 hr; from renal medulla: 156 hr. The distribution into the kidney has been studied more extensively, and drug is shown to be present in Bowman's capsule, the proximal convoluted tubule, the bush border membrane, and within renal tubular epithelial cells.[109] Data suggest that the oligonucleotides are filtered by the glomerulus and then reabsorbed by the proximal convoluted tubule epithelial cells.

---

[108] P. Iversen, *Anticancer Drug Des.* **6**, 531 (1991).

[108a] Leeds, unpublished data.

[109] J. Rappaport, B. Hanss, J. B. Kopp, T. D. Copeland, L. A. Bruggeman, T. M. Coffman, and P. E. Klotman, *Kidney Int.* **47**, 1462 (1995).

Moreover, the authors suggested that reabsorption might be mediated by interactions with specific proteins in the bush border membranes.

At relatively low doses, the clearance of phosphorothioate oligonucleotides is due primarily to metabolism.[11,12,108] Metabolism is mediated by exo- and endonucleases that result in shorter oligonucleotides and, ultimately, nucleosides that are degraded by normal metabolic pathways. Although no direct evidence of base excision or modification has been reported, these are theoretical possibilities that may occur. In one study, a larger molecular weight radioactive material was observed in urine, but was not fully characterized.[89] Clearly, the potential for conjugation reactions and the extension of oligonucleotides via these drugs serving as primers for polymerases must be explored in more detail. In a very thorough study, 20 nucleotide phosphodiester and phosphorothioate oligonucleotides were administered intravenously at a dose of 6 mg/kg to mice. The oligonucleotides were labeled internally with $[^3H]CH_3$ by methylation of an internal deoxycytidine residue using HhaI methylase and S-$[^3H]$adenosylmethionine.[95] Observations for the phosphorothioate oligonucleotide were entirely consistent with those made in our studies. Additionally, in this paper, autoradiographic analyses showed drug in renal cortical cells.[95]

One study of prolonged infusions of a phosphorothioate oligonucleotide to humans has been reported.[110] In this study, five patients with leukemia were given 10-day intravenous infusions at a dose of 0.05 mg/kg/hr. Elimination half-lives reportedly varied from 5.9 to 14.7 days. Urinary recovery of radioactivity was reported to be 30–60% of the total dose, with 30% of the radioactivity being intact drug. Metabolites in urine included both higher and lower molecular weight compounds. In contrast, when GEM-91 (a 25-mer phosphorothioate oligodeoxynucleotide) was administered to humans as a 2-hr intravenous infusion at a dose of 0.1 mg/kg, a peak plasma concentration of 295.8 mg/ml was observed at the cessation of the infusion. Plasma clearance of total radioactivity was biexponential with initial and terminal eliminations half-lives of 0.18 and 26.71 hr, respectively. However, degradation was extensive and intact drug pharmacokinetic models were not presented. Nearly 50% of the administered radioactivity was recovered in urine, but most of the radioactivity represented degradates. In fact, no intact drug was found in the urine at any time.[111]

In a more recent study in which the level of intact drug was evaluated

[110] E. Bayever, P. L. Iversen, M. R. Bishop, J. G. Sharp, H. K. Tewary, M. A. Arneson, S. J. Pirruccello, R. W. Ruddon, A. Kessinger, and G. Zon, $Antisense\ Res.\ Dev.$ **3**, 383 (1993).
[111] R. Zhang, J. Yan, H. Shahinian, G. Amin, Z. Lu, T. Liu, M. S. Saag, Z. Jiang, J. Temsamani, R. R. Martin, P. J. Schechter, S. Agrawal, and R. B. Diasio, $Clin.\ Pharmacol.\ Ther.$ **58**, 44 (1995).

carefully using capillary gel electrophoresis, the pharmacokinetics of ISIS 2302, a 20-mer phosphorothioate oligodeoxynucleotide, after a 2-hr infusion, were determined. Doses from 0.06 to 2.0 mg/kg were studied, and the peak plasma concentrations were shown to increase linearly, with the 2-mg/kg dose resulting in peak plasma concentrations of intact drug of approximately 9.5 μg/ml. Clearance from plasma, however, was dose dependent, with the 2-mg/kg dose having a clearance of 1.28 ml min^{-1} kg^{-1}, whereas that of 0.5 mg/kg was 2.07 ml min^{-1} kg^{-1}. Essentially, no intact drug was found in urine.

Clearly, the two most recent studies differ from the initial report in several facets. Although a number of factors may explain the discrepancies, the most likely explanation is related to the evolution of assay methodology, not a difference between compounds. Overall, the behavior of phosphorothioates in the plasma of humans appears to be similar to that in other species.

In addition to the pharmacological effects that have been observed after phosphorothioate oligonucleotides have been administered to animals (and humans), a number of other lines of evidence show that these drugs enter cells in organs. Autoradiographic, fluorescent, and immunohistochemical approaches have shown that these drugs are localized in endopromal convoluted tubular cells, various bone marrow cells, cells in the skin, and liver.[109,112,113]

Perhaps more compelling and of more long-term value are studies showing the distribution of phosphorothioate oligonucleotides in the liver of rats treated intravenously with the drugs at various doses.[114] This study showed that the kinetics and extent of the accumulation into the Kuppfer, endothelial, and hepatocyte cell population varied and that as doses were increased, the distribution changed. Moreover, the study showed that subcellular distribution also varied.

We have also performed oral bioavailability experiments in rodents treated with an H$_2$ receptor antagonist to avoid acid-mediated depurination or precipitation. In these studies, very limited (<5%) bioavailability was observed.[114a] However, it seems likely that a principal limiting factor in the oral bioavailability of phosphorothioates may be degradation in the gut rather than absorption. Studies using everted rat jejunum sacs demonstrated

[112] Y. Takakura, R. I. Mahato, M. Yoshida, T. Kanamaru, and M. Hashida, *Antisense Nucleic Acid Drug Del.* **6**, 177 (1996).
[113] M. Butler, K. Stecker, and C. F. Bennett, *Lab. Invest.* **77**, 379 (1997).
[114] M. J. Graham, S. T. Crooke, D. K. Monteith, S. R. Cooper, K. M. Lemonidis, K. K. Stecker, M. J. Martin, and R. M. Crooke, *J. Pharmacol. Exp. Ther.* **286**, 447 (1998).
[114a] D. K. Crooke, unpublished observations.

passive transport across the intestinal epithelium.[115] Further, studies using more stable 2′-methoxy phosphorothioate oligonucleotides showed a significant increase in oral bioavailability that appeared to be associated with the improved stability of the analogs.[116]

In summary, pharmacokinetic studies of several phosphorothioates demonstrate that they are well absorbed from parenteral sites, distribute broadly to all peripheral tissues, do not cross the blood–brain barrier, and are eliminated primarily by slow metabolism. In short, once a day or every other day systemic dosing should be feasible. Although the similarities between oligonucleotides of different sequences are far greater than the differences, additional studies are required before determining whether there are subtle effects of sequence on the pharmacokinetic profile of this class of drugs.

Pharmacological Properties

Molecular Pharmacology. Antisense oligonucleotides are designed to bind to RNA targets via Watson–Crick hybridization. As RNA can adopt a variety of secondary structures via Watson–Crick hybridization, one useful way to think of antisense oligonucleotides is as competitive antagonists for self-complementary regions of the target RNA. Obviously, creating oligonucleotides with the highest affinity per nucleotide unit is pharmacologically important, and a comparison of the affinity of the oligonucleotide to a complementary RNA oligonucleotide is the most sensible comparison. In this context, phosphorothioate oligodeoxynucleotides are relatively competitively disadvantaged as the affinity per nucleotide unit of oligomer is less than RNA ($>-2.0°$ T_m per unit).[117] This results in a requirement of at least 15–17 nucleotides in order to have sufficient affinity to produce biological activity.[81]

Although multiple mechanisms by which an oligonucleotide may terminate the activity of an RNA species to which it binds are possible, examples of biological activity have been reported for only three of these mechanisms. Antisense oligonucleotides have been reported to inhibit RNA splicing, affect translation of mRNA, and induce degradation of RNA by RNase H.[17,22,27] Without question, the mechanism that has resulted in the most

[115] J. A. Hughes, A. V. Avrutskaya, K. L. R. Brouwer, E. Wickstrom, and R. L. Juliano, *Pharm. Res.* **12,** 817 (1995).

[116] S. Agrawal, X. Zhang, Z. Lu, H. Zhao, J. M. Tamburin, J. Yan, H. Cai, R. B. Diasio, I. Habus, Z. Jiang, R. P. Iyer, D. Yu, and R. Zhang, *Biochem. Pharmacol.* **50,** 571 (1995).

[117] P. D. Cook, *in* "Antisense Research and Applications" (S. T. Crooke and B. Lebleu, eds.), p. 149. CRC Press, Boca Raton, FL, 1993.

potent compounds and is best understood is RNase H activation. To serve as a substrate for RNase H, a duplex between RNA and a "DNA-like" oligonucleotide is required. Specifically, a sugar moiety in the oligonucleotide that induces a duplex conformation equivalent to that of a DNA–RNA duplex and a charged phosphate are required.[118] Thus, phosphorothioate oligodeoxynucleotides are expected to induce RNase H-mediated cleavage of the RNA when bound. As discussed later, many chemical approaches that enhance the affinity of an oligonucleotide for RNA result in duplexes that are no longer substrates for RNase H.

Selection of sites at which optimal antisense activity may be induced in a RNA molecule is complex, dependent on terminating mechanism, and influenced by the chemical class of the oligonucleotide. Each RNA appears to display unique patterns of sites of sensitivity. Within the phosphorothioate oligodeoxynucleotide chemical class, studies in our laboratory have shown that antisense activity can vary from undetectable to 100% by shifting an oligonucleotide by just a few bases in the RNA target.[17,78,119] Although significant progress has been made in developing general rules that help define potentially optimal sites in RNA species, to a large extent, this remains an empirical process that must be performed for each RNA target and every new chemical class of oligonucleotides.

Phosphorothioates have also been shown to have effects inconsistent with the antisense mechanism for which they were designed. Some of these effects are due to sequence or are structure specific. Others are due to nonspecific interactions with proteins. These effects are particularly prominent in *in vitro* tests for antiviral activity as often high concentrations of cells, viruses, and oligonucleotides are coincubated.[15,120] The human immune deficiency virus is particularly problematic as many oligonucleotides bind to the gp 120 protein.[3] However, the potential for confusion arising from the misinterpretation of an activity as being due to an antisense mechanism when, in fact, it is due to nonantisense effects is certainly not limited to antiviral or just *in vitro* tests.[121–123] Again, these data simply urge caution

[118] C. K. Mirabelli and S. T. Crooke, *in* "Antisense Research and Applications" (S. T. Crooke and B. Lebleu, eds.), p. 7. CRC Press, Boca Raton, FL, 1993.

[119] C. F. Bennett, and S. T. Crooke, *in* "Ther. Modulation Cytokines" (B. B. Henderson and W. Mark, eds.), p. 171. CRC, Boca Raton, FL, 1996.

[120] R. W. Wagner, M. D. Matteucci, J. G. Lewis, A. J. Gutierrez, C. Moulds, and B. C. Froehler, *Science* **260**, 1510 (1993).

[121] C. M. Barton and N. R. Lemoine, *Br. J. Cancer* **71**, 429 (1995).

[122] T. L. Burgess, E. F. Fisher, S. L. Ross, J. V. Bready, Y. Qian, L. A. Bayewitch, A. M. Cohen, C. J. Herrera, S. F. Hu, T. B. Kramer, F. D. Lott, F. H. Martin, G. F. Pierce, L. Simonet, and C. L. Farrell, *Proc. Natl. Acad. Sci. U.S.A.* **92**, 4051 (1995).

[123] M. Hertl, L. M. Neckers, and S. I. Katz, *J. Invest. Dermatol.* **104**, 813 (1995).

and argue for careful dose–response curves, direct analyses of target protein or RNA, and inclusion of appropriate controls before drawing conclusions concerning the mechanisms of action of oligonucleotide-based drugs. In addition to protein interactions, other factors, such as overrepresented sequences of RNA and unusual structures that may be adopted by oligonucleotides, can contribute to unexpected results.[3]

Given the variability in cellular uptake of oligonucleotides, the variability in potency as a function of the binding site in an RNA target, and potential nonantisense activities of oligonucleotides, careful evaluation of dose–response curves and clear demonstration of the antisense mechanism are required before drawing conclusions from *in vitro* experiments. Nevertheless, numerous well-controlled studies have been reported in which antisense activity was demonstrated conclusively. As many of these studies have been reviewed previously, suffice it to say that antisense effects of phosphorothioate oligodeoxynucleotides against a variety of targets are well documented.[1,9,56,78,124]

In Vivo Pharmacological Activities. A relatively large number of reports of *in vivo* activities of phosphorothioate oligonucleotides have now appeared documenting activities after both local and systemic administration (for reviews see Crooke[125]). However, for only a few of these reports have sufficient studies been performed to draw relatively firm conclusions concerning the mechanism of action. Consequently, this article review, in some detail only a few reports that provide sufficient data to support a relatively firm conclusion regarding a mechanism of action. Local effects have been reported for phosphorothioate and methylphosphonate oligonucleotides. A phosphorothioate oligonucleotide designed to inhibit c-*myb* production and applied locally was shown to inhibit intimal accumulation in the rat carotid artery.[126] In this study, a Northern blot analysis showed a significant reduction in c-*myb* RNA in animals treated with the antisense compound, but no effect by a control oligonucleotide. In another study, the effects of the oligonucleotide were suggested to be due to a nonantisense mechanism.[12] However, only one dose level was studied, so much remains to be done before definitive conclusions are possible. Similar effects were reported for phosphorothioate oligodeoxynucleotides designed to inhibit cyclin-dependent kinases (CDC-2 and CDK-2). Again, the antisense oligonucleotide inhibited intimal thickening and cyclin-dependent kinase activity, while a control oligonucleotide had no effect.[127] Additionally, local

[124] K. M. Nagel, S. G. Holstad, and K. E. Isenberg, *Pharmacotherapy* **13**, 177 (1993).
[125] S. T. Crooke, ed., *in* "Handbook of Experimental Pharmacology." Springer, Berlin, 1998.
[126] M. Simons, E. R. Edelman, J.-L. DeKeyser, R. Langer, and R. D. Rosenberg, *Nature* **359**, 67 (1992).
[127] J. Abe, W. Zhou, J. Taguchi, N. Takuwa, K. Miki, H. Okazaki, K. Kurokawa, M. Kumada, and Y. Takuwa, *Biochem. Biophys. Res. Commun.* **198**, 16 (1994).

administration of a phosphorothioate oligonucleotide designed to inhibit n-*myc* resulted in reduction in n-*myc* expression and slower growth of a subcutaneously transplanted human tumor in nude mice.[128]

Antisense oligonucleotides administered intraventricularly have been reported to induce a variety of effects in the central nervous system. Intraventricular injection of antisense oligonucleotides to neuropeptide-y-y1 receptors reduced the density of the receptors and resulted in behavioral signs of anxiety.[129] Similarly, an antisense oligonucleotide designed to bind to NMDA-R1 receptor channel RNA inhibited the synthesis of these channels and reduced the volume of focal ischemia produced by occlusion of the middle cerebral artery in rats.[129]

In a series of well-controlled studies, antisense oligonucleotides administered intraventricularly selectively inhibited dopamine type 2 receptor expression, dopamine type 2 receptor RNA levels, and behavioral effects in animals with chemical lesions. Controls included randomized oligonucleotides and the observation that no effects were observed on dopamine type 1 receptor or RNA levels.[130–132] This laboratory also reported the selective reduction of dopamine type 1 receptor and RNA levels with the appropriate oligonucleotide.[133]

Similar observations were reported in studies on AT-1 angiotensin receptors and tryptophan hydroxylase. In studies in rats, direct observations of AT-1 and AT-2 receptor densities in various sites in the brain after the administration of different doses of phosphorothioate antisense, sense, and scrambled oligonucleotides were reported.[134] Again, in rats, intraventricular administration of the phosphorothioate antisense oligonucleotide resulted in a decrease in tryptophan hydroxylase levels in the brain, while a scrambled control did not.[135]

Injection of antisense oligonucleotides to synaptosomal-associated protein-25 into the vitreous body of rat embryos reduced the expression of the protein and inhibited neurite elongation by rat cortical neurons.[136]

Aerosol administration to rabbits of an antisense phosphorothioate oligodeoxynucleotide designed to inhibit the production of the antisense

[128] L. Whitesell, A. Rosolen, and L. M. Neckers, *Antisense Res. Dev.* **1,** 343 (1991).

[129] C. Wahlestedt, E. M. Pich, G. F. Koob, F. Yee, and M. Heilig, *Science* **259,** 528 (1993).

[130] B. Weiss, L.-W. Zhou, S.-P. Zhang, and Z.-H. Qin, *Neuroscience* **55,** 607 (1993).

[131] L.-W. Zhou, S.-P. Zhang, Z.-H. Qin, and B. Weiss, *J. Pharmacol. Exp. Therap.* **268,** 1015 (1994).

[132] Z. H. Qin, L. W. Zhou, S. P. Zhang, Y. Wang, and B. Weiss, *Mol. Pharmacol.* **48,** 730 (1995).

[133] S.-P. Zhang, L.-W. Zhou, and B. Weiss, *J. Pharmacol. Exp. Therap.* **271,** 1462 (1994).

[134] P. Ambuhl, R. Gyurko, and M. I. Phillips, *Regul. Pept.* **59,** 171 (1995).

[135] M. M. McCarthy, D. A. Nielsen, and D. Goldman, *Regul. Pept.* **59,** 163 (1995).

[136] A. Osen-Sand, M. Catsicas, J. K. Staple, K. A. Jones, G. Ayala, J. Knowles, G. Grenningloh, and S. Catsicas, *Nature* **364,** 445 (1993).

A_1 receptor has been reported to reduce receptor numbers in the airway smooth muscle and to inhibit adenosine, house dust mite allergen, and histamine-induced bronchoconstriction.[137] Neither control nor oligonucleotide complementary to bradykinin B_2 receptors reduced the density of adenosine A_1 receptors, although the oligonucleotides complementary to bradykin in B_2 receptor mRNA reduced the density of these receptors.

In addition to local and regional effects of antisense oligonucleotides, a growing number of well-controlled studies have demonstrated systemic effects of phosphorothioate oligodeoxynucleotides. The expression of interleukin-1 in mice was inhibited by the systemic administration of antisense oligonucleotides.[138] Oligonucleotides to the NF-κB p65 subunit administered intraperitoneally at 40 mg/kg every 3 days slowed tumor growth in mice transgenic for human T-cell leukemia viruses.[139] Similar results with other antisense oligonucleotides were shown in another *in vivo* tumor model after either prolonged subcutaneous infusion or intermittent subcutaneous injection.[140]

Several reports further extend the studies of phosphorothioate oligonucleotides as antitumor agents in mice. In one study, a phosphorothioate oligonucleotide directed to inhibition of the *bcr-abl* oncogene was administered at a dose of 1 mg/day for 9 days intravenously to immunodeficient mice injected with human leukemic cells. The drug was shown to inhibit the development of leukemic colonies in mice and to selectively reduce *bcr-abl* RNA levels in peripheral blood lymphocytes, spleen, bone marrow, liver, lungs, and brain.[19] However, it is possible that the effects on RNA levels were secondary to effects on the growth of various cell types. In the second study, a phosphorothioate oligonucleotide antisense to the proto-oncogene *myb* inhibited the growth of human melanoma in mice. Again, *myb* mRNA levels appeared to be selectively reduced.[141]

A number of studies from our laboratories that directly examined target RNA levels, target protein levels, and pharmacological effects using a wide range of control oligonucleotides and examination of the effects on closely related isotypes have been completed. Single and chronic daily administration of a phosphorothioate oligonucleotide designed to inhibit mouse protein kinase C-a (PKC-a) selectively inhibited expression of PKC-a RNA

[137] J. W. Nyce and W. J. Metzger, *Nature* **385,** 721 (1997).
[138] R. M. Burch and L. C. Mahan, *J. Clin. Invest.* **88,** 1190 (1991).
[139] I. Kitajima, T. Shinohara, J. Bilakovics, D. A. Brown, X. Xiao, and M. Nerenberg, *Science* **258,** 1792 (1992).
[140] K. A. Higgins, J. R. Perez, T. A. Coleman, K. Dorshkind, W. A. McComas, U. M. Sarmiento, C. A. Rosen, and R. Narayan, *Proc. Natl. Acad. Sci. U.S.A.* **90,** 9901 (1993).
[141] N. Hijiya, J. Zhang, M. Z. Ratajczak, J. A. Kant, K. DeRiel, M. Herlyn, G. Zon, and A. M. Gewirtz, *Proc. Natl. Acad. Sci. U.S.A.* **91,** 4499 (1994).

in mouse liver without effects on any other isotype. The effects lasted at least 24 hr after a dose, and a clear dose–response curve was observed with a dose of 10–15 mg/kg intraperitoneally reducing PKC-a RNA levels in liver by 50% 24 hr after a dose.[18]

A phosphorothioate oligonucleotide designed to inhibit human PKC-a expression selectively inhibited the expression of PKC-a RNA and PKC-a protein in human tumor cell lines implanted subcutaneously in nude mice after intravenous administration.[142] In these studies, effects on RNA and protein levels were highly specific and were observed at doses lower than 6 mg/kg/day. A large number of control oligonucleotides failed to show activity.

In a similar series of studies, Monia et al.[143,144] demonstrated a highly specific loss of human c-raf kinase RNA in human tumor xenografts and antitumor activity that correlated with the loss of RNA.

Finally, a single injection of a phosphorothioate oligonucleotide designed to inhibit c-AMP-dependent protein kinase type 1 was reported to selectively reduce RNA and protein levels in human tumor xenografts and to reduce tumor growth.[145]

Thus, there is a growing body of evidence that phosphorothioate oligonucleotides can induce potent systemic and local effects in vivo. More importantly, there are now a number of studies with sufficient controls and direct observation of target RNA and protein levels to suggest highly specific effects that are difficult to explain via any mechanism other than antisense. As would be expected, the potency of these effects varies depending on the target, the organ, and the end point measured as well as the route of administration and the time after a dose when the effect is measured.

In conclusion, although it is of obvious importance to interpret in vivo activity data cautiously, and it is clearly necessary to include a range of controls and to evaluate effects on target RNA and protein levels and control RNA and protein levels directly, it is difficult to argue with the conclusion that some effects observed in animals are most likely primarily due to an antisense mechanism.

Additionally, in studies on patients with cytomegalovirus-induced retinitis, local injections of ISIS 2922 have resulted in impressive efficacy, although it is obviously impossible to prove that the mechanism of action is

[142] N. Dean, R. McKay, L. Miraglia, R. Howard, S. Cooper, J. Giddings, P. Nicklin, L. Meister, R. Ziel et. al., Cancer Res. 56, 3499 (1996).
[143] B. P. Monia, J. F. Johnston, T. Geiger, M. Muller, and D. Fabbro, Nature Med. 2, 668 (1995).
[144] B. P. Monia, J. F. Johnston, H. Sasmor, and L. L. Cummins, J. Biol. Chem. 271, 14533 (1996).
[145] M. Nesterova and Y. S. Cho-Chung, Nature Med. 1, 528 (1995).

antisense in these studies.[146] This drug has now been approved for commercialization by the FDA. More recently, ISIS 2302, an ICAM-1 inhibitor, was reported to result in statistically significant reductions in steroid doses and prolonged remissions in a small group of steroid-dependent patients with Crohn's disease. As this study was randomized, double-blinded, and included serial colonoscopies, it may be considered the first study in humans to demonstrate the therapeutic activity of an antisense drug after systemic administration.[147] Finally, ISIS 5132 has been shown to reduce c-*raf* kinase message levels in peripheral blood mononuclear cells of patients with cancer after intravenous dosing.[148]

Toxicological Properties

In Vitro. In our laboratory, we have evaluated the toxicities of scores of phosphorothioate oligodeoxynucleotides in a significant number of cell lines in tissue culture. As a general rule, no significant cytotoxicity is induced at concentrations below 100 μM oligonucleotide. Additionally, with a few exceptions, no significant effect on macromolecular synthesis is observed at concentrations below 100 μM.[79,100]

Polynucleotides and other polyanions have been shown to cause the release of cytokines.[149] Also, bacterial DNA species have been reported to be mitogenic for lymphocytes *in vitro.*[150] Furthermore, oligodeoxynucleotides (30–45 nucleotides in length) were reported to induce and enhance the natural killer (NK) cell activity of interferon.[151] In the latter study, oligonucleotides that displayed NK cell stimulating activity contained specific palindromic sequences and tended to be guanosine rich. Collectively, these observations indicate that nucleic acids may have broad immunostimulatory activity.

It has been shown that phosphorothioate oligonucleotides stimulate B-lymphocyte proliferation in a mouse splenocyte preparation (analogous to bacterial DNA), and the response may underlie the observations of

[146] S. L. Hutcherson, A. G. Palestine, H. L. Cantrill, R. M. Lieberman, G. N. Holland, and K. P. Anderson, *35th ICAAC*, 204 (1995).
[147] B. R. Yacyshyn, M. B. Bowen-Yacyshyn, L. Jewell, J. A. Tami, C. F. Bennett, D. L. Kisner, and W. R. Shanahan, Jr., *Gastroenterology* **114,** 1133 (1998).
[148] P. J. O'Dwyer, J. P. Stevenson, M. Gallagher, E. Mitchell, D. Friedland, L. Rose, A. Cassella, J. Holmlund, N. Dean, A. Dorr, J. Geary, and K.-S. Yao, *in* "34th Annual Meeting of the American Society of Clinical Oncology." Los Angeles, CA, 1998.
[149] C. J. Colby, *Prog. Nucleic Acid Res. Mol. Biol.* **11,** 1 (1971).
[150] J. P. Messina, G. S. Gilkeson, and D. S. Pisetsky, *J. Immunol.* **147,** 1759 (1991).
[151] E. Kuramoto, O. Yano, Y. Kimura, M. Baba, T. Makino, S. Yamamoto, T. Yamamoto, T. Kataoka, and T. Tokunaga, *Jpn. J. Cancer Res.* **83,** 1128 (1992).

lymphoid hyperplasia in the spleen and lymph nodes of rodents caused by repeated administration of these compounds (see later).[152] We also have evidence of enhanced cytokine release by immunocompetent cells when exposed to phosphorothioates *in vitro*.[153] In this study, both human keratinocytes and an *in vitro* model of human skin released interleukin 1a (IL-1a) when treated with 250 μM–1 mm of phosphorothioate oligonucleotides. The effects seemed to be dependent on the phosphorothioate backbone and independent of sequence or 2'-modification. In a study in which murine B lymphocytes were treated with phosphodiester oligonucleotides, B-cell activation was induced by oligonucleotides with unmethylated CpG dinucleotides.[154] This has been extrapolated to suggest that the CpG motif may be required for the immune stimulation of oligonucleotide analogs such as phosphorothioates. This is clearly not the case with regard to release of IL-1a from keratinocytes[153] nor is it the case with regard to *in vivo* immune stimulation (see later).

Genotoxicity. As with any new chemical class of therapeutic agents, concerns about genotoxicity cannot be dismissed as little *in vitro* testing has been performed and no data from long-term studies of oligonucleotides are available. Clearly, given the limitations in our understanding about the basic mechanisms that might be involved, empirical data must be generated. We have performed mutagenicity studies on two phosphorothioate oligonucleotides, ISIS 2105 and ISIS 2922, and found them to be nonmutagenic at all concentrations studied.[8]

Two mechanisms of genotoxicity that may be unique to oligonucleotides have been considered. One possibility is that an oligonucleotide analog could be integrated into the genome and produce mutagenic events. Although integration of an oligonucleotide into the genome is conceivable, it is likely to be extremely rare. For most viruses, viral DNA integration is itself a rare event and, of course, viruses have evolved specialized enzyme-mediated mechanisms to achieve integration. Moreover, preliminary studies in our laboratory have shown that phosphorothioate oligodeoxynucleotides are generally poor substrates for DNA polymerases, and it is unlikely that enzymes such as integrases, gyrases, and topoisomerases (that have obligate DNA cleavage as intermediate steps in their enzymatic processes) will accept these compounds as substrates. Consequently, it would seem that the risk of genotoxicity due to genomic integration is no greater and proba-

[152] D. S. Pisetsky and C. F. Reich, *Life Sci.* **54,** 101 (1994).

[153] R. M. Crooke, S. T. Crooke, M. J. Graham, and M. E. Cooke, *Toxicol. Appl. Pharmacol.* **140,** 85 (1996).

[154] A. M. Krieg, A.-K. Yi, S. Matson, T. J. Waldschmidt, G. A. Bishop, R. Teasdale, G. A. Koretzky, and D. M. Klinman, *Nature* **374,** 546 (1995).

bly less than that of other potential mechanisms, e.g., alteration of the activity of growth factors, cytokine release, nonspecific effects on membranes that might trigger arachidonic acid release, or inappropriate intracellular signaling. Presumably, new analogs that deviate significantly more from natural DNA would be even less likely to be integrated.

A second concern that has been raised about possible genotoxicity is the risk that oligonucleotides might be degraded to toxic or carcinogenic metabolites. However, metabolism of phosphorothioate oligodeoxynucleotides by base excision would release normal bases, which presumably would be nongenotoxic. Similarly, oxidation of the phosphorothioate backbone to the natural phosphodiester structure would also yield nonmutagenic (and probably nontoxic) metabolites. Finally, it is possible that phosphorothioate bonds could be hydrolyzed slowly, releasing nucleoside phosphorothioates that presumably would be oxidized rapidly to natural (nontoxic) nucleoside phosphates. However, oligonucleotides with modified bases and/or backbones may pose different risks.

In Vivo. The acute LD_{50} in mice of all phosphorothioate oligonucleotides tested to date is in excess of 500 mg/kg.[154a] In rodents, we have had the opportunity to evaluate the acute and chronic toxicities of multiple phosphorothioate oligonucleotides administered by multiple routes.[155,156] The consistent dose-limiting toxicity was immune stimulation manifested by lymphoid hyperplasia, spelnomegaly, and a multiorgan monocellular infiltrate. These effects occurred only with chronic dosing at doses >20 mg/kg and were dose dependent. The liver and kidney were the organs affected most prominently by monocellular infiltrates. All of these effects appeared to be reversible, and chronic intradermal administration appeared to be the most toxic route, probably because of high local concentrations of the drugs resulting in local cytokine release and initiation of a cytokine cascade. There were no obvious effects of sequence. At doses of 100 mg/kg and greater, minor increases in liver enzyme levels and mild thrombocytopenia were also observed.

In monkeys, however, the toxicological profile of phosphorothioate oligonucleotides is quite different. The most prominent dose-limiting side effect is sporadic reductions in blood pressure associated with bradycardia. When these events are observed, they are often associated with activation

[154a] D. J. Kornbrust, unpublished observations.
[155] S. P. Henry, L. R. Grillone, J. L. Orr, R. H. Brunner, and D. J. Kornbrust, *Toxicology* **116,** 77 (1997).
[156] S. P. Henry, J. Taylor, L. Midgley, A. A. Levin, and D. J. Kornbrust, *Antisense Nucleic Acid Drug Dev.* **7,** 473 (1997).

of the C-5 complement and they are dose related and peak plasma concentration related. This appears to be related to the activation of the alternative pathway.[157] All phosphorothioate oligonucleotides tested to date appear to induce these effects, although there may be slight variations in potency as a function of sequence and/or length.[156,158,159]

A second prominent toxicologic effect in the monkey is the prolongation of activated partial thromboplastin time. At higher doses, evidence of clotting abnormalities is observed. Again, these effects are dose and peak plasma concentration dependent.[156,159] Although no evidence of sequence dependence has been observed, there appears to be a linear correlation between the number of phosphorothioate linkages and the potency among 18–25 nucleotides.[159a] Mechanisms responsible for these effects are likely very complex, but preliminary data suggest that direct interactions with thrombin may be at least partially responsible for the effects observed.[160]

In humans, again the toxicological profile differs a bit. When ISIS 2922 is administered intravitreally to patients with cytomegalovirus retinitis, the most common adverse event is anterior chamber inflammation, which is managed easily with steroids. A relatively rare and dose-related adverse event is morphological changes in the retina associated with loss in peripheral vision.[146]

When ISIS 2105, a 20-mer phosphorothioate designed to inhibit the replication of human papilloma viruses that cause genital warts, is administered intradermally at doses as high as 3 mg/wart weekly for 3 weeks, essentially no toxicities have been observed, including, remarkably, a complete absence of local inflammation.

Every other day administration of 2-h intravenous infusions of ISIS 2302 at doses as high as 2 mg/kg resulted in no significant toxicities, including no evidence of immune stimulation and no hypotension. A slight subclinical increase in APTT was observed at the 2-mg/kg dose.[161]

[157] S. P. Henry, P. C. Giclas, J. Leeds, M. Pangburn, C. Auletta, A. A. Levin, and D. J. Kornbrust, *J. Pharmacol. Exp. Ther.* **281,** 810 (1997).
[158] K. G. Cornish, P. Iversen, L. Smith, M. Arneson, and E. Bayever, *Pharmacol. Comm.* **3,** 239 (1993).
[159] W. M. Galbraith, W. C. Hobson, P. C. Giclas, P. J. Schechter, and S. Agrawal, *Antisense Res. Dev.* **4,** 201 (1994).
[159a] P. Nicklin, unpublished observation.
[160] S. P. Henry, W. Novotny, J. Leeds, C. Auletta, and D. J. Kornbrust, *Antisense Nucleic Acid Drug Dev.* **7,** 503 (1997).
[161] J. M. Glover, J. M. Leeds, T. G. K. Mant, D. Amin, D. L. Kisner, J. E. Zuckerman, R. S. Geary, A. A. Levin, and W. R. Shanahan, Jr., *J. Pharmacol. Exp. Ther.* **282,** 1173 (1997).

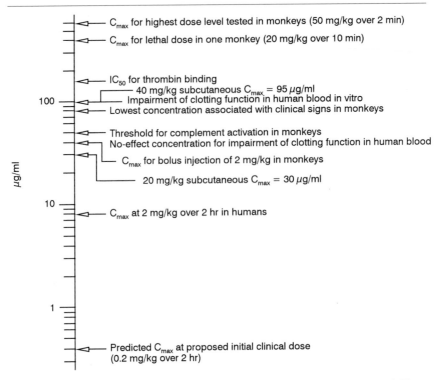

FIG. 2. Plasma concentrations of ISIS 2302 at which various activities are observed. These concentrations are determined by extracting plasma and analyzing by capillary gel electrophoresis and represent intact ISIS 2302.

Therapeutic Index

Figure 2 attempts to put the toxicities and their dose–response relationships in a therapeutic context. This is particularly important as considerable confusion has arisen concerning the potential utility of phosphorothioate oligonucleotides for selected therapeutic purposes deriving from the unsophisticated interpretation of toxicological data. As can be seen readily, the immune stimulation induced by these compounds appears to be particularly prominent in rodents and is unlikely to be dose limiting in humans. No hypotensive events have been observed in humans to date. Thus, this toxicity appears to occur at lower doses in monkeys than in humans and is certainly not dose limiting in humans.

Based on our experience to date, we believe that the dose-limiting toxicity in humans will be clotting abnormalities, which will be associated with peak plasma concentrations well in excess of 10 μg/ml. In animals, pharmacological activities have been observed with intravenous bolus doses from 0.006 to 10–15 mg/kg, depending on the target, the end point, the organ studied, and the time after a dose when the effect is measured. Thus, it would appear that phosphorothioate oligonucleotides have a therapeutic index that supports their evaluation for a number of therapeutic indications.

Conclusions

Phosphorothioate oligonucleotides have perhaps outperformed many expectations. They display attractive parenteral pharmacokinetic properties. They have produced potent systemic effects in a number of animal models and, in many experiments, the antisense mechanism has been directly demonstrated as the hoped-for selectivity. Further, these compounds appear to display satisfactory therapeutic indices for many indications.

Nevertheless, phosphorothioates clearly have significant limits. Pharmacodynamically, they have relatively low affinity per nucleotide unit. This means that longer oligonucleotides are required for biological activity and that invasion of many RNA structures may not be possible. At higher concentrations, these compounds inhibit RNase H as well. Thus, the higher end of the pharmacologic dose–response curve is lost. Pharmacokinetically, phosphorothioates do not cross the blood–brain barrier, are not significantly orally bioavailable, and may display dose-dependent pharmacokinetics. Toxicologically, clearly the release of cytokines, activation of complement, and interference with clotting will pose dose limits if they are encountered in the clinic.

As several clinical trials are in progress with phosphorothioates and others will be initiated shortly, we shall soon have more definitive information about the activities, toxicities, and value of this class of antisense drugs in human beings.

Medicinal Chemistry of Oligonucleotides

Introduction

The core of any rational drug discovery program is medicinal chemistry. Although the synthesis of modified nucleic acids has been a subject of interest for some time, the intense focus on the medicinal chemistry of oligonucleotides dates perhaps to no more than the early 1990s. Consequently, the scope of medicinal chemistry has expanded enormously, but biological data to support conclusions about synthetic strategies are only beginning to emerge.

Modifications in the base, sugar, and phosphate moieties of oligonucleotides have been reported. The subjects of medicinal chemical programs include approaches to create enhanced affinity and more selective affinity for RNA or duplex structures, the ability to cleave nucleic acid targets, enhanced nuclease stability, cellular uptake and distribution, and *in vivo* tissue distribution, metabolism, and clearance.

Heterocycle Modifications

Pyrimidine Modifications. A relatively large number of modified pyrimidines have been synthesized and are now incorporated into oligonucleotides and are being evaluated. The principal sites of modification are C-2, C-4, C-5, and C-6. These and other nucleoside analogs have been reviewed thoroughly.[162] Consequently, a very brief summary of the analogs that displayed interesting properties is incorporated here.

Inasmuch as the C-2 position is involved in Watson–Crick hybridization, C-2-modified pyrimidine-containing oligonucleotides have shown unattractive hybridization properties. An oligonucleotide containing 2-thiothymidine was found to hybridize well to DNA and, in fact, even better to RNA ΔT_m 1.5° modification.

In contrast, several modifications in the 4 position that have interesting properties have been reported. 4-Thiopyrimidines have been incorporated into oligonucleotides with no significant negative effect on hybridization.[163] A bicyclic and an N^4-methoxy analog of cytosine were shown to hybridize with both purine bases in DNA with T_m values approximately equal to natural base pairs.[164] Additionally, a fluorescent base has been incorporated into oligonucleotides and shown to enhance DNA–DNA duplex stability.[165]

A large number of modifications at the C-5 position have also been reported, including halogenated nucleosides. Although the stability of duplexes may be enhanced by incorporating 5-halogenated nucleosides, the occasional mispairing with G and the potential that the oligonucleotide might degrade and release toxic nucleosides analogs cause concern.[162]

Furthermore, oligonucleotides containing 5-propynylpyrimidine modifications have been shown to enhance the duplex stability ΔT_m 1.6°/modifi-

[162] Y. S. Sanghvi, *in* "Antisense Research and Applications" (S. T. Crooke and B. Lebleu, eds.), p. 273. CRC Press, Boca Raton, FL, 1993.

[163] T. T. Nikiforov and B. A. Connolly, *Tetrahed. Lett.* **32**, 3851 (1991).

[164] P. K. T. Lin and D. M. Brown, *Nucleic Acids Res.* **17**, 10373 (1989).

[165] H. Inoue, A. Imura, and E. Ohtsuka, *Nucleic Acids Res.* **13**, 7119 (1985).

cation and support the RNase H activity. 5-Heteroarylpyrimidines were also shown to influence the stability of duplexes.[120,166] A more dramatic influence was reported for the tricyclic 2'-deoxycytidine analogs, exhibiting an enhancement of 2–5°/modification depending on the positioning of the modified bases.[167] It is believed that the enhanced binding properties of these analogs are due to extended stacking and increased hydrophobic interactions.

In general, as expected, modifications in the C-6 position of pyrimidines are highly duplex destabilizing.[168] Oligonucleotides containing 6-aza pyrimidines not only reduce T_m by 1–2° per modification, but enhance the nuclease stability of oligonucleotides and support RNase H-induced degradation of RNA targets.[162]

Purine Modifications. Although numerous purine analogs have been synthesized, when incorporated into oligonucleotides, they usually have resulted in the destabilization of duplexes. However, there are a few exceptions where a purine modification had a stabilizing effect. A brief summary of some of these analogs is discussed.

Generally, N-1 modifications of purine moiety have resulted in destabilization of the duplex.[169] Similarly, C-2 modifications have usually resulted in destabilization. However, 2,6-diaminopurine has been reported to enhance hybridization by approximately 1° per modification when paired with T.[170] Of the 3-position substituted bases reported to date, only the 3-deazaadenosine analog has been shown to have no negative effect on hybridization.

Modifications at the C-6 and C-7 positions have likewise resulted in only a few interesting bases from the point of view of hybridization. Inosine has been shown to have little effect on duplex stability, but because it can pair and stack with all four normal DNA bases, it behaves as a universal base and creates an ambiguous position in an oligonucleotide.[171] Incorporation of 7-deazainosine into oligonucleotides was destabilizing, and this was considered to be due to its relatively hydrophobic nature.[172] 7-Deazaguanine was

[166] A. J. Gutierrez, T. J. Terhorst, M. D. Matteucci, and B. C. Froehler, *J. Am. Chem. Soc.* **116,** 5540 (1994).

[167] K.-Y. Lin, R. J. Jones, and M. Matteucci, *J. Am. Chem. Soc.* **117,** 3873 (1995).

[168] Y. S. Sanghvi, G. D. Hoke, S. M. Freier, M. C. Zounes, C. Gonzalez, L. Cummins, H. Sasmor, and P. D. Cook, *Nucleic Acids Res.* **21,** 3197 (1993).

[169] M. Manoharan, *in* "Antisense Research and Applications" (S. T. Crooke and B. Lebleu, eds.), p. 303. CRC Press, Boca Raton, FL, 1993.

[170] B. S. Sproat, A. M. Iribarren, R. G. Garcia, and B. Beijer, *Nucleic Acids Res.* **19,** 733 (1991).

[171] F. H. Martin, M. M. Castro, F. Aboul-ela, and I. J. Tinoco, *Nucleic Acids. Res.* **13,** 8927 (1985).

[172] J. SantaLucia, Jr., R. Kierzek, and D. H. Turner, *J. Am. Chem. Soc.* **113,** 4313 (1991).

similarly destabilizing, but when 8-aza-7-deazaguanine was incorporated into oligonucleotides, it enhanced hybridizations.[173] Thus, on occasion, introduction of more than one modification in a nucleobase may compensate for the destabilizing effects of some modifications. Interestingly, the 7-iodo-7-deazaguanine residue has been incorporated into oligonucleotides and shown to enhance the binding affinity dramatically (ΔT_m 10.0°/modification compared to 7-deazaguanine).[174] The increase in the T_m value was attributed to (1) the hydrophobic nature of the modification, (2) increased stacking interaction, and (3) favorable pK_a of the base.

In contrast, some C-8-substituted bases have yielded improved nuclease resistance when incorporated in oligonucleotides, but seem to be somewhat destabilizing.[162]

Oligonucleotide Conjugates. Although the conjugation of various functionalities to oligonucleotides has been reported to achieve a number of important objectives, data supporting some of the claims are limited and generalizations are not possible based on data presently available.

NUCLEASE STABILITY. Numerous 3'-modifications have been reported to enhance the stability of oligonucleotides in serum.[169] Both neutral and charged substituents have been reported to stabilize oligonucleotides in serum and, as a general rule, the stability of a conjugated oligonucleotide tends to be greater as bulkier substituents are added. Inasmuch as the principle nuclease in serum is a 3'-exonuclease, it is not surprising that 5'-modifications have resulted in significantly less stabilization. Internal modifications of base, sugar, and backbone have also been reported to enhance nuclease stability at or near the modified nucleoside.[169] Thiono triester (adamantyl, cholesteryl, and others)-modified oligonucleotides have shown improved nuclease stability, cellular association, and binding affinity.[175]

The demonstration that modifications may induce nuclease stability sufficient to enhance activity in cells in tissue culture and in animals has proven to be much more complicated because of the presence of 5'-exonucleases and endonucleases. In our laboratory, 3'-modifications and internal point modifications have not provided sufficient nuclease stability to demonstrate pharmacological activity in cells.[96] In fact, even a 5-nucleotide-long phosphodiester gap in the middle of a phosphorothioate oligonucleotide resulted in sufficient loss of nuclease resistance to cause a complete loss of pharmacological activity.[81]

---

[173] F. Seela, K. Kaiser, and U. Binding. *Helv. Chim. Acta* **72,** 868 (1989).

[174] F. Seela, N. Ramzaeva, and Y. Chen, *Bioorgan. Med. Chem. Lett.* **5,** 3049 (1995).

[175] R. Zhang, Z. Lu, X. Zhang, H. Zhao, R. B. Diasio, T. Liu, Z. Jiang, and S. Agrawal, *Clin. Chem.* **41,** 836 (1995).

In mice, neither a 5'-cholesterol nor a 5'-C-18 amine conjugate altered the metabolic rate of a phosphorothioate oligodeoxynucleotide in liver, kidney, or plasma.[83] Furthermore, blocking the 3'- and 5'-termini of a phosphodiester oligonucleotide did not markedly enhance the nuclease stability of the parent compound in mice.[13] However, 3'-modification of a phosphorothioate oligonucleotide was reported to enhance its stability in mice relative to the parent phosphorothioate.[176] Moreover, a phosphorothioate oligonucleotide with a 3'-hairpin loop was reported to be more stable in rats than its parent.[175] Thus, 3'-modifications may enhance the stability of the relatively stable phosphorothioates sufficiently to be of value.

ENHANCED CELLULAR UPTAKE. Although oligonucleotides have been shown to be taken up by a number of cell lines in tissue culture, with perhaps the most compelling data relating to phosphorothioate oligonucleotides, a clear objective has been to improve the cellular uptake of oligonucleotides.[2,8] Inasmuch as the mechanisms of cellular uptake of oligonucleotides are still very poorly understood, the medicinal chemistry approaches have been largely empirical and based on many unproven assumptions.

Because phosphodiester and phosphorothioate oligonucleotides are water soluble, the conjugation of lipophilic substituents to enhance membrane permeability has been a subject of considerable interest. Unfortunately, studies in this area have not been systematic and, at present, there is precious little information about the changes in physicochemical properties of oligonucleotides actually affected by specific lipid conjugates. Phospholipids, cholesterol and cholesterol derivatives, cholic acid, and simple alkyl chains have been conjugated to oligonucleotides at various sites in the oligonucleotide. The effects of these modifications on cellular uptake have been assessed using fluorescent, or radiolabeled, oligonucleotides or by measuring pharmacological activities. From the perspective of medicinal chemistry, very few systematic studies have been performed. The activities of short alkyl chains, adamantine, daunomycin, fluorescein, cholesterol, and porphyrin-conjugated oligonucleotides were compared in one study.[177] A cholesterol modification was reported to be more effective at enhancing uptake than the other substituents. It also seems likely that the effects of various conjugates on cellular uptake may be affected by the cell type and target studied. For example, we have studied cholic acid conjugates of phosphorothioate deoxyoligonucleotides or phosphorothioate 2'-methoxy oligonucleotides and observed enhanced activity against HIV and no effect on the activity of ICAM-directed oligonucleotides.

[176] J. Temsamani, J. Tang, A. Padmapriya, M. Kubert, and S. Agrawal, *Antisense Res. Dev.* **3**, 277 (1993).

[177] A. Boutorine, C. Huet, and T. Saison, *Nucleic Acid Therap.* 1991.

Additionally, polycationic substitutions and various groups designed to bind to cellular systems have been synthesized. Although many compounds have been synthesized, data reported to date are insufficient to draw firm conclusions about the value of such approaches or structure–activity relationships.[169]

RNA CLEAVING GROUPS. Oligonucleotide conjugates have been reported to act as artificial ribonucleases, albeit in low efficiencies.[178] The conjugation of chemically reactive groups such as alkylating agents, photoinduced azides, prophine, and psoralene has been utilized extensively to effect a cross-linking of oligonucleotide and the target RNA. In principle, this treatment may lead to translation arrest. In addition, lanthanides and complexes thereof have been reported to cleave RNA via a hydrolytic pathway. A novel europium complex has been linked covalently to an oligonucleotide and shown to cleave 88% of the complementary RNA at physiological pH.[179]

IN VIVO EFFECTS. To date, relatively few studies have been reported *in vivo*. Properties of a 5'-cholesterol and 5'-C_{18} amine conjugates of a 20-mer phosphorothioate oligodeoxynucleotide have been determined in mice. Both compounds increased the fraction of an intravenous bolus dose found in the liver. The cholesterol conjugate, in fact, resulted in more than 80% of the dose accumulating in the liver. Neither conjugate enhanced stability in plasma, liver, or kidney.[83] Interestingly, the only significant change in the toxicity profile was a slight increase in effects on serum transamineses and histopathological changes indicative of slight liver toxicity associated with the cholesterol conjugate.[180] A 5'-cholesterol phosphorothioate conjugate was also reported to have a longer elimination half-life, to be more potent, and to induce greater liver toxicity in rats.[181]

Sugar Modifications. The focus of second-generation oligonucleotide modifications has centered on the sugar moiety. In oligonucleotides, the pentofuranose sugar ring occupies a central connecting manifold that also positions the nucleobases for effective stacking. A symposium series has been published on the carbohydrate modifications in antisense research, which covers this topic in great detail.[182] Therefore, the content of the

[178] A. De Mesmaeker, R. Haener, P. Martin, and H. E. Moser, *Acc. Chem. Res.* **28,** 366 (1995).
[179] J. Hall, D. Hüsken, U. Pieles, H. E. Moser, and R. Haner, *Chem. Biol.* **1,** 185 (1994).
[180] S. P. Henry, J. E. Zuckerman, J. Rojko, W. C. Hall, R. J. Harman, D. Kitchen, and S. T. Crooke, *Anti-Cancer Drug Design* **12,** 1 (1997).
[181] J. Desjardins, J. Mata, T. Brown, D. Graham, G. Zon, and P. Iversen, *J. Drug Targeting* **2,** 477 (1995).
[182] Y. S. Sanghvi and P. D. Cook, *in* "ACS Symposium Series No. 580." American Chemical Society, Washington, DC, 1994.

following discussion is restricted to a summary of the main events in this area.

A growing number of oligonucleotides in which the pentofuranose ring is modified or replaced have been reported.[183] Uniform modifications at the 2′-position have been shown to enhance hybridization to RNA and, in some cases, to enhance nuclease resistance.[183] Chimeric oligonucleotides containing 2′-deoxyoligonucleotide gaps with 2′-modified wings have been shown to be more potent than parent molecules.[61]

Other sugar modifications include α-oligonucleotides, carbocyclic oligonucleotides, and hexapyranosyl oligonucleotides.[183] Of these, α-oligonucleotides have been studied the most extensively. They hybridize in parallel fashion to single-stranded DNA and RNA and are nuclease resistant. However, they have been reported to be oligonucleotides designed to inhibit Ha-ras expression. All these oligonucleotides support RNase H and, as can be seen, a direct correlation between affinity and potency exists.

A growing number of oligonucleotides in which the C-2′ position of the sugar ring is modified have been reported.[169,178] These modifications include lipophilic alkyl groups, intercalators, amphipathic aminoalkyl tethers, positively charged polyamines, highly electronegative fluoro or fluoroalkyl moieties, and sterically bulky methylthio derivatives. The beneficial effects of a C-2′ substitution on the antisense oligonucleotide cellular uptake, nuclease resistance, and binding affinity have been well documented in the literature. In addition, excellent review articles have appeared on the synthesis and properties of C-2′-modified oligonucleotides.[178,184–186]

Other modifications of the sugar moiety have also been studied, including other sites, as well as more substantial modifications. However, much less is known about the antisense effects of these modifications.[16]

2′-Methoxy-substituted phosphorothioate oligonucleotides have been reported to be more stable in mice than their parent compounds and to display enhanced oral bioavailability.[116,175] The analogs displayed tissue distribution similar to that of the parent phosphorothioate.

Similarly, we have compared the pharmacokinetics of 2′-propoxy-modified phosphodiester and phosphorothioate deoxynucleotides.[83] As expected, the 2′-propoxy modification increased lipophilicity and nuclease resistance. In fact, in mice the 2′-propoxyphosphorothioate was too stable in liver or kidney to measure an elimination half-life.

[183] K. J. Breslauer, R. Frank, H. Blocker, and L. A. Marky, *Proc. Natl. Acad. Sci. U.S.A.* **83,** 3746 (1986).
[184] A. I. Lamond and B. S. Sproat, *FEBS Lett.* **325,** 123 (1993).
[185] B. S. Sproat and A. I. Lamond, *in* "Antisense Research and Applications" (S. T. Crooke and B. Lebleu, eds.), p. 351. CRC Press, Boca Raton, FL, 1993.
[186] G. Parmentier, G. Schmitt, F. Dolle, and B. Luu, *Tetrahedron* **50,** 5361 (1994).

Interestingly, the 2'-propoxy phosphodiester was much less stable than the parent phosphorothioate in all organs except the kidney in which the 2'-propoxy phosphodiester was remarkably stable. The 2'-propoxy phosphodiester did not bind to albumin significantly, whereas the affinity of the phosphorothioate for albumin was enhanced. The only difference in toxicity between the analogs was a slight increase in renal toxicity associated with the 2'-propoxy phosphodiester analog.[180]

Incorporation of the 2'-methoxyethoxy group into oligonucleotides increased the T_m by 1.1°/modification when hybridized to the complement RNA. In a similar manner, several other 2'-O-alkoxy modifications have been reported to enhance the affinity.[187] The increase in affinity with these modifications was attributed to (1) the favorable gauche effect of the side chain and (2) additional solvation of the alkoxy substituent in water.

More substantial carbohydrate modifications have also been studied. Hexose-containing oligonucleotides were created and found to have very low affinity for RNA.[188] Also, the 4'-oxygen has been replaced with sulfur. Although a single substitution of a 4'-thio-modified nucleoside resulted in destabilization of a duplex, incorporation of two 4'-thio-modified nucleosides increased the affinity of the duplex.[189] Finally, bicyclic sugars have been synthesized with the hope that preorganization into more rigid structures would enhance hybridization. Several of these modifications have been reported to enhance hybridization.[182]

Backbone Modifications. Substantial progress in creating new backbones for oligonucleotides that replace the phosphate or the sugar–phosphate unit has been made. The objectives of these programs are to improve hybridization by removing the negative charge, enhance stability, and potentially improve pharmacokinetics.

For a review of the backbone modifications reported to date, see Refs. 16, 182. Suffice it to say that numerous modifications have been made that replace phosphate, retain hybridization, alter charge, and enhance stability. Because these modifications are now being evaluated *in vitro* and *in vivo,* a preliminary assessment should be possible shortly.

Replacement of the entire sugar–phosphate unit has also been accomplished and the oligonucleotides produced have displayed very interesting characteristics. Pestide nucleic acid oligonucleotides have been shown to bind to single-stranded DNA and RNA with extraordinary affinity and

[187] P. Martin, *Helv. Chim. Acta* **78,** 486 (1995).

[188] S. Pitsch, R. Krishnamurthy, M. Bolli, S. Wendeborn, A. Holzner, M. Minton, C. Lesueur, I. Schloenvogt, B. Jaun *et. al., Helv. Chim. Acta.* **78,** 1621 (1995).

[189] L. Bellon, C. Leydier, and J. L. Barascut, *in* "Carbohydrate Modifications in Antisense Research" (Y. S. Sanghvi and P. D. Cook, eds.), p. 68. American Chemical Society, Washington, DC, 1994.

high sequence specificity. They have been shown to be able to invade some double-stranded nucleic acid structures. Peptide nucleic acid oligonucleotides can form triple-stranded structures with DNA or RNA.

Peptide nucleic acid oligonucleotides were shown to be able to act as antisense and transcriptional inhibitors when microinjected in cells.[190] Peptide nucleic acid oligonucleotides appear to be quite stable to nucleases and peptidases as well.

In summary, enormous advances in the medicinal chemistry of oligonucleotides have been reported since the early 1990s. Modifications at nearly every position in oligonucleotides have been attempted, and numerous potentially interesting analogs have been identified. Although it is far too early to determine which of the modifications may be most useful for particular purposes, it is clear that a wealth of new chemicals is available for systematic evaluation and that these studies should provide important insights into the structure–activity relationship of oligonucleotide analogs.

Conclusions

Although many more questions about antisense remain to be answered than are answered, progress has continued to be gratifying. Clearly, as more is learned, we will be in the position to perform progressively more sophisticated studies and to understand more of the factors that determine whether an oligonucleotide actually works via an antisense mechanisms. We should also have the opportunity to learn a great deal more about this class of drugs as additional studies are completed in humans.

Acknowledgment

The author thanks Donna Musacchia for excellent typographic and administrative assistance.

[190] J. C. Hanvey, N. C. Peffer, J. E. Bisi, S. A. Thomson, R. Cadilla, J. A. Josey, D. J. Ricca, C. F. Hassman, M. A. Bonham, K. G. Au, S. G. Carter, D. A. Bruckenstein, A. L. Boyd, S. A. Noble, and L. E. Babiss, *Science* **258**, 1481 (1992).

[2] Basic Principles of Using Antisense Oligonucleotides In Vivo

By M. Ian Phillips and Y. Clare Zhang

Introduction

Antisense (AS) inhibition has been developed, particularly in cell culture applications, to the point where it is being tested in clinical trials for human immunodeficiency virus (HIV) and cancer.[1,2] The advantages and disadvantages of its use have been reviewed elsewhere.[3] Before 1992, however, antisense had not been applied *in vivo* with any success. There was much concern about the efficiency of the cellular uptake of oligonucleotides (ODNs). In 1992–1993, three or four laboratories simultaneously and independently made and tested AS–ODNs in the brain.[4–8] Cellular uptake was not a limiting factor in the central nervous system. Receptor binding reduction in the brain with antisense to the angiotensin AT_1 receptor, neuropeptide Y, and NMDA receptors showed that AS delivered centrally was an effective inhibitior.[5,6,8,9]

AS–ODNs have many potential attractive features as a new class of therapeutic agents. They can be studied in the same way as drugs for dose–response effects, length of time of activity, and pharmacokinetics. They have longer lasting actions than current drugs and, because of their molecular specificity, fewer side effects. To prolong the effect of antisense inhibition for weeks or months, DNA (partial or full-length) can be inserted in the antisense direction in viral vectors.[10]

[1] R. W. Wagner, *Nature* **372,** 333 (1994).
[2] S. T. Crooke, *Annu. Rev. Pharmacol. Toxicol.* **32,** 329 (1992).
[3] C. A. Stein and Y.-C. Cheng, *Science* **261,** 1004 (1993).
[4] B. J. Chiasson, M. L. Hooper, P. R. Murphy, and H. A. Robertson, *Eur. J. Pharmacol.* **277,** 451 (1992).
[5] C. Wahlestedt, E. M. Pich, G. F. Koob, F. Yee, and M. Heilig, *Science* **259,** 528 (1993).
[6] C. Wahlestedt, E. Golanov, S. Yamamoto, F. Yee, H. Ericson, H. Yoo, C. E. Inturrisi, and D. J. Reis, *Nature* **363,** 260 (1993).
[7] M. M. McCarthy, D. B. Masters, K. Rimvall, S. Schwartz-Giblin, and D. W. Pfaff, *Brain Res.* **636,** 209 (1994).
[8] R. Gyurko, D. Wielbo, and M. I. Phillips, *Reg. Pep.* **49,** 167 (1993).
[9] P. Ambuhl, R. Gyurko, and M. I. Phillips, *Reg. Pep.* **59,** 171 (1995).
[10] M. I. Phillips, *Hypertension* **29,** 177 (1997).

TABLE I
CONDITIONS FOR ANTISENSE OLIGONUCLEOTIDE INHIBITION

DNA sequence is specific and unique
Uptake into cells is efficient
Effect in cells is stable
No nonspecific binding to proteins
Hybridization of ODN is specific to target mRNA
Targeted protein and/or mRNA level is reduced
ODN is not toxic
No inflammatory or immune response is induced
ODN is effective compared to appropriate sense and mismatch ODN controls

Antisense Oligonucleotides

Designing Antisense Molecules

Table I lists the characteristics of AS–ODNs that need to be incorporated into their design and use. The concept of antisense inhibition assumes that a short DNA sequence in the antisense direction binds to the specific mRNA of the target protein in the cytoplasm and prevents either ribosomal assembly or read-through of the message.[11] There are several potential target sites,[12] but AS–ODNs are generally targeted to the gene initiation codon (AUG) or part of the coding region downstream from it. Other approaches involved the formation of triple helix with antisense constructed to the promoter region of a specified DNA.[13]

AS–ODNs are short, single stranded, and commonly 15–20 bases long, but longer or full-length DNA in the antisense direction is used in viral vectors. The length of 15–20 bases is optimal because shorter sequences are more likely to be nonspecific and sequences longer than 25 are less able to enter cells and are also more probable to contain a repeat sequence of purines, which can result in binding to proteins. When designing antisense molecules, one has to consider two antagonistic factors: (1) the affinity of oligonucleotide to its target sequence, which is dependent on the number and composition of complementary bases, and (2) the availability of the target sequence, which is dependent on the folding of the mRNA molecule.[14]

---

[11] R. W. Simons, *Gene* **72,** 35 (1988).
[12] M. I. Phillips and R. Gyurko, *Reg. Pep.* **59,** 131 (1995).
[13] C. Helene, *Anticancer Drug Res.* **6,** 569 (1991).
[14] R. A. Stull, L. A. Taylor, and F. C. Szoka, *Nucleic Acids. Res.* **20,** 3501 (1992).

Several reports suggested that AS–ODNs targeted to different regions of RNA have unequal efficiencies.[15,16] These differences may be related to the predicted secondary structure of the target mRNA.[17–19] The folding of mRNA influences target sequence availability. RNA double helices, which are responsible for the secondary structure of the mRNA, incorporate a weaker G–U base pairing next to A–U and G–C and are generally short and rarely perfect. Therefore the design should avoid G repeats. Burgess and Farrell[20] showed that when repeated G sequences appear in the oligonucleotide, the effects that are produced can be due to nonantisense mechanisms.

Controls. The proper testing of antisense requires a sense ODN and a mismatch ODN control for every antisense ODN. An ODN that has strong (Watson–Crick) base pairing with 100% complementariness will form the more thermodynamically favorable structure with its target RNA.[21]

Sites of Action. AS–ODNs have several potential sites of action. AS–ODNs inhibit translation by hybridizing to the specific mRNA that they are designed for and the hybridization prevents either ribosomal assembly or ribosomal sliding along the mRNA. This kind of action assumes that AS–ODNs are acting in the cytosol and do not affect measurable mRNA levels. Indeed, there are several articles reporting antisense effects without a detectable change in target mRNA levels.[6]

Alternatively, the mechanism can be by reduction of mRNA. Decreased mRNA levels can occur by RNase H digestion of the RNA portion of the mRNA–antisense DNA hybrid. RNase H is found in the cytoplasm as well as in the nucleus and it is normally involved in DNA duplication. The role of RNase H is to cleave RNA that has bound to DNA. The activation of RNase H is advantageous because the enzyme leaves the AS–ODN intact so it is free to hybridize with another mRNA, making the reaction catalytic rather than stoichiometric. In addition to inhibition of translation, other possible antisense mechanisms of action have been proposed. Based on studies of cellular uptake of labeled ODNs, the picture emerged that most AS–ODNs migrate quickly to the cell nucleus, suggesting an intranuclear

[15] L. M. Cowsert, M. C. Fox, G. Zon, and C. K. Mirabelli, *Antimicrob. Agents Chemother.* **37**, 171 (1993).

[16] T. Wakita and J. R. Wands, *J. Biol. Chem.* **269**, 14205 (1994).

[17] W. F. Lima, B. P. Monia, D. J. Ecker, and S. M. Freier, *Biochemistry* **31**, 12055 (1992).

[18] K. Rittner and G. Sczakiel, *Nucleic Acid Res.* **19**, 1421 (1991).

[19] J. W. Jaroszevski, J. L. Syi, M. Ghosh, K. Ghosh, and J. S. Cohen, *Antisense Res. Dev.* **3**, 339 (1993).

[20] T. Burgess and C. Farrell, *Proc. Natl. Acad. Sci. U.S.A.* **92**, 4051 (1995).

[21] M. Singer and P. Berg, "Genes and Genomes," p. 54. University Science Books, Mill Valley, CA, 1992.

site of action.[22,23] Antisense DNA can hybridize to its target mRNA or pre-mRNA in the nucleus, forming a partially double-stranded structure that would inhibit its transport out of the nucleus into the cytoplasm, thus preventing translation. AS–ODNs targeted to intron–exon junction sites prevent the splicing process and consequently the maturation of the transcript. Therefore, antisense molecules might inhibit pre-RNA splicing or the transport of mRNA from the nucleus to the cytoplasm. An alternate antigene strategy is to target the DNA with triplex forming ODNs to block DNA transcription. Effective AS–ODNs have been designed targeting exon–intron splicing sites.[24] AS–ODNs have been targeted to the major groove of the DNA[13] but triplex formation is corrected by DNA repair mechanisms.

General Principles. An AS–ODN specifically against rat β_1-adrenoceptor mRNA is used here as an example to illustrate the general principles we follow in the design of antisense ODN.

1. *Identify the sequence:* Check GenBank for the mRNA sequence of the target protein in the species to be studied. If there is more than one laboratory cloning the same protein, compare the homology of different reports. These sequences will point out the controversial bases, which can be due to natural mutations in variant strains of the same species or sequencing errors. Try to avoid these debatable regions. Target the ODN to the identical regions, which ensures reliability of the sequence being used. Figure 1 shows the homology comparison of the rat β_1-adrenoceptor mRNA sequence reported by two groups.

2. *Target the sites:* Although there are several possible sites within the DNA sequence to target for antisense design, the three regions that are considered to be the best targets for constructing effective AS–ODNs are the 5′ cap region, the AUG translation initiation codon, and the 3′-untranslated region of the mRNA.[25–27] In our experience, the AUG start codon and nearby bases downstream of AUG in the coding region are the most promising sites for antisense inhibition. Usually the length of 15- to 20-mer is sufficient to guarantee sequence

[22] P. L. Iversen, S. Zhu, A. Meyer, and G. Zon, *Antisense Res. Dev.* **2,** 211 (1992).

[23] B. Li, J. A. Hughes, and M. I. Phillips, *Neurochem. Intl.* **31,** 393 (1996).

[24] A. Colige, B. P. Sokolov, P. Nugent, R. Baserge, and D. J. Prockop, *Biochemistry* **32,** 7 and 569 (1993).

[25] C. F. Bennett, T. P. Condon, S. Grimm, H. Chan, and M. Y. Chiang, *J. Immunol.* **152,** 3530 (1994).

[26] T. A. Bacon and E. Wickstrom, *Oncogene Res.* **6,** 13 (1991).

[27] C. Wahlestedt, *TIPS* **15,** 42 (1994).

```
   1 TCCTGGGGTGCTTCCCAGGCGCGGCCCAGTCCCGCCACACCCCCCGCCCC 50
     ||||||||||||||||||||||||||||||||||||||||||||||| ||
1190 TCCTGGGGTGCTTCCCAGGCGCGGCCCAGTCCCGCCACACCCCCCGCGCC 1239

  51 CGGCCTCCGAAGCTCGGCATGGGCGCGGGGGCGCTCGCCCTGGGCGCCTC 100
     |||||||||||| |||||||||||||||||||||||||||||||||||||
1240 CGGCCTCCGAAG.TCGGCATGGGCGCGGGGGCGCTCGCCCTGGGCGCCTC 1288

 101 CGAACCCTGCAACCTGTCGTCGGCCGCGCCGCTGCCCGACGGCGCGGCCA 150
     ||||||||||||||||||||||||||||||||||||||||||||||||||
1289 CGAACCCTGCAACCTGTCGTCGGCCGCGCCGCTGCCCGACGGCGCGGCCA 1338

 151 CCGCGGCACGACTGCTGGTGCTCGCGTCGCCTCCCGCCTCGCTGCTGCCT 200
     ||||||||||||||||||||||||||||||||||||||||||||||||||
1339 CCGCGGCACGACTGCTGGTGCTCGCGTCGCCTCCCGCCTCGCTGCTGCCT 1388

 201 CCAGCCAGCGAGGGCTCAGCGCCGCTGTCGCAGCAGTGGACCGCGGGTAT 250
     ||||||||||||||||||||||||||||||||||||||||||||||||||
1389 CCAGCCAGCGAGGGCTCAGCGCCGCTGTCGCAGCAGTGGACCGCGGGTAT 1438
```

FIG. 1. Homology comparison of rat β_1-adrenoceptor cDNA sequence (partial) reported by two groups [Machida *et al.*[41] upper lane, gene accession number J05561) and Shimomura *et al.*[42] (lower lane, gene accession number D00634)]. The ATG translation initiation codon is highlighted in boldface fonts and inconsistent bases between two reports in italic fonts. The target region of a 15-mer AS–ODN is underlined.

specificity, which must be confirmed by checking with GenBank for existing sequences to avoid any significant homology with other mRNAs. Figure 2 shows the BLAST search result of a 15-mer AS–ODN designed specifically for the rat β_1-adrenoceptor, which turns out to be conserved in β_1-adrenoceptor mRNA of rat, mouse, human, monkey, and dog, but has no appreciable overlapping with other rat mRNAs.

   3. *Modify the backbone:* Phosphorothiate linkage is currently the analog linkage of choice for antisense studies to improve ODN stability. However, phosphorothiation decreases the melting temperature (T_m) of ODNs significantly and reduces binding affinity and stability of DNA–DNA and DNA–RNA duplexes. For 15- to 25-mer, T_m of phosphorothiate ODN (S–ODN) is usually 7–12° lower than normal ODN, with AT bases showing more T_m depression than GC.[27,28] This fact must be taken into account in the design of short AS–ODN because 15-mer S–ODN–RNA hybrids may have T_m close to the body temperature of 37°.

[28] J. M. Campbell, T. A. Bacon, and E. Wickstrom, *J. Biochem. Biophys. Methods* **20,** 259 (1990).

Sequences producing significant alignments:

Gene accession	Gene description	Identical bases
gb|J05561|RATB1AR	R.norvegicus beta-1-adrenergic receptor gene,...	15
dbj|D00634|RATB1ARA	Rattus norvegicus gene for beta-1 adrenergi...	15
gb|AC005886|AC005886	Homo sapiens chromosome 10 clone CIT-HSP-1...	15
gb|M17350|KPNUSV	K.pneumoniae nitrogen fixation genes nifU, nif...	15
emb|X75540|MMB1AR	M.mullata beta 1 adrenergic receptor gene	15
gb|L10084|MUSADRR	Mus musculus beta-1 adrenergic receptor gene,...	15
gb|U73207|CFU73207	Canis familiaris beta1 adrenergic receptor (...	15
gb|AF072433|AF072433	Ovis aries beta 1 adrenergic receptor (BAR...	15
gb|J03019|HUMADRB1	Human beta-1-adrenergic receptor mRNA, compl...	15
emb|X82929|CGPROAGEN	C.glutamicum proA gene	14
gb|S69150|S69150	ALDC=alpha-acetolactate decarboxylase [Acetoba...	14
emb|AL023702|SC1C3	Streptomyces coelicolor cosmid 1C3	14
gb|M83095|BPESODB	Bordetella pertussis superoxide dismutase (so...	14
gb|AF035395|AF035395	Pseudomonas aeruginosa serine/threonine pr...	14
emb|AJ001084|RSPRHA	Ralstonia solanacearum prhA gene	14
gb|AF031406|AF031406	Rhodobacter capsulatus NADPH dependent glu...	14
gb|U75215|MMU75215	Mus musculus neutral amino acid transporter ...	14
gb|AF082100|AF082100	Streptomyces sp. MA6548 FK506 peptide synt...	14
dbj|AB013077|AB013077	Alcaligenes xylosoxidans az1 gene for azu...	14
emb|X15286|HVDHN17	Barley mRNA for dehydrin (dhn17)	13
emb|AL031350|SC1F2	Streptomyces coelicolor cosmid 1F2	13
gb|U08229|CLU08229	Columba livia carnitine acetyltransferase mR...	13
gb|AC004441|AC004441	Drosophila melanogaster DNA sequence (P1 D...	13
gb|U04874|GLU04874	Giardia lamblia Portland 1 cytoplasmic 70 kD...	13
emb|X51542|CSIG18SR	Cucumber intergenic spacer DNA and 18S rRNA...	13
gb|AF010496|AF010496	Rhodobacter capsulatus strain SB1003, part...	13
emb|X68444|RCNIF	R.capsulatus genes nifU, nifS, nifV, nifW and ...	13
gb|U25811|CVU25811	Chromatium vinosum cytochrome c' precursor (...	13
gb|M73546|BACTENAI	Bacillus subtilis transcription activator (t...	13
gb|L78817|MSGB27CS	Mycobacterium leprae cosmid B27 DNA sequence.	13
emb|AL008635|HS510H16	Homo sapiens DNA sequence from PAC 510H16...	13
gb|M26595|TRBHSP70AB	T.cruzi heat shock protein (HSP70) gene, c...	13
gb|L37440|SYNSUICA	Expression vector pZEO-SG2 cytosine deaminas...	12
dbj|AB015509|AB015509	Aspergillus aculeatus gene for beta-manno...	12
emb|Y00556|BPFIMX	Bordetella pertussis fimX gene for fimbrial p...	12
gb|U94825|ASU94825	Actinomyces sp. 40 endoglucanase gene, compl...	12
emb|Z95120|MTCY7D11	Mycobacterium tuberculosis H37Rv complete g...	12
emb|X76532|TAPWR5PI	Triticum aestivium pWR5 RNA for protochloro...	12
emb|Z18284|HSOCT6TR	Homo sapiens Oct-6 transcription factor.	12
emb|X84895|AGCODA	A.globiformis codA gene	12
gb|U79264|HSU79264	Human clone 23814 mRNA sequence	12
gb|U58949|ASU58949	Allium sativum lectin related protein mRNA, ...	12
gb|AF029673|AF029673	Pseudomonas aeruginosa HexR (hexR), glucos...	12
gb|AC004770|AC004770	Homo sapiens chromosome 11, BAC CIT-HSP-31...	12
gb|U94899|SMU94899	Sinorhizobium meliloti dissimilatory nitrous...	12
emb|X98826|RVSSURRNA	R.venosa small subunit ribosomal RNA gene	12
emb|AJ001848|TAAJ1848	Thauera aromatica tdiSR and bssDCAB opero...	12

FIG. 2. BLAST search results (partial) of a 15-mer AS–ODN against rat β_1-adrenoceptor mRNA (see Fig. 3). BLAST is an on-line program at *http://www.ncbi.nlm.nih.gov*. Sequences sharing significant homology with AS–ODN are aligned in the order of identical bases. (Left) Gene accession number of sequences. (Center) Gene name and species. (Right) Number of identical bases between AS–ODN and corresponding genes.

1. Identify AUG initiation
   codon in mRNA

2. Complementary DNA

   |
   ↓

   Reverse sequence

   |
   ↓

3. Antisense ODN
4. Purify and modify
   (e.g., phosphorothioation)

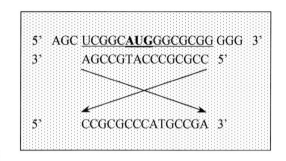

FIG. 3. AS–ODN is complementary to the mRNA sequence and displayed 5′ → 3′. The 15-mer AS–ODN shown here is targeted to rat β_1-adrenoceptor mRNA, spanning the AUG start codon (see Fig. 1).

4. *Avoid pitfalls:* Some other basic rules applicable in the design of polymerase chain reaction primers also hold true for AS–ODN,[29] e.g., avoidance of palindromic structure and primer–dimer formation and choosing a sequence with a balanced AT/GC ratio to minimize toxicity and nonspecific binding caused by high GC content. Although a high GC/AT ratio should be avoided generally, the danger of toxicity and nonspecific binding can be reduced by phosphorothioation. For example, the AS–ODN targeted to rat β_1-adrenoceptor mRNA as shown in Fig. 3, which has a relatively high GC content, has been used successfully without any nonspecific effects.

Lacking a sure-fire method of oligonucleotide prediction, antisense design boils down to trial-and-error testing in a model first. A general rule suggested by our own experience is that for a 15-mer oligonucleotide, test three different sites with the expectation that at least one will work. Obviously, it is desirable to have a rapid screening test *in vitro* or *in vivo* for the specific protein that the AS–ODN has been designed to inhibit. Controls are sense ODN and mismatch (one or more nucleotides different from AS) or scrambled where the entire sequence is random.

Stability of Oligonucleotides

Oligonucleotides in their natural form as phosphodiesters are subject to rapid degradation in the blood, intracellular fluid, or cerebrospinal fluid

[29] H. Takaku, *Neuropeptide* **15**, 519 (1996).

by exo- and endonucleases. The half-life of phosphodiester oligonucleotides is in the range of minutes in blood and tissue culture media. The half-life of ODNs is somewhat longer in cerebrospinal fluid, and intact ODNs can be detected 24 hr after injection into the cerebral ventricles.[27] Effects of a single injection of antisense in the cerebroventricles produce a physiological change for up to 7 days.[30] Several chemical modifications have been proposed to prolong the half-life of ODNs in biological fluids and enhance uptake while retaining their activity and specificity.[31]

Phosphorothioates. The most widely used modified ODNs are phosphorothioates, where one of the oxygen atoms in the phosphodiester bond between nucleotides is replaced with a sulfur atom. These phosphorothioate ODNs have greater stability in biological fluids than normal oligonucleotides. The half-life of a 15-mer phosphorothioate ODN is 9 hr in human serum, 4 days in tissue culture media,[23] and 19 hr in cerebrospinal fluid.[28] Phosphorothioate ODNs can be synthesized with automated DNA synthesizers, but the product may contain impurities unless purified on an affinity gel. ODNs used *in vivo* should be checked to ensure that they are pure.

One or more of the oxygen atoms in the phosphodiester bond can be replaced with a variety of other compounds, such as methyl groups (methylophosphonate), alkyl phosphotriester, phosphoramidate, or boranophosphate, all of which expand the half-life of ODNs in *in vivo* experiments. Newer designs include a dumbbell-shaped ODN, produced by a hairpin extension at the 3' end.[29] It is hoped that these third- or fourth-generation ODNs will provide longer-lasting stability, enhanced uptake kinetics, and affinity for the target.[32–34]

Cellular Uptake of Oligonucleotides

In order to hybridize with the target mRNA, AS–ODNs have to cross the cell membrane. The saturable uptake of ODNs reaches a plateau within 50 hr, occurs rapidly, and, depending on the cells, the uptake can be efficient.[35] Uptake is faster for shorter ODNs than for longer ones.[35] Decreasing the temperature prevents oligonucleotide uptake, indicating that there is an active uptake mechanism. An 80-kDa ODN-binding protein has been proposed to be the receptor moleculer for ODN uptake.[35] An efflux mecha-

[30] R. Gyurko, D. Tran, and M. I. Phillips, *Am. J. Hypertens.* **10,** 56S (1997).
[31] R. W. Wagner, M. D. Matteucci, J. G. Lewis, A. J. Gutierrez, C. Moulds, and B. C. Froehler, *Science* **260,** 1510 (1993).
[32] C. A. Stein, *Chem. Biol.* **3,** 319 (1996).
[33] C. A. Stein, K. Mori, S. L. Loke *et al., Gene* **72,** 333 (1988).
[34] C. L. Clark, P. K. Cecil, D. Singh, and D. M. Gray, *Nucleic Acids Res.* **25,** 4098 (1997).
[35] S. L. Loke, C. A. Stein, X. H. Zhang, K. Mori, M. Nakanishi, C. Subasinghe, J. S. Cohen, and L. M. Neckers, *Proc. Natl. Acad. Sci. U.S.A.* **86,** 3474 (1998).

nism has also been described indicating temperature-dependent secretion of the ODNs from the cells to the extracellular space.[23] Thus, uptake depends on a two-component process of uptake versus efflux until an equilibrium is reached. In adrenal cells, uptake was rapid (<60 min) and efflux was slow to begin (74 hr) so that initially ODNs were able to reach nuclei and have antisense effects before efflux began.[23] This may account for the effectiveness of a single dose of AS–ODN. The initial dose is unopposed by efflux mechanisms but repeated doses would be less effective.

Pharmacology of AS–ODNs

Antisense inhibition can be considered pharmacologically a drug–receptor interaction, where the oligonucleotide is the drug and the target sequence is the receptor. For binding to occur between the two, a minimum level of affinity is required, which is provided by hydrogen bonding between Watson–Crick base pairs and base stacking in the double helix that is formed. In order to achieve pharmacological activity, a minimum number of 12–15 bases can provide the minimum level of affinity.[31] Longer sequences may increase specificity, but above 20 bases, problems of cell uptake begin to reduce the effectiveness of ODNs.[35]

One of the main advantages of antisense inhibition is the specificity of the AS–ODN target sequence interaction provided by Watson–Crick base pairing. An oligonucleotide 12–15 nucleotides long is specific enough statistically to be complementary to a single sequence.[2] A different approach is to use antisense vectors. With viral vectors, the uptake problem is overcome because the virus enters cells freely by binding to viral receptors on cell membranes. Therefore, in a viral vector a full-length DNA antisense sequence can be used. The mechanism of action of antisense DNA is different from that of AS–ODN. Antisense DNA produces an antisense mRNA that competes negatively with mRNA in the cytoplasm. The problem then becomes which vector is the most appropriate.

Toxicity

AS–ODNs can inhibit protein synthesis in cultured cells in nanomoles per liter doses. The therapeutic window for AS–ODNs is rather narrow[31]: when testing for the optimal dose, small increments in the high nanomole per liter range should be tested.[27] High concentrations may produce nonspecific binding to cytosolic proteins and give misleading results.

Because phosphodiester oligonucleotides are degraded to their naturally occurring nucleotide-building blocks relatively quickly, no toxic reaction is expected from even high doses of phosphodiester. However, because they are short lasting, they have little therapeutic use. Studies on phosphorothioated oligonucleotides in rats show that following intravenous injection, phosphorothioated oligonucleotides are taken up from the plasma mainly by the liver, fat, and muscle tissues. Phosphorothioate oligonucleotides are excreted through the urine in 3 days mainly in their original form. An apparent mild increase in plasma LDH, and indicators of a possible transient liver toxicity are found with very high doses of phosphorothioated oligonucleotides.[36] Whole new classes of oligonucleotide backbone modifications are being developed to avoid the possible liver toxicity in humans with phosphorothioates.[30,34]

Delivery of Antisense

Naked DNA. Direct injection of the antisense DNA has been used successfully in numerous experiments.[37] For injections into the brain, naked DNA appears to be very effective[4–8] and because liposomes are probably toxic in the brain, naked ODNs are satisfactory. In a number of studies, using different AS–ODNs in the brain, uptake is efficient and effective for reduction in protein and inhibition of physiological events without liposomes.[38] The uptake of ODNs in brain is so avid that one difficulty with intracerebroventricular injections is that the DNA tends to be taken up close to the site of injection and does not spread evenly to other parts of the brain. This is an important consideration in antisense strategies for the treatment of brain diseases such as Parkinson's thalamic pain, Alzheimer's disease, and gliomas.

Liposomes. Liposomes that are self-assembling particles of bilipid layers have been used for encapsulating antisense ODN for delivery in blood. Antisense directed to angiotensinogen mRNA in liposomes has had successful results. Tomita *et al.*[39] used liposome encapsulation of angiotensinogen antisense and a Sendai virus injected in the portal vein. Blood pressure decreased for several days. However, they did not compare their results to naked DNA effects. Wielbo *et al.*[40] compared liposome-encapsulated antisense and naked DNA given intraarterially. They found that only liposome encapsulation was effective, whereas naked DNA was not, under the same conditions. Twenty-four hours after injection of 50 μg of liposome-encapsulated antisense ODN, blood pressure decreased to 25 mm Hg. Empty liposomes showed no effect and liposomes encapsulating scrambled

[36] P. L. Iversen, J. Mata, W. G. Tracewell, and Zon, G. *Antisense Res. Dev.* **4**, 43 (1994).
[37] M. I. Phillips, D. Wielbo, and R. Gyurko, *Kid. Intl.* **46**, 1554 (1994).
[38] R. C. Mulligan, *Science* **260**, 926 (1993).
[39] N. Tomita, R. Morishita, J. Higaki, Y. Kaneda, H. Mikami, and T. Ogihara, *Hypertension* **24**, 397 (1994).
[40] D. Wielbo, A. Simon, M. I. Phillips, and S. Toffolo, *Hypertension* **28**, 147 (1996).

ODN had no significant action.[41] Unencapsulated antisense ODN also had no significant effect on blood pressure. Confocal microscopy of rat liver tissue 1 hr after an intraarterial injection of 50 μg of unencapsulated fluorescein (FITC) antisense or liposome-encapsulated FITC-conjugated antisense showed intense fluorescence in liver tissue sinusoids with the liposome-encapsulated ODN.[42] Levels of protein (angiotensinogen and angiotensin II in the plasma) were reduced significantly in the liposome-encapsulated ODN group. Antisense alone, lipids alone, and scrambled ODN in liposomes did not affect protein levels.

Liposome development with cationic lipids allow high transfection efficiency of plasmid DNA. Short, single-stranded AS–ODNs are not actually encapsulated but are complexed with milamellar vesicles by electrostatic interactions. This simplifies the production of the antisense delivery system and allows for a variety of routes of delivery, including aerosol nasal sprays and parenteral injections.

[41] C. A. Machida, J. R. Bunzow, R. P. Searles, T. H. Van, B. Tester, K. A. Neve, P. Teal, V. Nipper, and O. Civelli, *J. Biol. Chem.* **265,** 12960 (1990).
[42] H. Shimomura and A. Terada, *Nuclei Acids Res.* **18,** 4591 (1990).

[3] Polyethyleneimine-Mediated Transfection to Improve Antisense Activity of 3'-Capped Phosphodiester Oligonucleotides

By Sonia Dheur and Tula E. Saison-Behmoaras

Introduction

Antisense technology can knock out selected genes, including disease-causing genes in cultured cells. A number of first-generation antisense compounds have entered human clinical trials.[1] Most of these oligonucleotides are nuclease-resistant phosphorothioates (PS). However, it has been recognized for some years that these compounds may have nonsequence-specific effects on cellular functions.[2] Compared to phosphodiester oligonucleotides (PO), phosphorothioates have a decreased affinity for complementary RNA sequences and a generally increased affinity for cellular proteins. Usually, in cell culture experiments, oligonucleotides with a phosphoro-

[1] S. Akhtar and S. Agrawal, *Trends Pharmacol. Sci.* **18,** 12 (1997).
[2] C. A. Stein, *Nature Med.* **1,** 1119 (1995).

thioate backbone are delivered with a cationic lipid. Cationic lipids increase cellular uptake and facilitate nuclear accumulation of the oligonucleotides.

Phosphodiester oligonucleotides are degraded rapidly by exo- and endonucleases. The stability of phosphodiester oligonucleotides can be increased by blocking the 3'-hydroxyl function. There are few examples in the literature relating the biological activity of phosphodiester oligonucleotides.[3,4] It is commonly thought that even 3' end-protected phosphodiester oligonucleotides are not active as antisense because they are not stable enough within cells. Few studies, however, have investigated the efficacy of cationic lipids to transfect oligonucleotides bearing different backbone modifications. This article shows that the transfection efficacy of a carrier depends on the chemical form of the nucleic acid being transfected. We have compared the antisense activity of PS and 3'-capped PO oligomers delivered by cationic lipids (Lipofectin) or a cationic polymer, polyethyleneimine (PEI), in inhibiting Ha-*ras* expression. Four hours after the application of different oligomer–carrier combinations to the cells, almost complete depletion of the mRNA was obtained with both PS–Lipofectin and PO–PEI formulations. No antisense effect was observed with the formulation PS–PEI or PO–Lipofectin. It is also often reported that the intranuclear accumulation of oligonucleotides is a prerequisite in achieving an efficient antisense effect. Confocal microscopy experiments carried out with fluorescein-labeled oligomers showed that a high nuclear accumulation of PO–ODN is not required to achieve efficient antisense activity.

Materials and Assays Procedures

Modified Oligonucleotides

Phosphorothioates and 3'-capped oligonucleotides are from Eurogentec (Belgium). The 3' end of PO oligonucleotides is protected with a C_3 spacer. Similar results are obtained when a longer spacer such as C_6 or a different protecting group as an amine is used to protect the 3' end of the PO oligonucleotide. The 20-mers used in this study are targeted to the initiation codon region of Ha-*ras* mRNA. Antisense activity of the PS–ODN delivered by cationic lipids has already been demonstrated in cultured cells.[5] As a control, oligomers with the same sequence but inverse polarity are used.

[3] G. Schwab, C. Chavany, I. Duroux, G. Goubin, J. Lebeau, C. Hélène, and T. Saison-Behmoaras, *Proc. Natl. Acad. Sci. U.S.A.* **91,** 10460 (1994).

[4] O. Zelphati, J. L. Imbach, N. Signoret, G. Zon, B. Rayner, and L. Leserman, *Nucleic Acids Res.* **22,** 4307 (1994).

[5] B. P. Monia, J. F. Johnston, D. J. Ecker, M. A. Zounes, W. F. Lima, and S. M. Freier, *J. Biol. Chem.* **267,** 19954 (1992).

Oligonucleotides are resuspended in PBS (Dulbecco's phosphate-buffered saline without calcium chloride and magnesium chloride, Sigma, St. Louis, MO) and precipitated overnight at −20° with 0.3 M sodium acetate and 10 volumes of ethanol. After centrifugation at 12,000 rpm for 30 min, the ODN pellet is resuspended in PBS. Quality of the oligonucleotides is checked after 5′ labeling (see Intracellular Stability of Oligonucleotides for details) and denaturing polyacrylamide gel electrophoresis (PAGE) analysis.

Cationic Lipids and Polymers

The prototype cationic lipid reagent for transfection applications is available commercially under the trade name Lipofectin Reagent (Life Technologies) as a 1:1 (w/w) liposome formulation of the cationic lipid N-(1-(2,3-dioleoyloxy)propyl)-N,N,N-trimethylammonium chloride (DOTMA) and the fusion-enhancing neutral lipid dioleoyl phosphatidylethanolamine (DOPE) (for chemical structure, see Zelphati and Szoka[6]). Many new reagents acting through a similar mechanism have become widely available, and because additional improved reagents are being developed rapidly, the term "cytofectin" is used to specify the class of positively charged lipid molecules that can facilitate the entry of oligo- and polynucleotides into cells. Cytofectin lipids can be obtained from three commercial sources: Invitrogen, Life Technologies, and Promega (Madison, WI). The different cationic lipids are presented singly, in combination with a second cationic lipid, or in combination with the fusion-enhancing neutral lipid DOPE. We use the term "Lipofectin" for the Life Technologies lipid containing DOTMA/DOPE in a 1:1 ratio.

PEI exists as 22-, 25-, 50-, and 800-kDa polymers. We use the commercially available (Aldrich, Milwaukee, WI) 25-kDa molecule, which is the branched form of the polymer.[7,8] The PEI stock solution contains 100 mM monomer in water, neutralized to pH 7 with HCl. This solution is diluted in 150 mM NaCl before transfection.

Cell Culture

The human bladder carcinoma cell line (T24) is from the American Type Culture Collection (Rockville, MD). Cells are grown in Dulbecco's

---

[6] O. Zelphati and F. C. Szoka, Jr., *J. Cont. Rel.* **41**, 99 (1996).

[7] O. Boussif, F. Lezoualćh, M. A. Zanta, M. D Mergny, D. Scherman, B. Demeneix, and J. P. Behr, *Proc. Natl. Acad. Sci. U.S.A.* **92**, 7297 (1995).

[8] H. Pollard, J. S. Remy, G. Loussouarn, S. Demolombe, J.-P. Behr, and D. Escande, *J. Biol. Chem.* **273**, 7507 (1998).

modified Eagle's medium (DMEM containing 4.5 g/liter glucose and sodium bicarbonate, Sigma) supplemented with 7% heat-inactivated fetal calf serum (30 min at 56°), antibiotics (50 U/ml penicillin and 50 U/ml streptomycin), and 4 mM glutamine.

Laser Scanning Confocal Microscopy

Fluorescent Carriers. N-(Lissamine rhodamine B sulfonyl)-1,2-dihexa-decanoyl-*sn*-glycero-3-phosphoethanolamine, triethylammonium salt (rhodamine-DHPE), is obtained from Molecular Probes (Eugene, OR). Rhodamine-DHPE is solubilized in chloroform. The Lipofectin reagent is an aqueous solution. Before the addition of rhodamine-DHPE, Lipofectin is lyophilized and resuspended in chloroform. Rhodamine-labeled Lipofectin is obtained by the addition of 1 mole percent rhodamine-DHPE to Lipofectin. Chloroform is partially evaporated under an argon atmosphere for a few minutes, and lipids are placed under vacuum for 2 hr to eliminate any trace of solvent to achieve solubilization of the lipids in water (final concentration of 1 μg/μl).

In order to prepare rhodamine-labeled PEI, tetramethylrhodamine isothiocyanate (TRITC) is purchased from Molecular Probes. To a solution containing 100 μmol of PEI (100 μl of 1 M PEI at pH 7) in 900 μl borate buffer (200 mM, pH 8.9), a solution of DMSO (50 μl) containing 0.44 mg of TRITC (1 μmol) is added drop by drop. Eighteen hours after incubation in the dark at room temperature, rhodamine-labeled PEI is purified by molecular filtration on Sephadex G-25 (5 $\times$ 130 mm) equilibrated previously with 10 mM NaCl. Fractions containing fluorescent PEI are collected. The determination of amine concentration using the method of Snyder and Sobocinski[9] and measurement of absorbance at 544 nm (ε = 90,000 M^{-1} cm^{-1} in methanol) allowed the determination of the number of tetramethylrhodamine residues attached to PEI (four rhodamines per polymer).

Cell Transfection Conditions. T24 cells are seeded into eight-chamber glass slides (Lab Tek from Nalge Nunc International, Naperville, IL), 18 hr before transfection. Cells are treated with 200 nM fluorescein-labeled oligomers complexed with rhodamine-labeled carriers (10 equivalents of PEI or 5 μg/ml Lipofectin) in serum-free medium for 4 hr (see Table I for details).

Analysis. After three washes in PBS (Sigma), cells are fixed with 2% paraformaldehyde (Sigma) and mounted with Mowiol 4-88 (Hoechst, F, Frankfurt) as described.[10] Cells are analyzed with a confocal imaging system (MRC-600, Bio-Rad Richmond, CA) equipped with a Nikon Optiphot epifluorescence microscope (Nikon, Tokyo, Japan) and a 50 and 100$\times$

---

[9] S. L. Snyder and P. Z. Sobocinski, *Anal. Biochem.* **64**, 284 (1975).
[10] G. V. Heimer and C. E. Taylor, *J. Clin. Pathol.* **27**, 254 (1974).

TABLE I
TRANSFECTION CONDITIONS ACCORDING TO DIFFERENT METHODS OF ANALYSIS

Analysis	Cell culture vessel (cm²)	Cell number	PS or PO (100 μM)	PEI (10 mM)	Lipofectin (1 μg/μl)	Serum-free medium
Northern blot	10-cm petri dish (56 cm²)	8×10^5	6 μl	12 μl	15 μl	3 ml
FACS	6-well plate (9.5 cm²)	2×10^5	1 μl	2 μl	2.5 μl	500 μl
Confocal microscopy	8-chamber glass slide (1 cm²)	3×10^4	0.4 μl	0.8 μl	1 μl	200 μl

Planapo objective (numerical aperture, 1.4). A krypton/argon laser tuned to produce both 488-nm fluorescein excitation and 568-nm rhodamine excitation wavelength beams allows simultaneous reading of both fluorescent signals and image merging. Diaphragm and fluorescence detection levels are adjusted to reduce to a minimum any interference between fluorescein and rhodamine channels. Pictures are recorded with a Kalman filter (average of five images).

Northern Blot

Cell Transfection Conditions. T24 cells are seeded 18 hr before treatment (for details, see Table I). Cells are washed twice with PBS and treated with serum-free medium containing 200 nM oligonucleotide and the carrier. As cationic lipid carrier, 5 μg/ml Lipofectin solution is used.

Different ratios of PEI nitrogen per DNA phosphate (equivalents) are tested (4, 6, 10, and 20) with PO and PS oligonucleotides. Similar efficiency in mRNA depletion is obtained with the different ratios of nitrogen/phosphate. Ten equivalents are used for further experiments (for details, see Table I). Practically, oligonucleotide and PEI are first diluted separately in 50 μl of 150 mM NaCl. PEI solution is added to the oligonucleotide solution. This solution is mixed gently every 2 min for 10 min and is incubated at room temperature for another 10 min. Three milliliters of DMEM without serum is added to the PEI–oligomer complexes, and the cells are incubated with this mixture for 4 hr.

Northern Blot Conditions. After 4 hr, total RNA is prepared from oligonucleotide-treated cells using the RNeasy Mini Kit (Qiagen). RNA samples are electrophoresed through 0.8% agarose–formaldehyde gels and transferred to a nylon hybridization membrane (Hybond N⁺, Amersham) by capillary diffusion over a 12- to 18-hr period. RNA is fixed by alkali treatment of the membrane (0.05 M NaOH for 20 min) and hybridized to random-primed, ^{32}P-labeled, full-length cDNA probes corresponding to human Ha-*ras* (Oncogene Research Products, Calbiochem, La Jolla, CA) or human glyceraldehyde-3-phosphate dehydrogenase (AMBION). RNA is quantified using a Molecular Dynamics phosphorimager.

FACS Analysis

Transfection Conditions. Cells are seeded in six-well plates at a density of 2×10^5 and treated with 200 nM oligonucleotide in 500 μl serum-free medium for 4 hr (for details, see Table I).

Monensin Treatment. Monensin is from Sigma. A stock solution (25 mM) is prepared in ethanol. After a 4 hr treatment with carrier-complexed oligonucleotides, the medium is discarded and cells are washed twice with PBS. Cells are treated with 50 μM monensin diluted in PBS (10 ml) for 30 min at 4°.

Analysis. Transfected cells are rinsed three times with PBS, released from substrate by incubation with a solution of 0.05% trypsin–EDTA, and resuspended in DMEM supplemented with 7% serum. After centrifugation at 3000 rpm for 5 min, pellets were washed with PBS and submitted to another centrifugation. The supernatant is discarded and cells are resuspended in 500 μl PBS. Fluorescence from 10^4 individual cells is analyzed (FACSort, Becton-Dickinson, San Jose, CA).

Intracellular Stability of Oligonucleotides

5′-Radiolabeling of PO and PS Oligonucleotides. Twenty picomoles of oligonucleotide is first heat denatured for 5 min at 95° in 100 mM Tris–HCl, pH 7.6, 100 mM MgCl$_2$, and 50 mM dithiothreitol (DTT) and equilibrated at 37° before the addition of 10 units of T4 polynucleotide kinase and 10 μCi [γ^{32}-P]ATP (3000 Ci/mmol). PO oligomers are incubated for 30 min at 37°. Phosphorothioate oligonucleotides interact with and inhibit T4 polynucleotide kinase activity. Therefore, labeling of PS oligonucleotides is carried out in the following conditions: 3 hr of incubation at 37° and 18 hr at 4° after the addition of 10 additional units of enzyme. Radiolabeled oligomers are purified through a Micro Bio Spin-6 chromatography column (Bio-Rad) to remove unincorporated nucleotides.

Cell Transfection Conditions. Cells are seeded in six-well plates at a density of 2×10^5 and treated with 200 nM oligonucleotide (100 pmol containing 5% radiolabeled oligonucleotide, about 1×10^5 cpm) in 500 μl of serum-free medium for 4 hr. Cells are collected, washed twice with PBS, and lysed by sonication in a buffer containing 10 mM Tris–HCl (pH 7.4), 10 mM NaCl, 3 mM MgCl$_2$, 0.5% NP-40, and 5% sodium dodecyl sulfate (SDS). Cell lysates are phenol extracted and precipitated in the presence of 0.3 M sodium acetate. Precipitated oligonucleotides extracted from cells and from 100 μl culture medium (1/5) are subjected to electrophoresis on 20% polyacrylamide gels containing 7 M urea.

Fluorescence Measurements

Fluorescence measurements are carried out on a SPEX Fluorolog F1T11T spectrofluorimeter, and the cell holder is maintained at a constant temperature of 23° with a Huber Ministat circulating water bath. For all samples, the excitation path length is 0.2 cm, whereas the emission path length is 1 cm. Experimental solutions contain either PO or PS oligonucleotides at a concentration of 6 μM and PEI at 10 equivalents. The solutions contain 10 mM sodium cacodylate (pH 7) and 150 mM NaCl. Fluorescence emission spectra of these solutions are recorded. The bandwidth of the excitation and emission slits is 1.9 nm. The excitation wavelength is 480 nm.

Results and Discussion

Northern Blot Analysis

The limiting steps in mRNA degradation when oligonucleotides that induce RNase H activity are used as antisense are essentially the uptake kinetics of the oligonucleotide, the escape kinetics from endosomes, and the hybridization kinetics to the cytosolic or nuclear target sequence. Depletion of the messenger RNA of interest can be followed by Northern blot analysis for several hours after oligonucleotide application to the cells in culture. However, the corresponding protein inhibition will depend on the half-life of the protein and will be deleted with respect to messenger RNA depletion. Once released in the cytoplasm, oligonucleotides reach the targeted messenger RNA in the different compartments very rapidly. Therefore, if the cationic lipid formulation and cationic polymer formulation deliver the oligonucleotide with the same kinetics into the cytoplasm and have the same rate of degradation in this compartment, identical results should be obtained with respect to RNA degradation. It is noteworthy that the mechanism by which cationic lipids and polymers increase the activity of antisense oligonucleotides is poorly understood. Mechanisms by which the oligonucleotide is released from the endocytic compartments in order to gain access to its RNA target is not yet fully understood, although different models have been proposed.[7,11]

Functional properties that characterize cationic lipid formulations are (i) 100% spontaneous capture of negatively charged polynucleotides with

[11] O. Zelphati and F. C. Szoka, Jr., *Proc. Natl. Acad. Sci. U.S.A.* **93,** 11493 (1996).

cationic lipids by a condensation reaction, (ii) increased cellular uptake due to the interaction of positively charged complexes with negatively charged biological surfaces, and (iii) membrane fusion (or transient membrane destabilization) with the endosome in order to achieve delivery into the cytoplasm.

Functional properties that characterize PEI formulations are (i) 100% capture of negatively charged polynucleotides with the protonated amino groups in PEI, (ii) the cationic particles bind ionically to the cell surface and are taken up by spontaneous endocytosis, and (iii) endosome disruption and delivery into the cytoplasm.

There is no apparent difference between cationic lipid- and cationic polymer-mediated transfection mechanisms.

We have used PEI and Lipofectin as carriers to enhance the uptake and antisense activities of phosphodiester and phosphorothioate 20-mers (5′ TCC GTC ATC GCT CCT CAG GG 3′) targeted to the initiation codon (AUG) of Ha-*ras* mRNA. We have used as a control sequence an oligonucleotide with the same sequence as antisense but with inverted polarity. To determine the ability of different oligonucleotides to elicit antisense effects, T24 cells were treated in culture with oligonucleotides at a final oligonucleotide concentration of 200 nM and antisense activity was assessed by analysis of Ha-*ras* mRNA expression. Four hours after treatment, Ha-*ras* mRNA was reduced selectively and efficiently in two cases: PO antisense delivered with PEI and PS oligonucleotide delivered with Lipofectin. There was no significant change in the level of Ha-*ras* mRNA following treatments with all other combinations (Fig. 1).

The kinetics of antisense inhibition of Ha-*ras* mRNA was determined for PEI–PO. In this analysis, T24 cells were treated with 200 nM oligonucleotide, and Ha-*ras* mRNA levels were determined 4, 8, 10, 22, and 28 hr following the initiation of oligonucleotide treatment. The degree of antisense activity induced by PO antisense delivered with PEI was found to be highly time dependent. After 10 hr, oligonucleotide activity diminished substantially and was completely lost at 24 hr. The PS-modified 20-mer delivered with Lipofectin was as active at 24 hr as at 4 hr.

Agarose Gel Analysis of Oligonucleotide–Carrier Complex Formation

In theory, the carrier–oligonucleotide entity should be stable in the culture medium, but once within the cells, the oligonucleotide should be released from the carrier in order to reach its target. Therefore, there are at least two cases that are not favorable for the activity of antisense oligonucleotides: (a) if the oligonucleotide–carrier complex is not stable and dissociates rapidly in the culture medium and (b) if the oligonucleotide–

PEI   Lipofectin

PO PS   PO PS

C A I A I   C A I A I

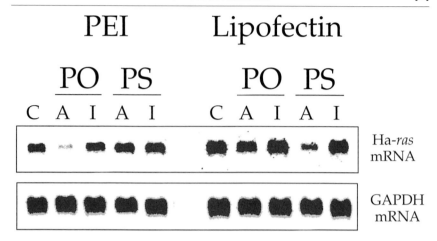

Ha-*ras*
mRNA

GAPDH
mRNA

FIG. 1. Downregulation of Ha-*ras* mRNA by antisense oligonucleotides. Northern blot analysis was performed on RNA samples isolated from T24 cells treated with 200 nM PO– or PS–ODN in the presence of 10 equivalents of PEI or 5 μg/ml Lipofectin reagent for 4 hr in serum-free medium. Medium was removed and total cellular RNA was prepared and separated on an agarose gel. Blots were hybridized with a Ha-*ras* cDNA probe, stripped, and rehybridized with the GAPDH probe to test equal loading. C refers to no oligonucleotide treatment. A refers to treatment with antisense ODN. I refers to treatment with inverse ODN.

carrier complex is too stable, the oligonucleotide will not be released from the carrier.

To investigate if backbone modification influences carrier–oligonucleotide interactions, fluorescently labeled oligonucleotides were complexed to the carriers and subjected to agarose gel electrophoresis (Fig. 2). 5′-Fluorescein–PO–20-mer and 5′-fluorescein–PS–20-mer were equally labeled and loaded into the gel. Three hundred picomoles of 5′-fluorescein-conjugated ODNs (PS and PO–20-mers) was added to the PEI or Lipofectin. The PEI concentration was adjusted in order to reach a molar ratio of PEI nitrogen/phosphate equal to 5, 10, and 20. PEI–ODN complexes were retained in the top of the gel and no band was observed where free ODN migrated. At a charge ratio of 20, the highly positive charge excess in the complex explains its migration to the negative pole. The same amount of fluorescent PO and PS oligomers was complexed to 7.5 μg of Lipofectin. These complexes were also retained. In contrast, with PO–20-mer, a band corresponding to the free oligonucleotide migration was detected. Quantification of the free oligonucleotide fluorescence shows that 30% of the oligonucleotide was not associated with Lipofectin. This implies that backbone modification of the PS oligonucleotide augmented the stability of the

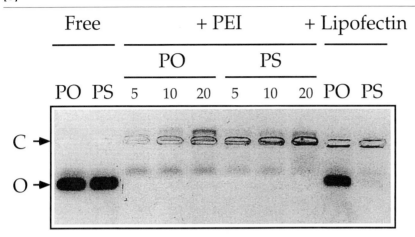

FIG. 2. *In vitro* stability of ODN–carrier complexes. The same amount of fluorescent ODN (2 μg ODN) was loaded in each well. ODNs were mixed with different equivalents (5, 10, and 20) of PEI or 7.5 μg of Lipofectin. Complexes were loaded on a 1% agarose gel and run for a few minutes. The gel was exposed to UV irradiation and photographed. A negative picture is represented. C refers to carrier-complexed ODN and O to free ODN.

complex. The instability of PO–20-mer–Lipofectin complexes could explain the lack of biological activity of the antisense oligonucleotide.

Laser Scanning Confocal Microscopy Analysis

To investigate the time course of distribution of the ODN and carrier (Lipofectin, PEI) in cells, the complex was prepared using rhodamine-labeled carrier and fluorescein-labeled oligonucleotide. T24 cells were treated for 4 hr with the complex (200 n*M* 5′-FITC-oligonucleotide + carrier) before the addition of 7% FCS containing medium. The following results were obtained (Fig. 3).

Lipofectin + PS–20-mer. This combination led to a complete inhibition of Ha-*ras* mRNA, after 4 hr of treatment, up to 24 hr.

At 4 hr, almost all cells incorporated the ODN into the nucleus. Oligonucleotide is also localized into cytoplasm and cytoplasmic punctate structures; however, a large amount of the PS–ODN is found in the nucleus. The cationic lipid could be seen localized around the nucleus and in some larger punctate structures. Some of these structures contain the PS–ODN. This pattern of distribution was observed up to 10 hr.

Lipofectin + PO–20-mer. Up to 24 hr no inhibition of Ha-*ras* mRNA was obtained with this complex.

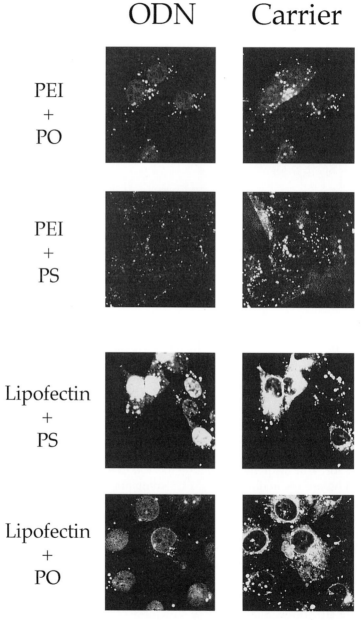

Fig. 3. Intracellular distribution of fluorescein-labeled ODNs and rhodamine-labeled carriers. T24 cells were treated with 200 n*M* fluorescently labeled PO– or PS–ODN carried by 10 equivalents of rhodamine-labeled PEI or 5 μg/ml rhodamine-stained Lipofectin for 4 hr. Cells were then rinsed, fixed with paraformaldehyde, and mounted with Mowiol before examination under a laser scanning confocal microscope.

ODN is nearly equally distributed between the cytoplasm and the nucleus. Nuclear accumulation was not as important as with PS oligonucleotides. The cationic lipid stays outside of the nucleus. At 10 hr, Lipofectin was accumulated in large punctate structures into the cytoplasm.

PEI + PO–20-mer. Application of this complex to the cells for 4 hr resulted in the depletion of Ha-*ras* mRNA. This inhibition decreased after 10 hr and was lost at 24 hr.

At 4 hr, ODN was found in fine speckles in the cytoplasm and only a small amount of ODN staining could be seen in the nucleus. Nuclear staining increased over time. PEI was found in large punctate structures. PEI was never found within the nuclei.

PEI + PS–20-mer. No inhibition of Ha-*ras* mRNA was detected with this complex. At 4 hr, PS–20-mer was almost undetectable in the cells. This very weak staining pattern did not change appreciably after an additional 20 hr of incubation. However, PEI was detected in large punctate structures at 4 hr and this staining increased with a longer incubation period.

Using gel electrophoresis, PS–PEI complexes are very stable. However, confocal microscopy experiments show that PEI is internalized into cells. Therefore, it seems likely that PS–ODN is internalized, but is not detected because its fluorescence is quenched.

Measurement of Fluorescence Emission of Free and PEI-Complexed ODNs

Fluorescence emission of oligonucleotides complexed to transporters is generally quenched. Fluorescence emission quenching will depend on oligonucleotide interaction with the transporter. The fluorescence emission of free and PEI-complexed FITC–PS and FITC–PO oligonucleotides has been measured. Figure 4 shows that PO and PS have the same fluorescence quantum efficiency as free agents and this fluorescence emission is reduced to the same extent when they are carrier bound.

Flow Cytometric Analysis

5'-Fluorescein-labeled oligonucleotides were used for cytofluorimetric cellular uptake studies. Table II gives the fluorescence intensity associated with cells exposed to 200 n*M* of free or carrier-associated FITC–ODNs for 4 hr at 37°. The fluorescence intensity associated with cells exposed to 200 n*M* PO oligomer increased by 7.5-fold when the oligonucleotide was complexed to either PEI or Lipofectin. In contrast, no enhancement of cell-associated fluorescence was observed when PS–ODN was complexed to PEI. The highest enhancement of cell-associated fluorescence intensity was recorded with PS–ODN complexed to Lipofectin (9 → 600).

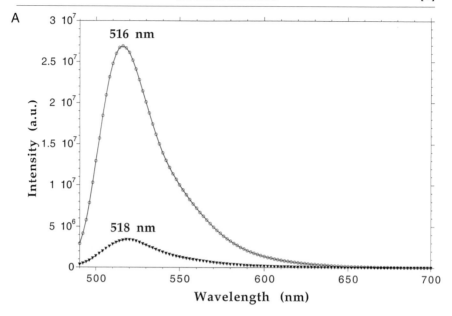

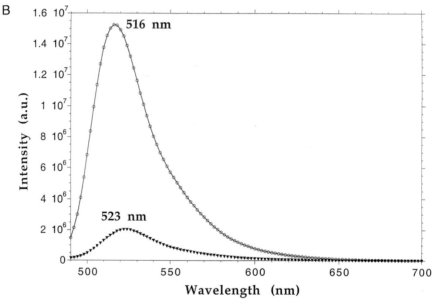

FIG. 4. Fluorescence emission spectra of free (o) and PEI-complexed (▼) FITC–PO (A) and FITC–PS (B) oligonucleotides. Oligonucleotides (6 μM) were mixed with 10 equivalents of PEI in 10 mM sodium cacodylate (pH 7), 150 mM NaCl. Fluorescence emission spectra were recorded at an excitation wavelength of 480 nm.

TABLE II
FLOW CYTOMETRIC QUANTIFICATION OF
FITC–ODN INTERNALIZATION[a]

Compound	Cell fluorescence intensity (arbitrary units)
PO	4
PO + PEI	33
PO + Lipofectin	28
PS	9
PS + PEI	8
PS + Lipofectin	600

[a] T24 cells were treated with free or transporter-complexed ODNs (200 nM) for 4 hr.

The fluorescence intensity was also measured after monensin treatment. This ionophore neutralizes endocytic compartments, allowing the retrieval of the fluorescence quenched in an acidic environment. A twofold increase in the cell fluorescence intensity after monensin treatment was measured only in the case when cells were treated with PO–PEI complexes, indicating that these complexes were within acidic intracellular compartments.

These results are in agreement with those obtained with confocal microscopy. Indeed, an enhancement of intracellular fluorescence was observed when ODNs were delivered with a carrier, except when PS–ODN was delivered with PEI. The highest intracellular fluorescence was obtained in the nuclear compartment when PS–ODN was delivered with Lipofectin. A similar intensity of nuclear fluorescence was observed when PO–ODN was delivered with either PEI or Lipofectin.

Analysis of Intracellular Stability of Oligonucleotides

Confocal microscopy and flow cytometric experiments cannot be interpreted properly if the 5'-FITC–ODNs are degraded within cells. If the oligonucleotide is degraded by intracellular exo- or endonucleases, the recorded fluorescence intensity will reflect the fluorescence of free and not ODN-associated fluorophore.

The intracellular stability of oligonucleotides can be determined after the extraction of ^{32}P-labeled ODNs from cells followed by denaturing PAGE analysis. (Fig. 5) Not only the intracellular stability but also the quantity of the oligonucleotide taken up by cells can be calculated. It is noteworthy that cell lysates were phenol extracted. Phenol extraction dissociates oligonucleotides from carrier. Therefore, undegraded oligonucleotide detected in the gel comes from free and carrier-associated oligonucleotide.

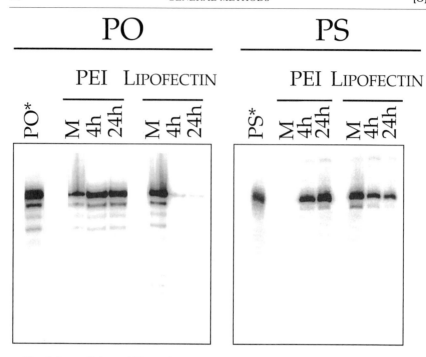

Fig. 5. Intracellular stability and penetration of radiolabeled PO– and PS–ODNs carried by PEI or cationic lipids. T24 cells were treated for 4 hr in serum-free medium with 200 nM ODN containing 5% ^{32}P-labeled ODN carried by 10 equivalents of PEI or 5 μg/ml lipofectin. Cells were collected and sonicated. Nucleic acids were phenol extracted, precipitated, and loaded onto a 20% polyacrylamide gel containing 7 M urea. PO* and PS* refer to aliquots of radiolabeled ODNs. M refers to precipitated ODN present in one-fifth of the culture medium at 4 hr. 4h refers to precipitated nucleic acids extracted from cells treated for 4 hr with radiolabeled ODN. 24h refers to nucleic acids extracted from cells treated for 4 hr with radiolabeled ODN and cultured in 7% FCS containing medium for an additional 20 hr.

Fully degraded radiolabeled oligonucleotides cannot be precipitated efficiently. Therefore, if the oligonucleotide is degraded very rapidly, we must observe only a disappearance of full-length radiolabeled oligonucleotide.

Analysis of Two Combinations That Result in Similar Depletion of Messenger RNA after 4 hr of Treatment: PO–PEI and PS–Lipofectin. PO–ODN (37%) was incorporated into the cells when delivered with PEI, whereas only 6% PS–ODN was incorporated when the Lipofectin carrier was used (Table III). However, confocal microscopy images show a much higher intranuclear fluorescence intensity when fluorescein-labeled PS–ODN was delivered with Lipofectin compared to the intranuclear intensity of fluorescein-labeled PO–ODN delivered with PEI. These two observations are

TABLE III

DISTRIBUTION OF OLIGONUCLEOTIDE-ASSOCIATED RADIOACTIVITY[a]

Carrier	Oligonucleotide	Medium	Intracellular
PEI	PO	63%	37%
	PS	7%	93%
Lipofectin	PO	99.5%	0.5%
	PS	94%	6%

[a] T24 cells were treated for 4 hr at 37° with 200 nM of ^{32}P-labeled ODNs complexed to PEI or Lipofectin. Quantification of oligonucleotide-associated radioactivity in the medium and within cells is given as a percentage of total radioactivity.

compatible if we assume that the majority of the 37% radioactive PO–ODN found in the cells are complexed to the PEI (the fluorescence of this fraction will be quenched) and only a small fraction is released. We detect the fluorescence associated with the free fraction. In contrast, it is likely that the totality of PS–ODN delivered with Lipofectin (6%) was released very rapidly from the carrier. The high fluorescence intensity observed in the nucleus comes from this free fraction (6%). Taking into account the intracellular fluorescence intensities in both cases (Table II) and the uptake percentage, we can roughly estimate that the intranuclear-free PS–ODN concentration was 20-fold higher than the PO–ODN concentration. It is probable that a large amount of intranuclear PS–ODN interacts very tightly with nuclear proteins and only a small fraction is available for nucleic acid interaction. These observations imply that with PO–ODNs, which do not elicit high affinity for proteins, low intranuclear concentrations are sufficient to exert an efficient antisense effect. It is also obvious that intranuclear fluorescence intensity cannot be directly related to a biological effect.

Analysis of Two Combinations That Do Not Allow Messenger RNA Depletion: PS–PEI and PO–Lipofectin. The highest cellular incorporation was found when PS–ODN was delivered with PEI (93%). However, FITC-labeled PS–ODN was not detected within cells when confocal microscopy or flow cytometry analysis was used. The explanation is that the majority of the intracellular PS–ODN is complexed to the carrier. The fluorescence of carrier-associated oligonucleotide is quenched as discussed earlier. Therefore, almost the totality of the radiolabeled PS–ODN extracted from the cells was released from the carrier after phenol extraction. PS–ODN forms very stable complexes with PEI (see agarose gel electrophoresis, Fig. 2). These complexes are taken up very efficiently by cells (93% of uptake), but the oligonucleotide is not released from the carrier and therefore cannot hybridize to messenger RNA.

PO–ODN formed unstable complexes with Lipofectin (see agarose gel electrophoresis). The very low incorporation into the cells (0.5%) can be explained by the fact that PO–ODN was dissociated from the carrier in the culture medium. No intact radiolabeled PO–ODN was detected. It is probable that the intranuclear fluorescence intensity comes from the fluorophore detached from the oligonucleotide after the intracellular degradation of PO–ODN.

Conclusions

Selection of Appropriate Carrier

The selection of an appropriate carrier cannot be based on fluorescence detection of oligonucleotide incorporation into the cells. Indeed, we show that a 20-fold higher nuclear fluorescence was recorded when cells were treated with PS–Lipofectin compared to PO–PEI. The efficiency of both oligonucleotides to deplete Ha-*ras* mRNA was similar. It is likely that the majority of nuclear PS–ODNs are trapped by nuclear proteins. The fraction of free nuclear PS–ODN that interacts with nucleic acids is probably equivalent to the PO–ODN.

The *in vitro* stability of the ODN–carrier complex is not predictive of its delivery efficiency. We show that the most stable complex is formed between PEI and PS–ODN. The highest oligonucleotide incorporation into the cells was achieved with this combination (93%). However, no biological activity was obtained, probably because the oligonucleotide was not released from the carrier.

The intracellular stability of the carrier-delivered oligonucleotide is not predictive of its antisense efficacy. Indeed, after cell extraction of the radio-labeled oligonucleotide, the step of phenol extraction before PAGE analysis cannot be avoided. Phenol extraction dissociates the oligonucleotide from the carrier. Because usually ODNs are protected from nucleases when complexed to a carrier, the proportion of intact ODN detected on PAGE analysis reflects the carrier-associated fraction of ODN instead of the undegraded free fraction of ODN that is implicated on antisense activity.

Our results suggest that each nucleic acid must be evaluated empirically to determine the most effective carrier for the enhancement of its antisense activity. The earliest detectable biological activity is the depletion of mRNA when oligonucleotides inducing RNase H activity are used as antisense. Only the biological activity of the carrier-delivered oligonucleotide can predict if the uptake, escape from endosomes, and interaction with messenger RNA have been maximized.

Advantages of Using PO–ODNs instead of PS–ODNs

To identify effective antisense sequences capable of inhibiting one gene expression, usually 50 different oligonucleotides targeted to different sites of messenger RNA have to be tested. For example, of the 34 antisense ODNs that were evaluated for their ability to inhibit c-*raf* mRNA expression in cultured tumor cells, only 1 was identified as a potent antisense inhibitor.[12] The selection consists of treating cells in culture with an oligonucleotide delivered with a carrier and determining the level of messenger RNA after a few hours by Northern blot analysis. Phosphodiester oligonucleotides can be incorporated efficiently into cells in culture using the PEI carrier. The advantages of using PO–ODNs instead of PS–ODNs in the selection experiments are multiple: (a) PO–ODNs have an increased affinity for hybridization to complementary mRNA sequences compared to PS–ODNs. (b) PO–ODNs have a decreased affinity toward proteins compared to PS–ODNs. Therefore, less nonantisense effects are generated with PO–ODNs compared to PS–ODNs. (c) The cost of PO–ODNs is much less than PS–ODNs. (d) PEI is less toxic to cells compared to cationic lipids. (e) One transfection experiment mediated by Lipofectin reagent is much more costly than one transfection experiment with PEI.

However, only a short-term antisense effect can be obtained with PO–ODNs. We show that a single treatment (4 hr) with 200 nM PO–ODN delivered with PEI allows downregulation of the messenger RNA of interest during 6 hr. After that time, the messenger RNA level increases. The half-life of the carrier-released oligonucleotide determines roughly the duration of mRNA depletion. However, we have determined that only a small fraction of PO–ODN was released from PEI within cells. Therefore, there is a pool of undegraded PO–ODN–carrier complexes in the cells (PO–ODNs complexed to carrier, which is protected efficiently from nuclease degradation). Therefore, future studies on the mechanisms by which ODNs are released from the carrier and endocytic compartments will probably allow the control of release of the PO–ODN from the ODN–PEI pool and will permit the antisense activity to be useful for inhibition of the expression of the protein product whose mRNA is targeted by the antisense.

Acknowledgments

The authors express their gratitude to Pr. Claude Hélène for his support of the research forming the basis of this article, Dr. Jean-Paul Behr for helpful suggestions, Dr. Thierry Bettinger for the preparation of rhodamine-labeled PEI, and Dr. Jean-Louis Mergny for fluorescence spectroscopy experiments.

[12] B. P. Monia, J. F. Johnston, T. Geiger, M. Muller, and D. Fabbro, *Nature Med.* **2,** 668 (1996).

[4] Design of Antisense and Triplex-Forming Oligonucleotides

By Jean-Christophe François, Jérome Lacoste, Laurent Lacroix, and Jean-Louis Mergny

Introduction

Antisense oligonucleotides bind to their complementary sequence on messenger RNAs and inhibit translation of the mRNA into the protein. Short oligonucleotides may also be used to inhibit gene transcription via triple-helix formation on a double-stranded DNA gene. These oligonucleotides are called triplex-forming oligonucleotides (TFO). Certain TFOs (also called clamp oligonucleotides) are designed to bind to single-stranded nucleic acids and may inhibit mRNA translation by physical blockade. The numerous applications of triplex-based strategies have been well described.[1–5]

Triplex-forming oligonucleotides and antisense oligonucleotides share the same requirements for cellular applications, such as high binding affinity for their targets, nuclease resistance, and efficient cell penetration. Very similar physicochemical methods are used to study antisense and triplex-forming oligonucleotides *in vitro* before considering their utilization in cell culture. This article focuses on some of the methods that have been employed to determine the critical parameters that help in designing an efficient antisense or triplex-forming oligonucleotide.

Design of Antisense and Triplex-Forming Oligonucleotides

Choice of Target Sequence

The choice of a target sequence for antisense or triplex-forming oligonucleotides depends greatly on the gene context. Obviously, the uniqueness of the targeted sequence and of the oligonucleotide sequence should be checked by searching for homologies in available DNA sequence data banks. A 17-nucleotide-long oligonucleotide should find a unique target

[1] N. T. Thuong and C. Hélène, *Angew. Chem. Int. Ed.* **32**, 666 (1993).
[2] L. J. Maher III, *Cancer Invest.* **14**, 66 (1996).
[3] P. P. Chan and P. M. Glazer, *J. Mol. Med.* **75**, 267 (1997).
[4] C. Giovannangéli and C. Hélène, *Antisense Nucleic Acid Drug Dev.* **7**, 413 (1997).
[5] K. M. Vasquez and J. H. Wilson, *Trends Biochem. Sci.* **23**, 4 (1998).

within the human genome, assuming a random distribution of base pairs (for discussion about the selectivity of oligonucleotides, see Ref. 1). Antisense oligonucleotides were often targeted to coding regions of messenger RNA and especially to the initiation codon, arguing that these regions are accessible to hybridization with oligonucleotides. Nevertheless, antisense oligonucleotides targeted to the 5'- or to the 3'-untranslated regions were also able to specifically inhibit translation.[6]

Target sequences for triplex-forming oligonucleotides, which are located on the double-stranded DNA, should contain long stretches of purines (or pyrimidines). These polypurine–polypyrimidine sequences, which are overrepresented in regulatory regions of eukaryotic genes, are able to bind a third oligonucleotide via the formation of base triplets. Different triplets may be formed depending on the purine or pyrimidine composition of the third strand.[7] Briefly, the CT-containing TFO binds to the polypurine strand of the duplex DNA in a parallel orientation through the formation of base triads, $C \cdot G*C^+$ and $T \cdot A*T$. A GA-containing TFO forms an antiparallel triplex through the formation of $C \cdot G*G$ and $T \cdot A*A$ base triads (Fig. 1). GT-containing TFO binds to the oligopurine sequence of the DNA duplex in an orientation that depends both on the number of GpT and TpG steps and on the length of G and T tracts.[8] Acidic pH is required to protonate cytosines within the standard pyrimidine-rich triplex-binding motif, limiting the potential use of unmodified CT–TFOs for cellular studies. The formation of triple helices involving purine-rich third strands has been shown to be pH independent and is stabilized by multivalent cations such as magnesium ions (for details, see Refs. 1–3).

In order to control gene expression of specific genes, it was sometimes necessary to expand the library of triplex-binding sites by either alternate strand recognition or targeting interrupted polypurine–polypyrimidine sequences using modified nucleotides.[4,9] A tentative triplex-binding code showing the different modified nucleotides known to stabilize TFOs is available.[3,5]

Regulatory regions such as binding sites for transcriptional activators or for RNA polymerases are often chosen as targets for triplex-forming oligonucleotides. A TFO that recognizes the target sequence of a regulatory protein located in the promoter region may inhibit the binding of this protein, leading to transcriptional inhibition (reviewed in Refs. 1–5). It has

[6] B. P. Monia, J. F. Johnston, T. Geiger, M. Muller, and D. Fabbro, *Nat. Med.* **2,** 668 (1996).
[7] V. N. Soyfer and V. N. Potaman, "Triple-Helical Nucleic Acids." Springer-Verlag, New York, 1996.
[8] T. De Bizemont, G. Duval-Valentin, J.-S. Sun, E. Bisagni, T. Garestier, and C Hélène, *Nucleic Acids Res.* **24,** 1136 (1996).
[9] J. S. Sun, T. Garestier, and C. Hélène, *Curr Opin. Struct. Biol.* **6,** 327 (1996).

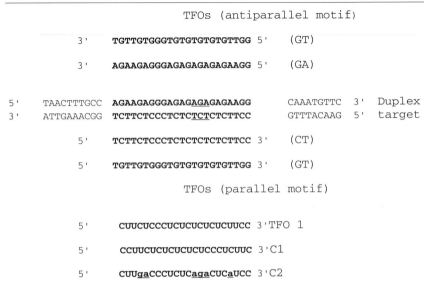

FIG. 1. Two binding motifs for triple-helix formation. The duplex fragment corresponds to a 42-bp sequence of the rat insulin-like growth factor I gene with the polypurine–polypyrimidine target site for triplex-forming oligonucleotides shown in bold. TFOs form base triads with the purine-rich strand of the duplex. TFOs designed to bind in the antiparallel or in the parallel orientation with respect to the oligopurine sequence of the duplex are shown above and below the 42-mer duplex. The GA-containing TFO designed to bind in the antiparallel orientation is used for the gel retardation experiment shown in Fig. 2. TFO1, a modified oligopyrimidine containing 5Me-dC and C-5 propyne(dU) with phosphorothioate linkages used in transient transfection, is shown. Underlined nucleotides AGA/TCT in the duplex are changed to CTC/GAG in a mutated target for TFO1. Control oligonucleotides C1 and C2 correspond to the reverse sequence and to a six base mismatch sequence of TFO1, respectively.

been demonstrated that *in vitro* transcription elongation could be inhibited using a TFO containing phosphoramidate linkages.[10] Different mechanistic models of triplex-mediated gene repression have been proposed.[2]

Nuclease-Resistant Antisense or Triplex-Forming Oligonucleotides

Unmodified antisense oligonucleotides are susceptible to degradation by extra- and intracellular nucleases. Different chemical modifications of antisense oligomers have been used to confer resistance to degradation. Oligonucleotides containing phosphorothioate linkages are used routinely

---

[10] C. Escudé, C. Giovannangéli, J. S. Sun, D. H. Lloyd, J. K. Chen, S. M. Gryaznov, T. Garestier, and C. Hélène, *Proc. Natl. Acad. Sci. U.S.A.* **93,** 4365 (1996).

in antisense strategy. Like antisense oligomers, natural triplex-forming oligonucleotides are sensitive to nuclease degradation. The oligonucleotide stability toward exonucleases can be increased slightly when extremities are blocked via chemical linkers. The introduction of these blocking groups usually does not affect the binding affinity of TFOs for their target. Conventional backbone modifications, such as RNA, 2'-OMe-RNA, phosphorothioate, α-DNA, and phosphoramidate studied in the context of antisense oligonucleotides, are also used in the design of TFOs.[9–22] Although most of these modifications confer increased nuclease resistance, some have a deleterious effect on triplex-binding affinities (Table I). This destabilization depends on the purine or pyrimidine motif of the third strand. Different strategies have been proposed to increase the binding affinities of nuclease-resistant TFOs. For example, covalent linkage of an intercalating agent such as an acridine derivative to a GA-containing triplex-forming oligophosphorothioate could partially restore binding affinities compatible with physiological conditions.[18] It has been shown that the introduction of C-5 propyne pyrimidines in nuclease-resistant oligophosphorothioate increased its binding affinity for the mRNA target, thereby decreasing its effective inhibitory concentration.[23–25] Introduction of C-5 propyne uracil

[11] C. Escudé, J. C. François, J. S. Sun, G. Ott, M. Sprinzl, T. Garestier, and C. Hélène, Nucleic Acids Res. 21, 5547 (1993).
[12] L. Lacroix, J. L. Mergny, J. L. Leroy, and C. Hélène, Biochemistry 35, 8715 (1996).
[13] M. Alluni-Fabbroni, G. Manfioletti, G. Manzini, and L. E. Xodo, Eur. J. Biochem. 226, 831 (1994).
[14] L. Xodo, M. Alunni-Fabbroni, G. Manzini, and F. Quadrifoglio, Nucleic Acids Res. 22, 3322 (1994).
[15] J. G. Hacia, B. J. Wold, and P. B. Dervan, Biochemistry 33, 5367 (1994).
[16] M. Musso and M. W. Van Dyke, Nucleic Acids Res. 23, 2320 (1995).
[17] G. Tu, Q. Cao, and Y. Israel, J. Biol. Chem. 270, 28402 (1995).
[18] J. Lacoste, J. C. François, and C. Hélène, Nucleic Acids Res. 25, 1991 (1997).
[19] S. B. Noonberg, J. C. François, D. Praseuth, A. L. Guieysse-Peugeot, J. Lacoste, T. Garestier, and C. Hélène, Nucleic Acids Res. 25, 4042 (1995).
[20] C. Giovannangéli, S. Diviacco, V. Labrousse, S. Gryaznov, P. Charneau, and C. Hélène, Proc. Natl. Acad. Sci. U.S.A. 94, 79 (1997).
[21] M. J. Ferber and L. J. Maher III, Anal. Biochem. 244, 312 (1997).
[22] M. Shimizu, A. Konishi, Y. Shimada, H. Inoue, and E. Ohtsuka, FEBS Lett. 302, 155 (1992).
[23] R. Wagner, M. Matteucci, J. Lewis, A. Gutierrez, C. Moulds, and B. Froehler, Science 260, 1510 (1993).
[24] J. G. Lewis, K. Y. Lin, A. Kothavale, W. M. Flanagan, M. D. Matteucci, R. B. DePrince, R. A. J. Mook, R. W. Hendren, and R. W. Wagner, Proc. Natl. Acad. Sci. U.S.A. 93, 3176 (1996).
[25] J. O. Ojwang, S. D. Mustain, H. B. Marshall, T. S. Rao, N. Chaudhary, D. A. Walker, M. E. Hogan, T. Akiyama, G. R. Revankar, A. Peyman, E. Uhlmann, and R. F. Rando, Biochemistry 36, 6033 (1997).

TABLE I

BASE MODIFICATIONS AND PHOSPHATE–SUGAR BACKBONES USED IN ANTISENSE STRATEGY TO DESIGN NUCLEASE-RESISTANT TFOs[a]

Modifications[b]	Nuclease resistance	Py-rich TFO	Pu-rich TFO	i motif[c]	G quartet[d]	GA duplex[d]	Comments on triplex formation	Refs.
5Me-dC	No	+	o	=	o	o	Reduce pH requirement	26, 27
C-5 propyne (dU)	No	+	o	?	o	o	Gain of +2° per substitution in ΔT_m	26, 27
C-5 propyne (dC)	No	−	o	?	o	o	Loss of −3° per substitution in ΔT_m	26, 27
RNA	No	++	− −	−	=	− −	UC-RNA forms more stable triplex than TC-DNA. No triplex observed with GA and GU RNA	11, 12
2'-OMe-RNA	Yes	++	− −	− −	?	− −	Sequence-dependent stabilization with 2'OMeUC RNA. No triplex was observed with antiparallel 2'OMe GA TFO. No data available on 2'OMe GU-TFO	21, 22
Phosphorothioate (PS-DNA)	Yes	−	=/−	=	=	−	In TC-PS TFO, destabilization depends on the number of PS linkages. GT-PS TFOs exhibit similar affinities than GT-PO. GA-PS TFOs form triplex less stable than GA-PO	13–18
α-Anomers (α-DNA)	Yes	=	− −	=	?	− −	Similar binding affinities with CT-TFOs. GT-α exhibits lower affinities than GT-β. No triplex observed with antiparallel or parallel GA-α TFOs	9, 19
Phosphoramidate (N3' → 5' PN)	Yes	++	− −	− −	?	?	Parallel N3' → 5' PN TC and PN TCG exhibit higher affinities compatible with physiological pH. No triplex was observed with antiparallel N3' → 5' PN GA TFO	10, 20

[a] The effects of oligonucleotide substitutions on the stability of DNA triple helices and three competing structures (i-DNA, G quartets, and GA duplexes) are given. o, not applicable; ?, not determined; + or ++, moderate or high stabilization; − or − −, moderate or high destabilization; =, not significant effect; − or − −, moderate or high destabilization.

[b] Modified bases and backbone modifications may be combined. For example, pyrimidine oligophosphorothioate containing 5Me-dC and C-5 propyne (dU) forms a stable triplex.

[c] For pyrimidine, C-rich TFO.

[d] For purine-rich TFO.

but not C-5 propyne cytosine was also shown to stabilize triplex formation with oligodeoxypyrimidines.[26,27]

Design of Control Oligonucleotides

A general consensus to demonstrate antisense mechanisms involved in translation inhibition has been proposed.[28] Briefly, the demonstration of a specific recognition of a target sequence by an antisense oligonucleotide should be provided, at least two control oligonucleotides having the same base composition (a mutated and a reverse version of the specific oligonucleotide) should be tested, and their inhibition activities compared with the specific oligomer. The same rules for the design of control sequences in the antisense strategy may apply for the triplex-mediated strategy. Additionally, when specific motifs such as GA repeats, C, or G blocks are present in third strands, these motifs should be included in control oligonucleotides. Self-association or even aggregation may occur with oligonucleotides containing these motifs, therefore decreasing their apparent binding constant for their targets. As oligonucleotide degradation products could perturb gene expression,[29] control oligonucleotides should also bear 5' and 3' ends similar to the specific oligonucleotides.

Monitoring Binding of Antisense and Triplex-Forming Oligomers to Their Targets by Gel Electrophoresis

The binding of antisense oligonucleotides to their target sequence can be studied by performing RNase H-induced cleavage of the *in vitro*-transcribed mRNA,[30] provided that the antisense–mRNA complex is still recognized by RNase H. Gel mobility shift experiments were also used to determine binding affinities of antisense oligomers for their complementary sequence using short synthetic oligoribonucleotide targets.[31,32]

Similar experiments may be done to demonstrate binding of TFOs to duplex DNA. Interference with a restriction endonuclease that recognizes

---

[26] B. C. Froehler, S. Wadwani, T. J. Terhorst, and S. R. Gerrard, *Tetrah. Lett.* **33,** 5307 (1992).

[27] N. Colocci and P. B. Dervan, *J. Am. Chem. Soc.* **116,** 785 (1994).

[28] S. T. Crooke, *Antisense Nucleic Acid Drug Dev.* **6,** 145 (1996).

[29] J. L. Vaerman, P. Moureau, F. Deldime, P. Lewalle, C. Lammineur, F. Morschhauser, and P. Martiat, *Blood* **90,** 331 (1997).

[30] T. E. Saison-Behmoaras, I. Duroux, N. T. Thuong, U. Asseline, and C. Hélène, *Antisense Nucleic Acid Drug Dev.* **7,** 361 (1997).

[31] B. P. Monia, E. A. Lesnik, C. Gonzales, W. F. Lima, D. McGee, C. J. Guinosso, A. M. Kawasaki, P. Dan Cook, and S. M. Freier, *J. Biol. Chem.* **268,** 14514 (1993).

[32] G. Godard, J. C. François, I. Duroux, U. Asseline, M. Chassignol, T. Nguyen, C. Hélène, and T. Saison-Behmoaras, *Nucleic Acids Res.* **22,** 4789 (1994).

totally or even partially the binding site of the triplex-forming oligonucleotide on the DNA duplex has been performed to demonstrate specific binding to DNA. The recognition sites of these restriction endonucleases may be either palindromic or nonpalindromic sequences and must contain a short (2–6 bp) polypurine–polypyrimidine sequence $R(n)Y(m)$ (Table II). The specificity of this cleavage assay is demonstrated either through the utilization of DNA targets containing cleavage sites for the restriction enzyme but no binding sites for the TFO or through the utilization of restriction enzymes that cleave at more than two sites on the DNA target containing only one TFO target site. *In vitro* footprinting experiments with DNase I nuclease or copper–phenanthroline as cleaving reagents have been used to evidence triplex-forming oligonucleotide-binding sites on the DNA double helix.[11,21,33–35]

Gel retardation assays with synthetic duplexes, restriction fragments of a plasmid, and even comigration assays with complete plasmids are popular in studying the binding of TFOs. Experimental procedures depend greatly on the nature of the third strand (purine or pyrimidine), as acidic pH and magnesium ions have dramatic effects on TFO-binding affinities.[1,3] An interesting comparison of electrophoretic mobility shift assays (EMSA) and quantitative DNase I footprinting assays to determine the affinities of pyrimidine-rich oligonucleotides for duplex DNA has been performed.[21] Briefly, it was shown that both techniques are in good agreement when low K^+ concentrations are used. In the presence of physiological concentrations of potassium ions (100 mM), EMSA slightly overestimates triplex stabilities relative to footprinting analysis.[21] Apparent equilibrium dissociation constants (K_d) differed by 4- to 11-fold in buffer containing KCl.

Method: A Typical Gel Shift Experiment to Evidence TFO Binding

A short double-stranded DNA target (or a restriction fragment) containing the target sequence may be used for electrophoretic mobility shift assays. Either strand of the synthetic duplex is labeled at its 5′ end with T4 polynucleotide kinase and [γ-^{32}P]ATP. Unlabeled larger targets may also be used in combination with radiolabeled TFOs to perform comigration assays (see later). The synthetic duplex target is formed first by heating briefly (1 min) both complementary strands at 80° in a buffer containing 100 mM NaCl and then cooling (overnight) the sample slowly to room temperature. Increasing concentrations of TFOs (from 10 nM to 10 μM) are then added to the labeled preformed duplex (10 nM) and incubated in

[33] J. C. François, T. Saison-Behmoaras, and C. Hélène, *Nucleic Acids Res.* **16,** 11431 (1988).
[34] B. Faucon, J. L. Mergny, and C. Hélène, *Nucleic Acids Res.* **24,** 3181 (1996).
[35] H. M. Paes and K. R. Fox, *Nucleic Acids Res.* **25,** 3269 (1997).

TABLE II
RESTRICTION ENZYME INHIBITION ASSAYS TO STUDY TFO BINDING

Restriction enzyme	Recognition sequence[a]	Refs.
AciI	CCGC(−3/−1)	c
AvaI	C/YCGRG	d
BamHI	G/GATCC	e
BsgI[b]	GTGCAG(16/14)	f
DraI	TTT/AAA	g
Eco57I	CTGAAG(16/14)	h
EcoRI	G/AATTC	i
HaeIII	GG/CC	j
HindIII	A/AGCTT	e
HinfI	G/ANTC	k
Ksp632I, EarI	CTCTTC(1/4)	l, m
MnlI	CCTC(7/6)	k
SacI	GAGCT/C	e
SmaI	CCC/GGG	e

[a] All recognition sequences are written 5' to 3' using the single letter code nomenclature with the point of cleavage indicated by a "/". Numbers in parentheses indicate point of cleavage for nonpalindromic enzymes.

[b] Utilization in a combinatorial approach to design TFOs.

[c] C. A. Hobbs and K. Yoon, *Antisense Res. Dev.* **4**, 1 (1994).

[d] L. J. Maher III, B. Wold, and P. B. Dervan, *Science* **245**, 725 (1989).

[e] R. Kiyama and M. Oishi, *Nucleic Acids Res.* **23**, 452 (1995).

[f] P. Hardenbol and M. W. Van Dyke, *Proc. Natl. Acad. Sci. U.S.A.* **93**, 2811 (1996).

[g] C. Giovannangéli, N. T. Thuong, and C. Hélène, *Nucleic Acids Res.* **20**, 4275 (1992).

[h] B. Ward, *Nucleic Acids Res.* **24**, 2435 (1996).

[i] J. C. Hanvey, M. Shimizu, and R. D. Wells, *Nucleic Acids Res.* **18**, 157 (1990).

[j] N. Bianchi, C. Rutigliano, M. Passadore, M. Tomasetti, L. Pippo, C. Mischiati, G. Feriotto, and R. Gambari, *Biochem. J.* **326**, 919 (1997).

[k] M. Grigoriev, D. Praseuth, P. Robin, A. Hemar, T. Saison-Behmoaras, A. Dautry-Varsat, N. T. Thuong, C. Hélène, and A. Harel-Bellan, *J. Biol. Chem.* **267**, 3389 (1992).

[l] J. C. François, T. Saison-Behmoaras, N. T. Thuong, and C. Hélène, *Biochemistry* **28**, 9617 (1989).

[m] D. A. Collier, N. T. Thuong, and C. Hélène, *J. Am. Chem. Soc.* **113**, 1457 (1991).

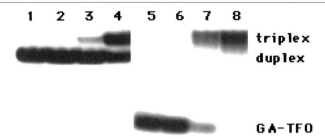

FIG. 2. A typical gel retardation profile showing a triplex-forming oligonucleotide binding to a double-stranded target. The radiolabeled duplex target (10 nM, lane 1) was incubated with increasing concentrations of unlabeled GA-TFO (see GA-TFO sequence in Fig. 1). Lanes 2–4 correspond to 10 nM, 100 nM, and 1 μM of added GA-TFO. Incubation was performed at 37° in a buffer containing 50 mM HEPES, pH 7.2, 100 mM NaCl, 10% sucrose, 0.5 μg/ml tRNA, 10 mM MgCl$_2$. In lanes 5–8, the GA-TFO is labeled, and increasing concentrations of an unlabeled duplex target were added (lanes 5–8: 0, 10 nM, 100 nM, 1 μM duplex). Electrophoresis was performed at 37° in a nondenaturing polyacrylamide gel (10%) run in a 50 mM HEPES buffer, 10 mM MgCl$_2$. The apparent dissociation constant (K_d) of the GA-TFO/duplex is approximately 10^{-6} M when the duplex is labeled, in contrast with an estimated value of 10^{-7} M when the TFO is labeled.

an appropriate buffer, close to physiological conditions. The incubation may be done at 37° in a 50 mM, pH 7.2, HEPES buffer containing magnesium cations (1–10 mM), potassium ions (100 mM), and spermine (0.2–1 mM). It should be noted that the presence of physiological concentrations of monovalent potassium ions during incubation may promote G-tetrads or aggregates in oligomers.[3,36] Sucrose (10%) is added just prior to loading on a polyacrylamide gel in native conditions or during the incubation of the TFOs and the duplex target (Fig. 2). Nonspecific DNA or tRNA can be used as competitors to avoid loss of material by adsorption to the tubes. It should be noted that these competitors may also trap oligonucleotides, as we have observed with TFOs containing phosphoramidate linkages that interacted strongly with tRNA.

Kinetics of triplex formation depend on the sequence of the third strand. Pyrimidine-containing TFOs bind more slowly to DNA than purine-containing TFOs and therefore need longer incubation times.[34,35] Electrophoresis is performed on a nondenaturing polyacrylamide gel (10–15%) containing 1–10 mM MgCl$_2$ and 50 mM HEPES, pH 7.2, at the incubation temperature or at 4°. For studies of CT-containing TFOs at a more acidic pH, MES buffer (at pH 6) can be used for sample incubation and electrophoresis. Apparent dissociation constants, K_d, for triplex formation are estimated as the added TFO concentration required to displace 50% of the

[36] A. J. Cheng, J. C. Wang, and M. W. Van Dyke, *Antisense Nucleic Acid Drug Res.* **8,** 215 (1998).

radiolabel from a double-strand state to a triplex state based on gel mobility. These apparent K_d values depend greatly on the experimental procedure (incubation and electrophoresis conditions).

Triplex formation may also be analyzed using a radiolabeled TFO and increasing concentrations of a duplex target (a synthetic short duplex, a restriction fragment, or a full plasmid). Dissociation constants obtained with this procedure are usually more accurate, as intermolecular self-structures of the TFOs are minimized. A near 10-fold difference in apparent K_d values was observed comparing the addition of increasing GA-TFO to a radiolabeled duplex and the reverse experiment, the addition of increasing duplex to a radiolabeled GA-TFO (Fig. 2).[37] Unfortunately, the labeling of some modified TFOs, such as phosphorothioate-containing C-5 propyne by standard techniques, is sometimes not readily achievable. Nevertheless, a study of triplex formation at different temperatures may be performed to demonstrate that certain TFOs exhibit intermolecular secondary structures that prevent the oligomers from binding to their target.

A tract of d(GA) repeats in TFOs can result in a substantial decrease in triplex formation due to competing stable oligonucleotide self-association.[37] To identify the potential formation of intermolecular self-structures (GA homoduplex, G-tetrads, i-DNA) in TFOs, single-stranded gel retardation could be performed with radiolabeled oligonucleotide and increasing concentrations of the same unlabeled oligonucleotide.[19,36,37] Melting experiments to study intramolecular and intermolecular self-structures could be done provided that the formation of these structures leads to a change in UV absorption (hyperchromism).

Identification of Secondary Structures in Antisense and Triplex-Forming Oligonucleotides by UV-Melting Experiments

Ultraviolet spectroscopy may be used to demonstrate the binding of an antisense oligomer to its RNA target as heating of the oligomer–RNA complex results in an increase in UV absorbance. Melting temperatures corresponding to half-dissociation of the antisense from its target may be compared between various modified oligomers.[25,31]

UV-melting curves are obtained routinely in the case of pyrimidine triplexes. The hypochromism associated with pyrimidine triplex formation is large enough to provide a satisfactory signal in the micromolar strand concentration range. Fewer examples of triplex formation monitored by UV-melting experiments have been reported for GA- and GT-containing

[37] S. B. Noonberg, J. C. François, T. Garestier, and C. Hélène, *Nucleic Acids Res.* **23,** 1956 (1995).

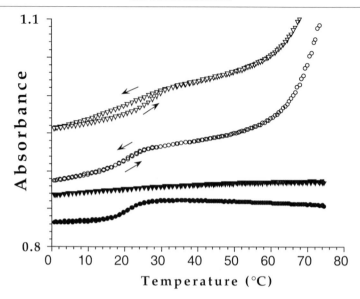

Fig. 3. Effect of oligonucleotide substitutions on the stability of triple helices. Sequence of the TFOs: 18D,$^{5'}C_3TC_3T_4C_3TC_3^{3'}$; 18R,$^{5'}C_3UC_3U_4C_3UC_3^{3'}$. An 18-bp complementary oligopurine–oligopyrimidine target is inserted in the center of a 40-bp DNA duplex. Triplex formation is analyzed by a UV-melting experiment. ●, DNA third strand (18D) alone (+0.6 OD offset); ▼, RNA third strand (18R) alone (+0.6 OD offset); ○, 40-bp duplex + 18D (DNA third strand); ▽, 40-bp duplex + 18R (RNA third strand). Buffer conditions: 10 mM sodium-cacodylate, pH 6.4, 0.14 M KCl, 10 mM MgCl$_2$. Concentrations of oligomers and duplex are 1.8 and 1.5 μM, respectively.

TFOs.[38,39] This section presents results highlighting the necessity of performing appropriate controls in the case of T_m measurements. Melting curve profiles should be analyzed very carefully to avoid the possible causes of errors illustrated in Fig. 3 (hysteresis in triplex formation, self-structures of TFOs).

Method to Monitor Melting Experiments

Absorbance versus temperature heating and cooling curves are obtained using a KONTRON-UVIKON 940 spectrophotometer. The temperature of the six-cells holder is regulated by a circulating liquid (80% water–20% glycerol) using a HAAKE cryothermostat and is monitored by a thermoresistance immersed in an accompanying cell containing only buffer. Constant

[38] F. Svinarchuk, J. Paoletti, and C. Malvy, *J. Biol. Chem.* **270**, 14068 (1995).
[39] M. Alluni-Fabbroni, G. Manzini, F. Quadrifoglio, and L. E. Xodo, *Eur. J. Biochem.* **238**, 143 (1996).

heating or cooling rates (dT/dt) are obtained using a HAAKE PG20 temperature programmer. The rates of temperature changes range from 0.1 to 0.3°/min. Absorbance and temperature are recorded at appropriate constant time intervals, usually 4–6 min, depending on the conditions. For fast systems, cooling and heating profiles are superimposed. Water condensation on the cell walls at low temperatures is prevented by gently blowing a stream of dry air in the cell compartment. All strands are mixed at a high temperature (>80°) in a 10-mm pathway cuvette. The experiment starts with a cooling cycle and is followed by a heating cycle.

Triplex Formation

A prerequisite for the recovery of thermodynamic, i.e., equilibrium, parameters from these curves is that they are true equilibrium curves. This implies that the sample must be heated or cooled at a rate that is slow with respect to the rate of the association–dissociation reaction under study. A simple and useful criterion for this is the coincidence of the heating and cooling curves. This is not always the case, and we have observed in many cases that, at pH 6–7 and moderate ionic strength (0.1–0.2 M NaCl), the thermal dissociation (heating) curves of the triplex are largely shifted toward higher temperatures as compared to the association (cooling) curves. Such a behavior is the result of slow association and dissociation kinetics and has already been described in detail.[40] It is rather obvious from the melting profiles presented in Fig. 3 (open triangles) that the heating and cooling cycles are not superimposable. This hysteresis phenomenon is perfectly reproducible: two successive heating cycles are perfectly superimposable, provided they were performed using the same temperature gradient. Therefore, a *chemically* reversible reaction, such as the folding/unfolding of a triple helix, may be *thermodynamically* not reversible (as shown by a hysteresis). A confusion could therefore arise from "reversible" or "irreversible" terms.

Many reports in the literature only analyze heating cycles, which leads to an overestimation of the true equilibrium melting temperature, as well as to completely inaccurate thermodynamic parameters. It is therefore not surprising that large discrepancies are obtained with thermodynamic parameters deduced from calorimetry experiments. A method has been described to extract kinetic and thermodynamic data from such a hysteresis phenomenon.[40] It is striking that such a hysteresis phenomenon almost never happens in the case of duplex formation between antisense oligonu-

[40] M. Rougée, B. Faucon, J. L. Mergny, F. Barcelo, C. Giovannangéli, T. Garestier, and C. Hélène, *Biochemistry*, **31**, 9269 (1992).

cleotide and RNA target, except at very low ionic strength, as a consequence of faster kinetics of association and dissociation of a duplex.

Self-Structure Formation

The second possible cause of error is shown on Fig. 3 (circles). By mixing equal amounts (1 : 1 strand stoichiometry) of the target duplex with a pyrimidine oligodeoxynucleotide (18D), a transition (midpoint = 21°) is obtained on raising or decreasing the temperature.[12] As expected, this transition is absent in the case of the duplex alone, which melts at 70° (not shown), and the T_m is strongly pH dependent. This transition is also evidenced at 295 nm, where cytosine protonation/deprotonation can be followed.[41] The T_m values obtained at 265 and 295 nm are the same, suggesting that melting of the triplex occurs at 21°. However, a similar transition is obtained with the third strand 18D alone, in the absence of any target duplex (Fig. 3, filled circles). This control shows that the transition is the result of the folding of the third strand rather than of triplex formation. Several other observations indicate that this transition does not correspond to triple-helix formation: this transition is concentration independent, and the T_m is reduced slightly on the addition of 10 mM MgCl$_2$.

Triplex formation was confirmed by a gel retardation assay (data not shown).[12] No retarded band was obtained when increasing concentrations (0–10 μM) of the 18D oligodeoxynucleotide were added to the labeled duplex. This was in clear contrast with the retarded band obtained in the presence of increasing amounts of an 18-nucleotide-long oligoribonucleotide (18R) or of a 2'-O-methyl oligoribonucleotide (18M). Apparent K_d values of 0.06 and 0.1 μM were determined for 18M and 18R. The gel retardation experiment shows that triplex formation is possible with a cytosine-rich RNA third strand, but not with a DNA oligonucleotide (18D): in none of the experimental conditions tested (pH 5–7; MgCl$_2$, 0–20 mM) was any triplex evidenced with the 18D oligonucleotide.

We have shown that the folding of the CT-rich third strand can interfere with triplex formation. This folded structure, which includes stretches of cytidines, forms a tetrameric structure involving C · C$^+$ base pairs in a so-called i motif.[42] Several parameters have a differential impact on triplex and i-DNA stability. To design a triplex-forming oligodeoxynucleotide, the choice of the target sequence is of course the first constraint, as triplex formation is mostly limited to oligopurine–oligopyrimidine stretches. Any sequence containing at least four repeats of two cytosines (or more) is

[41] J. L. Mergny, L. Lacroix, X. Han, J. L. Leroy, and C. Hélène, *J. Am. Chem. Soc.* **117**, 8887 (1995).
[42] K. Gehring, J. Leroy, and M. Gueron, *Nature* **363**, 561 (1993).

compatible with i-motif formation. Sequence requirements for a bimolecular i motif are even less stringent, as only two stretches of two cytosines are required. Many of the potential binding sites for triplex-forming oligodeoxynucleotides belong to this family. The presence of a repeated motif of several C · G base pairs will be deleterious to triplex formation by oligodeoxynucleotides because the third strand is likely to self-associate and thus be blocked in an inactive conformation: i-motif formation was preferred over triplex formation in all the experimental conditions tested. The addition of 10–20 mM magnesium ions, known to favor triplex formation, was insufficient to promote triplex formation. Fortunately, this limitation can be overcome by CU-rich RNA and 2'-O-methyl oligoribonucleotides, which were able to bind to the C · G-rich target with submicromolar dissociation constants: the presence of adjacent C · G*C$^+$ triplets is not a major limitation to triplex formation. It should be noted that this DNA to RNA change in the third strand chemistry, which is doubly advantageous in the pyrimidine motif, is of little use in the purine motif. As a matter of fact, the G quartet is effectively formed with RNA strands[43] and, thus, guanine-rich RNA third strands could also be trapped in G quartet structures.

Formation of undesired structures should not only be taken into account for triplex formation, but also for antisense applications. A quick overview of the literature on antisenses revealed that many oligodeoxynucleotides studied are potentially able to form the i motif. For example, several antisense oligonucleotides directed against HIV have a sequence compatible with i-motif formation,[44] and we have shown that such oligonucleotides, whether phophodiester or phosphorothioate, effectively fold *in vitro*.[45] Cell uptake may also be affected by the folded conformation. In fact, it has been shown that a cytosine-rich oligonucleotide had a different fate in HL60 cells as compared with a control oligonucleotide of similar length[46] and was trapped in acidic compartments where i-motif formation is the most likely.

Many research groups are designing oligonucleotide analogs to improve the efficacy of such molecules as antisense or antigene agents. Such chemical modifications could also have an impact on i-motif stability or, more generally, on any type of competing self-structure. Thus, the choice of an oligodeoxynucleotide for triplex formation must take into account not only its ability to form a triplex, but also its ability to form an undesired, competing

[43] C. J. Cheong and P. B. Moore, *Biochemistry* **31**, 8406 (1992).
[44] L. Perlaky, Y. Saijo, R. K. Busch, C. F. Bennett, C. K. Mirabelli, S. T. Crooke, and H. Busch, *Anticancer Drug Des.* **8**, 3 (1993).
[45] J. L. Mergny and L. Lacroix, *Nucleic Acids Res.* **26**, 4797 (1998).
[46] J. Tonkinson and C. A. Stein, *Nucleic Acids Res.* **22**, 4268 (1994).

structure such as G quartets,[47] GA-parallel duplexes,[37] or i-DNA.[42] Table I summarizes the effect of oligonucleotide subtitutions on the stability of DNA triple helices. Several conclusions may be drawn from Table I: some modifications (RNA, 2′-O-methyl-RNA, and phosphoramidate) are promising for triplex formation with CT-rich oligonucleotides. From the number of question marks in Table I, it is clear that, at least for some modifications, the occurrence of competing G-quartet structures has not yet been explored.

Delivery of Antisense and Triplex-Forming Oligonucleotides into Cells

The introduction of antisense or triplex-forming oligonucleotides into cells represents an important challenge for the utilization of oligonucleotides in the control of gene expression. Depending on the cell type, cell penetration of the oligonucleotides could be increased by various chemical or physical methods. Numerous formulations of cationic lipids have been employed successfully for *in vitro* oligonucleotide delivery. For example, the inhibition of gene expression using a low nanomolar concentration of C-5 propyne containing antisense oligonucleotides could be achieved with the GS cytofectin (Fig. 4).[24,25]

In the presence of cationic lipids or cationic polymers, the structure of the antisense oligomers (phosphodiester, phosphorothioate) seems to play an important role in achieving efficient penetration.[48] We have also observed that GS cytofectin increased the cell penetration of phosphorothioates but not of phosphodiesters. Despite the real progress in methods in DNA transfection, the choice of cationic lipids for an efficient delivery of antisense or triplex-forming oligonucleotides into cells remains sometimes empirical.

Delivery of C5-Propyne-Containing Oligonucleotides in Cells with GS Cytofectin

Efficient cell delivery of C-5 propyne-containing oligophosphorothioate was obtained using the cationic lipid, GS cytofectin (Glen Research, Sterling, VA) (Fig. 4). Dilution of antisense oligomers in Opti-MEM (Life Technologies, Rockville, MA) was done in polystyrene tubes. GS cytofectin was also diluted in another tube with Opti-MEM. Cytofectin was then combined with an equal volume of a diluted oligonucleotide solution. Cytofectin/oligonucleotide mixtures were incubated 10 to 15 min at room

[47] W. M. Olivas and L. J. Maher III, *Biochemistry* **34**, 278 (1995).
[48] S. Dheur and T. E. Saison-Behmoaras, *Methods Enzymol.* **313** [3] 1999 (this volume).

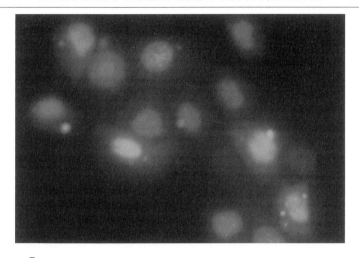

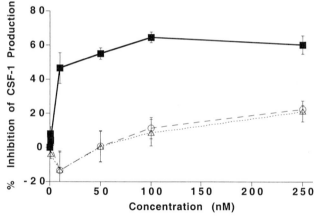

FIG. 4. Effect of the antisense oligonucleotide on CSF-1 expression. (Top) Representation of the antisense distribution after 24 hr in the presence of 2.5 μg/ml GS cytofectin. (Bottom) Percentage inhibition of CSF-1 levels in NS2T2A1 cell supernatant after 24 hr treatment by antisense and control oligonucleotides. Sequences of the antisense and control PS oligomers (mutated and reverse) containing C-5 propyne(dU) and C-5 propyne(dC) are $^{5'}$d(UA$_2$U$_3$G$_2$ CACGAG$_2$UCU)$^{3'}$ (antisense, ■), $^{5'}$d(UA$\underline{UA}$U$\underline{GU}$GCC$\underline{AG}$AGA$\underline{GU}$GCU)$^{3'}$ (control 1, △), and $^{5'}$d(UCUGGAGCACGGUUUAAU)$^{3'}$ (control 2, ○), respectively. Measurement of CSF-1 production was done using an ELISA test (R&D).

temperature to allow complex formation. Complete growth media were then added to dilute GS cytofectin and oligonucleotides before adding them to cells. The GS cytofectin could be used without exhibiting any toxicity at a final concentration of 1 to 10 μg/ml, depending on the cell type.

Uptake studies of the oligonucleotide were essentially done as described in this volume.[48] Briefly, a 5'-fluorescein-labeled C-5 propyne oligophosphorothioate (100 nM) in combination with GS cytofectin was added to cells grown on cover glasses. After different transfection times (2 to 24 hr), cells were washed with phosphate-buffered saline (PBS) and fixed with paraformaldehyde (2% in PBS) (no difference in cellular distribution was observed when this fixation step was omitted). To detect dead cells or permeabilized cells, propidium iodide (10 μg/ml) was added to cover glasses for 10 min before mounting. A Zeiss microscope with an epifluorescence illumination system (with fluorescein and propidium iodide filters) and a cooled three charged-coupled device (3CCD) camera (LHESA France) equipped with a triple band pass was used to analyze the subcellular localization of fluorochrome-conjugate oligomers.

The GS cytofectin (2.5 μg/ml) was able to transfect efficiently anti-CSF-1 oligonucleotides into the nuclei of NS2T2A1 cells in the presence of calf serum (10%). This cell line, which is able to induce tumors in nude mice, derived from normal mammary primary cell cultures following SV40 T antigen immortalization.[49,50] The transfected antisense oligonucleotide inhibited specifically CSF-1 production by NS2T2A1 cells as measured by a standard ELISA assay (Fig. 4).

Rapid Analysis of Gene Inhibition by Antisense and Triplex-Forming Oligonucleotides

The effect of antisense oligonucleotides on the expression of the targeted protein has been measured by various techniques, including Western blot, ELISA assays, flow cytometric analysis, and semiquantitative immunocytochemistry depending on the targeted gene. Determination of mRNA levels after oligonucleotide treatment is usually performed by Northern blot experiments or by reverse-transcribed polymerase chain reaction (RT-PCR).[17,30]

The effect of oligonucleotide treatment on the expression of the targeted protein could be analyzed more quickly using reporter gene strategies. Cotransfections of oligonucleotides and plasmids containing reporter genes and target sequences for antisense or TFOs have been done to optimize the oligonucleotide uptake and the choice of the oligonucleotide sequence and also to measure rapidly the effect on protein synthesis. Such reporter genes, such as chloramphenicol acetyltransferase, luciferase, and alkaline

[49] P. Berthon, G. Goubin, B. Dutrillaux, A. Degeorges, A. Faille, C. Gespach, and F. Calvo, *Int. J. Cancer* **52,** 92 (1992).
[50] L. Ma, C. Gauvillé, Y. Berthois, A. Desgeorges, G. Millot, P. Martin, and F. Calvo, *Int. J. Cancer* **78,** 112 (1998).

phosphatase, were used previously to determine the most efficient antisense or triplex-forming oligonucleotides in cells and also to demonstrate the specificity of oligonucleotide treatment.[31,51–58]

Analysis of Inhibition of Gene Transcription Using Transient Transfection Assays

We used a rat hepatocarcinoma cell line for the cotransfection of antisense or triplex-forming oligonucleotides and reporter vectors containing targeted sequences for the oligomers. Antisense (or TFOs) must inhibit the reporter protein synthesis in a sequence-specific manner, provided that these targeted sequences play important roles in gene expression. A plasmid (p1711b/luc) containing the *Pyralis* firefly luciferase gene (*luc*) under the control of the promoter region of the rat insulin-like growth factor I gene (IGF-I) was used. The pRL family of *Renilla* luciferase control vectors (Promega, Madison, WI) was used to correct for transfection efficiency, as oligonucleotide transfection may inhibit gene expression nonspecifically. β-Galactosidase-containing vectors may also be used.[51,54,56,59,60] The pRL-TK plasmid contains the herpes simplex virus thymidine kinase (HSV-TK) promoter region controlling the transcription of the sea pansy luciferase gene (*Rluc*). All plasmids used in the transfection experiments were first propagated in JM109 bacterial strain (Promega), purified by Qiagen Maxipreps kits (Qiagen, Hilden, Germany), and ethanol precipitated.

The transfections of plasmids plus oligonucleotides were carried out using the TFX50 reagent (Promega). A typical transfection experiment with TFX50 is performed as follows. Various amounts of oligonucleotides are mixed with 6 μg of firefly luciferase-containing plasmid DNA (p1711b/luc) and 0.1 μg of *Renilla* luciferase-containing plasmid DNA (pRL-TK)

[51] M. Grigoriev, D. Praseuth, A. L. Guieysse, P. Robin, N. T. Thuong, C. Hélène, and A. Harel-Bellan, *Proc. Natl. Acad. Sci. U.S.A.* **90,** 3501 (1993).
[52] N. H. Ing, J. M. Beekman, D. J. Kessler, M. Murphy, K. Jayaraman, J. G. Zendegui, M. E. Hogan, B. W. O'Malley, and M. J. Tsai, *Nucleic Acids Res.* **21,** 2789 (1993).
[53] C. Roy, *Nucleic Acids Res.* **21,** 2845 (1993).
[54] F. Svinarchuk, A. Debin, J.-R. Bertrand, and C. Malvy, *Nucleic Acids Res.* **24,** 295 (1996).
[55] C. Delporte, I. G. Panyutin, O. A. Sedelnikova, C. D. Lillibridge, B. C. O'Connell, and B. J. Baum, *Antisense Nucleic Acid Drug Dev.* **7,** 523 (1997).
[56] J. Joseph, J. C. Kandala, D. Veerapanane, K. T. Weber, and R. V. Guntaka, *Nucleic Acids Res.* **25,** 182 (1997).
[57] H. G. Kim, J. F. Reddoch, C. Mayfield, S. Ebbinghaus, N. Vigneswaran, S. Thomas, D. E. Jones, and D. M. Miller, *Biochemistry* **37,** 2299 (1998).
[58] H. G. Kim and D. M. Miller, *Biochemistry* **37,** 2666 (1998).
[59] M. Grigoriev, D. Praseuth, P. Robin, A. Hemar, T. Saison-Behmoaras, A. Dautry-Varsat, N. T. Thuong, C. Hélène, and A. Harel-Bellan, *J. Biol. Chem.* **267,** 3389 (1992).
[60] C. A. Hobbs and K. Yoon, *Antisense Res. Dev.* **4,** 1 (1994).

in a total volume of 1.5 ml (medium without serum). The TFX50 reagent (final concentration 18 μM) is then added to the mixture to obtain a 3:1 cationic lipid/DNA ratio. The mixture is split in quadruplicates and then added to cells (5×10^4 cells/well spread in 24-well plates, 350 μl per well). Cells are then incubated overnight and harvested 24 hr after transfection. Both luciferase activities (*Pyralis* and *Renilla*) are measured in cell extracts using the dual-luciferase assay kit (Promega). Briefly, transfected LFCL2A cells are lysed directly with 150 μl/well of passive lysis buffer (Promega) for 30 min. Equal amounts of proteins measured by the Bradford reagent (Protein Assay kit, Bio-Rad, Richmond, CA) are then analyzed sequentially for both luciferase gene expression levels with a Microbeta Trilux luminescent counter (EGG Wallac). The second luciferase (*Rluc*) gene expression provides an internal control value to which expression of the experimental firefly reporter gene is normalized. The luciferase activity corresponds to the ratio of the relative light units detected for the *Pyralis* luciferase and the *Renilla* luciferase. All transfection experiments should be done in quadruplicate and repeated at least twice. When the cotransfection experiments are performed with the plasmid containing its complementary target sequence, TFO1 inhibits luciferase enzyme activity by 60%, whereas control oligonucleotides C1 and C2 do not (see TFO sequences in Fig. 1).

Triplex-Mediated Mechanims Involved in Gene Inhibition

Transient Transfection

The demonstration that gene inhibition obtained by the oligonucleotide treatment of cells involves antisense- or triplex-mediated mechanisms is often difficult to provide. The synthesis of control oligonucleotides that do not bind *in vitro* to the target sequence contributes partially to the answer.[28,61] Many unexpected effects of antisense oligonucleotides unrelated to their intended antisense activity have been described.[61,62] Scrambled sequences of antisense and triplex-forming oligonucleotides used as controls cannot completely rule out nonspecific effects of the antisense or triplex-forming oligonucleotides.

Sequence specificity of transcription inhibition by oligomers in cells can be demonstrated using a plasmid carrying the gene of interest. The target site present on this plasmid may be mutated by *in vitro* mutagenesis to provide evidence for the involvement of triple-helix formation in the observed inhibition. This could also be true for antisense oligonucleotides

[61] C. A. Stein, *Antisense Nucleic Acid Drug Dev.* **8,** 129 (1998).
[62] S. T. Crooke, *Antisense Nucleic Acid Drug Dev.* **8,** 115 (1998).

targeted to mRNA. Mutations of the target sequences of antisense or TFOs help demonstrate that sequence-specific binding of the oligonucleotide is required to observe gene repression, provided that these mutations do not affect gene regulation and expression level. It is also possible to analyze the effect of specific TFOs on the reporter expression driven by an unrelated plasmid that do not contain a target sequence for these oligonucleotides. Plasmids bearing small deletions or mutations of the target sequences are easily achievable by *in vitro* mutagenesis and should be preferred to unrelated plasmids.

The introduction of mutations in the target sequence for a pyrimidine-containing TFO designed to bind to the promoter region of the α subunit of the interleukin-2 receptor demonstrated that the inhibitory effect of this TFO is due to the binding to the targeted DNA sequence.[51,59] Three nucleotides (AGA) in the polypurine–polypyrimidine sequence of the IGF-1 gene were changed to (CTC) (see Fig. 1) by a standard *in vitro* site-directed mutagenesis.[63] The mutated plasmid was then used in transient transfection assays with oligonucleotides as mentioned previously. The expression of the luciferase gene was not changed by the introduction of mutations. We found that the specific triplex-forming oligonucleotide (TFO1) inhibited the mutated plasmid at 30% and the wild-type plasmid at 60%, in agreement with an inhibitory mechanism involving the oligonucleotide binding to the targeted sequence. Although TFO1 did not bind *in vitro* to the mutated duplex, only a small difference in inhibition levels was observed. Further studies should be provided to demonstrate clearly that this triplex-forming oligonucleotide inhibited luciferase gene expression through a triplex-mediated mechanism.

Demonstration of TFO Binding by Cleaving or Cross-Linking DNA Targets in Cells

In vivo DMS footprinting showed that conditions inside the cells are favorable with intermolecular triplex formation.[54] An indication of triplex formation may also be obtained based on the inhibition of DNase I hypersensitive sites in nuclei.[4,64]

TFOs linked covalently to psoralen, which induces specific cross-links to DNA, have been used to detect triple-helix formation inside the cell. The binding of an oligopyrimidine to its duplex target via triple helix formation inside the cell has been demonstrated using psoralen–oligo-

[63] L. Zhu, *Methods Mol. Biol.* **57**, 13 (1995).
[64] E. H. Postel, S. J. Flint, D. J. Kessler, and M. E. Hogan, *Proc. Natl. Acad. Sci. U.S.A.* **88**, 8227 (1991).

nucleotide conjugates.[65] The nuclear accessibility of target sequences for TFOs within the chromatin structure has been shown using quantitative PCR methods.[20] Psoralen-oligomers have also been used to induce specific mutations via triplex-mediated site-directed mutagenesis in cell culture on polypurine–polypyrimidine sequences located in plasmids[3] and even in endogenous genes such as hprt.[66] All these methods suggested strongly that specific inhibition of gene expression may be achieved via a triplex-mediated mechanism.

Concluding Remarks

Theoretically, the inhibition of gene expression using triplex-forming oligonucleotides that are targeted to the two copies of a double-stranded gene could be an obvious advantage as compared to antisense oligonucleotides that are targeted to numerous messenger RNAs. Nevertheless, whether there is any advantage in targeting the duplex gene rather than its messenger RNA still remains an open question to be determined in each particular system. In the antisense and antigene strategy, two scrambled control oligonucleotides are required to demonstrate the sequence specificity of the observed biological effect. Reverse sequence and several base mismatch sequences could be used as controls, provided that their structures are similar to those of the antisense sequence (G or C blocks, GA repeats). Self-structures of oligonucleotides can be analyzed easily by gel shift assays and UV-melting experiments. The utilization of transient cotransfection experiments with plasmids and oligonucleotides is a convenient method to demonstrate the sequence specificity of antisense or antigene effects. Oligonucleotides and plasmids containing the target sequence could be transfected separately to show that the oligomers inside the cells could encounter target sequences and inhibit specifically gene expression. Obviously, the conditions of these transfection experiments cannot resemble the real physiological conditions. For the antisense approach, mRNAs were often synthesized easily from heterologous promoters in plasmids such as viral promoters.[31] For the triple-helix approach, the accessibility of regulatory regions in plasmids is likely to be different from that of endogenous genes, which bind to proteins in chromatin. Nevertheless, these transient cellular assays might give precious answers about the mechanisms of inhibition induced by TFOs and even by antisense oligomers.

[65] A. L. Guieysse, D. Praseuth, M. Grigoriev, A. Harel-Bellan, and C. Hélène, *Nucleic Acids Res.* **24,** 4210 (1996).
[66] A. Majumdar, A. Khorlin, N. Dyatkina, F. L. Lin, J. Powell, J. Liu, Z. Fei, Y. Khripine, K. A. Watanabe, J. George, P. M. Glazer, and M. M. Seidman, *Nat. Genet.* **20,** 212 (1998).

Furthermore, studies on plasmids with mutated targets for antisense or TFOs can be performed easily to demonstrate that inhibitory effects are mediated through sequence-specific binding via duplex or triplex formation. From the progress in the past decade, triplex-based strategies became a powerful approach to selectively inhibit gene expression, despite sharing similar properties and limitations with antisense strategies.

Acknowledgments

This work was supported in part by grants from the Ligue Nationale contre le Cancer. The authors thank all the scientists of the Laboratoire de Biophysique who contributed by their work and by their useful comments to this manuscript. The authors also thank members of the Experimental and Clinical Pharmacologics laboratory and Dr. F. Calvo for helpful discussion in conducting CSF-1 experiments. The authors gratefully acknowledge Dr. P. Rotwein for providing the IGF-I plasmids and Dr. S. Scholl for help and support.

[5] Chimeric Oligodeoxynucleotide Analogs: Chemical Synthesis, Purification, and Molecular and Cellular Biology Protocols

By RICHARD V. GILES, DAVID G. SPILLER, and DAVID M. TIDD

Introduction

The consequences of forming an intracellular antisense oligodeoxynucleotide (ON)–RNA heteroduplex may include inhibition of splicing,[1] or other RNA processing events,[2] and inhibition of translation.[3] Perhaps the most widely anticipated result is cleavage of the RNA, within the heteroduplex region, resulting from the action of the apparently ubiquitous endogenous enzyme ribonuclease H (RNase H).[4] RNases H have been convincingly demonstrated to mediate many of the antisense effects ob-

[1] R. Kole, R. R. Shulka, and S. Akhtar, *Adv. Drug Delivery Rev.* **6**, 271 (1991).

[2] M.-Y. Chiang, H. Chan, M. A. Zounes, S. M. Freier, W. F. Lima, and C. F. Bennett, *J. Biol. Chem.* **266**, 18162 (1991).

[3] J. Summerton, D. Stein, S. B. Huang, P. Matthews, D. D. Weller, and M. Partridge, *Antisense Nucleic Acid Drug Dev.* **7**, 63 (1997).

[4] R. J. Crouch and M. L. Dirksen, *in* "Nucleases" (S. M. Linn and R. J. Roberts, eds.), p. 211. Cold Spring Harbor Laboratory Press, Cold Spring Harbor, NY, 1982.

served in *in vitro* systems,[5–9] following ON microinjection into *Xenopus* oocytes, eggs, and embryos[10,11] and following delivery of antisense effectors into human cells in culture.[12,13] Clearly, if the transcript becomes degraded through the action of RNase H, then translation of the cognate protein is impossible.

Whereas the theory of using antisense ON may have an attractive simplicity, the practice is complicated by numerous technical difficulties. Unmodified phosphodiester ON are degraded rapidly in a variety of biological fluids,[14–17] particularly by 3' to 5' (3' → 5') exonucleases, and release potentially toxic deoxynucleosides and deoxynucleotides. Certain sequence-specific nonantisense effects[18] may be related to the release of nuclease degradation products, which distort intracellular deoxynucleotide pools.

A variety of analog structures have been designed that possess the beneficial characteristic of nuclease resistance. The most commonly used nuclease-resistant ON structure is the phosphorothioate form, generated by replacement of one of the acidic hydroxyl oxygen atoms in the phosphodiester linkage by sulfur. However, phosphorothioate ON have been widely reported to interact with both intracellular and extracellular proteins and to induce a variety of nonantisense biological effects (reviewed by Stein and Cheng[19] and Stein[20,21]). Moreover, both phosphodiester and phosphorothioate antisense ON are taken up by endocytosis, remain trapped in endo-

[5] J. Minshull and T. Hunt, *Nucleic Acids Res.* **14**, 6433 (1986).

[6] R. Y. Walder and J. A. Walder, *Proc. Natl. Acad. Sci. U.S.A.* **85**, 5011 (1988).

[7] P. J. Furdon, Z. Dominski, and R. Kole, *Nucleic Acids Res.* **17**, 9193 (1989).

[8] R. V. Giles and D. M. Tidd, *Anti-Cancer Drug Des.* **7**, 37 (1992).

[9] R. V. Giles and D. M. Tidd, *Nucleic Acids Res.* **20**, 763 (1992).

[10] J. Shuttleworth and A. Colman, *EMBO J.* **7**, 427 (1988).

[11] J. Shuttleworth, G. Matthews, L. Dale, C. Baker, and A. Colman, *Gene* **72**, 267 (1988).

[12] R. V. Giles, D. G. Spiller, and D. M. Tidd, *Antisense Res. Dev.* **5**, 23 (1995).

[13] R. V. Giles, C. J. Ruddell, D. G. Spiller, J. A. Green, and D. M. Tidd, *Nucleic Acids Res.* **23**, 954 (1995).

[14] D. M. Tidd and H. M. Warenius, *Br. J. Cancer* **60**, 343 (1989).

[15] T. M. Woolf, C. G. B. Jennings, M. Rebagliati, and D. A. Melton, *Nucleic Acids Res.* **18**, 1763 (1990).

[16] G. D. Hoke, K. Draper, S. M. Freier, C. Gonzalez, V. B. Driver, M. C. Zounes, and D. J. Ecker, *Nucleic Acids Res.* **19**, 5743 (1991).

[17] A. R. Sburlati, R. E. Manrow, and S. L. Berger, *Proc. Natl. Acad. Sci. U.S.A.* **88**, 253 (1991).

[18] J. L. Vaerman, L. P. Moureau, P. Lewalle, P. Deldime, M. Blumenfeld, and P. Martiat, *Blood* **86**, 3891 (1995).

[19] C. A. Stein and Y.-C. Cheng, *Science* **261**, 1004 (1993).

[20] C. A. Stein, *Nature Med.* **1**, 1119 (1995).

[21] C. A. Stein, *Trends Biotechnol.* **14**, 147 (1996).

somes, and so do not gain access to the cytoplasmic or nuclear compartments of cells in culture following simple topical application.[22–24]

Another nuclease-resistant ON analog,[14,25,26] the methylphosphondiester structure, is created by replacement of one of the nonbridging oxygen atoms in a phosphodiester linkage with a nonionic methyl group. ON composed entirely of methylphosphonodiester linkages do not direct the activity of RNase H[7,8] but were reported to enter cells by passive diffusion.[27]

We hoped that combining methylphosphonodiester and phosphodiester linkages in a single compound would provide a chimeric ON structure, similar to that originally described for 2′-O-methyl/2′-deoxy phosphodiester,[28,29] but which combined the beneficial characteristics of exonuclease resistance, the ability to direct RNase H and passive diffusion across the cytoplasmic membrane into cells in culture. In fact, neither all-methylphosphonodiester ON nor chimeric structures that possess a central phosphodiester, or phosphorothiodiester, region large enough to efficiently direct human RNase H accumulate in the cytoplasm or nucleus of cells in the absence of specific treatments that permeabilize the plasma membrane to ON.[30] However, we found that such structures were resistant to the exonucleases found in cell culture medium,[14] in cell extracts,[31] and in living cells. Moreover, chimeric methylphosphonodiester–phosphodiester ON have been found to direct RNase H cleavage of target RNA with high efficiency and enhanced specificity in vitro[8,9,32] and in living cells in culture,[12,13,31] resulting in reproducible antisense effects.

This article focuses on chimeric ON constructed with terminal methylphosphonodiester sections and a central phosphodiester region. Procedures for the synthesis, fluorescent labeling, purification, and use of chimeric ON are provided. All of the procedures described have been found to be robust and most are generally applicable to a wide range of antisense ON structures.

[22] S. L. Loke, C. A. Stein, X. H. Zhang, K. Mori, M. Nakanishi, C. Subasinghe, J. S. Cohen, and L. M. Neckers, Proc. Natl. Acad. Sci. U.S.A. 86, 3474 (1989).
[23] L. A. Yakubov, E. A. Deeva, V. F. Zarytova, E. M. Ivanova, A. S. Ryte, L. V. Yurchenko, and V. V. Vlassov, Proc. Natl. Acad. Sci. U.S.A. 86, 6454 (1989).
[24] D. G. Spiller and D. M. Tidd, Anti-Cancer Drug Des. 7, 115 (1992).
[25] S. Agrawal and J. Goodchild, Tetrah. Lett. 28, 3539 (1987).
[26] S. Akhtar, R. Kole, and R. L. Juliano, Life Sci. 49, 1793 (1991).
[27] P. S. Miller, K. B. McParland, K. Jayaraman, and P. O. P. Ts'o, Biochemistry 20, 1874 (1981).
[28] H. Inoue, Y. Hayase, A. Imura, S. Iwai, K. Miura, and E. Ohtsuka, Nucleic Acids Res. 15, 6131 (1987).
[29] H. Inoue, Y. Hayase, S. Iwai, and E. Ohtsuka, FEBS Lett. 215, 327 (1987).
[30] D. G. Spiller and D. M. Tidd, Antisense Res. Dev. 5, 13 (1995).
[31] R. V. Giles, D. G. Spiller, J. A. Green, R. E. Clark, and D. M. Tidd, Blood 86, 744 (1995).
[32] R. V. Giles, D. G. Spiller, and D. M. Tidd, Anti-Cancer Drug Des. 8, 33 (1993).

Oligodeoxynucleotide Synthesis

Chimeric Methylphosphonodiester/Phosphodiester Oligodeoxynucleotides

Methylphosphonate ON analogs are synthesized in an analogous fashion to normal phosphodiester ON on automatic DNA synthesizers, except that 5'-O-dimethoxytrityl-2'-deoxynucleoside 3'-O-methylphosphonamidite synthons (available commercially from Glen Research Corporation, Sterling, VA; UK supplier Cambio Ltd., Cambridge) replace the 5'-O-dimethoxytrityl-2'-deoxynucleoside 3'-O-(β-cyanoethyl) phosphoramidites. Methylphosphonamidites are less reactive than phosphoramidites and coupling times need to be extended to 5 min. Apart from that, the same cycle and reagents as used for phosphodiester ON synthesis may be employed. In our hands, an optimal final yield of pure product from syntheses on an Applied Biosystems Model 381A synthesizer (Applied Biosystems, Warrington, Cheshire) was achieved using the slower cycle, version 1.23 software, 1 μmol cycle, and the "improved 10-μmol synthesis cycle Model 381" (ABI User Bulletin No. 15, August 12, 1988).

Methylphosphonamidites are made up in solution at a concentration of 0.1 M. Thymidine, N^4-isobutyryl protected deoxycytidine and N^6-benzoyl protected deoxyadenosine methylphosphonamidites are dissolved in the anhydrous acetonitrile supplied for ON synthesis. However, N^2-isobutyryl protected deoxyguanosine methylphosphonamidite is insoluble in this solvent. The manufacturer recommends that this synthon be dissolved in anhydrous tetrahydrofuran. In our experience, coupling efficiencies for deoxyguanosine methylphosphonamidite were suboptimal even when dissolved in the most anhydrous commercially available tetrahydrofuran, suggesting that further purification of the solvent was required. However, the deoxyguanosine methylphosphonamidite gave good coupling efficiencies when dissolved in a 1:1 (v/v) mixture of anhydrous acetonitrile and analytical grade dichloromethane, dried over a 4-Å molecular sieve, or commercially available anhydrous dichloromethane (Aldrich Chemical Company, Gillingham, Dorset). Deoxycytidine methylphosphonamidite is also available in the N^4-acetyl form, which is deprotected rapidly during the ammonium hydroxide cleavage step and, therefore, is not susceptible to the transamination reaction with ethylenediamine during deprotection (see later). The compound is readily soluble in anhydrous acetonitrile and gives good coupling efficiencies in this solvent.

With five-port synthesizers, a batch of several different methylphosphonodiester/phosphodiester chimeric ON is synthesized by first installing the methylphosphonamidite solutions on the synthesizer and running syntheses of the 3'-methylphosphonate end sections of each consecutively before replacing the methylphosphonamidites with phosphoramidite solutions and continuing syntheses of the central phosphodiester sections, and so on. The final trityl group should be left on the ON to aid in identification and purification of the full-length product as, because of the reduced coupling efficiencies of methylphosphonamidites, relative to those of phosphoramidites, failure sequences can be quite abundant in the crude deprotection reaction mixture.

Chimeric Methylphosphonodiester/Phosphorothiodiester ON

Methylphosphonodiester/phosphorothiodiester chimeric ON are synthesized in analogous fashion using the synthesizer manufacturer's recommended protocol for phosphorothioate ON analog synthesis. Successful results were achieved on an Applied Biosystems Model 381A synthesizer using tetraethylthiuram disulfide in acetonitrile as the sulfurizing reagent and the Applied Biosystem's recommended phosphorothioate cycle (ABI User Bulletin Number 58, February 1991).

Oligodeoxynucleotide Purification

Deprotection of Chimeric ON

Unlike the normal phosphodiester linkage, the methylphosphonodiester internucleoside linkage is highly susceptible to base-catalyzed hydrolysis, and the usual treatment with concentrated ammonia at 55° for 16 hr used to deprotect phosphodiester and phosphorothioate ON cannot be employed. A modification of the one-pot deprotection procedure for methylphosphonate ON analogs recommended by Hogrefe et al.[33,34] has consistently given the best results with methylphosphonate containing chimeric ON. The ON is cleaved from the support by placing the controlled pore glass support from a 1-μmol synthesis in 0.5 ml of an ammonium hydroxide solution consisting of acetonitrile/ethanol/concentrated ammonium hydroxide (45:45:10, by volume) in a sealed screw cap vial for 30 min at room temperature. This treatment also removes the N^4-acetyl-protecting group from cytosine residues, thereby preventing the formation of N^4-ethylamino adducts by transamination during full deprotection with ethyl-

[33] R. I. Hogrefe, M. M. Vaghefi, M. A. Reynolds, K. M. Young, and L. J. Arnold, Jr., Nucleic Acids Res. 21, 2031 (1993).

[34] R. I. Hogrefe, M. A. Reynolds, M. M. Vaghefi, K. M. Young, T. A. Riley, R. E. Klein, and L. J. Arnold, Jr., in "Methods in Molecular Biology" (S. Agrawal, ed.), Vol. 20, p. 143. Humana Press, Totowa, NJ, 1993.

enediamine.[33] Full deprotection is achieved by adding 0.5 ml ethylene-diamine to the vial and leaving it at room temperature overnight. The extended incubation time required for full deprotection, relative to that recommended for all-methylphosphonate ON, may reflect a slower rate of deprotection of the phosphodiester ON section in the chimeric molecules under these conditions. The supernatant is collected and the support is washed twice with 0.5 ml acetonitrile in water (1:1, v/v). The solvent is removed on a rotary evaporator with a high vacuum pump and a liquid nitrogen trap, and the final traces of ethylenediamine are cleared by co-evaporation with ethanol. The residue is extracted with 1 ml water containing 0.1% pyridine (v/v) to prevent premature detritylation of the 5'-dimethoxytrityl-ON, which may occur slowly in aqueous solutions even at neutral or slightly alkaline pH. After centrifugation to remove insoluble material, the ON is precipitated by the addition of 0.1 ml 3 M sodium acetate and 10 ml ethanol in a 14-ml polypropylene centrifuge tube and cooling to $-80°$ for 1 hr. The precipitate is pelleted and washed twice with 3 ml ethanol before being dissolved in 10 ml 0.1% pyridine in 0.1 M triethylammonium acetate, pH 8.0 (v/v, Py/TEAA). It is recommended that the triethylammonium acetate used for purification and high-performance liquid chromatography (HPLC) analysis of ON is prepared from triethyl-amine that has been refluxed for 4 hr with potassium hydroxide pellets and redistilled to remove UV-absorbing impurities.

One milliliter of 70% perchloric acid/absolute ethanol (1:1, v/v) is added to a sample of the crude ON solution (50 μl), and the absorbance of the released dimethoxytrityl cation is recorded at 498 nm to give a rough estimate of the trityl yield of the synthesis (dimethoxytrityl cation millimolar absorptivity at 498 nm = 71.7). A second sample is subjected to analysis by reversed phase HPLC (see later) to check the ratio achieved in the synthesis of trityl-on product, the longest retained major peak, to failure sequences.

Primary, Trityl-On, Purification by Reversed-Phase, Solid-Phase Extraction on C$_{18}$ Sep-Pak Cartridges

C$_{18}$ Sep-Pak solid-phase extraction cartridges (Waters Chromatography Division of Millipore, Watford, Hertfordshire) and all glassware (Pasteur pipettes, beakers, etc.) contacting the ON solutions during purification are siliconized to prevent loss of ON through nonspecific adsorption. This is achieved by treatment with "Repelcote(VS)" (BDH/Merck, Lutterworth, Leicestershire) for 5 min, followed by successive washes with dichloromethane, acetonitrile, and water. C$_{18}$ Sep-Pak cartridges are given a final wash with Py/TEAA.

The crude, deprotected ON solution in Py/TEAA is applied to the siliconized C_{18} Sep-Pak cartridge and allowed to drip through under gravity using a polypropylene disposable syringe barrel as a reservoir (e.g., Plasti-pak, Becton-Dickinson, ON may adsorb to other types of plastic). A 50-μl sample of the effluent is mixed with 1 ml perchloric acid/absolute ethanol for trityl analysis at 498 nm as described earlier. In the event that not all trityl-containing material has bound to the cartridge, the cartridge is washed with 5 ml Py/TEAA, 10 ml 8% acetonitrile in Py/TEAA, and 5 ml Py/TEAA, and the initial effluent is diluted further with an excess of Py/TEAA and reapplied to the same cartridge. When all trityl is bound, the cartridge is washed with 10 ml Py/TEAA and developed with a step gradient of acetonitrile in Py/TEAA starting at 10 ml 8% acetonitrile (v/v).

Elution of failure sequences is monitored by spectrophotometric examination (270–290 nm, away from the absorption maximum of pyridine) of cartridge effluents against a Py/TEAA blank. Two hundred-microliter samples of the effluents are also assayed for trityl as described earlier to test for elution of the product ON. Generally, failure sequences will begin to elute at around 10% acetonitrile and the trityl-on product will begin to appear in the effluent at or above 15% acetonitrile. Trityl-on methylphosphonodiester/phosphorothiodiester chimeric ON are more highly retained than their phosphodiester counterparts. A precise concentration of acetonitrile, which elutes failure sequences but permits retention of trityl-on product, cannot be provided because the degree of substitution of phosphodiester linkages with the more lipophilic, nonionic methylphosphonate groups in the chimeric ON will affect retention, with more highly substituted molecules requiring higher concentrations of the acetonitrile organic modifier to effect their elution from the cartridge. Furthermore, the characteristics of C_{18} Sep-Pak cartridges tend to vary slightly from batch to batch. Once substantial amounts of failure sequences begin to elute, it is advisable to hold the acetonitrile concentration of the eluent constant until the absorbance of the effluent at 270–290 nm reduces to a low constant value before increasing the concentration further.

Detritylation

When trityl is detected in the effluent, the fraction is diluted with TEAA and reapplied to the cartridge, which is then washed with 10 ml TEAA and 10 ml water. Pyridine is not required in the TEAA at this stage to protect against premature detritylation. Detritylation is performed *in situ* by passing 10 ml 0.5% trifluoroacetic acid in water (v/v) through the cartridge for 30 min before neutralizing with 10 ml TEAA and washing twice with 10 ml water. The partly purified chimeric ON product is eluted from the cartridge with 2 ml 50% acetonitrile/water, and the total number of A_{260} units recovered is determined from the absorbance at 260 nm of a diluted sample.

Samples of the eluate are analyzed by anion-exchange HPLC and reversed phase HPLC (see later). Cleavage of methylphosphonodiester internucleoside linkages may occur to a certain degree during deprotection to yield fragments carrying a 5'-dimethoxytrityl group, which will copurify with the product during trityl selection on C_{18} Sep-Pak cartridges. Whereas reversed phase HPLC is useful for distinguishing between trityl-on ON and trityl-off failure sequences, this technique may give inadequate or no separation between trityl-off product and some trityl-off ON fragments. In this case, anion-exchange HPLC, where the separation is not based solely on the overall negative charge on the molecule, provides a useful crosscheck of the purity of chimeric ON preparations. It has generally been found that this technique is more discriminating than reversed-phase HPLC.

Secondary, Trityl-Off, Purification by Reversed-Phase, Solid-Phase Extraction on C_{18} Sep-Pak Cartridges

Further cleanup of the product is achieved by diluting the ON solution with 10 volumes of TEAA and applying it to a new siliconized C_{18} Sep-Pak cartridge, which is then eluted with a step gradient of acetonitrile in TEAA. The major ON containing fractions are identified rapidly by UV absorption spectra between 220 and 300 nm and their purity is assessed by HPLC analysis. The appropriate fractions are pooled and desalted by diluting with TEAA and applying the solution to a siliconized C_{18} Sep-Pak cartridge, which is then washed exhaustively with water. The desalted product is eluted with 2 ml 50% acetonitrile/water. This technique may also be used to rapidly concentrate and desalt fractions collected from HPLC purification of the ON, as discussed later.

When the second-stage, trityl-off separation is reiterated with exhaustive elution at an acetonitrile concentration just slightly less than that required to start eluting the product, the Sep-Pak purification scheme is generally adequate to provide a product that is sufficiently pure for most purposes.

At this stage the ON is in the form of its triethylammonium salt. The triethylammonium cation is quite toxic to mammalian cells, and for biological applications it is advisable to convert the ON to the sodium salt. This may be achieved by passing the ON solution through a small column of Dowex 50WX8-400 ion-exchange resin (Aldrich), sodium form, held in a 10-ml polypropylene syringe barrel with a plug of siliconized glass wool, and eluting with water until all UV-absorbing material is recovered. The Dowex 50 resin is supplied in the hydrogen form and is converted into the sodium form by washing the column with 1 M sodium chloride solution until the effluent is no longer acidic, followed by a water wash to remove the salt. Alternatively, the sodium form of the ON may be generated by three consecutive ethanol precipitations through the addition of 0.1 volume 3 M sodium acetate plus 10 volumes ethanol followed by 1 hr at $-80°$. The final precipitate is washed twice with ethanol.

The ON pellet from ethanol precipitation is dissolved in 1 ml water, or the Dowex 50 ON eluate is concentrated in a Savant Speed-Vac concentrator, and adjusted to a volume of 1 ml with water. The product is cleared of all residual small molecule contaminants by passage through a gel filtration NAP 10 column (Pharmacia, St. Albans, Hertfordshire), eluting with 1.5 ml water. The final product solution is evaporated to dryness in a Speed-Vac concentrator and is stored dessicted at $-20°$.

Analysis of Chimeric ON by HPLC

Reversed-Phase HPLC. Chimeric methylphosphonodiester/phosphodiester and methylphosphonodiester/phosphorothiodiester ON may be analyzed by reversed-phase HPLC on Brownlee Aquapore RP-300 7-μm (Applied Biosystems) and Asahipak C8P-50 (UK supplier Prolabo/Rhone-Poulenc, Manchester) columns. A generally useful, steep gradient from 5 to 70% acetonitrile (HPLC grade) in 0.1 M TEAA, pH 7.0, in 20 min at a flow rate of 1 ml/min serves to separate 5′-dimethoxytrityl-ON from failure sequences in all applications. Shallower gradients over shorter concentration ranges may be used for a more stringent analysis of final products. The TEAA buffer is best prepared from triethylamine that has been refluxed with potassium hydroxide pellets for 4 hr and redistilled to remove UV-absorbing impurities that can interfere with the analyses. Installing a column dry packed with a silica precolumn gel (37–53 μm, Whatman, Maidstone, Kent) upstream of the sample injector to saturate the buffer with silicate prior to contact with the analytical column enhances the lifetime of the silica-based Aquapore column. In the case of analyzing fluorescein tagged ON (see later), where optimum signal from the fluorescence detector is achieved by using a slightly alkaline TEAA buffer pH of 8.0, use of a silica-packed column in the solvent delivery line is highly recommended.

Anion-Exchange HPLC. Excellent separations of chimeric ON were achieved by anion-exchange HPLC on HRLC MA7Q columns (Bio-Rad, Hemel Hempstead, Hertfordshire) at a column temperature of 65° using a 60-min gradient from 0 to 1.5 M potassium chloride in 20 mM potassium phosphate, 50% formamide (v/v), pH 7.5, and a flow rate of 1 ml/min, with the UV detector set at 280 nm, off the absorption of formamide. In the case of chimeric ON with only four phosphodiester linkages, and therefore carrying only four negative charges, the composition of the weak buffer was changed to 1 mM potassium phosphate in 50% formamide, pH 7.5, to achieve retention on the column. The use of low UV-absorbing reagents ("HiPerSolv" grade, potassium dihydrogen orthophosphate and potassium chloride, BDH/Merck and spectrophotometric grade formamide, Aldrich) is recommended for the preparation of buffers. Nevertheless, UV-absorbing impurities in these constituents can still interfere with the analyses by concentrating on the column at low ionic strength and then eluting during the gradient as the salt concentration is increased. Installation of a column dry packed with Partisil-10 SAX, a strong ion exchanger (Whatman), in the solvent line on the instrument upstream of the sample injector largely eliminates such problems. The packing material in this column is replaced periodically as it becomes exhausted.

   HPLC Column Regeneration. Polymer-based C8P-50 and HRLC columns have the advantage that they may be regenerated by flushing with a 0.2 M solution of sodium hydroxide, followed by 0.2 M acetic acid, whereas such a treatment would be incompatible with silica-based chromatographic supports. Therefore, these columns are suitable for "dirty" applications, such as the analysis of ON in cell and tissue extracts, which would rapidly and irreversibly poison the latter. Column performance is monitored periodically with a cocktail of ON standards, and the regeneration procedure is applied when a deterioration in the separation is observed.

Purification of Chimeric ON by HPLC

   The product may be subjected to a final purification step on the analytical HRLC anion-exchange column when highly pure chimeric ON are required. This is particularly appropriate for phosphorothioate containing ON, where the preparation will contain substantial amounts of material seen as an apparent "*n*-1" peak on anion-exchange HPLC analysis, but which coelutes with the true product on reversed-phase HPLC analysis and during C$_{18}$ Sep-Pak purification. In this case, the impurity is a mixture of full-length ON in which one of the internucleoside linkages is phosphodiester rather than the desired phosphorothiodiester as a result of failure in the synthesis sulfurization step or oxidation occurring during deprotection.[35]

   Preparative Isocratic Anion-Exchange HPLC. The injector should be fitted with a large-volume sample loop. Significant amounts of ON may be purified relatively quickly by running the HPLC isocratically and injecting

---

[35] G. Zon, *in* "Methods in Molecular Biology" (S. Agrawal, ed.), Vol. 20, p. 165. Humana Press, Totowa, NJ, 1993.

the ON in batches of 5–10 A_{260} units. The same conditions and buffers as used for analytical separation are used.

The nominal percentage of strong buffer in the 60-min 0–100% strong buffer gradient at which the product elutes is determined during an analytical separation. The column is then equilibrated at this constant eluent composition. The ON should elute in the void volume of the column at this concentration under isocratic conditions. That this occurs is verified by injection of an analytical sample of ca. 0.05 A_{260} units. The column is reequilibrated at progressively reduced percentages of strong buffer until injection of an analytical sample of the ON produces a product peak retention time of about 20 min. This eluent composition should suffice for preparative separation as the retention time reduces when vastly greater amounts of ON are injected. The HPLC UV detector is offset from the ON absorption maximum to 290 nm and the range is set to 2 absorbance units full scale. The separation may be fine-tuned by reducing the percentage of strong buffer still further during the preparative runs. Inadequately purified fractions may be rerun under the improved separation conditions after being rapidly concentrated and desalted. This is achieved by dilution with 10 volumes of TEAA to permit binding of the ON to a siliconized C_{18} Sep-Pak cartridge, which is then washed exhaustively with water and the product eluted with 2 ml 50% acetonitrile/water. Product fractions are pooled and desalted on a C_{18} Sep-Pak cartridge prior to removal of all small molecule contaminants by passage through a NAP 10 gel filtration column, as described earlier.

After completion of the preparative separation the column must be regenerated by flushing with 0.2 M sodium hydroxide followed by 0.2 M acetic acid before being used again for analytical work, as memory effects can be quite serious.

Preparative Gradient Anion-Exchange HPLC. An alternative approach is to load up to 120 A_{260} units onto the HRLC column under isocratic conditions of 100% weak buffer and a flow rate of 1 ml/min using multiple injections if necessitated by the volume of sample. A manual step gradient is applied cautiously until UV-absorbing material appears in the effluent. The elution is allowed to run isocratically until a constant baseline is reestablished. This process is repeated until all the ON that was applied to the column is eluted. The UV absorption trace obtained from this approach is uninterpretable as a chromatograph, as fluctuations in the effluent concentration of pure product give the appearance of multiple peaks. This effect is particularly noticeable when increasing the concentration of strong buffer over the range required to elute all product from the column. The column is regenerated and then used analytically to identify the range of fractions containing pure ON.

Fluorescein-Labeled Chimeric Oligodeoxynucleotides:
  Synthesis and Purification

It is desirable to have a readily detectable, fluorescent reporter group attached to ON for monitoring intracytoplasmic delivery (see later) and metabolism against a background of UV-absorbing biomolecules. This function is fulfilled adequately by fluorescein, which, at the same time, does not appear to affect unduly the biochemical and biological properties of antisense ON.

ON may be most conveniently labeled with fluorescein by introducing an amino-linker group as the last cycle on the synthesizer, followed by postsynthetic derivatization. A number of amino-linker phosphoramidites are available commercially, but for creating a linkage to methylphosphonodiester/phosphodiester and methylphosphonodiester/phosphorothiodiester chimeric ON, a degree of selectivity is required. Aminolink 2 (Applied Biosystems), for example, is protected on phosphorus with a methoxy group, and the extended deprotection treatment with ammonia at room temperature required for this function may lead to loss of ON product through hydrolysis of methyphosphonodiester internucleoside linkages. However, use of the β-cyanoethyl-protected 5'-amino-modifier C$_6$-TFA (Glen Research) is compatible with the deprotection conditions for methylphosphonate containing chimeric ON, as described earlier, provided that the deprotection time is extended to 72 hr to ensure complete removal of the trifluoroacetyl amino-protecting group.

In order to confirm that efficient coupling of the amino linker to the ON has been achieved, an interrupt is programmed into the DNA synthesizer immediately prior to the capping step of this cycle. The capping step is omitted by jumping to the next command. A further two cycles are executed for coupling an arbitrary deoxynucleoside phosphoramidite. The trityl fraction of the first supplementary cycle gives an indication of the efficiency of coupling of the amino linker and ideally should be colorless. If coupling has been less than ideal, the detritylation step of the final cycle may be omitted, as for trityl-on ON synthesis. ON molecules that failed to couple with the amino linker may then be removed during the initial C$_{18}$ Sep-Pak cleanup of deprotected product through a reversal of the strategy used to select the trityl-on product from failure sequences.

In the absence of a 5'-dimethoxytrityl group, the complete separation of 5'-amino ON from all end-capped failure sequences cannot be achieved on C$_{18}$ Sep-Pak cartridges. However, significant enhancement of the purity of the preparation is accomplished by exhaustive elution of the cartridge with a concentration of acetonitrile in TEAA just slightly less than that required to displace the product. This is determined with an exploratory step gradient, monitoring fractions by weak anion-exchange HPLC, following which product containing fractions are diluted with TEAA and returned to the cartridge. The 5'-amino ON is eluted with an appropriate concentration of acetonitrile in TEAA, desalted on a new C_{18} Sep-Pak as described earlier, and dried down in a Speed-Vac concentrator.

5'-Amino oligodeoxynucleotide (1 μmol, very approximately 200 A_{260} units of a 20-mer) is dissolved in 150 μl 0.6 M sodium bicarbonate/sodium carbonate buffer, pH 8.5. A useful working means of converting A_{260} units into approximate numbers of micromoles of ON is provided by calculating the millimolar absorptivity, ε_{260}, of an ON at 260 nm. This is achieved by summing the values for each base[36]: 8.8 for T; 7.3 for C; 11.7 for G; and 15.4 for A. A solution of 5(6)-carboxyfluorescein N-hydroxysuccinimide ester (Molecular Probes Inc., UK supplier, Cambridge BioScience, Cambridge) in dimethylformamide (12 μmol in 50 μl) is added and the pH of the solution is checked immediately with a semimicro electrode. If below 8, the pH is readjusted to 8.5 by addition of one or two grains of solid sodium carbonate. Reaction is complete in less than 30 min. A sample (1 μl) of the reaction mixture is precipitated by the addition of 1 ml 90% ethanol and analyzed by weak anion-exchange HPLC, with the column effluent being monitored with UV and fluorescence (excitation 494 nm, emission detection centered on 530 nm, Wratten 15 520-nm cut-off emission filter) detectors connected in series. The reaction proceeds with the disappearance of the 5'-amino ON UV absorption peak and the appearance of a UV absorption/fluorescence peak at longer retention time, consistent with an increase in the overall charge on the molecule due to the negative charges on fluorescein and removal of the positive charge at the amino function through formation of the amide. If unreacted 5'-amino ON remains at 30 min, it is necessary to add more carboxyfluorescein ester solution to achieve further reaction. Failure to observe additional conversion of 5'-amino ON to product may mean that the reaction is inhibited by some form of intermolecular secondary structure. In this case a complete reaction is secured by adding fresh carboxyfluorescein ester solution and heating the reaction mixture briefly to 90°.

To remove the bulk of the carboxyfluorescein impurity, 0.85 ml water is added to the reaction mixture and the product is precipitated by the addition of 0.1 ml 3 M sodium acetate and 10 ml ethanol, followed by 1 hr at −80°. The precipitate is redissolved in 1 ml water, applied to a NAP 10 gel filtration column, and eluted with 1.5 ml water. The eluate is diluted

[36] B. S. Sproat and M. J. Gait, in "Oligonucleotide Synthesis: A Practical Approach" (M. J. Gait, ed.), p. 83. IRL Press, Oxford, 1984.

with 10 ml TEAA, and the fluorescein-tagged product is purified further on a C_{18} Sep-Pak cartridge using the standard approach.

In Vitro Antisense Techniques Using Escherichia coli Ribonuclease H

Antisense ON Effects against Oligomeric RNA Targets Using E. coli Ribonuclease H

The intrinsic potential for ON of a given structure to direct the RNase H cleavage of target RNA sequences may be conveniently assayed using chemically synthesized RNA and E. coli RNase H. Using the following procedure, we showed that chimeric methylphosphonodiester/phosphodiester ON demonstrated enhanced activity, relative to phosphodiester congeners.[8] This effect is thought to be due to the reduced hybrid-melting temperature, conferred on the heteroduplex through the presence of methylphosphonodiester internucleoside linkages in the ON,[8,37] permitting enhanced dissociation of the quaternary ON–enzyme–cleavage product complex.

Assay Conditions. Assays are initiated by the addition of E. coli RNase H (final concentration: 0.0015 U/μl, Amersham International, Little Chalfont, Buckinghamshire or Cambio) to mixtures of target RNA (10 μM) and ON (10 μM) in RNase H digestion buffer (40 mM Tris–HCl, 4 mM MgCl$_2$, 1 mM dithiothreitol, 0.003% bovine serum albumen, pH 7.6, at 37°), and incubation at 37° for 0 to 120 min. Reactions are terminated by the addition of 1 volume of ice-cold 8 mM EDTA (pH 8.0), 0.02% bromphenol blue.

One valuable modification of the just-described assay is to place the RNA substrate in 100-fold molar excess over the ON (300 μM to 3 μM) in the presence of a greatly increased concentration of RNase H (0.1 U/μl). Reactions under such conditions particularly test the capacity of the ON to "catalytically" direct the RNase H-mediated destruction of target RNA.

PhastGel Electrophoresis and Silver Staining. ON, oligoribonucleotides, and RNase H cleavage products of oligoribonucleotides may be conveniently analyzed by fractionation through "Homogenous 20" PhastGels (Pharmacia), using "native" buffer strips and visualized by automated silver staining.[8]

Electrophoresis is carried out at 15° using the protocol of preseparation 400 V, 10 mA, 2.5 W, 100 Vhr; sample application 400 V, 1 mA, 2.5 W, 5 Vhr; separation 400 V, 10 mA, 2.5 W, 50 Vhr to 80 Vhr.[38] The values for

[37] D. M. Tidd, *AntiCancer Res.* **10**, 1169 (1990).
[38] R. V. Giles, Ph.D. Thesis, 1993.

voltage, current, and power represent upper limits whereas "Vhr" represents the actual duration.

Optimal silver staining was obtained using the following protocol: (1) Rinse in water 0.1 min, 20°; (2) fix gel with 20% trichloroacetic acid 5 min, 20°; (3) sensitize with 8.3% glutaraldehyde 5 min, 50°; (4) wash twice in water 2 min, 50° each; (5) stain with 0.5% $AgNO_3$ 10 min, 40°; (6) wash twice in water 0.5 min, 30° each; (7) develop with 2.5% Na_2CO_3, 0.04% HCHO 0.5 min, 30°; (8) as step 7 but 4.5 min, 30°; (9) reduce background with 2.5% $Na_2S_2O_3 \cdot 5H_2O$, 3.7% Tris–HCl 1 min, 30°; (10) stop further reaction with 5% acetic acid 2 min, 50°; (11) preserve gel with 10% acetic acid, 5% glycerol 3 min, 50°. This procedure was found to have a sensitivity limit of 0.1 pmol of ON and to stain maximally with 15 pmol of 15-mer ON.[38]

Antisense ON Activity against In Vitro-Transcribed RNA Targets Using E. coli RNase H

The efficiency and specificity of ON-directed RNase H scission of target and nontarget RNA may be conveniently assayed using *E. coli* RNase H and *in vitro*-transcribed RNA.[9] Arguably, the most meaningful results are obtained when the *in vitro*-transcribed RNA is as near full-length as possible, as its secondary structure may then more accurately reflect that observed for natural mRNA within living cells.

In Vitro Transcription of Target RNA. Target RNA is produced by a modification of standard *in vitro* transcription conditions [50-μl reaction containing ca. 2 pmol of template DNA (linearized if plasmid), 250 U of the appropriate bacteriophage RNA polymerase (Epicentre, UK Distributor Cambio), 100 U RNase Block I (Stratagene, Cambridge, Cambridgeshire), and 1 mM of each NTP in 40 mM Tris–HCl, pH 7.5, 10 mM NaCl, 2 mM spermidine, 6 mM $MgCl_2$, and 10 mM dithiothreitol, incubate at 37° to 40° for 2 hr] by replacing 1 mM UTP with 0.1 mM digoxigenin-11-UTP (Boehringer Mannheim, Lewes, East Sussex) and 0.9 mM UTP. For this application, template DNA should not be removed by DNase I digestion and further manipulation of the RNA product is not required.

Assay Conditions. Reactions containing 25 ng/μl of digoxigenin-labeled RNA (from unpurified transcription reactions), *E. coli* RNase H (0.025 U/μl), and 1 μM ON (where present) in RNase H digestion buffer containing 2 U/μl RNase Block I are incubated at 37° for 0 to 60 min. Reactions are terminated at intervals by the removal of samples (~75 ng transcript) to 16 volumes of ice-cold 100 ng/μl *E. coli* ribosomal RNA in 1 mM EDTA, pH 8.0. Proteins in the reaction mixture are removed by 1 round of phenol/chloroform/isoamyl alcohol extraction (25 : 24 : 1, v/v, pH 8.0). Reactants and products in the aqueous phase of such extractions (~3 ng *in vitro* transcript and ~200 ng rRNA) are fractionated by formaldehyde agarose gel electrophoresis, blotted onto nylon membrane, and visualized immunologically. Band intensities are quantified by densitometry.

Formaldehyde Gel Electrophoresis of RNA and Northern Blotting. This protocol is streamlined relative to, and more robust than, standard protocols that have been described.

SAMPLE PREPARATION. The solvent containing RNA samples for gel electrophoresis is adjusted to 50% formamide (v/v), 1× MOPS buffer (20 mM MOPS, 5 mM sodium acetate, 1 mM EDTA, pH 7.0), 6.6% formaldehyde, 50 μg/ml ethidium bromide, 0.5 mg/ml orange G dye, and 6.6% sucrose (w/v). Samples are vortexed and incubated at 65° for 10 min and then placed on ice. Moisture within the tubes is collected by brief centrifugation and the samples are mixed by vortexing. Undissolved solids are pelleted by centrifugation (12,000g, 5 min, room temperature) and the samples are returned to ice prior to loading on the gel.

GEL PREPARATION. Volumes for a 100-ml gel. Agarose (1–1.5 g) is dissolved in 77 ml diethyl pyrocarbonate (DEPC)-treated water by microwave heating, and the gel solution is then placed at 60° to equilibrate. When the agarose solution has cooled to 60°, add 5 ml 20× MOPS buffer and 18 ml 37% formaldehyde, swirl to mix, cast into suitably sealed, preheated (60°, prevents warping) gel-casting tray, and position comb. Allow gel to set for at least 1 hr.

When the gel has fully set, remove comb and immerse in running buffer (1× MOPS, 6.6% formaldehyde), load samples, and allow to settle for 10 min. Apply a potential difference of 6 V/cm of gel. The orange G dye runs just ahead of the position of tRNA. When the RNAs have migrated the desired distance, remove the gel and visualize abundant RNA species (>about 20 ng/band) by UV (330 nm) excitation of intercalated ethidium bromide.[39]

Nucleic acids contained in the gel may be capillary blotted onto amphipathic nylon membranes (Hybond N, Amersham International, or Nytran, Schleicher & Schuell, London) without further treatment. A standard capillary blotting procedure is used[40] with 20× saline sodium citrate (20× SSC, 3 M NaCl, 0.3 M sodium citrate, pH 7.0) as the mobile phase. Nytran membranes do not wet efficiently in high salt buffers and should be floated on distilled water prior to use. Replacing the "stack" of paper towels with 10 to 15 sheets of QuickDraw paper (Sigma, Poole, Dorset) allows quantitative transfer of RNAs < about 5 kb in length to be achieved in 2

[39] R. A. Kroczek, *Nucleic Acids Res.* **17**, 9497 (1989).
[40] J. Sambrook, E. F. Fritsch, and T. Maniatis, *in* "Molecular Cloning: A Laboratory Manual." Cold Spring Harbor Laboratory Press, Cold Spring Harbor, NY, 1989.

hr. The membrane should then be allowed to dry thoroughly before UV (254 nm) cross-linking of the nucleic acids to the membrane. The duration of UV exposure, required for optimal cross-linking of nucleic acids to membrane, should be established empirically for each application. These optimizations should be repeated occasionally to maintain a maximal signal-to-noise ratio.

Immunological Detection of Membrane-Bound Nonradioactive Nucleic Acids. Nonradioactively labeled RNA may be detected using antidigoxigenin (or antifluorescein)–alkaline phosphatase-conjugated Fab fragments (Boehringer) essentially as described by the manufacturer.[41] Briefly, the membrane is equilibrated for 5 min in wash buffer [0.1 M maleic acid, 0.15 M NaCl, pH 7.5, 0.3% (v/v) Tween 20] and then blocked for 30 min to 1 hr in blocking buffer [1% (w/v) blocking reagent (Blocking Reagent, Boehringer) in wash buffer, freshly diluted from a DEPC-treated 10% solution of blocking reagent in wash buffer]. Antibody is applied (20 ml/ 100 cm^2 of membrane), diluted 1 : 5000 (v/v) in blocking buffer, for 30 min to 1 hr. The membrane is then washed once for 5 min and twice more for 15 min each in wash buffer. It is important to change the vessel between the first and second postantibody washes. The membrane is then equilibrated in 0.1 M diethanolamine, 5 mM MgCl$_2$, pH 10.0, for 5 min. These steps should be carried out at room temperature with constant gentle agitation. The membrane is drained extensively and 10 ml/100 cm^2 chromogen [10 ml diethanolamine/MgCl$_2$ buffer, 90 μl 37.5 mg/ml nitro blue tetrazolium chloride (NBT) in 70% dimethyformamide, and 70 μl 25 mg/ml 5-bromo-4-chloro-3-indolyl phosphate p-toluidine salt (BCIP) in 100% dimethylformamide] is added. The chromogen-covered membrane should be placed in the dark at room temperature with minimal disturbance until the required bands are the desired intensity. The blot should be washed extensively in water, dried, and stored in the dark.

Results. Phosphodiester[9] and phosphorothioate[32] ON induced substantial RNase H-dependent scission of nontarget RNA sequences in this assay, whereas a progressively reduced undesired cleavage was observed using chimeric analogs with increasing methylphosphonodiester substitution.[9,32] Essentially identical results were obtained using similar protocols that replaced the *in vitro*-transcribed RNA with total RNA extracted from cell lines (500 ng/μl) or A$^+$ selected RNA (25 ng/μl).[32] The enhanced specificity of chimeric analogs is thought to result from a combination of two mechanisms. (1) The reduced hybrid stability observed in heteroduplexes between RNA and methylphosphonodiester-containing ON,[8,37] relative to parent phosphodiester congeners, affects an increased stringency of hybridization

---

[41] Boehringer, DIG nucleic acid detection kit.

under conditions of constant temperature and ionic strength. (2) Partial complementarities contained entirely within the methylphosphonodiester section of the ON will not result in RNase H cleavage of the hybridized RNA.[7,8]

Identification of Open Loop (Accessible) Regions of RNA. It occurred to us that the undesired cleavage, observed in the experiments described earlier, may provide for the identification of regions of RNA that are accessible to ON hybridization.[42,43] Consequently, a computer-fitting algorithm was developed that returns all regions of contiguous partial complementarity (of a user-designated minimum length) between input RNA and ON sequences. MSDOS executable is available from our web site.[44] By matching the observed cleavage fragments to the computed potential cleavage sites it was noted that (1) undesired cleavage occurs at only a subset of the possible sites and (2) undesired RNase H activity could be supported by partial complementarities that extended over just six contiguous base pairs.

This system was investigated for the c-*myc* oncogene. The starting ON was directed to the translation initiation codon. ON complementary to regions of RNA observed to be efficient substrates for undesired cleavage were synthesized. Parent and derived ON were assayed for antisense activity in living cells, following streptolysin O (SLO)-mediated transfection (see later). A chimeric ON with a sequence derived by this procedure was found to be a much more potent effector than a chimeric ON with the parent sequence.[45]

Coupled In Vitro Transcription/RNase H Cleavage Reactions. The most convenient method to assay ON-directed RNase H cleavage in *in vitro*-transcribed RNA is to dope standard (nonlabeling) *in vitro* transcription reactions (see earlier discussion) with 1 μM ON and 0.02 U/μl *E. coli* RNase H.[13] RNA products in samples taken from such reactions may be fractionated most simply by electrophoresis through native agarose gels and visualized by ethidium bromide staining and UV illumination. However, because the secondary structure influences electrophoretic mobility in native agarose gels, apparent fragment sizes may differ significantly from their true lengths. If accurate determination of transcript RNA and RNase H cleavage product lengths is required, coupled transcription/cleavage reactions should be stopped and products concentrated by ethanol precipitation

[42] R. V. Giles, D. G. Spiller, J. Grzybowski, R. E. Clark, P. Nicklin, and D. M. Tidd, *Nucleic Acids Res.* **26**, 1567 (1998).

[43] R. V. Giles, D. G. Spiller, R. E. Clark, and D. M. Tidd, in press.

[44] http://www.liv.ac.uk/~giles/

[45] D. G. Spiller, R. V. Giles, C. M. Broughton, J. Grzybowski, C. J. Ruddell, R. E. Clark, and D. M. Tidd, *Antisense Nucleic Acid Drug Dev.* **8**, 281 (1998).

prior to formaldehyde gel electrophoresis and visualization of the ethidium bromide-stained RNA by UV illumination (see earlier discussion). We have noted that phosphorothioate ON and $2'$-O-allyl-protected ribozymes may not be used in this system as minimal transcription occurs in the presence of such compounds.

Antisense Oligodeoxynucleotide Activity in Total Cell Protein Extracts with *In Vitro*-Transcribed RNA Targets

Whole cell extracts have proved invaluable as a source of human RNase H, which allow predictions to be made regarding the intracellular antisense efficacy of ON structures. We have shown previously that chimeric ON with a central gap as small as two phosphodiesters can efficiently direct the action of *E. coli* RNase H.[32] Activity from such ON can also be obtained in human cell extracts under conditions of low salt. However, when the experimental conditions were altered to reflect the intracellular concentrations of ions, no antisense activity was observed from chimeric ON with central regions containing three phosphodiesters,[31] even when the target RNA possessed a relatively favorable secondary structure. Further experimentation, under physiological conditions, revealed that ON with central regions containing four, or fewer, phosphodiester internucleoside linkages were significantly less active than chimeric ON with larger phosphodiester regions (unpublished data). Moreover, chimeric 15-mer ON with eight central phosphodiester linkages were shown to provide point mutation selectivity for human RNase H cleavage in cell extract experiments.[13]

The point mutation specificity obtained in cell extracts has been duplicated in intact cells.[13] Furthermore, the minimum phosphodiester requirements for ON activity in cell extracts have been largely duplicated in living cells following ON introduction using SLO. However, target site accessibility clearly has an impact on the observed activity of antisense ON. Chimeric ON targeted to exceptionally accessible sites (e.g., the new c-*myc* site derived earlier) were found to retain substantial activity when the central region was reduced to four phosphodiester internucleoside linkages (unpublished data).

Preparation of Whole Cell Extract

Human leukemia cells are maintained in exponential growth in 90% RPMI 1640 medium (GIBCO, Paisley, Renfrewshire), 10% fetal calf serum (SeraLab, Crawley Down, West Sussex) by incubation at 37° in an atmosphere of 5% CO_2, 95% air. Fetal calf serum is heat-inactivated by incubation at 56° for 30 min prior to use.

Total Cell Count by Flow Cytometry. The flow cytometric procedures described here were derived using an Ortho Cytoron Absolute flow cytometer (Ortho, High Wycombe, Buckinghamshire). This machine uses a 15-mW argon ion laser and will record the variables of forward and side scatter, which provide information on cell size and cell size/morphology, respectively. The remaining two photomultiplier tubes are equipped with 515- to 548-nm band pass and 620-nm long pass filters to provide analysis of green and red fluorescence, respectively.

A protocol is first established on the flow cytometer that passes the sample through the laser beam at the known rate of 1 μl/sec for a known duration (10 sec). Forward versus side scatter dot-plot output is selected, and forward and side scatter detector gains are optimized, using linear amplification, to obtain a tight population distribution somewhat away from the axes. This population is selected (gated) so that subsequent measurements only include single cells. A simple cell count may be obtained by taking an undiluted sample of cell culture through this protocol. Cells per milliliter in the original culture = selected cell count $\times$ 1000 μl/ml $\div$ 10 μl.

Preparation of Whole Cell Extract. The extract is obtained as described by Manley.[46] All procedures are carried out on ice and with ice-cold buffers. Briefly, 2.5×10^8 exponentially growing cells are collected and washed twice in phosphate-buffered saline. The cells are then resuspended in 4 ml 10 mM Tris–HCl (pH 7.9), 1 mM EDTA, 5 mM dithiothreitol and incubated on ice for 20 min. The swollen cells are then lysed by Dounce homogenization (eight strokes, "B" pestle"). Four milliliters of 50 mM Tris–HCl (pH 7.9), 10 mM MgCl$_2$, 2 mM dithiothreitol, 25% (w/v) sucrose, 50% glycerol is added to the suspension and the suspension is mixed by gentle stirring. The stirring is continued as 1 ml of saturated ammonium sulfate (pH 7.0) is added dropwise and extended for an additional 30 min. Centrifuge the suspension for 3 hr at 175,000g. Collect the supernatant ($\sim$10 ml) and precipitate the remaining proteins and nucleic acids by adding 3.5 g of solid ammonium sulfate. When the ammonium sulfate has dissolved, add 3.5 μl of 1 M NaOH and stir for 30 min. Collect the precipitate by centrifugation (15,000g, 20 min) and resuspend in 500 μl of 40 mM Tris–HCl (pH 7.9), 0.1 M KCl, 10 mM MgCl$_2$, 0.2 mM EDTA, 15% glycerol. Dialyze the suspension against 100 ml of the same buffer overnight. Divide the suspension ($\sim$800 μl) into 100-μl aliquots, snap freeze, and store in liquid nitrogen. The extracts retain RNase H activity through at least three rounds of thawing and snap freezing.

---

[46] J. L. Manley, *in* "Transcription and Translation: A Practical Approach" (B. D. Hames and S. J. Higgins, eds.), p. 71. IRL Press, Oxford/Washington, DC, 1984.

Cell Extract RNase H Assay

Reactions were performed by suspending 5 ng/μl to 10 ng/μl *in vitro*-transcribed RNA (labeled, 1:9 digoxigenin-11-UTP:UTP, see earlier discussion). 1 μg/μl yeast tRNA, 1 μM ON, and 1 U/μl RNase Block I in intracellular buffer (11 m*M* potassium phosphate, pH 7.4, 108 m*M* KCl, 22 m*M* NaCl, 1 m*M* dithiothreitol, 3 m*M* MgCl$_2$, 1 m*M* ATP) at 37°. Cell extract, preheated to 37°, is added to a final concentration of 10% (v/v) and incubation is continued at 37° for 0 to 60 min. Vanadyl ribonucleoside complexes (20 m*M* final concentration) may be substituted for the human placental ribonuclease inhibitor in these assays. Reactions are stopped by guanidine thiocyanate/acid phenol extraction of the RNA as if preparing total cell RNA.

Purification of Total RNA from Enzyme Assays and Human Cells. A streamlined adaptation of the guanidine thiocyanate/acid phenol method of Chomcyznski and Sacchi[47] is used routinely.

ENZYME ASSAY. Add 500 μl of lysis buffer (4 *M* guanidine thiocyanate, 5 m*M* sodium citrate, pH 7.0, 0.5% sodium sarkosyl, 0.2 *M* sodium acetate, pH 4.2, and 0.1 *M* 2-mercaptoethanol) directly to the enzyme assay. Vortex to ensure complete mixing.

CELLS. Samples of 0.5 to 1 × 10^6 cells are pelleted by centrifugation (800g, 4 min) and the supernatant is discarded. The cells are agitated gently to form a cell "slurry" from the packed pellet, and 500 μl of lysis buffer is squirted directly onto the slurry. Vortex the suspension vigorously to ensure complete lysis.

Add 500 μl water-saturated phenol (pH 4.0 to 4.3) and vortex to obtain a single clear phase. Add 130 μl chloroform:isoamyl alcohol (24:1, v/v) and vortex to obtain a cloudy suspension. Incubate on ice for 10 to 15 min and centrifuge (room temperature, 12,000g or higher for 10 min) to separate the phases. At this stage it is important to not allow the contents of the tube to cool below ca. 20° or gross DNA contamination will result. Remove 400 μl of the upper aqueous phase to a fresh tube and precipitate RNA by adding 500 μl isopropanol and incubating at −20° for 2 hr or overnight. Do not attempt to increase the yield of RNA by removing more than 400 μl of the aqueous phase as significant DNA contamination may result. Pellet the RNA by centrifugation (room temperature, 12,000g or higher for 30 min) and remove and discard the supernatant. Subject the tubes to a brief second centrifugation (room temperature, 12,000g for 10 sec) to collect the fluid adhering to the tube walls, remove, and discard. For subsequent RT-PCR applications, resuspend the pellet in DEPC-treated water;

[47] P. Chomczynski and N. Sacchi, *Anal. Biochem.* **162,** 156 (1987).

for subsequent electrophoresis applications, resuspend in DEPC-treated, deionized formamide. In both cases, resuspension can be facilitated by heating to 65° and vortexing. Resuspension in formamide is preferable as it is a good solvent for RNA, a poor solvent for DNA, and inhibits the action of ribonucleases.[48]

Spectrophotometric analysis in pH 8.0 buffer at 230, 260, and 280 nm of RNA extracted from MOLT4 cells by this method indicates a recovery of ca. 10–15 μg per 10^6 cells devoid of overt protein ($A_{260}/A_{280} > 1.8$) or guanidine thiocyanate ($A_{260}/A_{230} > 2.0$) contamination.

Analysis of Substrate and Product RNAs. Lengthy *in vitro*-transcribed RNA substrate and RNase H cleavage product fragments are fractionated most conveniently by formaldehyde–agarose gel, capillary blotted onto nylon membrane, and immunologically visualized, as described earlier. Shorter RNAs (*in vitro* transcript <ca. 1000 nucleotides) should be analyzed by urea–polyacrylamide gel electrophoresis (urea–PAGE). Resuspend the RNA pellet in 90% formamide, 0.03% xylene cyanol, 0.03% orange G, denature by heating to 90° for 5 min, and quench on ice prior to loading on the gel. Fractionate through a 0.75-mm-thick, 20-cm-long gel composed of 7 M urea, 1× TBE (90 mM Tris–borate, 2 mM EDTA), and 4 to 6% acrylamide (19:1 acrylamide:bisacrylamide) maintained thermostatically at 50°, with a potential difference of 25 V/cm of gel using 1× TBE as running buffer. When the orange G band has migrated ~19 cm, dismantle the electrophoresis apparatus and soak the gel, still attached to one glass plate, for 15 min in 0.5× TBE. Nucleic acids contained in the polyacrylamide gel may be semidry electroblotted onto a nylon membrane using a TransBlot semidry cell (Bio-Rad) as described by the manufacturer using 0.5× TBE as transfer buffer with maximal current and voltage of 2 mA/cm^2 of membrane and 25 V, respectively, for 30 min. We have found that other makes of semidry blotter, which possess carbon electrodes, gave irreproducible results in this, and similar, applications. The membrane should then be dried and UV cross-linked. Labeled RNA present on the membrane may be detected immunologically, as described earlier.

Identification of RNase H Class in Cell Extracts

The ionic constituents of the cell extract RNase H assays may be varied, in line with previous work,[49] in an effort to identify the type of RNase H present. The RNases H present in extract from human acute lymphocytic leukemia MOLT4 and human chronic myeloid leukemia K562 cells were sensitive to increased concentrations of salt (as intimated earlier) and activ-

---

[48] P. Chomczynski, *Nucleic Acids Res.* **20,** 3791 (1992).
[49] C. Cazenave, P. Frank, and W. Busen, *Biochimie* **75,** 113 (1993).

ity was maintained when Mn^{2+} ions replaced Mg^{2+}. These data are consistent with the majority of the RNase H activity in these cell lines being of class I type.

ON Degradation in Cell Extract RNase H Assays

The relative concentration and integrity of fluorescein end-labeled ON in standard cell extract RNase H assays may be examined. In this case, vanadyl ribonucleoside complexes were avoided as it was uncertain whether they would also inhibit the action of deoxyribonucleases potentially involved in the degradation of ON. Two-microliter samples of reaction (containing 2 pmol ON) are removed at intervals to 10 μl of ice-cold 10 mM EDTA (pH 8.0). For anion exchange or reversed phase HPLC analysis (see earlier), 6-μl aliquots of the ON–EDTA solution (containing 1 pmol ON) are diluted to 100 μl with the appropriate weak buffer, heated to 95°, and immediately injected onto the column.

PAGE Analysis of ON. For PAGE analysis, 0.5-pmol (3 μl) aliquots of ON–EDTA are diluted by the addition of 7 μl 90% formamide, 0.03% xylene cyanol, 0.03% orange G, heated to 90°, quenched on ice, and loaded onto urea–PAGE as described earlier for short RNA transcripts, except that 15% polyacrylamide gels are used. Electrophoresis and semidry electroblotting conditions identical to those described previously for short RNA transcripts are used. Examination of the dried membrane under UV reveals the presence of fluorescent species. The signal may be "amplified" by immunologically detecting the fluorescein group on UV cross-linked membranes, as described earlier for digoxigenin-labeled RNAs.

Cautionary Remarks. It is strongly recommended that equal amounts of authentic stock ON are analyzed under identical conditions to the experimental points. When analysis proceeds by HPLC, one experimental sample of ON should be spiked with an equal amount of authentic stock and the resultant mixture analyzed to positively identify the observed ON peak, as retention time may be altered unpredictably by other biomolecules in the injected sample.

Results. In the main, the two analytical procedures give directly comparable results. End-protected phosphodiester ON, present in reactions containing 10% cell extract, are essentially fully degraded within 30 min. Chimeric ON with a central region of five phosphodiester linkages remain essentially fully intact over the same period in comparable reactions. Chimeric ON with larger central phosphodiester regions present intermediate stability.[31] Neither analytical procedure revealed significant levels of degradation intermediates with any of the structures described earlier. This indicates that the rate-limiting step of ON degradation in cell extracts is endonuclease cleavage, which is followed by rapid exonuclease degradation. That is, the bulk of the nuclease activity present in cell extracts is of the exonuclease type rather than the endonuclease class, consistent with what has been observed in other biological fluids.[14]

Phosphorothioate and chimeric methylphosphonate–phosphorothioate ON are not analyzed conveniently by the PAGE separation/immunological visualization method as very weak signals are obtained. We found that the intensity of UV-induced fluorescence on the membranes was similar to that observed with phosphodiester ON. Moreover, no evidence for preferential elution of the phosphorothioate containing ON from the membranes was obtained. These data were interpreted to indicate that the phosphorothioate backbone of the ON inhibits either the antibody-fluorescein label interaction or the activity of the antibody-conjugated alkaline phosphatase enzyme.

Antisense Oligodeoxynucleotide Activity in Nonidet P-40-Lysed Cells

RNase H activity can be observed in cell lysates,[12] generated by gentle removal of the cytoplasmic membrane using the nonionic detergent Nonidet P-40 (NP-40). The primary advantage of NP-40 lysis experiments is that the target RNA is the full-length endogenous mRNA. Such assays, therefore, provide a rapid and convenient mechanism to check that a given ON structure is antisense active when it gains access to its target mRNA.

Essentially similar experiments to those described earlier for cell extracts can be performed in NP-40 lysates: a variation of ON structure and ionic composition, followed by analysis of target RNA and input ON. For example, the addition of N-ethylmaleimide[49] to, and variation of monovalent cation concentration and divalent cation (Mg^{2+} and Mn^{2+}) in, NP-40 lysates of human chronic myeloid leukemia KYO1 cells, containing an appropriate antisense ON, produced data consistent with the majority of the RNase H being of the class I type.

NP-40 Lysis Assay

Using this procedure, 10^6 exponentially proliferating cells per assay are pelleted by centrifugation (800g, 4 min) and washed once by resuspension in an isotonic buffer, such as the intracellular buffer described earlier or HEPES-buffered saline (HBS, 10 mM HEPES, 137 mM NaCl, pH 7.4), and recentrifugation. One microliter of a 50 μM solution of ON is added to the packed cell pellet, which is then agitated gently to obtain a cell/ON "slurry." The cells are then resuspended in 50 μl of HBS containing 10% (v/v) NP-40 and 1.3 U/μl RNase Block I by gentle vortex mixing. The nonviscous cloudy suspension is incubated at 37° for 10 min. Samples for

ON and RNA analysis are taken, and RNA samples are processed as described earlier for cell extract assays, with the exception that an additional Northern hybridization step will be required to detect specific RNA species. ON samples should be diluted by the addition of 1 ml 20 mM EDTA (pH 8.0) filtered through Centricon 30 ultrafilters (Millipore, Watford, Hertfordshire) to remove high molecular weight biomolecules. The filters should be washed once with 1 ml water and the combined filtrates concentrated to dryness in a Speed-Vac concentrator. The ON-containing pellet should be resuspended in 10 μl water, diluted with HPLC weak buffer/ PAGE loading buffer, and analyzed by HPLC or PAGE, respectively, as described earlier.

Nonradioactive Northern Hybridization. Dried, UV cross-linked membranes are placed in a preheated hybridization tube, and the tube manufacturers' recommended volume of DEPC-treated hybridization buffer [50% formamide (v/v, deionized), 5× SSC, 5% blocking reagent (w/v, Boehringer), 0.1% sodium N-laurylsarcosine (w/v), 0.02% sodium dodecyl sulfate (SDS, w/v)] supplemented with 100 μg/ml yeast tRNA (Sigma) is added and incubation is continued for 2 hr. At the end of this period, 1 to 2 μl *in vitro*-transcribed nonradioactively labeled antisense probe per 10 ml of hybridization buffer is added and incubation is continued overnight.

The nonradioactively labeled antisense probe RNA may be synthesized in modified transcription reactions where 1 mM UTP is replaced with 0.35 mM digoxigenin[50] or fluorescein-UTP (using 0.65 mM UTP to maintain total UTP concentration at 1 mM). Transcripts with this high level of label may not be purified by extraction with phenol as the great majority will partition into the phenol rather than aqueous phase. The most convenient method we have found to preserve RNA for subsequent use as a probe is to add 0.1 volume of 10% SDS and 0.04 volume of 2-mercaptoethanol to the transcription reaction and incubate at 80° for 5 min. The reaction products should then be stored at −20° until required. Multiple rounds of thawing and refreezing do not adversely affect the performance of the probe.

Following hybridization, membranes are washed three times in at least 50 ml/100 cm^2 of membrane in high stringency buffer [0.1× SSC, 0.1% SDS (w/v)] at 65°. The nonradioactively labeled probe RNA is then detected immunologically as described earlier.

We have observed that the hybridization temperature is the most critical parameter for the efficient and specific detection of RNA species and that deficiencies in specificity during hybridization cannot be readily rectified by increasing the stringency of the wash conditions. Remarkably high hy-

[50] H.-J. Holtke and C. Kessler, *Nucleic Acids Res.* **18,** 5843 (1990).

bridization temperatures may be required. For example, a 1700 nucleotide human c-*myc* antisense RNA probe is used in the conditions described previously at 80°, and significantly lower hybridization temperatures result in mishybridization to the ribosomal (and preribosomal) RNAs.

Antisense Oligonucleotide Effects in Living Cells

Human leukemia cells are maintained in exponential growth as described previously. We have found that ON do not gain access to the cytoplasmic and nuclear compartments of human leukemia cells, either cultured or primary tissue, following simple addition to the culture medium. Accordingly, we have utilized two methods that do affect intracellular delivery and, therefore, permit subsequent investigation of ON antisense activity in living cells. It is important, however, that the cells entered into these procedures are proliferating exponentially and are as viable as possible. We only perform experiments on cultures assayed to >95% viability. Cell viability may be checked by the standard method of trypan blue exclusion or by incubation of the cells with propidium iodide (PI) followed by flow cytometry.

Viable Cell Count by Flow Cytometry. A viable cell count may be obtained by adding PI (final concentration 10 μg/ml) to 1 ml of culture. PI is a red fluorescent DNA stain that is excluded from healthy cells. The cells are incubated on ice for 5 min, pelleted by centrifugation ($\sim$800g, 4 min), resuspended in 1 ml ice-cold RPMl 1640, and passed through the flow cytometer using a modification of the total cell count protocol. The forward/side scatter selected population is used to generate a red fluorescent histogram. The red fluorescence detector gain is optimized, using log amplification, to obtain good separation between nonred (healthy, PI excluding) and red (dead, PI stained) subpopulations. Regions are set to select PI-stained and -unstained populations. This returns the percentage and the number of the unstained and PI-stained cells. Viable cells per milliliter in the original culture = PI unstained cell count $\times$ 100.

Streptolysin O Permeabilization

Antisense ON delivery by streptolysin O was first described by Barry *et al.*[51] and was subsequently adapted by us.[30,52] In addition to its utility for introducing ON into cells in culture, we have also found that SLO may be

[51] E. L. R. Barry, F. A. Gesek, and P. A. Friedman, *BioTechniques* **15,** 1016 (1993).
[52] R. V. Giles, J. Grzybowski, D. G. Spiller, and D. M. Tidd, *Nucleosides Nucleotides* **16,** 1155 (1997).

used to deliver ON into cells resulting from bone marrow and peripheral blood stem cell harvests from patients with chronic myeloid leukemia.[53]

Activation of SLO. SLO (Sigma) is resuspended at 1000 U/ml in phosphate-buffered saline containing 0.01% bovine serum albumen (Sigma), and is activated by the addition of freshly prepared dithiothreitol to 5 mM followed by incubation at 37° for 2 hr. The SLO is then tested routinely for permeabilization activity and, if found acceptable, is divided into small aliquots (500 to 1000 U) and stored at $-20°$ until required. Cycles of thawing and refreezing should be avoided or else reduced SLO activity will be observed.

Reversible Cell Permeabilization Using SLO. Wash 5×10^6 cells per permeabilization point twice by centrifugation at 800g and resuspension in serum-free RPMI 1640 medium. It is essential to remove the fetal calf serum for reproducible and efficient SLO reversible permeabilization. Pellet the cells and resuspend in 200 μl RPMI 1640 per point. Add 200 μl of cell suspension to a mixture of SLO and sufficient fluorescently labeled ON to achieve the desired final concentration and agitate the suspension gently. Incubate at 37° for 10 min agitating twice during this period. Reseal the cells by adding 1 ml of 90% RPMI 1640, 10% fetal calf serum followed by incubation at 37° for 20–30 min. Transfer the cell suspension to flasks containing a volume of warmed and gassed normal growth medium that permits exponential expansion of the culture. It is suggested that, in the first instance, a concentration of 10 to 20 μM ON is appropriate. The amount of SLO used in these experiments should be optimized for each cell line and batch of SLO. We usually find that between 5 and 20 U SLO per 10^6 cells provides optimal results.

ON delivery and culture viability should be assessed 30 min to 1 hr after the initiation of permeabilization by counterstaining $\sim 5 \times 10^5$ cells with PI and analyzing by two-color flow cytometry as described later. Typically, we observe >90% of the population reversibly permeabilized to ON and the remainder is split equally between dead and nonpermeabilized at optimal conditions. No overall reduction in cell number should be observed relative to control cells not permeabilized with SLO.

Electroporation

Human chronic myeloid leukemia KYO1 cells may be reversibly permeabilized to ON by electroporation using a method similar to that described by Bergan et al.[54] Minor modification of this procedure, particularly optimi-

---

[53] C. M. Broughton, D. G. Spiller, N. Pender, M. Komorovskaya, J. Grzybowski, R. V. Giles, D. M. Tidd, and R. E. Clark, *Leukemia* **11,** 1435 (1997).

[54] R. Bergan, Y. Connell, B. Fahmy, and L. Neckers, *Nucleic Acids Res.* **21,** 3567 (1993).

zation of permeabilization voltage, should allow a range of suspension cell lines to be similarly treated. Culture viability is assessed, and 5×10^6 cells per point are washed (once) as described earlier for SLO permeabilization. The cells are pelleted and resuspended in 800 μl ice-cold RPMI 1640 per point, fluorescent ON is added to the desired final concentration, and the culture of 5×10^6 cells is transferred to an ice-cold 4-mm gap electroporation cuvette (Bio-Rad). Cells are permeabilized by a single pulse from a Gene Pulser (Bio-Rad) attached to the optional capacitance extender set to 250 V and 960 μF. Following electroporation, the cells are left in the cuvette, which is placed on ice for 1 hr. At the end of this period the cells are transferred to flasks containing normal growth medium. Culture viability, ON delivery, and cell viability are assessed by two-color flow cytometry (later). Under optimal conditions, approximately 80% of the culture is reversibly permeabilized to ON and ~14% killed. A modest reduction in cell number, relative to unelectroporated controls, is observed.

Two-Color Flow Cytometry

When green fluorescent ON are introduced into cells that are subsequently stained with PI, two-color flow cytometry may be used to calculate the percentages of the population that are (1) living but did not take up the ON (neither green nor red), (2) living and took up the ON (green not red), and (3) dead (red only or red and green).

A modification of the flow cytometric protocols described earlier is generated. The forward/side scatter selected population is used to generate green fluorescence versus red fluorescence dot-plot output. Red fluorescence detector gain and amplification identical to that optimized for viable cell counting are used. Green fluorescence detector gain and amplification identical to that identified in "Calibration of the Flow Cytometer Mean Channel Green Fluorescence for Attomoles of Fluorescein" is used. When performing two-color flow cytometry, electronic compensation must be used to avoid the green channel signal spilling over into the red channel, and vice versa. This is achieved by running samples of cells containing similar amounts of the individual green and red fluorochromes as used in dual-labeling cell experiments. Compensation is then applied such that a population of cells containing different concentrations of an individual fluorochrome exhibit a spread parallel to the appropriate axis on a dual fluorescence dot plot. Alternate green- and red-only labeled cells are passed through the flow cytometer until conditions of electronic compensation are found such that each individually labeled population distributes parallel to the appropriate red or green axis. A final check for the validity of these settings is performed by passing a series of mixtures through the flow cytometer where the red-stained and green-stained cells are present in different proportions. Regions are set, using experimental samples of cells, that discriminate between PI-stained and -unstained cells, as described earlier, and green and nongreen cells. The green fluorescent/nonfluorescent boundary is set such that cells that have not received the green fluorescent compound are observed to be 1 to 2% positive for green fluorescence.

Calculation of Average Intracellular ON Concentration by Flow Cytometry

The average intracellular concentration of fluorescent ON in viable cells of the population, obtained following one of the previous delivery procedures, may be determined by flow cytometry.

Estimation of Average Cell Volume Using a Coulter Counter. A Model ZM Coulter Counter (Coulter Electronics Ltd., Luton, Bedfordshire) equipped with a 100-μm-diameter orifice tube aperture is calibrated using polystyrene beads of designated size using the half-count method.

A background count is taken to ensure that the electrolyte solution is free of particulates. Set the current to 2 mA, polarity to "+", lower and upper thresholds to 10 and 99.9 (out), respectively, attenuation to 1, and the preset gain to the machines' optimum, determined during installation. Ensure that the mean of five counts, each of 500 μl of electrolyte, is less than 400.

Dilute 2 drops of 10 μm (nominal diameter) polystyrene DVB latex bead suspension, supplied by Coulter, in 100 ml saline (9 g/liter) and place the orifice tube in the suspension. Use the stirrer to maintain the beads in suspension. Select corrected count (requires that the manometer volume and aperture diameter be entered) and turn the reset/count control to reset. Adjust the current to a value that results in most of the pulses on the oscilloscope screen being 15–20 mm high. Alter the lower threshold until the shadow line is roughly the same height as half of the pulses and note this value (T_0). Take a count with the threshold at $0.5 \times T_0$ (N_1). Take a count with the threshold at $1.5 \times T_0$ (N_2). Calculate the average of N_1 and N_2 (N). Take counts and adjust the lower threshold until the obtained count = N; note this lower threshold value (T_E). If T_E differs from T_0 by more than two threshold units, repeat the counting procedure, replacing T_0 with T_E. In this way one may rapidly identify the approximate T_0. When T_E differs from T_0 by less than two threshold units, repeat the procedure, but taking four counts at $0.5 \times T_0$ and $1.5 \times T_0$ and find the averages (A_1 and A_2, respectively). Calculate the average of A_1 and A_2 (A). Find the threshold value (T_F) where the average of four counts = A. Enter the singlet number mean diameter (on the beads' assay sheet) into the "DIA (μm)" register.

Set the Coulter Counter to "Mode VOL. T_L." Pass a suitable dilution of cells, known to be greater than 95% viable, through the Counter with the reset/count control in the reset position. Adjust current and attenuation so that the great majority of the pulses remains within the scale of the oscilloscope and adjust the lower threshold to exclude debris. Find the average of four counts at this setting (C). Take further counts, adjusting the lower threshold until the number of cells counted = $0.5 \times C$. The volume that this lower threshold value corresponds to may be read directly from the "data" digital display and is the median volume of the cells in that culture.

Calibration of Flow Cytometer Mean Channel Green Fluorescence for Attomoles of Fluorescein. As observed previously, ON bind irreversibly to APS-Hypersill, 5-μm Hypersphere HPLC support[24] (Shandon Southern Products Ltd., Runcorn, Merseyside). We exploit this feature to bind known amounts of fluorescently labeled phosphodiester ON to known numbers of hyperspheres. Spheres are wetted with isopropanol and washed with anion-exchange strong buffer and are then resuspended in saline (9 g/liter) at ca. 10^8 spheres/ml. The number of particles per milliliter in this suspension is obtained using a Coulter Counter. Aliquots of 10^8 particles are placed into 2-ml Eppendorf tubes, pelleted by centrifugation (800g, 4 min), and the supernatant is discarded. Care should be exercised in removing the supernatant as the pellet is disturbed easily. Indeed, removal of all of the supernatant is not advisable as significant numbers of hyperspheres are lost when the meniscus passes over the pellet. The spheres are resuspended by the dropwise addition of 1 ml of ON solution in phosphate-buffered saline with continuous vortex mixing. A range of concentrations of fluorescent phosphodiester ON, from 0.01 to 2.5 μM, are used to obtain a series of hypersphere populations where between 0.1 and 25 amol of ON are bound per sphere, on average. Confirmation of complete ON binding to the hyperspheres may be obtained by pelleting the cells and removing an aliquot of the supernatant for analysis by HPLC or PAGE. It is advisable to check a dilution of each fluorescent hypersphere suspension on the Coulter Counter and adjust the nominal amount of ON/sphere in the light of the obtained counts.

The flow cytometer is first checked for alignment (if required) using the common reference of gluteraldehyde-fixed chick red cells (Sigma), suspended in phosphate-buffered saline. Linear amplification of the green channel signal is set to $\times 2$ and detector gain is manipulated until the modal signal from chick red cells falls into channel 50 (eight-bit data collection). Under these conditions, the mean green fluorescent signal from hyperspheres with 1 amol ON/sphere falls into channel 100. Log green channel amplification is then selected and the populations of hyperspheres are passed through the flow cytometer. The mean channel green fluorescence for samples with each amount of ON/sphere is noted and used to construct the standard curve. The standard curve need not be offset to account for differences in autofluorescence between hyperspheres and human leukemia cells, as we have found that both produce similar autofluorescence readings under these flow cytometry conditions. We have found that the Ortho Cytoron Absolute flow cytometer remains in calibration for periods in excess of 24 months.

Calculation of Average Intracellular ON Concentration. Deliver ON into the cells using one of the methods described and counterstain with PI. Pass PI-treated cells through the flow cytometer, select the PI excluding population (see earlier), and generate a green fluorescence histogram. It is essential to use the same green fluorescence detector gain and amplification as used to obtain the standard curve. Identify the mean channel green fluorescence of this population. Relate the obtained mean channel green fluorescence to attomoles of fluorescein using the standard curve generated earlier and divide this value by the average volume of a cell as estimated by Coulter counting. Under optimal conditions of both SLO permeabilization and electroporation, the intracellular concentration of ON does not exceed the extracellular concentration (typically 10 to 30%) and SLO permeabilization delivers ca. 10-fold more ON than electroporation.[55]

Fluorescence Microscopy

The intracellular localization of fluorescent ON observed following one of the delivery methods described previously may be confirmed by fluorescent microscopy performed on the residue of PI counterstained cells that were not used in flow cytometry. The cells are pelleted by centrifugation (800*g*, 4 min) and most of the supernatant is removed and resuspended in the remaining fluid by gentle agitation to form a cell slurry. Five to 10 μl of the slurry is transferred to a clean slide and overlaid with a coverslip. It is important that the cells are not fixed prior to microscopy as we have observed that this alters the subcellular distribution of ON.

Fluorescent signals obtained following delivery of ON from 1 μM extracellular concentration, by either electroporation or SLO reversible permeabilization, are of sufficient intensity to be readily visible by fluorescence microscopy. Specifically, we examine cells at ×400 or ×630, under oil

[55] D. G. Spiller, R. V. Giles, J. Grzybowski, D. M. Tidd, and R. E. Clark, *Blood* **91,** 4738 (1998).

immersion, using 40/0.70 PL Fluotar and 63/1.30 Fluoreszenz Phaco3 objectives, respectively, on a Leitz Laborlux S fluorescence microscope (Leica, Milton Keynes, Buckinghamshire) using the filters supplied by the manufacturer for fluorescein. Images are captured using the automatic exposure settings on a Wild MPS 48 Photoautomat camera, using Kodachrome Elite or Fujichrome ASA 400 color slide film. Exposure times for cells loaded with ON using SLO are typically in the region of 10 sec, which allows simultaneous illumination with white light for phase contrast if the lamp intensity is reduced to, or near, the minimum.

Predominantly nuclear localization is observed for all ON structures that we have examined in this way. Minor differences in subcompartment distribution are observed between ON with differing backbone structures. Chimeric methylphosphonodiester/phosphodiester ON demonstrate even nuclear distribution, whereas phosphorothiodiester ON display punctate nuclear localization.

In contrast, cells exposed to less than 50 μM of fluorescently labeled ON for 4 hr, without manipulation to effect intracellular delivery, may be extremely difficult to visualize under the same conditions. Typical exposure times were found to be in excess of 150 sec, even in the presence of fluorescence-bleaching inhibitors. Images may be captured from cells incubated in the presence of higher concentrations of ON and demonstrate classic "punctate" endosomal ON localization.[30,32,55] If such cells are fixed with 70% ethanol, ON is observed to redistribute rapidly from the endosomes to the nucleus.

Antisense Effects in Living Cells

ON may be delivered into living human leukemia cells using SLO reversible permeabilization or electroporation as described earlier. Cells belonging to other lineages may be loaded with ON using cationic lipid delivery systems[56] such as Lipofectin. We have observed that methylphosphonodiester-containing ON are not delivered efficiently using Lipofectin because the reduced net negative charge on the ON limits interaction with the cationic lipid.

Samples of the culture are taken routinely at intervals for two-color flow cytometry, fluorescence or confocal microscopy, and Northern analysis of target and control mRNA expression, as described earlier, and Western analysis of protein expression (later).

Extraction of Cellular Protein for Analysis by SDS–PAGE. Samples of cells (1×10^6) should be pelleted by centrifugation (800g, 4 min) and

[56] C. F. Bennett, M.-Y. Chiang, H. Chan, J. E. E. Shoemaker, and C. K. Mirabelli, *Mol. Pharmacol.* **41**, 1023 (1992).

washed once by resuspension in RPMI 1640 and recentrifugation to remove excess fetal calf serum. A cell "slurry" should be made by gentle agitation of the packed cell pellet. Two hundred microliters of prewarmed lysis buffer (1% SDS, 10% glycerol, 10% 2-mercaptoethanol, 40 mM Tris–HCl, pH 6.8, 0.001% bromphenol blue) should be added directly to the slurry. The suspension should be vortexed vigorously and incubated at 100° for 10 min. The contents of the tube should be collected by brief centrifugation and mixed by vortexing. The protein samples may be stored at −20° until required.

SDS–PAGE and Western Blotting. Cellular proteins, solubilized in SDS/ 2-mecaptoethanol as described earlier, may be fractionated on the basis of size by electrophoresis through polyacrylamide gels containing SDS. Proteins contained within the gel may be transferred onto nitrocellulose membranes and specific proteins may be detected immunologically. The following protocol has been found to be very robust.

GEL PREPARATION. We routinely use the Bio-Rad Protean II vertical electrophoresis system with a 10% separation gel and a 5% stacking gel. Vacuum degas the main gel solution [10% polyacrylamide (w/v, 37.5:1 acrylamide:bisacrylamide), 0.5 M Tris–HCl, pH 8.8], add 10 μl 10% SDS, 10 μl 10% (w/v) ammonium persulfate, and 1 μl N,N,N',N',-tetramethylethylenediamine (TEMED) per milliliter of gel solution, and cast the gel. Immediately overlay the gel solution with tertiary amyl alcohol (water saturated) and allow the gel to polymerize for 30 min to 1 hr. The tertiary amyl alcohol should be added carefully so as to cause minimum disturbance to the gel surface. Discard the tertiary amyl alcohol, wash the gel surface with water, discard, and remove excess with a filter paper. Vacuum degas the stacking gel solution [5% polyacrylamide (w/v, 37.5:1 acrylamide:bisacrylamide), 70 mM Tris–HCl, pH 6.8], add 10 μl 10% SDS, 10 μl 10% ammonium persulfate, and 1 μl TEMED/ml of gel solution, and cast the gel. Position the comb and allow the gel to polymerize for 30 min to 1 hr. Aim to have ca. 1 to 1.5 cm of stacking gel between the bottom of the wells and the top of the main gel.

GEL RUNNING. Remove comb and flush the wells with water to remove unpolymerized acrylamide and drain briefly. Construct the gel electrophoresis apparatus, enable the cooling system, and fill the reservoirs with running buffer (3 g/liter Tris base, 14.4 g/liter glycine, 1 g/liter SDS). Flush wells with running buffer and load a 30- to 60-μl sample. Apply a potential difference of 10 V/cm of gel and run until the bromphenol blue dye band nears the lower end of the gel.

WESTERN BLOTTING. We have found that wet electroblotting, using Bio-Rad TransBlot (with plate electrodes) equipment, produces more reliable and reproducible transfers than semidry electroblotting for this application.

Remove a glass plate from the gel "sandwich" and soak the gel, still attached to the other plate, in blotting buffer [20% methanol (v/v), 3 g/liter Tris base, 14.4 g/liter glycine] for 30 min. Cut one sheet of nitrocellulose membrane (Optitran, Schleicher & Schuell) and two sheets of Whatman 3M paper (or equivalent) to the same size as the gel. Soak the membrane, chromotography paper, and four TransBlot sponge pads in blotting buffer. Construct the transfer cassette by positioning, in order, two sponge pads, one sheet of 3M paper, the gel, the nitrocellulose membrane, one sheet of 3M paper, and two sponge pads. Remove air bubbles gently after the addition of each layer. Position the membrane side of the cassette toward the positive electrode in a buffer-filled TransBlot cell. Blot for 3 hr, with cooling, using upper limits of 100 V and 0.8 A.

Specific proteins may be detected using a nonradioactive immunological detection procedure analogous to that described earlier for Northern blots. Recover the membrane from the transfer cassette and wash briefly in blot wash buffer. Incubate the membrane for 1 hr in blocking buffer. Incubate the membrane in primary antibody solution for 1 hr (or more). Commercial antibodies should be diluted by the manufacturer's recommended amount in blocking buffer. The supernatant from hybridoma cultures may be used without further manipulation. Wash the blots three times for 5 min each in wash buffer. Apply the secondary antibody diluted appropriately in blocking buffer (e.g., alkaline phosphatase-conjugated sheep antimouse, Sigma, 1:2500) for 1 hr. Wash membrane for 5 min in wash buffer, transfer to a clean vessel, and wash twice for 15 min each in wash buffer. Equilibrate the membrane in diethanolamine/MgCl$_2$ buffer for 5 min and apply chromogen. When the desired bands are the required intensities, wash the blot extensively in water, dry, and store in the dark.

Estimation of Intracellular ON Half-Life

In addition to removing samples of culture for analysis of mRNA and protein expression, samples may also be taken to analyze the relative concentration of intact intracellular ON, by PAGE or HPLC, as described earlier.

Extraction of Oligonucleotide from Cells. Samples containing 10^6 cells are taken at intervals, following ON delivery, and pelleted by centrifugation (800g, 4 min). The cells are washed five times in ice-cold phosphate-buffered saline by centrifugation and resuspension, and the resulting cell pellet is snap frozen in liquid nitrogen and stored at −20°. ON are extracted by resuspension and lysis of the cell pellets with 1 ml 0.5% Nonidet P-40 (NP-40) in 20 mM EDTA (pH 8.0), followed by incubation at 95° for 5 min. The lysates are filtered through Centricon 30 ultrafilters, filters are washed, and combined filtrates are concentrated as described for ON samples from NP-40 lysate experiments. Lyophilized material is resuspended in the appropriate HPLC weak buffer, or PAGE loading buffer, and analyzed by HPLC or PAGE, respectively, as described previously.

Phosphodiester ON are degraded essentially to completion within 4 hr following intracellular delivery by SLO. Chimeric molecules with increasing methylphosphonodiester substitution demonstrate increasing stability, such that the intracellular levels of intact effectors with very small phosphodiester sections appear to reduce only in proportion to proliferation of the cell population.

Data obtained using these direct methods have been compared with the apparent intracellular concentration obtained by flow cytometry. In many cases the latter reflects the actual intracellular concentration of intact ON accurately. Consequently, due to the relative simplicity and convenience of the procedures, we do not routinely assess intracellular ON concentration by extraction and PAGE/HPLC analysis but use the indirect measure provided by flow cytometry as an *estimate*. However, in the case of chimeric methylphosphonodiester/phosphodiester ON with methylphosphonodiester sections in excess of four linkages adjacent to the fluorescein label and with central phosphodiester regions in excess of three or four internucleoside bonds, the flow cytometrically determined value does not reflect the actual intracellular concentration of ON accurately. In this case, the phosphodiester region is sufficiently large to permit endonuclease cleavage, and the fluorescein-labeled methylphosphonodiester degradation product is not removed from cells rapidly, whereas smaller fluorescein-methylphosphonodiester fragments are.

Results

We have tested a variety of antisense ON analog structures in this system, including 3′-end protected phosphodiester, phosphorothioate, C5-propyne modified phosphodiester, and phosphorothioate analog, chimeric methylphosphonodiester/phosphodiester, chimeric methylphosphonodiester/phosphorothioate, chimeric methylphosphonodiester/phosphodiester/phosphorothioate, chimeric 2′ modified (O-methyl, O-allyl and methoxyethoxy) phosphodiester/2′-deoxyphosphodiester, chimeric 2′ modified (methoxyethoxy and methoxytriethoxy) phosphodiester, and phosphorothioate/2′-deoxyphosphorothioate. In every case, methylphosphonodiester/phosphodiester chimeric ON provided the most efficient inhibition of target mRNA expression when assayed 4 hr postdelivery. Identical rank potencies are obtained following both SLO-mediated and electroporation delivery with the structures that have been examined. Where suitable series of compounds are available, reduced antisense activity is observed to correlate with an increased phosphorothioate content.[42] In addition, no increase in antisense activity is observed to correlate with the increasing hybridization potential of the ON.

However, it may sometimes be necessary to compromise short-term antisense efficacy and use ON structures with greater biological stability such that a significant downregulation of proteins with long half-lives may be achieved.[55] For example, the mutant p53 protein in KYO1 cells is observed to possess a half-life of about 8 hr. Twenty-four hours after delivery of the chimeric methylphosphonodiester/phosphodiester chimeric antisense ON p53 protein, expression was reduced to ~60% of control levels and expression had returned to control levels another 24 hr later. In contrast, a chimeric methylphosphonodiester/phosphorothiodiester effector targeted to the same region of p53 mRNA reduced protein expression to ~30% of control levels both 24 and 48 hr after delivery.

Selection of Control ON

Experiments of such a nature require carefully selected controls to demonstrate that the observed biological consequences of ON delivery are, in fact, antisense in nature rather than sequence-dependent nonantisense. It has become widely accepted that, as a minimum, biological response should be preceded by a detectable downregulation of the target protein expression and, for ON structures capable of directing the action of RNase H, target mRNA expression. The active antisense effector should not inhibit the expression of nontarget genes. Further, at least two appropriate control ON of similar size and structure to the active compound should be demonstrated to be inactive at inhibiting target expression and inducing the biological response.

We suggest that inverse antisense (reverse polarity antisense) and scrambled antisense sequences provide the best controls for potential aptameric, sequence-specific and base composition nonantisense effects. In the case of chimeric ON, care should be taken to conserve the base composition of the methylphosphonodiester and phosphodiester (or phosphorothiodiester) sections of the antisense molecule in the control structures. Arguably, more convincing controls for an antisense mechanism of action than those described earlier are provided by antisense ON that efficiently downregulated the expression of other gene products. For example, we have shown that c-*myc* expression is inhibited following the delivery of *myc* antisense but not p53 or *bcr-abl* antisense effectors, that p53 expression is inhibited by p53 but not *myc* or *bcr-abl* antisense compounds, and that *bcr-abl* mRNA was depleted by *bcr-abl* but not *myc* or p53 antisense ON.[12]

In any event, caution is required in interpreting the molecular effects of antisense and control ON. Cleavage of nontarget RNA species, first described in *in vitro* systems,[9] is also observed following ON delivery into living cells.[42,55]

Mechanism of Oligodeoxynucleotide Action in Living Cells

Oligodeoxynucleotides capable of directing RNase H activities are observed to reduce the expression of target mRNA following the delivery into living cells. In some instances, short mRNA species are observed on Northern blots, which are the correct size to be fragments of the parent mRNA produced through RNase H cleavage at the site of ON hybridization.[12,13] Such data, however, fall short of providing definitive proof of RNase H mediating antisense effects in intact cells. Consequently, a protocol was developed, described as reverse ligation-mediated RT-PCR (RL-PCR), which amplifies and positively identifies the 3′ RNA products of RNase H cleavage reactions.[12,13] The procedure was originally used to identify *in vivo* protein–RNA interactions and ribozyme function.[57,58] An overview of the protocol for RL-PCR amplification of RNase H cleavage fragments is presented in Fig. 1. It may be seen that the 3′ RNase H cleavage product is detected. Human RNase H has been reported to produce 3′ RNA fragments with a 5′ phosphate group.[59] In agreement with this, we have found that polynucleotide kinase treatment of extracted RNA prior to linker ligation, described by Bertrand *et al.* in the original protocols, is not required.

RL-PCR Detection of RNase H-Generated mRNA Fragments

Synthesis of RNA Linker. A short 25-mer RNA linker is synthesized using the high yield T7 RNA polymerase transcription conditions recommended by Milligan *et al.*[60] Hybridize 10 nmol of each template ON (5′ . . . TTT CAG CGA GGG TCA GCC TAT GCC CTA TAG TGA GTC GTA TTA . . . 3′ and 5′. . . TAA TAC GAC TCA CTA TAG . . . 3′) by incubation in 250 μl of 30 mM NaCl, 2 mM Tris–HCl, 0.2 mM EDTA, pH 8.0 (0.2× TEN) at 95° for 10 min, allowing to cool slowly to room temperature. Analyze a small aliquot by PhastGel "Homogenous-20"

[57] E. Bertrand, M. Fromont-Racine, R. Pictet, and T. Grange, *Proc. Natl. Acad. Sci. U.S.A.* **90,** 3496 (1993).
[58] E. Bertrand, R. Pictet, and T. Grange, *Nucleic Acids Res.* **22,** 293 (1994).
[59] P. S. Eder, R. Y. Walder, and J. A. Walder, *Biochimie* **75,** 123 (1993).
[60] J. F. Milligan, D. R. Groebe, G. W. Witherall, and O. C. Uhlenbeck, *Nucleic Acids Res.* **15,** 8783 (1987).

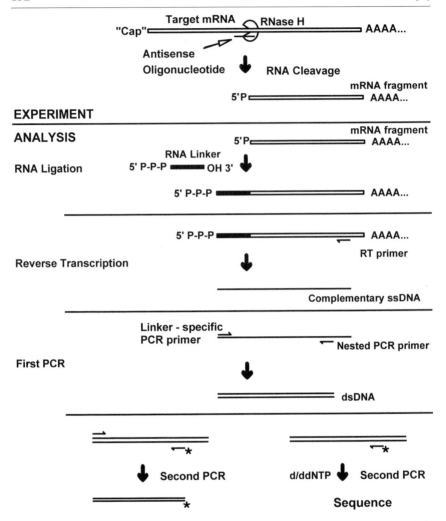

Fig. 1. Schematic representation of RNase H cleavage of target mRNA and the analysis of 3' mRNA cleavage fragments by reverse ligation-mediated RT-PCR. The asterisk in the final analytical stage represents a digoxigenin or fluorescein label.

electrophoresis to ensure hybridization and visualize by ethidium bromide staining and UV illumination. Dilute hybrid to 5 μM by adding 1750 μl 0.2× TEN. Transcribe the template by placing in a reaction tube in the following order: 153 μl water, 50 μl 0.1 M dithiothreitol, 7 μl 1 M MgCl$_2$, 100 μl 5× transcription buffer (0.2 M Tris–HCl, pH 7.5, 50 mM NaCl, 10 mM spermidine, 30 mM MgCl$_2$), 20 μl 100 mM ATP, 20 μl 100 mM CTP,

20 μl 100 mM GTP, 20 μl 100 mM UTP, 20 μl 5 μM template hybrid, 20 μl RNase Block I (40 U/μl), and 50 μl T7 RNA polymerase (200 U/μl) followed by incubation at 37° for 6 hr. Fractionate reaction template and products by electrophoresis through a 20 cm $\times$ 0.75 mm 20% nondenaturing polyacrylamide gel (19 : 1 acrylamide : bisacrylamide, 1$\times$ TBE) using 1$\times$ TBE running buffer. Identify the intense transcript band by UV shadowing over a fluorescent thin-layer chromatography plate. Excise the band, mash the gel slice, and extract the linker RNA transcript from the gel by constant agitation in ~2 volumes of the buffer used to lyse cells for RNA, overnight. Centrifuge (10,000–15,000g, 10 min) to pellet the mashed gel. Collect the supernatant and reextract the mashed gel slice pellet with the same volume of lysis buffer as just described, but for 2 hr. Pool the collected supernatant and precipitate RNA by adding 1 volume of isopropanol followed by incubation at $-20°$ overnight. Pellet the precipitated RNA linker by centrifugation at 10,000–15,000g for 30 min. Reextract the RNA as if purifying total RNA from cells. Resuspend the resultant RNA pellet in 100 μl water and divide into 10 aliquots of 10 μl. Store at $-20°$ until required.

RNA Ligation Reaction. ON are introduced into cells and total RNA extracted as described earlier. Resuspend the cellular total RNA pellet in distilled water at 250 ng/μl. Ligate the linker RNA onto all available 5' phosphate groups in 1 μg of the total RNA using T4 RNA ligase. Remove 4 μl of the total RNA to a fresh tube, add 1 μl gel-purified 25-mer RNA (~500 ng), 1 μl 10$\times$ ligase buffer, 1 μl RNase Block I (40 U), 3 U T4 RNA ligase (Boehringer, ~0.5 μl), and 0.5 MBU DNase I (~0.5 μl), and adjust volume to 10 μl/reaction with water. Incubate overnight at 17° followed by 30 min at 37° (to ensure complete DNase I digestion). Stop the ligation reaction by adding 5 μl 10 mM EDTA, pH 8.0, and 200 μl of modified lysis buffer (RNA extraction buffer : phenol, pH 4.3, 1 : 1, v/v), that contains 5 μg linear polyacrylamide (Aldrich, coprecipitant). Vortex briefly. Induce phase separation by the addition of 25 μl chloroform. Recover the upper aqueous phase to a thermal cycler tube and precipitate nucleic acids with 1 volume (ca. 130 μl) isopropanol and overnight incubation at $-20°$. Pellet precipitates at 10,000–15,000g for 30 min. Remove and discard supernatant. Perform a brief second spin, to collect fluid adhering to the vessel walls, and discard the 2–4 μl of supernatant so collected.

Reverse Transcription. Resuspend the precipitates in 4 μl water. Store on ice until ready to proceed. Make a mix with the following components: 0.9 μl water, 0.7 μl *Pfu* polymerase buffer 1 (Stratagene), 0.7 μl 0.1 M dithiothreitol, 0.2 μl 10 μM gene-specific RT primer, and 0.5 μl RNase Block I and place at 37°. Heat RNA solutions to 95° for 5 min to destroy RNA secondary structure and then place immediately at 37°. Incubate for 2–3 min. Hybridize the RT primer and the RNA by adding the mix to the

RNA solution followed by incubation at 37° for 45 min. Make a second mix with the following components: 1.4 μl water, 0.3 μl *Pfu* polymerase buffer 1, 0.3 μl 0.1 *M* dithiothreitol, 0.5 μl mix containing 10 m*M* dNTPs (each), and 0.5 μl Superscript reverse transcriptase (~100 U, GIBCO) and add to the RNA/primer hybridization reaction. Incubate at 37° for another 45 min. Stop the reaction by incubation at 95° for 10 min and place on ice. Add 1 μl RNase A (1 ng/μl, previously boiled to destroy any contaminating DNases) and incubate at 37° for 20 min. Place reactions on ice.

First-Round PCR Amplification. Amplify the reverse transcribed products using *Pfu* polymerase (Stratgene). Add to the RNase A-treated reverse transcription reactions: 4.8 μl water, 1 μl *Pfu* polymerase buffer 1, 2 μl 10 μ*M* linker-specific primer (5′ . . . GGG CAT AGG CTG ACC CTC GCT GAA A. . . 3′), 2 μl 10 μ*M* nested (position 5′ to the RT primer) gene-specific primer, and 0.2 μl native *Pfu* polymerase (~0.5 U). Overlay with 50 μl light mineral oil and cycle through 20 rounds of 95° 1 min, 55° to 65° (depending on the gene-specific primer) 1 min, and 72° 1 min.

Second-Round PCR Amplification. At this stage the first-round PCR products may be subamplified to provide highly sensitive detection of the RNase H cleavage products. To a fresh tube add 9.5 μl water, 1.5 μl *Pfu* polymerase buffer 1, 1.5 μl 10 μ*M* linker-specific primer, 2 μl 10 μ*M* 5′ nonradioactively labeled nested (position 5′ to the first-round PCR primer) primer, 0.3 μl mix containing 10 m*M* dNTPs (each), and 0.2 μl *Pfu* polymerase. To this reaction mixture add 5 μl of the first-round PCR, overlay with 50 μl light mineral oil, and perform 5 to 10 rounds of PCR using the conditions described earlier. Remove 2 μl of the second-round PCR to 8 μl sequencing gel-loading buffer (95% formamide, 10 m*M* NaOH, 0.05% bromphenol blue, 0.05% xylene cyanol), incubate at 95° for 10 min, and quench on ice. Fractionate through 7 *M* urea, 1× TBE, 6 to 8% polyacrylamide (19:1 acrylamide:bisacrylamide), semidry electroblot the nucleic acids contained in the gel onto nylon membrane, and visualize immunologically as described for RNA products of cell extract experiments.

Nonradioactive Sequencing. Alternatively, the first-round PCR products may be sequenced to positively identify the position at which the gene-specific sequence changes to RNA linker, which should occur within the region to which the antisense ON was targeted. An adaptation of the fmol (Promega, Southampton, Hampshire) cycle sequencing protocol is used. Into the four base-specific chain termination reactions tubes add 2 μl of dNTP/dideoxy-NTP mix. All d/ddNTP mixes contain 20 μ*M* ddATP, dCTP, dTTP, and 7-deaza-dGTP. In addition the A mix contains 350 μ*M* ddATP, the C mix 200 μ*M* ddCTP, the G mix 30 μ*M* ddGTP, and the T mix 600 μ*M* ddTTP. In a separate tube, place 5 μl of 5× sequencing buffer (250 mM Tris–HCl, 10 mM MgCl$_2$), 3 pmol of 5' nonradioactively labeled nested gene-specific primer (the same as that used in the second-round PCR), and 1 μl sequencing grade *Taq* (5 U/μl, Promega), adjust volume to 17 μl, and transfer 4 μl to each of the chain termination reaction tubes. Overlay with 25 μl light mineral oil and perform 30 cycles of PCR as described for first- and second-round PCRs. Reactions are stopped by the addition of 6 μl sequencing gel-loading buffer and heating to 95° for 10 min.

Sequencing reaction products are fractionated through a 0.4-mm-thick 7 M urea 6% polyacrylamide (w/v, 19:1 acrylamide:bisacrylamide) 1× TBE sequencing gel under standard conditions.[40] When the fragments have migrated the required distance, nucleic acids are contact blotted from the polyacrylamide gel onto a 0.2-μm pore Nytran membrane. Briefly, the gel is exposed by removing one of the plates, and a small volume of 1× TBE is pipetted onto the region of gel to be blotted. A membrane, wetted in 1× TBE, is placed over this region and air bubbles and exces buffer are removed gently. A stack of 5 to 10 sheets of absorbent chromatography paper (Whatman 3M or equivalent) is placed over the membrane and weighed down heavily. Transfer for at least 2 hr but preferably overnight. The membrane is then removed from the gel. Any fragments of polyacrylamide adhering to the membrane should be removed before the membrane is dried by briefly placing the membrane nucleic acid side down on chromatography paper. The membrane should then be dried and UV cross-linked. The sequence may be developed immunologically as described earlier.

Acknowledgments

Ongoing research described in this article is supported by the Leukaemia Research Fund of Great Britain. RVG is supported by a grant from The Liposome Company Inc., Princeton, New Jersey. We thank all members of this laboratory, past and present, for helping to bring these protocols to their current level of refinement.

[6] Evaluation of Antisense Mechanisms of Action

By Chandramallika Ghosh, David Stein,
Dwight Weller, and Patrick Iversen

Antisense inhibition of gene expression may provide for highly effective therapeutic applications as well as powerful tools in the exploration of gene function. This promise is currently facilitated by an ever-widening spectrum of oligonucleotide chemistries. However, key mechanistic questions need

$40S_{rib}$: eIF-2 : GTP : Met-tRNA
  $+ \ 5'$-UTR-mRNA $\rightleftharpoons [40S_{rib} : eIF\text{-}2 : GTP : Met\text{-}tRNA : 5'\text{-}UTR\text{-}mRNA]_{scanning} \rightarrow$
     $40S_{rib}$: eIF-2 : GTP : Met-tRNA : AUG-mRNA $+ \ 60S_{rib} \rightarrow$ translation begins[5] $\quad$ (1)
AS-ODN $+ \ 5'$-UTR-mRNA $\dashv$ compete with complex binding and scanning
  = noncompetitive inhibition of translation $\qquad\qquad\qquad\qquad\qquad\qquad$ (2)

SCHEME I. Noncompetitive inhibition of initiation of translation at $5'$-UTR.

to be addressed for each chemistry, for each target gene, and for each oligonucleotide sequence. Further, prevailing questions regarding appropriate negative control oligonucleotides can best be addressed when a definitive mechanism of action is known. This article has been prepared with the idea that evaluation of the inhibition of gene expression can be defined in quantitative terms consistent with inhibitors of enzymatic reactions and/or receptor antagonists. Hence, the V_{max}, K_m (alternatively B_{max}, K_d), and K_i terms may be defined and the comparison of oligonucleotide chemistries as well as comparisons with alternative methods of inhibition may be described. The site of oligonucleotide action within the cell may define different mechanisms such as interaction with transcription factors, formation of triple helix with DNA, interference with pre-mRNA splicing, or invasion of a DNA duplex. These mechanisms do not involve hybridization with mature mRNA and are not considered further in this article. This mechanistic and quantitative evaluation of oligonucleotide : RNA duplexes should allow comparisons from *in vitro* to cell culture and to *in vivo* applications. The potential for contributions by nonspecific or aptameric effects such as those defined for phosphorothioate oligonucleotides, including oligonucleotide interactions with extracellular and cellular proteins[1,2] and short heteroduplex recognition of RNase H,[3] can be observed. Finally, the quantitative agreement or potential disagreement with the rules of base pairing, including base mismatches described by Aboul-ela *et al.*,[4] can be evaluated experimentally.

Hypothesis for Antisense Mechanism of Action

The hypothesis for the mechanism of action of RNA oligonucleotide duplexes is shown in Scheme I. This hypothesis queries the oligonucleotide

[1] R. L. Shoeman, R. Hartig, Y. Huang, S. Grub, and P. Traub, *Antisense Nucleic Acid Drug Dev.* **7,** 291 (1997).
[2] L. Yakubov, Z. Khaled, L. M. Zhang, A. Truneh, V. Vlassov, and C. Stein, *J. Biol. Chem.* **268,** 18818 (1993).
[3] R. Crouch and M. Dirksen, *in* "Nucleases" (S. Linn and R. Roberts, eds.), p. 211. Cold Spring Harbor Laboratory Press, Cold Spring Harbor, NY, 1982.
[4] F. Aboul-ela, I. Tinoco, and F. H. Martin, *Nucleic Acids Res.* **13,** 4811 (1985).
[5] D. Voet and M. Voet, *in* "Biochemistry," 2nd Ed., p. 994. Wiley, New York, 1995.

$$40S_{rib} : eIF\text{-}2 : GTP : Met\text{-}tRNA : AUG\text{-}mRNA$$
$$+\ 60S_{rib} \rightarrow \text{translation begins} \tag{3}$$
$$AS\text{-}ODN + AUG\text{-}mRNA \dashv \text{compete with complex scanning}$$
$$+\ 60S_{rib} \text{ binding} = \text{inhibition} \tag{4}$$

SCHEME II. Mixed inhibition of initiation of translation at AUG.

as a noncompetitive inhibitor of translation involving the initiation complex recognition of the 5′-untranslated region (5′UTR) of the mRNA. This complex is composed of the 40S ribosomal subunit, initiation factor 2 (eIF-2), GTP, and the met-tRNA, which recognizes the mRNA at the 5′ cap and scans as an intermediate complex to the AUG sequence, which results in recruitment of the 60S ribosomal subunit so that translation may begin. The purpose of Eq. (2) in Scheme I is to indicate the oligonucleotide binding in the 5′-UTR. This interferes with the initiation complex recognition and downstream scanning in a noncompetitive manner, thus preventing the complex from identifying the AUG sequence in the mRNA. Equation (3) in Scheme II describes the unique phenomenon at the AUG start site recruitment of the 60S ribosomal subunit. Equation (4) indicates novel mechanistic properties of an oligonucleotide that binds at or near the AUG sequence, resulting in a biphasic mixed inhibition. Initially, it involves the same interference of the initiation complex recognition of the 5′-UTR and second interference with the recruitment of the 60S ribosomal subunit at the AUG sequence.

The initial experiments designed to test these hypotheses utilize an *in vitro* translation method because of its simplicity by not including the complexity of all cellular transcripts, mRNA-binding proteins, or RNA : RNA duplex interactions. The two most common methods for this assay are rabbit reticulocyte lysate and the wheat germ lysate.

These proposed mechanisms contrast Eq. (2) with antisense as a noncompetitive inhibitor of the initiation factor complex with the 5′-UTR and Eq. (4) with the antisense as a mixed inhibitor of both the initiation factor complex and the recruitment of the 60S ribosomal subunit assembly.

Vector Design Components for Evaluation of Antisense Action

Sample Promoters

Bacterial	T7
Eukaryotic Inducible	MMTV
Eukaryotic Constitutive	CMV

Sample Target α-Globin Fusion to Luciferase Reporter Gene:

5'-GACACUUCUGGTCCAGTCCGACTGAGAAGGAACCACC**ATG**GTGCTGTC-3'
3'-GUGAAGACCAGGUCAGGCUG-5'   antisense 1: −35 to −16
                    3'-AGGCAGACUCUUCCUUGGUGG-5' antisense 2: −21 to −1

Sample Reporter Genes

Reporter	Half-life (hr)	Detection methods
Green fluorescent protein (GFP)	24	Flow cytometry
dGFP	2	Spectrofluorometer
Luciferase	3	Luminometer
β-Galactosidase	20	Colorimetric
Chloramphenicol acetyltransferase	50	Antibody

An optimal vector provides for the preparation of RNA for *in vitro* translation and for expression in eukaryotic cells. Further, an inducible promoter provides a dynamic range of RNA in the cell. A constitutive promoter is useful for addressing steady-state inhibition and comparison of the time course of inhibition with the expected inhibition based on the protein half-life. The insert of antisense target as a fusion with the reporter gene can incorporate an AUG site, but it is important to note that even an out of frame AUG in this insert does not prevent reporter gene expression initiated at a downstream AUG. A convenient feature of this insert is that it can be small enough that synthetic oligonucleotides can be prepared and cloned into restriction sites between the promoter and the AUG of the reporter gene.

In Vitro Translations

RNA can be generated with a T7 polymerase following the protocol in the Ambion mMessage mMachine instruction manual (Ambion Inc., Austin, TX). Translations are carried out with rabbit reticulocyte lysate purchased from Promega Corp. (Madison, WI):

   4 μl oligonucleotide in water (10–3000 nM)

   4 μl RNA (0.1–50 nM)

   16 μl rabbit reticulocyte lysate solution: 12.0 μl nuclease-treated reticulocyte lysate, 1.0 μl amino acid mix (Promega Corp.), and 3.0 μl water or RNase H (0.67 units/μl).

Incubate this reaction for 120 min at 37° and stop by chilling to 4°. The luciferase assay shows both an extensive dynamic range and a high degree of linearity ($r^2 = 0.99$). Titration of mRNA indicates that light production is not linear to 10 nM mRNA, providing a working range for evaluation between 0.1 and 2.0 nM.

Ten microliters of this translation sample is then added to 50 μl of ambient temperature luciferase assay reagent (Promega Corp.) and mixed for 30 sec, and light emission is measured for 15 sec in a luminometer (Turner Model TD-20e, TurnerDesigns, Inc., Mountain View, CA). This luciferase assay reagent uses coenzyme A (CoA) as the energy source and affords higher and more extended light emission relative to the assay reagents that employ ATP. The luciferase detection is linear over 8 logs of enzyme concentration. The percentage inhibition is calculated as

% inhibition = 100*[1-(oligonucleotide treated/untreated control)]

Light output from the control reaction with no added antisense agents was 540 ± 17.3 (n = 47) whereas light output from control reactions in any single experiment varied between 3 and 13% (see Fig. 1).

The Lineweaver–Burke plots in Fig. 2 represent observed data for a phosphorodiamidate morpholine oligomer targeting in the 5′-UTR that is expected to be a noncompetitive inhibitor as per Eq. (2). The open squares are nonoligomer treated, the open circles represent 100 nM antisense oligomer, and the filled circles represent 300 nM antisense oligomer. The $-1/k_m$ is identical for all three lines, indicating that this is a noncompetitive inhibitor of translation. The bottom portion of Fig. 2 represents data for a phosphorodiamidate morpholine oligomer that hybridizes to the AUG site of the target mRNA and is expected to be a biphasic mixed inhibitor as per Eq. (4). Neither $-1/k_m$ or $1/V_{max}$ are equivalent for these three lines, indicating a mixed inhibitor consistent with a biphasic inhibition.

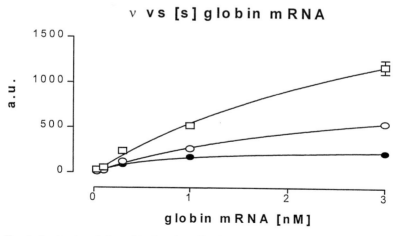

FIG. 1. *In vitro* translation with phosphorodiamidate morpholine antisense oligomer. The plot shows a velocity versus substrate curve for no antisense inhibitor (□) 100 nM antisense 2 (○), and 300 nM antisense 2 (●).

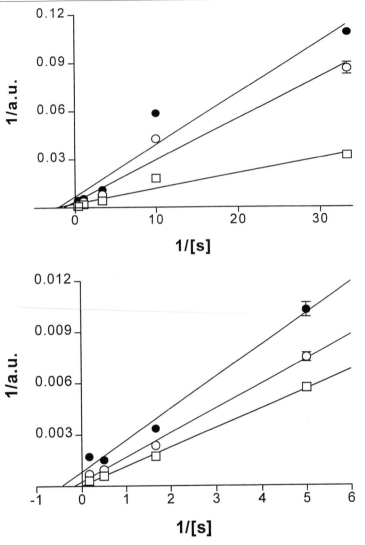

FIG. 2. The open squares are nonoligomer treated, the open circles represent 100 n*M* antisense oligomer, and the filled circles represent 300 n*M* antisense oligomer.

Cell Culture Studies

HeLa cells were stably transformed with a pAVI-3 vector derived from pMAMneo-LUC (Clonetech, Palo Alto, CA) containing luciferase and a 36-base insert for the oligomer target, rabbit globin immediately upstream.

The cells were grown in DME:F12 medium, 10% fetal bovine serum, and antibiotics (100 U/ml penicillin and 75 U/ml streptomycin) at 37° with 5% CO_2 in air.

One million cells per well of a 9.6-cm^2 plate were manipulated for oligomer entry by a scrape-loading method.[6] Figure 3 shows nonoligomer-treated cells (open squares) and 3 μM antisense oligomer-treated cultures (filled circles). Dexamethasone was employed to induce expression of a range of mRNA concentrations in the cells for the globin:luciferase fusion transcript. The sample antisense target is in the 5'-UTR of the mRNA and noncompetitive inhibition is expected. The $-1/k_m$ is identical for both lines, which is consistent with Eq. (2).

The cells were seeded onto plates 24 hr before the start of the experiment. Media were changed to include the antisense oligomer at a concentration of 3 μM in a total volume of 2 ml. The cells were scraped with a cell scraper (Sarstedt) and gently transferred with medium to a new plate with a sterile disposable 2-ml pipette. After 30 min, dexamethasone was added at a range of concentrations from 0.03 to 10 μM to induce the transcription of mRNA. The oligomer-loaded cells were then incubated for 16 to 18 hr at 37° and then harvested in 25 μl cell culture lysis reagent (Promega Corp.). Cell nuclei were pelleted by spinning at 16,000g at 4° for 15 sec. The amount of luciferase produced was determined by adding 10 μl of cell lysate to 50 μl of luciferase assay reagent and mixing, and light emission was measured as described earlier. The protein concentration was also determined for the cell lysates with 10 μl lysate in 2.5 ml of filtered Bio-Rad dye reagent (Hercules, CA) and incubation for 5 min at room temperature, and absorbance was determined at 595 nm. The luciferase activity was then adjusted by dividing light units by total protein. The mean luciferase activity of untreated samples in cells treated with 1 μM dexamethasone was 36.4 ± 1.5 light units/μg cellular protein.

Analytical evaluation of antisense oligonucleotides in cells in culture should include analysis of cell viability. There are a variety of methods, but the MTT ([3-(4,5-dimethylthiazol-2-yl)-2,5-diphenyltetrazolium bromide]) vital stain is convenient because it can be read on a 96-well plate reader. The assay involves the addition of 100 μl of 5 mg/ml filtered MTT to wells in a 96-well plate, incubation for 2 hr and removal of the solution. A 200-μl volume of dimethyl sulfoxide is added to the wells, the solution is mixed, and absorbance is read at 540 nm. The oligonucleotides should be evaluated at the highest concentrations to ensure that cell death is not an interfering feature of the evaluation of mechanism of action.

---

[6] M. Partridge, A. Vincent, P. Matthews, J. Puma, D. Stein, and J. Summerton, *Nucleic Acid Drug Dev.* **6,** 169 (1996).

Cell Culture ν vs. [S]

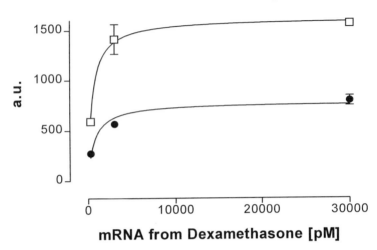

mRNA from Dexamethasone [pM]

Cell Culture Double Reciprocal Plot

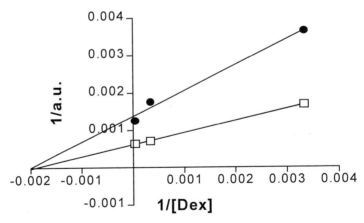

1/[Dex]

FIG. 3. (Top) Substrate versus activity and (bottom) double-reciprocal plot.

The fusion of target gene segment to reporter gene offers several advantages. First, studies can be conducted in a cell-free system as well as in cell culture. Second, in cell culture the reporter gene may act as an internal control for either observation of changes in cell phenotype or analysis of changes in target protein concentration. This strategy ensures that the antisense oligomer selected active in cell-free translation can be active in a more complex environment and that selected negative controls are indeed negative. Failure to observe inhibition in the reticulocyte lysate system would indicate a fundamental oligomer design problem. An observed inhibition in cell-free translation with failure to inhibit in the intact cell indicates that the failure is due to poor cellular availability or interference due to the complexity of the living system. Finally, inhibition observed in both *in vitro* translation and cell culture provides an opportunity to test the inhibition for equivalence between the two systems, e.g., noncompetitive inhibition. Data could be evaluated as an inhibitor in a saturation receptor-binding experiment or as an inhibitor in an enzymatic reaction.

The internal control feature in cell culture experiments may be of tremendous assistance to the mechanism of action studies in that rank order potency is expected. Further, temporal aspects of the degree of inhibition can be evaluated. Finally, the reporter gene may be employed with flow cytometry to select for a population that is nearly 100% inhibited for a more clear evaluation of the altered phenotype conferred by inhibition of the target mRNA translation.

[7] Physiology and Molecular Biology Brought to Single-Cell Level

By PIERRE-MARIE LLEDO, DAVID DESMAISONS, ALAN CARLETON, and JEAN-DIDIER VINCENT

Introduction

The traditional genetic approach uses chemicals or radiation to randomly delete or alter genes; interesting characteristics in the modified organisms or cells and their progeny are then selected by the experimenter. This approach has been highly successful in microorganisms, certain invertebrates, and even plants, but less successful for vertebrates. Their long generation times and the lethal nature of interesting mutations make vertebrates poor subjects for classical reverse genetic techniques (e.g., transgenesis and gene targeting).

METHODS IN ENZYMOLOGY, VOL. 313

Recent technological progress in cloning, or copying genes, makes inhibition or activation of any given gene in any given cell type a realistic option. One method, called antisense oligonucleotide strategy, is, in principle, remarkably simple. The basic idea underlying antisense approaches is to interfere at a specific point in the information flow from gene to protein. Thus, they offer the exciting possibility of selectively modifying the expression of a particular gene without affecting the function of others. As such, antisense oligonucleotides are useful tools in the study of gene function and have also aroused interest as possible therapeutic agents. Briefly, reagents that bind to single- or double-stranded nucleic acids can potentially target specific genes, either at the messenger RNA (mRNA) or at the double-stranded DNA stage. Using antisense DNA, an RNA duplex is formed that is either degraded rapidly (the mRNA is impaired in nuclear processing) or blocks RNA translation into protein.[1] In the case of double-stranded DNA, the oligonucleotide can form a triple helix by other forms of hydrogen bonding, such as Hoogsteen bonding, to interfere with the normal transcription of the gene. Triple-helix formation not only inhibits gene expression, but also offers the exciting potential of activating gene expression by targeting the binding sites of negatively acting transcription factors; such a strategy may be useful in activating tumor suppressor genes in the treatment of cancer. In fact, a natural biological function for antisense RNA molecules was first found in the regulation of certain genes during the life cycles of viruses and bacteria. Today, such an approach is practical enough for investigators to solve a broad range of problems.

This article reviews an antisense knockout approach that permits the direct correlation of a gene expression pattern, in a living cell and in its native cellular environment, with the functional properties expressed by the targeted protein. Our purpose is to discuss the strategy employed to assign biological functions to several native proteins found in excitable cells using a combination of antisense oligodeoxynucleotide approaches with electrophysiological techniques.

Oligodeoxynucleotides corresponding to the antisense orientation of messenger RNAs coding for the different proteins can be synthesized easily. These antisense probes specifically block the translation of the corresponding messages into proteins. For this approach to be effective, the oligonucleotide must be capable of crossing the cell membrane, of resisting nuclease degradation, and of hybridizing selectively to its target nucleotide sequence. Unmodified oligonucleotides are negatively charged and will not enter cells readily, but modification of the internucleoside phosphate linkages to uncharged methyl-phosphonates enables the oligonucleotides to gain access

[1] A. R. van der Krol, J. N. M. Mol, and A. R. Stuitje, *Biotechniques* **6**, 958 (1988).

to the cell interior when added to the incubation medium. An alternative mode of administration is by direct application of the oligonucleotides to the nucleus or the cytoplasm of individual cells either by microinjection using a sharp electrode (for examples, see Refs. 2 and 3) or by diffusion from the patch electrode into the cytoplasm.[4,5] In the latter approach, a sequential patch-clamp procedure allowed recording from the same cell, at several days' interval. Compared with the addition of antisense oligonucleotides to the cell medium, these two techniques are more direct and reliable. This article briefly reviews these two direct means of introducing the antisense probes into living cells.

Antisense Oligonucleotide Techniques

Principles of Antisense Techniques

Antisense RNA molecules were initially used to inactivate specific genes with higher selectivity than mutations do (for a review, see Ref. 6). However, antisense RNAs are not the only antisense molecules that can match onto messenger RNAs and prevent protein translation. Short complementary strands of DNA can also hybridize with messenger RNAs. Antisense oligonucleotides, which are strands of DNA only 15 to 20 bases long, can be introduced into cells where they will have inhibitory effects like those of antisense RNA. The antisense DNA technology that was used before the development of antisense RNA was first applied in an attempt to inhibit the *Rous sarcoma* virus from transforming cultured chicken cells into cancerous ones.[7]

The action of oligodeoxynucleotides can be rationalized by traditional receptor theory. The affinity of oligodeoxynucleotides for their nucleic acid receptors results from hybridization interactions that depend on hydrogen bonding and base stacking in the double helix that is formed. A minimum level of affinity is required for the desired interaction and this can be achieved with an oligodeoxynucleotide of at least 15 nucleotides in length. Several computational approaches have been designed to predict antisense

[2] C. Kleuss, J. Hescheler, C. Ewel, W. Rosenthal, G. Schultz, and B. Wittig, *Nature* **353,** 43 (1991).

[3] L. Johannes, P.-M. Lledo, M. Roa, J.-D. Vincent, J.-P. Henry, and F. Darchen, *EMBO J.* **13,** 2029 (1994).

[4] A. J. Baertschi, Y. Audigier, P. M. Lledo, J. M. Israel, J. Bockaert, and J.-D. Vincent, *Mol. Endocrinol.* **6,** 2257 (1992).

[5] P.-M. Lledo, P. Vernier, J.-D. Vincent, W. T. Mason, and R. Zorec, *Nature* **364,** 540 (1993).

[6] H. Weintraub, J. G. Izant, and R. M. Harland, *Trends Genet.* **1,** 23 (1985).

[7] P. C. Zamecnik and M. L. Stephenson, *Proc. Natl. Acad. Sci. U.S.A.* **75,** 280 (1978).

a. mRNA hydrolysis

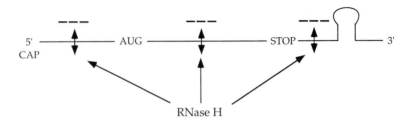

RNase H

b. Blockade of translation

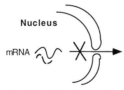

c. Change of the secondary structure

d. No exportation from the nucleus

FIG. 1. Mechanisms of action of antisense oligodeoxynucleotides. The antisense oligodeoxynucleotide probe is illustrated by the dashed line, AUG is the initiation codon, and STOP is the stop codon of a target messenger RNA. (a) Inhibition of target protein production may be initiated by RNase H cleavage of hybridized mRNA. RNase H acts as an amplifier of the antisense effect because it cleaves only duplex mRNA and not antisense DNA, which can therefore induce degradation of the multiple transcript. It is noteworthy that the majority of

oligodeoxynucleotide efficacy; before its use, it is imperative that every oligonucleotide sequence is checked for similarities with other sequences present in gene data bases.

The mechanisms of potential antisense oligonucleotide interactions with target nucleic acids are complex, as there is a long sequence of events leading from DNA to protein synthesis that is beyond the scope of this article. This article focuses on the experimental use of the antisense oligodeoxynucleotide targeted to messenger RNA. Figure 1 shows the possible mechanisms by which the association of messenger RNA with a complementary sequence could specifically block the synthesis of the encoded protein. Most of the results obtained with antisense DNA have been explained by the RNase H activity that induces the hydrolysis of messenger RNA bound to DNA (Fig. 1a). In the nucleus, this enzyme cleaves the RNA moiety of DNA/RNA hybrids during DNA replication. In the antisense approach, the cleaved messenger RNA can no longer support translation, whereas the antisense DNA will act catalytically for the RNase H activity by binding to a new messenger RNA, again inducing RNA hydrolysis. Another way to arrest the translation machinery results from a physical block to binding of the initiation complex (the 40S subunit) at the AUG initiation codon or at the region that covers the ribosome binding site, upstream of the initiation codon (close to the cap region; Fig. 1b). Binding of antisense oligodeoxynucleotides may also have two other consequences: an increase in the catabolism rate of the target messenger RNA due to a change of its secondary structure (Fig. 1c) and the arrest of the messenger RNA exportation from the nucleus to the cytoplasm (Fig. 1d).

A major advantage of this antisense oligodeoxynucleotide approach is the simplicity of the synthesis and testing procedures, meaning that a number of oligodeoxynucleotides can be screened before their use. Hence, because there are no *a priori* rules to predict the most efficient antisense sequence, we suggest using a "trial and error" approach to determine whether translated messenger RNA regions (including the initiation codon) as well as 5' (e.g., capping region)- or 3'-untranslated messenger RNA regions can be targeted. During our experiments, the most useful antisense oligodeoxynucleotides were found to be those binding to the translation

modified oligonucleotides are unable to elicit RNase H activity. (b) Physical block of the translation by impeding the binding of the initiation complex, which scans the 5' leader of the mRNA. It has been claimed that cap and AUG regions constitute the best targets for antisense oligonucleotides. (c) By changing the secondary structure of the messenger RNA, the antisense probe may induce an increase in messenger RNA degradation. (d) The duplex mRNA can no longer be exported from the nucleus. Note that mechanism (b) would not affect the total level of mRNA in the cytosol.

initiation codon or its immediate vicinity. Their efficiency was evaluated from an *in vitro* translation system using a rabbit reticulocyte lysate.

In Vitro Translation

mRNAs used in the *in vitro* translations were transcribed from cDNA that were subcloned into plBl31. Translation can be performed as following: mRNA ($0.3-1$ $\mu g/\mu l$) is diluted ($1:32-1:64$ in H_2O) and added ($0.5-1$ μl) to the vehicle ($1-2$ μl H_2O) or antisense solution ($1-2$ μl), heated for 1 min at 65° and cooled rapidly on ice. A mixture (8 μl) of rabbit reticulocyte lysate (Promega, 80%, v:v) and [^{35}S]methionine (NEN; 10 $\mu Ci/\mu l$, 13.3% v:v) is added to the mRNA/antisense, mixed gently, centrifuged for 1 min at 10,000g, and incubated for 45–60 min at 30° in a water bath. The translation product is run on 9 or 12% polyacrylamide–0.1% sodium dodecyl sulfate (SDS) gel electrophoresis. The gel autoradiograph is then scanned and analyzed on a Biolmage Analyzer (Visage 40000, Millipore, Bedford, MA). Integrated absorbance can be quantified for each band, and antisense effects are expressed as the percentage decrease relative to vehicle control.

Antisense Strategy: Technical Tips

Oligodeoxyribonucleotides were synthesized in our nucleotide sequencing center, purified on a diethylaminoethyl high-performance liquid chromatography (HPLC) ion-exchange column (Zorbax, Dupont), lyophilized, and reconstituted in distilled water at a concentration of 30 μM as verified by spectrophotometry. The specificity of an antisense can be optimized by designing antisense probes to nonconserved [e.g., 5'- or 3'-untranslated (UT) sequences] nonrepetitive sequences to minimize cross-hybridization. Specificity also depends on length. Indeed, the longer an oligonucleotide is, the more likely it is to bind to only one DNA or RNA target. To improve their efficiency, oligonucleotides can be modified in different ways by covalent linkage to various ligands. For example, it is possible to make phosphorothioate oligonucleotides in which sulfur atoms substitute for oxygen atoms in the sugar–phosphate backbone. We have found that such oligonucleotides are less vulnerable to nucleases and therefore remain inside the cell for longer periods. We have also synthesized oligonucleotides with uncharged sugar–phosphate backbones and have found that they enter a cell through its outer membrane more easily than ordinary oligonucleotides can. Finally, use of acridine attachment to an oligonucleotide has been shown to improve the rate at which cells absorb the oligonucleotides from the environment and also to improve the interaction of the oligonucleotide with the messenger RNA target (for a review, see Ref. 8).

After having loaded cells with the oligonucleotides, the extent of knock out should be investigated, preferably by the quantitative determination of target protein levels using binding or immunoblot analyses and the specificity can be assessed by monitoring the levels of proteins that are closely related to, or associated with, the target protein. One of the key aspects of protein depletion by antisense approaches is the rate of target protein turnover. Even with a 100% blockage of protein synthesis by antisense hybrids, no depletion of the target protein will occur until the remaining pool of previously synthesized protein is degraded. Thus, the half-life of degradation for the protein in question is an important consideration for acute, transient antisense experiments in which several days may be required for depletion of the target protein to produce optimal knockout conditions.

Patch-Clamp Technique for Loading Cells with Antisense Probes

The introduction of patch-clamp techniques[9] provided new possibilities for the study of single cell membrane conductances by measuring currents through single channels and through whole-cell membranes.[10] However, a further advantage of the patch-clamp techniques lies in providing the means to introduce substances into the cytosol. Molecules, ions, peptides, antibodies, and oligonucleotides, putatively interfering with cytosolic processes, can be dissolved in the electrode-filling solution, and on the formation of whole-cell configuration, the diffusional barrier between electrode lumen and cytosol is eliminated, allowing an exchange of substances between these two compartments.[10,11]

Loading of antisense probes into single cells maintained in culture can be accomplished by a sequential patch-clamp technique. This technique consists of a sequential patch-clamp procedure that allows recordings from the same injected cell, at intervals of several days. During the first whole-cell recording, the cell is loaded with the antisense DNA (the loading session) and then, after various periods of time, the same cell is recorded from again to determine the effects of protein absence on a specific cellular response (the test session).

[8] J.-J. Toulme and C. Helene, *Gene* **7,** 51 (1988).
[9] O. P. Hamill, A. Marty, E. Neher, B. Sakmann, and F. J. Sigworth, *Pflueg. Arch.* **391,** 85 (1981).
[10] A. Marty and E. Neher, *in* "Single-Channel Recording" (B. Sakmann and E. Neher, eds.), p. 107. Plenum Press, New York, 1983.
[11] M. Pusch and E. Neher, *Pflueg. Arch.* **411,** 204 (1988).

During the loading session, the time of loading can be determined from the empirical expression derived by Pusch and Neher[11]:

$$\tau = kR_aM^{1/3}$$

where τ is the time constant of loading; k is an empirical constant related to the size of the loaded cell (0.6 for sizes similar to endocrine cells), R_a is access resistance in ohms, and M is molecular weight of the antisense probe. The antisense probes used in our studies have molecular weights between 6000 and 6500, thus the time constant of loading can be determined by monitoring changes in R_a. For endocrine cells, we have used antisense probes at a concentration of about 2.7×10^{-8} pM. Taking an average access resistance of 12 MΩ and volume of a 4-pF cell to be 60% of an average chromaffin cell,[11] the time constant of loading for such a cell is 83 sec, and it can be estimated that over 11,000 molecules of the antisense probe were loaded into the cytoplasm after 120 sec of whole-cell recording.[4,5]

At the end of the loading session, the electrode is withdrawn carefully and slowly, and usually the cell reseals itself immediately as observed under Nomarski optics. Distinctive marks can be made next to the loaded cell. After this procedure, the petri dishes need to be placed back into the incubator, and the bathing medium for electrophysiological measurements is replaced by standard culture medium supplemented by antibiotics for 48 hr. All materials and solutions used in the loading procedure need to be sterilized. One of the major advantages of this loading technique is the possibility of recording electrical properties of interest during this session, before the antisense action.

During the test session, a second recording can usually be obtained in about 25% of loaded cells when performed after 3 days of antisense incubation. However, the type of antisense or vehicle dialyzed is not a factor in cell survival. In fact, about half of the loaded cells are lost during the washing of petri dishes with culture and bathing media. During the second recording, one would expect membrane capacitance (C_m, a parameter proportional to the membrane surface area) to be lower in comparison with the first recording, as one would expect a small part of the membrane to be removed during withdrawal of the pipette at the end of the probe loading. Using the patch-clamp technique to measure C_m, we have found a nonsignificant change in resting C_m (-0.04 ± 0.02 pF, mean $\pm$ SEM, $n = 40$) between the first and the second recording. This property may result from a rapid production of phospholipids that may counteract with the patch membrane lost. Similarly, the difference in resting conductance of -0.04 ± 0.04 nS ($n = 10$) between the first and the second recording was not significantly different from zero. Resting conductance was determined as the point conductance (I/V, I is current and V is driving potential) at the beginning of recordings. These results show that the double-patching procedure, and loading by antisense probes, does not affect the passive electrophysiological parameters of recorded cells.

We have used this technique to determine the nature of G proteins via which dopamine receptors can modulate five different ionic channels in pituitary cells[4] (for a review, see Ref. 12). On day 1, cells in culture were dialyzed with antisense oligodeoxynucleotides (15 nM–1.5 μM) or vehicle (control), and voltage-activated ionic currents were recorded in the whole-cell mode. For the loading session, the cell dialysis lasted an average of 150 sec. The effectiveness and specificity of action of six types of antisense were determined by *in vitro* translation. Table I shows the specificity of the antisense directed to the α_o subunit of G proteins (AS α_o) in reducing the dopamine-induced inhibition of calcium currents recorded from rat pituitary cells.

We have also used scrambled analogs of the presumably active oligodeoxynucleotide molecules, as well as mismatched analogs maintaining the same base composition as the antisense molecule, for more stringent control experiments. Another important control for specificity of the active antisense is the identification of any other nonoverlapping antisense oligodeoxynucleotides able to induce similar effects. During the course of this study and to check the effects of knocking out the expression of the target protein by the antisense approach, cells were visualized by immunofluorescence microscopy. Only cells loaded with antisense DNA, for periods of 42–52 hr, showed a dramatic reduction of the immunostaining for the target protein.

One of the major advantages of the antisense oligodeoxynucleotide strategy is the reversibility of the effect (other advantages include the possibility of studying the product of any cloned gene from any species and the low cost to the laboratory). Figure 2 illustrates that after loading a pituitary cell with α_o antisense DNA, inhibition of the calcium current induced by dopamine is reduced in a time-dependent manner. This reduction of dopamine response parallels the disappearance of α_o subunit immunoreactivity. Interestingly, 48 hr after cell loading, the calcium current response to dopamine tends to be restored, indicating that antisense suppression of gene expression can be overcome. The 48-hr period required for almost abolishing the calcium current response would be compatible with a half-life for α_o of 28 hr as measured in the GH$_4$ pituitary tumor cell line.[13]

[12] P.-M. Lledo, P. Vernier, J.-D. Vincent, W. T. Mason, and R. Zorec, *Ann. N.Y. Acad. Sci.* **710**, 301 (1994).
[13] S. Silbert, T. Michel, R. Lee, and E. J. Neer, *J. Biol. Chem.* **265**, 3102 (1990).

TABLE I

CALCIUM CURRENT INHIBITION BY DOPAMINE IN
LACTOTROPHS LOADED WITH ANTISENSE
OLIGODEOXYNUCLEOTIDE DIRECTED AGAINST
mRNAs ENCODING α SUBUNITS[a]

Treatment	Mean ± SEM (%) dopamine-induced inhibition	Number of cells tested
Control	54 ± 9	18
+AS α_o	15 ± 8	7
+AS α_s	50 ± 12	6
+AS α_{i1}	51 ± 8	5
+AS α_{i2}	49 ± 14	6
+AS α_{i3}	57 ± 11	8

[a] The voltage-activated calcium current was recorded every 5 sec in response to 200-msec depolarization steps from a holding potential of -80 to -10 mV. To isolate calcium current, pipette-filling solution contained (in mM) CsCl (130), tetraethylammonium chloride (20), MgCl$_2$ (1), HEPES (10), EGTA (10.25), CaCl$_2$ (1), cAMP (0.1), ATP (2), GTP (0.4), pH 7.25/CsOH. Control denotes vehicle-dialyzed cells; AS α_o (CCC CGG TGG TAC CCT ACA) is an antisense probe overlapping the translation initiation codon of mRNA α_o; AS α_s (CGG CGG CGG TAC CCG ACG) is an antisense probe overlapping the translation initiation codon of mRNA α_s; AS α_{i1} (TGG TAC CCG ACG TGT GAC) is an antisense probe overlapping the translation initiation codon of mRNA α_{i1}; AS α_{i2} (TCC TAC CCG ACG TGG CAC) is an antisense probe overlapping the translation initiation codon of mRNA α_{i2}; and AS α_{i3} (CAG TAC CCG ACG TGC AAC) is an antisense probe overlapping the translation initiation codon of mRNA α_{i3}. All of these sequences are oriented 3' to 5'.

Using a panel of different antisense DNA, it is also possible to establish a correlation between the percentage inhibition of calcium currents induced by dopamine and the percentage inhibition of *in vitro* translation of the α_o subunit. The effects of antisenses on *in vitro* translation can be quantified by image analysis of the gel autoradiographs. We have found that in a linear regression fitting the data, the percentage inhibition of calcium currents correlates inversely with inhibition of the *in vitro* translation of α_o.

The use of antibodies is also an interesting and alternative approach to characterize the function of a specific protein (for a review, see Ref. 14).

[14] G. Milligan, *Biochem. I.* **255,** 204 (1988).

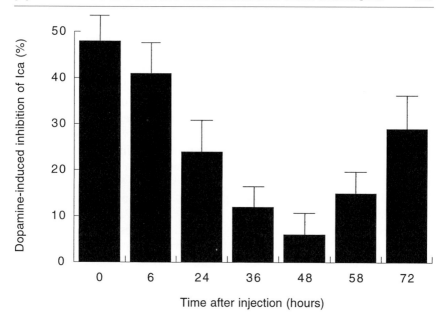

FIG. 2. Time course of calcium current inhibition by 10 nM dopamine in lactotroph cells injected with antisense α_o. Mean values are shown from seven different cells, and bars represent standard error of the mean.

However, there are inherent problems in the use of antibodies, which include the difficulty of introducing them into the cell, the low affinity of many antibodies for the peptides, and uncertainty as to their specificity for preventing molecular interactions because of their large size. There are three advantages of the antisense technique over antibody quenching of a target protein function. First, the risks of effector run-down such as ionic channels, for example, are much lower with the sequential patch-clamp technique, as at least 10 min more are required for the dialysis of antibody proteins than antisense molecules. Second, oligonucleotides are far easier to produce than antibodies. Finally, distinction between protein subtypes is easier at the nucleic acid than at the protein level.

Microinjection Loading Technique

Because of the technical difficulty of our double patching procedure and because of its poor success rate (about 25%), we have since established a microinjection technique using intracellular sharp electrodes. Antisense or control oligonucleotides (10 μg/ml) are injected in a buffer containing (in mM): sodium glutamate, 135; NaCl, 20; MgCl$_2$, 4; EGTA, 0.5; GDP, 0.05; HEPES, 10, pH 7.2, and 1 mg/ml fluorescein isothiocyanate (FITC)-dextran (Sigma). Just before injection, samples are cleared by centrifugation at 140,000g for 15 min. Injections can be performed easily with an Eppendorf microinjector at constant pressure. The immunofluorescent signal of coinjected FITC-dextran allowed the unambiguous identification of loaded cells. As during the sequential patch-clamp procedure, to check whether the antisense did suppress the target protein, cells were monitored by immunofluorescence microscopy following the microinjection procedure. Using this technique, more than 70% of cells were seen to survive microinjection loading after an incubation period of up to 9 days. Cells were subjected to electrophysiological experiments between several days after the loading session.

To demonstrate further the specificity of the antisense approach, rescue experiments can be designed with the microinjection procedure. If the observed effects of the antisense probe are due to suppression of a specific protein, reintroduction of the same purified protein should restore the original phenotype. For this purpose, cells are first injected with antisense oligonucleotides and, about 5 days later, with 0.5 mg/ml protein to which the antisense was directed in buffer containing rhodamine isothiocyanate (RITC)-dextran instead of FITC-dextran. Thus, cells injected with antisense oligonucleotides and subsequently with protein are identified positively by their fluorescence in the fluorescein and rhodamine channels.[3] In these conditions, cells show normal electrical membrane properties, indicating that the integrity of their membranes is preserved.

The advantages of patch pipette dialysis of antisense over nuclear microinjections are twofold. First, the cell under study may be functionally identified during the first recording. Second, the cell is dialyzed using a known concentration of antisense molecules in the patch electrode. The obvious disadvantages of the sequential patch-clamp technique are that it is more time-consuming, technically more demanding, and is endowed with a moderate rate of success, at least in normal pituitary cells. This technique would be particularly powerful in more hardy cells, such as neurons, which withstand the procedure much better. However, because the microinjection system with sharp electrodes can be applied easily to round cultured cells (such as endocrine cells), the microinjection procedure described in this article should prove generally useful and convenient.

Discussion and Concluding Remarks

Assigning biological function to proteins at the cellular level is a difficult task because cells are small and experimentally relatively inaccessible.

Moreover, some proteins are present in the cell as several isoforms, showing very similar structure, but with a distinctly different function. Thus, the best approach to distinguishing between the function of two isomeric proteins is to employ antisense probes directed to messenger RNA sequences unique to one of the two species. However, oligonucleotide uptake by cultured cells may be slow and highly inefficient, with only a small fraction of the added oligonucleotide actually entering cells. The extent of uptake and potency can be improved, however, by at least 1000-fold by applying the oligonucleotides to the cells in the presence of a liposome formulation containing cationic lipids. Furthermore, not all of the internalized oligonucleotide is necessarily available to interact with the intended subcellular targets. The intracellular bioavailability of oligonucleotides may be reduced by nuclease degradation, protein and lipid binding, other nonspecific binding, and localization within cellular organelles such as endosomes. Our delivery strategies circumvent some of these problems. Moreover, with electrophysiological recording techniques, this approach can be used at the single-cell level as the antisense probes can be loaded into cells and then function can be assayed at this same level.

Many important refinements of antisense technology are still needed, and many important questions must still be answered. It should be possible, for example, to chemically modify antisense oligonucleotides so that they can be introduced into cells more efficiently or be bound to their targets more effectively. In addition, certain parts of messenger RNA may be more susceptible than other parts to inhibition by antisense DNA. As these parameters become better understood, it should be possible to design more effective antisense DNA molecules. New methods should also provide antisense oligonucleotides more resistant to degradation inside the cells.

A greater understanding of the precise mechanisms whereby antisense DNA inhibits the production of proteins is also essential. Research suggests that antisense RNA can act both within the nucleus and in the cytoplasm of the cell and that it may arrest protein translation by doing more than hybridizing with messenger RNAs, but also preventing it from being spliced and modified in essential ways. A difficulty in all studies with antisense oligonucleotides is the independent demonstration of their effectiveness and specificity in single cells. Theoretically, *in situ* hybridizations could be carried out to detect the quantity of mRNA remaining after antisense treatment. However, the inhibition of target protein production may be initiated not only by RNase H cleavage of hybridized mRNA or by inhibition of transcription and splicing, but also by hybridization arrest of translation at the ribosomes. The quantification of remaining mRNA would therefore be inappropriate.

The challenge of finding new ways to manipulate the activity of cellular genes is spurring on the development of antisense technology and other techniques, such as gene targeting. No doubt with more information, more sophisticated technology, and additional imagination, new approaches will emerge that will complement those already in use. The new field of reverse genetics is rapidly providing inroads into the understanding of gene function; with luck it will eventually enhance the ability of medicine to understand and treat disease. In theory, the antisense approaches outlined in this article should be applicable for the stable or transient knock out of any class of signaling molecule for which cDNA sequences are available. Low stringency approaches to molecular cloning have led to the identification of multiple subtypes of signaling molecules: the use of antisense approaches to exploit the differences between these molecules should allow for the identification of specific roles for individual signaling molecules triggering specific cellular responses.

[8] Antisense Properties of Peptide Nucleic Acid

By Peter E. Nielsen

Introduction

Peptide nucleic acid (PNA) is a DNA mimic with a pseudopeptide backbone (Fig. 1). It was introduced in 1991,[1] and because of very favorable hybridization properties[2,3] and high chemical and, most importantly, biological stability,[4] it was regarded as a very promising lead for developing efficient antisense reagents and medical drugs.[5,6] *In vitro* translation[7-11] and

[1] P. E. Nielsen, M. Egholm, R. H. Berg, and O. Buchardt, *Science* **254,** 1497 (1991).
[2] M. Egholm, O. Buchardt, L. Christensen, C. Behrens, S. M. Freier, D. A. Driver, R. H. Berg, S. K. Kim, B. Norden, and P. E. Nielsen, *Nature* **365,** 566 (1993).
[3] K. K. Jensen, H. Ørum, P. E. Nielsen, and B. Norden, *Biochemistry* **36,** 5072 (1997).
[4] V. V. Demidov, V. N. Potaman, M. D. Frank-Kamenetskii, M. Egholm, O. Buchardt, S. H. Sönnichsen, and P. E. Nielsen, *Biochem. Pharmacol.* **48,** 1310 (1994).
[5] L. Good and P. E. Nielsen, *Antisense Nucleic Acid Drug Dev.* **7,** 431 (1997).
[6] A. De Mesmaeker, K. H. Altmann, A. Waldner, and S. Wendeborn, *Curr. Opin. Struct. Biol.* **5,** 343 (1995).
[7] J. C. Hanvey, N. J. Peffer, J. E. Bisi, S. A. Thomson, R. Cadilla, J. A. Josey, D. J. Ricca, C. F. Hassman, M. A. Bonham, K. G. Au et al., *Science* **258,** 1481 (1992).
[8] H. Knudsen and P. E. Nielsen, *Nucleic Acids Res.* **24,** 494 (1996).
[9] M. A. Bonham, S. Brown, A. L. Boyd, P. H. Brown, D. A. Bruckenstein, J. C. Hanvey, S. A. Thomson, A. Pipe, F. Hassman, J. E. Bisi et al., *Nucleic Acids Res.* **23,** 1197 (1995).

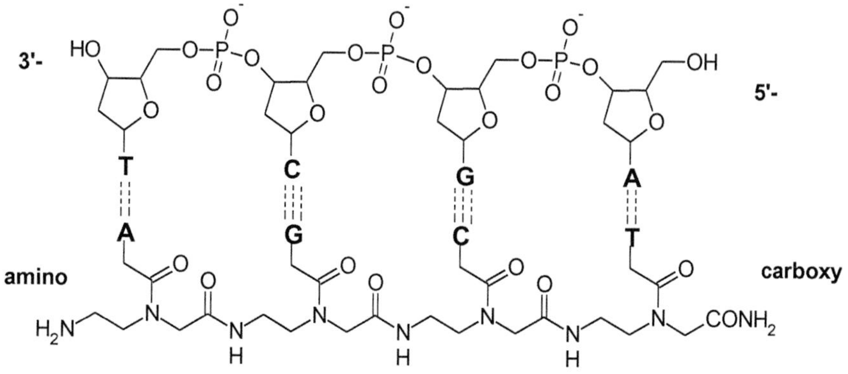

FIG. 1. An antiparallel PNA–DNA duplex showing the chemical structure of PNA (lower strand) as compared to that of DNA (upper strand).

cellular microinjection experiments[7] supported the enthusiasm, but progress was impeded by the low lipid membrane penetration[12] and inefficient cellular uptake of PNA.[9] However, a series of recent reports using both mammalian and bacterial systems have illustrated ways of delivering PNAs to cells in culture[13–21] as well as in animals,[14–16] and have also demonstrated antisense effects of such PNAs.[14,15,20] Therefore, development, studies, and application of PNA antisense reagents have gained new momentum.

[10] C. Gambacorti-Passerini, L. Mologni, C. Bertazzoli, P. le Coutre, E. Marchesi, F. Grignani, and P. E. Nielsen, *Blood* **88,** 1411 (1996).

[11] L. Mologni, P. leCoutre, P. E. Nielsen, and C. Gambacorti-Passerini, *Nucleic Acids Res.* **26,** 1934 (1998).

[12] P. Wittung, J. Kajanus, K. Edwards, G. Haaima, P. E. Nielsen, B. Norden, and B. G. Malmstrom, *FEBS Lett.* **375,** 27 (1995).

[13] C. G. Simmons, A. E. Pitts, L. D. Mayfield, J. W. Shay, and D. R. Corey, *Bioorg. Med. Chem. Lett.* **7,** 3001 (1997).

[14] M. Pooga, U. Soomets, M. Hällbrink, A. Valkna, K. Saar, K. Rezaei, U. Kahl, J.-X. Hao, X.-J. Xu, Z. Weisenfeld-Hallin, T. Hökfelt, T. Bartfai, and Ü. Langel, *Nature Biotechnol.* **16,** 857 (1998).

[15] B. M. Tyler, D. J. McCormick, C. V. Hoshall, C. L. Douglas, K. Jansen, B. W. Lacy, B. Cusack, and E. Richelson, *FEBS Lett.* **421,** 280 (1998).

[16] W. M. Pardridge, R. J. Boado, and Y. S. Kang, *Proc. Natl. Acad. Sci. U.S.A.* **92,** 5592 (1995).

[17] H. Knudsen, Ph.D. Thesis University of Copenhagen, 1997; T. Ljungstrøm, H. Knudsen, and P. E. Nielsen (in preparation).

[18] G. Aldrian-Herrada, M. G. Desarménien, H. Orcel, L. Boissin-Agasse, J. Méry, J. Brugidou, and A. Rabie, *Nucleic Acids Res.* **26,** 4910 (1998).

[19] S. Basu and E. Wickstrom, *Bioconj. Chem.* **8,** 481 (1997).

[20] L. Good and P. E. Nielsen, *Nature Biotechnol.* **16,** 355 (1998).

[21] A. F. Faruqi, M. Egholm, and P. M. Glazer, *Proc. Natl. Acad. Sci. U.S.A.* **95,** 1398 (1998).

Cellular Uptake

Peptide nucleic acid antisense compounds are rather large (~15 nucleo-bases, molecular weight ~3000), predominantly hydrophilic molecules with physical properties mostly reminiscent of peptides. It is therefore not sur-prising that mammalian cells generally do not take up such PNAs readily. However, certain peptides are internalized very efficiently by cells[22] and these may even "carry a load" such as a PNA into the cells.[13–15] Several such peptides have now been identified, some of natural origin (e.g., the third helix of the homeo domain of the antennepedia protein from *Drosoph-ila*) and others fully synthetic.[23] However, they share a common feature thought to reflect the mechanism of action. They are amphipatic α helices with a high content of basic amino acid residues (Lys, Arg), and it is speculated that the cationic face of the α helix interacts with the anionic surface of the cellular membrane and induces "inverted liposomes" that can transverse the membrane or just transiently destabilize it. This type of peptide is conjugated easily to PNAs either by continuous peptide synthe-sis[13,15] or by fragment coupling conjugation chemistry, e.g., via a disulfide linkage[14] (Fig. 2). Although the results on cellular uptake of such PNA–peptide conjugates are impressive, it is still too early to judge how general the principle is in terms of cell types and PNA reagent (sequence, length).

Oligonucleotides are not taken up very efficiently by mammalian cells either, and in general, good antisense efficacy is only achieved by employing cationic liposome preparations as carriers.[24] In principle, PNAs may also be delivered via liposomes, but because PNAs are inherently electrostatically neutral molecules, the electrostatic attraction to the cationic liposomes that make loading with the anionic oligonucleotides very efficient does not exist, and loading liposomes with PNA is difficult. In an effort to equip PNAs with a lipophilic tail that would confer liposome affinity, we have prepared PNA–lipid conjugates (Fig. 2) and studied the (liposome mediated) cellular uptake of these.[17] Although the results showed that the properties of such conjugates are influenced by the PNA sequence, the results were encourag-ing in the sense that in at least some cell types uptake of the PNA–lipid conjugates themselves were much enhanced, and in several cases diffuse cytoplasmic uptake was observed when using cationic liposomes as car-riers.[17]

---

[22] D. Derossi, A. H. Joliot, G. Chassaing, and A. Prochiantz, *J. Biol. Chem.* **269,** 10444 (1994).

[23] A. Prochiantz, *Curr. Opin. Neurobiol.* **6,** 629 (1996); E. Vives, P. Brodin, and B. Lebleu, *J. Biol. Chem.* **272,** 16010 (1997).

[24] J. G. Lewis, K. Y. Lin, A. Kothavale, W. M. Flanagan, M. D. Matteucci, R. B. DePrince, R. A. Mook, Jr., R. W. Hendren, and R. W. Wagner, *Proc. Natl. Acad. Sci. U.S.A.* **93,** 3176 (1996).

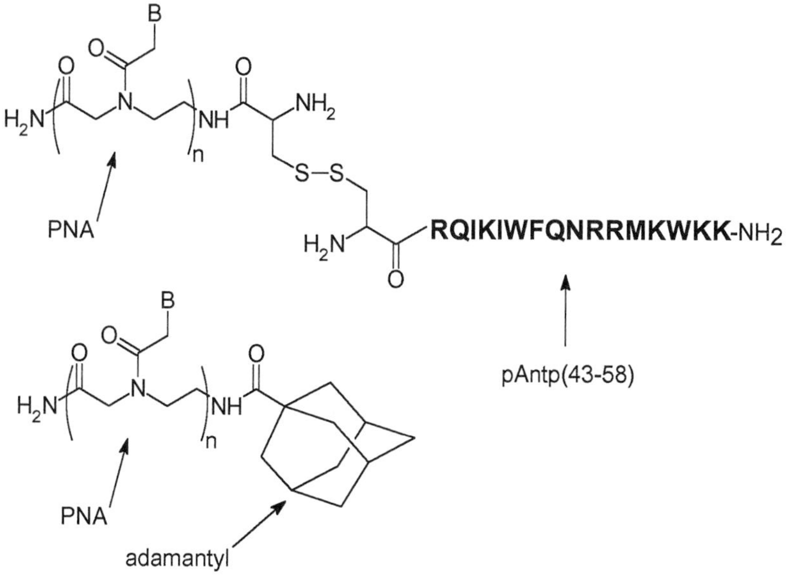

Fig. 2. Examples of PNA conjugates showing increased cellular uptake. A disulfide (Cys–Cys)-linked conjugate of PNA to the third helix of the homeo domain of the antennapedia protein (pAntp)[14] from *Drosophila* is shown on top, whereas a PNA–lipid (adamantyl) conjugate[17] is shown below. B is a nucleobase and *n* is 12–15.

Antisense

Target Selection

PNA–RNA hybrids are not substrates for RNase H,[7,8] the enzyme that is inferred as the mediater of antisense effects of phosphodiester and phosphorothioate oligonucleotides, and therefore PNAs must exert their antisense effect by other mechanisms, such as direct steric blocking of ribosomes or essential translation factors. Thus sensitive targets that have been optimized for phosphorothioates most likely are not good targets when using PNA (or other non-RNase H-activating antisense reagents, such as the 2'-substituted oligonucleotides). Although it is much too early for general rules concerning target selection for PNA antisense, targets around the AUG initiation codon in general seem sensitive.[8,10,11] In contrast, it has been found by *in vitro* translation that mixed purine/pyrimidine sequence PNAs, which form PNA–RNA duplexes with the target, did not arrest translation elongation, indicating that the PNA was displaced by the moving ribosome.[8] However, when homopurine targets are present in the translated region of the gene, these can be targeted by homopyrimidine (bis) PNAs, which form extremely stable PNA_2–DNA triplexes that are indeed capable of arresting the ribosome during elongation.[7,8] It should be stressed, however, that several mixed purine/purimidine sequence, and thus only duplex forming, PNAs that exhibit antisense gene inhibition, but do not target the AUG region, have been identified.[9,11,14,18] Therefore, as with antisense gene targeting in general, several sequence targets should be examined and, if possible a mRNA gene walk[25] is advisable. Nonetheless, a sensible first choice would be targets close to or overlapping the AUG initiation site.

Mammalian Cells

Until 1998, antisense studies with PNA had been confined to *in vitro* translation[8,10] or microinjection[7,9] experiments. The employment of peptide carriers[14,18] and the choice of nerve cells have now allowed cellular[14,15,18] and even animal studies.[14,15] Very interestingly and quite surprisingly, it seems that (at least some) nerve cells do indeed take up "naked" PNA quite efficiently,[15,18] and although the experiments reported so far do not entirely live up to the very stringent requirement now expected for antisense claims[26] (*vide infra*), the results are fully compatible with an antisense interpretation. Two of the studies even report a reduction of the activity of the targeted neuronal receptors on the injection of naked PNA or a peptide–PNA conjugate into the brain of rats.[14,18]

Bacteria

PNAs targeted to the AUG initiation codon of bacterial genes are very potent inhibitors of translation of these genes in cell extracts and, most surprisingly, also caused downregulation of the expression of the genes at micromolar PNA concentrations in a leaky mutant of *Escherichia coli* (AS19).[20] Moreover, when the β-lactamase gene, which is responsible for penicillin resistance, was targeted, bacteria were resensitized to ampicillin by at least six orders of magnitude. A slight, but still significant antisense effect was also demonstrated on wild-type *E. coli* K12.[20] These results open possibilities of performing antisense experiments in bacteria, both with the aim of developing new antibacterial drugs, and as a means of studying gene function.

[25] B. P. Monia, J. F. Johnston, T. Geiger, M. Muiler, and D. Fabbro, *Nature Med.* **2,** 668 (1996).
[26] F. Eckstein, A. M. Krieg, C. A. Stein, S. Agrawal, S. Beaucage, P. D. Cook, S. Crooke, M. J. Gait, A. Gewirtz, C. Helene, P. Miller, R. Narayanan, A. Nicolin, P. Nielsen, E. Ohtsuka, H. Seliger, W. Stec, D. Tidd, R. Wagner, and J. Zon, *Antisense Nucleic Acid Drug Dev.* **6,** 149 (1996).

Other RNA Targets

Peptide nucleic acid may also be used as specific inhibitors of the function of other cellular RNA molecules. For instance, PNAs targeted to the RNA of telomerase are efficient inhibitors of this enzyme *in vitro.*[27,28] It is even possible to inhibit the function of bacterial ribosomes by targeting ribosomal RNA of the peptidyl transferase center or the α-sarcin loop, and these PNAs exhibit bacteriostatic activity.[29] Finally, PNAs bound to the RNA template of reverse trancriptase, such as that from the HIV virus, cause translation arrest of this enzyme.[30,31]

Antigene

Homopurine regions of 8 bp or more in length in double-stranded DNA (dsDNA) can be targeted by homopyrimidine PNAs via the formation of extremely stable PNA triplex strand displacement complexes.[1,32] Because two PNAs—one Watson–Crick bound and the other Hoogsteen bound—are required to form these complexes, most often bis-PNAs in which the two parts are chemically linked are employed for dsDNA targeting.[33] PNA triplex invasion complexes occlude protein binding (transcription factors, restriction enzymes) to proximal or overlapping DNA sites and have sufficient stability to arrest the elongating RNA polymerase[7,34] or DNA polymerase,[35] thereby making PNAs good candidates for antigene reagents.[36,37] Binding to dsDNA is, however, very sensitive to even moderate ionic strength,[38] and only employing optimized bis-PNAs is stable binding possi-

[27] J. C. Norton, M. A. Piatyszek, W. E. Wright, J. W. Shay, and D. R. Corey, *Nature Biotechnol.* **14**, 615 (1996).

[28] S. E. Hamilton, A. E. Pitts, R. R. Katipally, X. Jia, J. P. Rutter, B. A. Davies, J. W. Shay, W. E. Wright, and D. R. Corey, *Biochemistry* **36**, 11873 (1997).

[29] L. Good and P. E. Nielsen, *Proc. Natl. Acad. Sci. U.S.A.* **95**, 2073 (1998).

[30] U. Koppelhus, V. Zachar, P. E. Nielsen, X. Liu, J. Eugen-Olsen, and P. Ebbesen, *Nucleic Acids Res.* **25**, 2167 (1997).

[31] R. Lee, N. Kaushik, M. J. Modak, R. Vinayak, and V. N. Pandey, *Biochemistry* **37**, 900 (1998).

[32] P. E. Nielsen, M. Egholm, and O. Buchardt, *J. Mol. Recogn.* **7**, 165 (1994).

[33] M. Egholm, L. Christensen, K. L. Dueholm, O. Buchardt, J. Coull, and P. E. Nielsen, *Nucleic Acids Res.* **23**, 217 (1995).

[34] P. E. Nielsen, M. Egholm, and O. Buchardt, *Gene* **149**, 139 (1994).

[35] R. W. Taylor, P. F. Chinnery, D. M. Turnbull, and R. N. Lightowlers, *Nature Genet.* **15**, 212 (1997).

[36] T. A. Vickers, M. C. Griffith, K. Ramasamy, L. M. Risen, and S. M. Freier, *Nucleic Acids Res.* **23**, 3003 (1995).

[37] D. Praseuth, M. Grigoriev, A. L. Guieysse, L. L. Pritchard, A. Harel-Bellan, P. E. Nielsen, and C. Helene, *Biochim. Biophys. Acta* **1309**, 226 (1996).

[38] N. J. Peffer, J. C. Hanvey, J. E. Bisi, S. A. Thomson, C. F. Hassman, S. A. Noble, and L. E. Babiss, *Proc. Natl. Acad. Sci. U.S.A.* **90**, 10648 (1993).

ble at 140 mM KCl.[39] Thus the possibilities of using antigene PNAs *in vivo* could seem slim. Nonetheless, it has been reported that PNAs targeted to a mouse gene cause a 10-fold increase in the mutation rate at the PNA target in this gene in cells in culture, strongly suggesting that significant PNA binding to the nuclear DNA had taken place.[21] Such binding may be catalyzed by active transcription, which has been shown to dramatically accelerate PNA binding *in vitro* either directly via the single-stranded transcription bubble[40] or indirectly via (transcription induced) DNA negative supercoiling.[41]

Sequence Considerations

Apart from the target selection discussed previously, a few sequence considerations relating to the PNA itself are also required. As mentioned earlier, homopyrimidine PNAs form extremely stable PNA_2–DNA (RNA) triplexes with their homopurine targets. Typically, the T_m of a 10-mer would be around 70°. In contrast, pyrimidine-rich PNAs (containing a few purines) form relatively unstable PNA–DNA duplexes (T_m of a 10-mer is ca. 40°C), whereas purine-rich PNAs form extremely stable duplexes (T_m of a 10-mer can be above 70°). This pronounced asymmetry in thermal stability is reflected in an empirical formula for T_m prediction of PNA–DNA duplexes based on more than 300 measured T_m values[42]:

$$T_{m_{\text{pred}}} = 20.79 + 0.83 T_{m_{\text{nnDNA}}} - 26.13 f_{\text{pyr}} + 0.44 \, \text{length}$$

in which $T_{m_{\text{nnDNA}}}$ is the melting temperature as calculated using a nearest neighbor model for the corresponding DNA/DNA duplex applying ΔH^0 and ΔS^0 values as described by SantaLucia *et al.*[43] f_{pyr} denotes the fractional pyrimidine content, and length is the PNA sequence length in bases.

Finally, it is worth noting that solubility/aggregation problems are sometimes encountered with purine-rich and especially guanine-rich PNAs. Such problems may, however, be solved by including 10–30% lysine backbone-modified PNA units[44] or by including a few hydrophilic "linker units" at the C or N-terminal of the PNA.[45]

[39] M. C. Griffith, L. M. Risen, M. J. Greig, E. A. Lesnik, K. G. Sprangle, R. H. Griffey, J. S. Kiely, and S. M. Freier, *J. Am. Chem. Soc.* **117**, 831 (1995).

[40] H. J. Larsen and P. E. Nielsen, *Nucleic Acids Res.* **24**, 458 (1996).

[41] T. Bentin and P. E. Nielsen, *Biochemistry* **35**, 8863 (1996).

[42] U. Giesen, W. Kleider, C. Berding, A. Geiger, H. Ørum, and P. E. Nielsen, *Nucleic Acids Res.* **26**, 5004 (1998).

[43] J. SantaLucia, H. T. Allawi, and P. A. Seneviratne, *Biochemistry* **35**, 3555 (1995).

[44] G. Haaima, A. Lohse, O. Buchardt, and P. E. Nielsen, *Ange. Chem.* **35**, 1939 (1996).

[45] B. D. Gildea, S. Casey, J. MacNeill, H. Perry-O'Keefe, D. Sørensen, and J. M. Coull, *Tetrah. Lett.* **39**, 7255 (1998).

Designing PNA Antisense Experiment

Although it is not sensible at this stage to provide a general procedure for performing PNA antisense experiments on cells in culture or in animals, the experience so far does allow for some guidelines:

1. Naked PNAs may show activity in some cells (bacteria, nerve cells), but PNA conjugates (e.g., peptides, or lipids) are often more potent. High nanomolar to low micromolar concentrations of PNA show activity, depending on the efficiency of the cellular uptake.
2. PNA size should be around 15 nucleobases with a sequence containing ca. 50% purines. A high pyrimidine content results in low hybrid stability and high purine (especially guanine) may cause solubility problems.
3. Several targets on the mRNA should be tested, but initially sites around AUG on the 5'-UTR may be chosen.
4. As in any other antisense experiment, stringent assay conditions and proper controls are of the utmost importance.[26] Assays should include quantification of the protein level (Western blot), whereas a decrease in the mRNA level (Northern) may not result.

Controls

a. The protein level of a gene with a similar or shorter half-life than that of the targeted protein should be monitored as a control.
b. A (PNA) dose–response curve should be produced.
c. Mismatched PNAs with identical nucleobase composition, but with interchanged bases should be used. Ideally, a series of PNAs with an increasing number of mismatches should exhibit corresponding decreased potency.
d. Alternatively, if a mutant is available for the gene in question, the wild-type PNA should be inactive on the mutant, whereas an analogous PNA with the mutant sequence should reestablish activity.

Because it is not known if PNAs have yet undiscovered (sequence dependent) biological nonantisense biological effects on cells or on animals, such stringent controls are required prior to claiming that any observed biological effect is caused by an antisense mechanism.

Technical Hints

PNA oligomers are synthesized conveniently by (automated) solid-phase synthesis using Boc-, Fmoc-, or Mmt-protected monomers[46] and are

---

[46] P. E. Nielsen and M. Egholm, eds., "Peptide Nucleic Acids (PNA): Protocols and Applications." Horizon Scientific Press, 1999.

purified easily by reversed-phase high-performance liquid chromatography using water/acetonitrile gradients (0.1% TFA). PNA monomers and oligomers are also available commercially (PerSeptive Biosystems). PNAs are best stored as dry powders, but stock solutions in pure water (which become acidic due to TFA from the purification) are also stable if kept at −20°. For some PNAs, aggregation becomes a problem on prolonged storage in solution.

For still unexplained reasons, PNA oligomers show decreased solubility in phosphate buffers. Thus PBS should be avoided if possible for cellular delivery of PNA. It is also important to note that PNA oligomers adsorb very strongly to polystyrene. Thus containers of this material should be used with caution.

[9] Synthesis of Oligonucleotide Conjugates in Anhydrous Dimethyl Sulfoxide

By David Milesi, Igor Kutyavin, Eugene A. Lukhtanov, Vladimir V. Gorn, and Michael W. Reed

Introduction

The performance of DNA probes or therapeutic oligonucleotides can often be improved by adding conjugate groups that provide unique biophysical properties. Conjugate groups for oligonucleotides have been classified into three major types[1]: chemically reactive groups, fluorescent or chemiluminescent groups, and groups promoting intermolecular interactions. Conjugate groups that can survive the automated DNA synthesis conditions can be added at the 3' terminus by using modified solid supports or into the growing oligonucleotide chain using modified phosphoramidites.[2] Those conjugate groups that are not compatible with phosphoramidite chemistry can often be added to the oligonucleotide postsynthetically. Various linker chemistries have been described for conjugating groups to oligonucleotides.[1] A common approach is to introduce nucleophilic alkylamine groups into the desired position of the oligonucleotide and to react these with an electrophilic derivative of the desired conjugate group. These methods generally require an aqueous base to dissolve the nucleic acid component and to increase the nucleophilicity of the alkylamine linker and an organic

[1] J. Goodchild, *Bioconj. Chem.* **1,** 165 (1990).
[2] S. L. Beaucage and R. P. Iyer, *Tetrahedron* **49,** 6123 (1993).

cosolvent to dissolve lipophilic electrophiles. Although these conditions are often successful, competing side reactions such as hydrolysis of the electrophilic conjugate group can give low yields and require large excess of the conjugate group. Extremely lipophilic conjugate groups are also a problem if they are insoluble in aqueous solutions.

This article describes two general methods for attaching conjugate groups to oligonucleotides containing alkylamine linker arms. We have used a simple acylation reaction with a variety of activated esters for the preparation of hundreds of oligonucleotide conjugates. The first method (Method A) describes reaction of high-performance liquid chromatography (HPLC)-purified amine-modified oligonucleotides under anhydrous conditions. Triethylammonium (TEA) salts of oligonucleotides are soluble in dimethyl sulfoxide (DMSO) and allow the successful conjugation of lipophilic and hydrolytically unstable groups in homogeneous organic solution. The anhydrous conditions also protect the electrophile from hydrolysis and require only a small excess of conjugate group. This versatile method has been used to prepare all three major types of oligonucleotide conjugates. The second method (Method B) describes acylation of 5′-alkylamine-modified oligonucleotides while still immobilized on the solid support. This simpler method has advantages for conjugate groups that can survive oligonucleotide deprotection conditions.

The two general methods just described are compared here by preparing a specific oligonucleotide conjugate: an oligodeoxynucleotide (ODN) with a 5′-minor groove-binding (MGB) conjugate group (Fig. 1). The solution structure of a 10-bp DNA duplex formed with MGB-ODN (3) has been solved by high-field nuclear magnetic resonance (NMR) analysis.[3] MGB–ODN conjugates have dramatically increased the affinity for complementary DNA targets.[4] A nonconjugated, alkylating minor groove binder (CC-1065) has been used to enhance suppression of translation by antisense ODNs,[5] but MGB–ODN conjugates were not prepared. Solution-phase synthesis of MGB–ODN conjugates has been described using cetylammonium salts of ODNs to provide organic solubility.[6] The MGB–ODN conjugates described here can stabilize DNA–RNA hybrids,[6] but have not been examined as antisense agents. In any event, the methods presented here

[3] S. Kumar, M. W. Reed, H. B. Gamper, V. V. Gorn, E. A. Lukhtanov, M. Foti, J. West, R. B. Meyer, and B. I. Schweitzer, *Nucleic Acids Res.* **26**, 831 (1998).

[4] A. N. Sinyakov, S. G. Lokhov, I. V. Kutyavin, H. Gamper, and R. B. Meyer, *J. Am. Chem. Soc.* **117**, 4995 (1995).

[5] D.-Y. Kim, D. H. Swenson, D.-Y. Cho, W. Taylor, and D. S. Shih, *Antisense Res. Dev.* **5**, 149 (1995).

[6] E. A. Lukhtanov, I. V. Kutyavin, H. B. Gamper, and R. B. Meyer, *Bioconj. Chem.* **6**, 418 (1995).

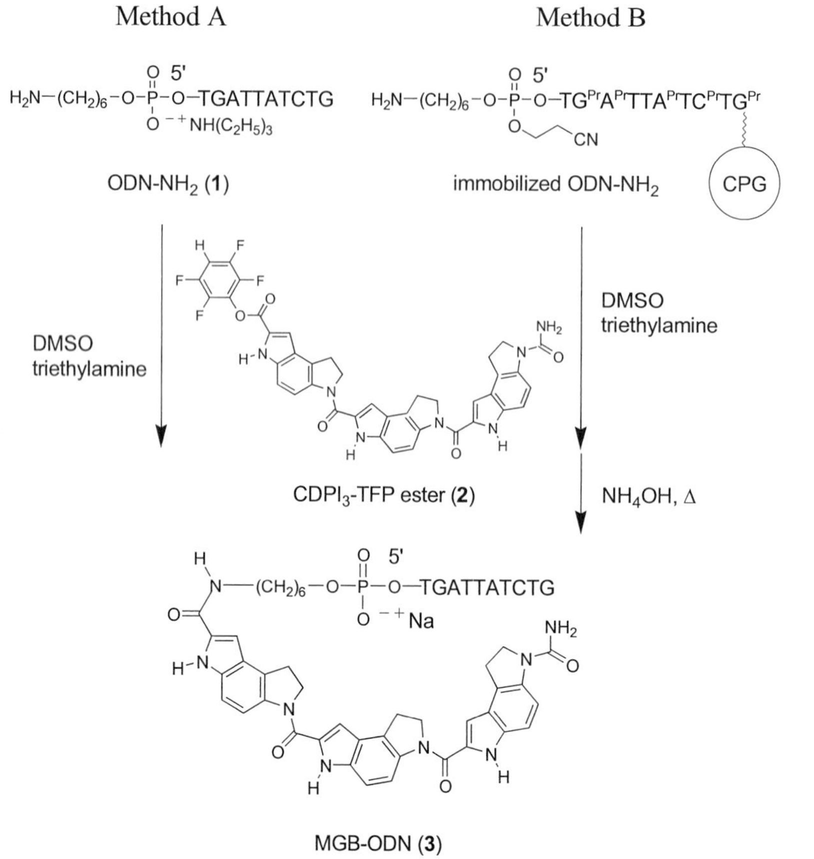

Method A

H₂N—(CH₂)₆—O—P(=O)—O—TGATTATCTG
with O⁻ ⁺NH(C₂H₅)₃ and O at 5'

ODN-NH₂ (1)

Method B

H₂N—(CH₂)₆—O—P(=O)—O—TG^Pr A^Pr TTA^Pr TC^Pr TG^Pr
with O—CH₂CH₂—CN and O at 5'

immobilized ODN-NH₂ (CPG)

DMSO
triethylamine

DMSO
triethylamine

CDPI₃-TFP ester (2)

NH₄OH, Δ

MGB-ODN (3)

FIG. 1. Two methods for synthesis of a minor groove binder–oligodeoxynucleotide conjugate (MGB–ODN).

are intended to provide a general method for conjugate preparation. The MGB conjugate group is a good case study because its lipophilicity prevents efficient conjugation in aqueous media. The anhydrous (DMSO) conditions of Method A were originally developed for the preparation of hydrolytically sensitive mustard–ODN conjugates.[7] Despite the fact that some of these conjugates have half-lives in aqueous solution of less than 30 min, HPLC purity of >90% was routine. The ability to precipitate these reactive ODNs rapidly from solution was critical for successful synthesis and is an important technique described here.

[7] M. W. Reed, E. A. Lukhtanov, V. Gorn, I. Kutyavin, A. Gall, A. Wald, and R. B. Meyer, *Bioconj. Chem.* **9,** 64 (1998).

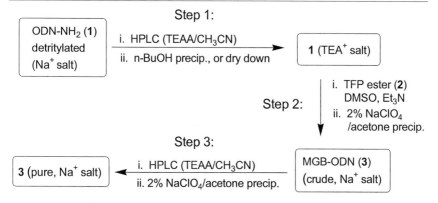

FIG. 2. Synthesis and purification flow chart for Method A.

Description of Methodology

Method A

A flow chart (Fig. 2) shows the synthetic and purification operations for Method A. A 1-μmol scale synthesis is used in the specific example described here, but the methods are scaled easily. The general methods rely heavily on reversed-phase HPLC for analysis, identification, and purification of the ODN conjugates. Analytical chromatograms in this article are shown for key steps during synthesis. These in-process controls are important for developing new types of ODN conjugates, but are not needed for an established method.

Step 1. The starting ODN-NH₂ (**1**) is a single peak by HPLC (Fig. 3A). More complex oligonucleotides may have unusual peak shape, or have impurities that coelute with the product. Impure **1** can be used in Method A, but yields suffer and product quality becomes suspect. The ODN-NH₂ used here is prepared using *N*-monomethoxytrityl (MMT)-protected hexanolamine phosphoramidite. Conditions for synthesis and HPLC purification of 5'-hexylamine-modified ODNs have been described.[8] Because sodium salts of ODNs are not soluble in DMSO, the counterion is exchanged by HPLC purification of the detritylated **1** using standard acetonitrile/ triethylammonium acetate (TEAA) mobile phases. Any residual trityl-on product is also removed at this stage. The fraction containing **1** can either be dried on a centrifugal evaporator or be concentrated and precipitated with 1-butanol to provide the required TEA salt of the ODN. The more volatile triethylammonium bicarbonate (TEAB) is advantageous because

[8] M. W. Reed, A. D. Adams, J. S. Nelson, and R. B. Meyer, *Bioconj. Chem.* **2**, 217 (1991)

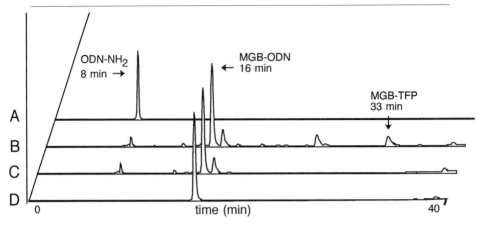

FIG. 3. HPLC analysis showing preparation of MGB–ODN **3** (Method A).

it leaves a more easily dissolved residue on drying. Mixed sequence ODNs are soluble in DMSO as their TEA salts (>10 mg/ml). Vigorous vortexing or sonication of the mixture for several minutes is often helpful in breaking up the dried ODN. Some modified purine-rich ODNs are sparingly soluble and this problem can be solved by using the corresponding tributylammonium (TBA) salts (using tributylammonium bicarbonate as the mobile phase).[9] Dried aqueous solutions of TEA salts are not strictly anhydrous because the ODN remains hydrated.

Concentrating aqueous solutions with butanol is a common biochemical technique. However, butanol concentration and precipitation of ODNs from the reversed-phase HPLC eluent are not trivial because the eluent contains variable amounts of acetonitrile as a cosolvent (depending on the lipophilicity of the ODN or conjugate). If the acetonitrile concentration is high, then the aqueous and organic phases will not separate. We have developed a "rule of thumb" that calls for adding 3 ml of *n*-butanol for each milliliter of collected eluent. If the lower aqueous layer is not apparent, then another 0.5 ml of water is added for each milliliter of eluent. After the aqueous layer is removed to another tube, further concentration with butanol is routine. For isolation of TEA salts as described earlier, the aqueous solution is concentrated to ~0.1 ml and then the ODN is precipitated with excess butanol, collected by centrifugation, washed with ethanol, and dried under vacuum.

[9] M. W. Reed, E. A. Lukhtanov, V. V. Gom, D. D. Lucas, J. H. Zhou, S. B. Pai, Y.-C. Cheng, and R. B. Meyer. *J. Med. Chem.* **38,** 4587 (1995).

Step 2. Reaction of the TEA salt of **1** with an active ester derivative of the desired conjugate group is accomplished easily, and the progress of the reaction can be followed by HPLC as shown in Fig. 3. The MGB conjugate group is especially easy to follow because it is lipophilic and has a unique absorbance maximum at 350 nm. The MGB also has an aborbance at 260 nm of 68,000 M^{-1} cm^{-1} that is almost as high as the ODN portion of the conjugate (96,000 M^{-1} cm^{-1}). Generally, the ODN is the more valuable component in the conjugate, and a large excess of the electrophile is used. For valuable conjugate groups (such as the MGB-TFP ester **2**), only a small excess is required. Only 2.5 molar equivalents are used in the example shown here, but 20 equivalents are generally used. Rigorously anhydrous DMSO (dried over calcium hydride) is particularly important for some aqueous-sensitive groups. More stable activated esters can tolerate some water, especially if a large excess is used. The excess triethylamine (or ethyldiisopropylamine) required in the reaction is also dried over calcium hydride before use. As shown in Fig. 3B, the crude reaction mix contains only a trace of starting **1** (8 min) and a major lipophilic conjugate peak corresponding to MGB-ODN **3** (16 min). Excess **2** elutes at 33 min, and a number of smaller side products are also seen. For novel conjugation reactions, identification of the desired conjugate peak can be challenging. Poor conversion, impurities from the excess activated ester, or small changes in retention time can be misleading. A dual wavelength UV or photodiode-array HPLC detector simplifies analysis, especially if the conjugate group has a unique chromophore.

Identification of the desired ODN conjugate is clarified during the isolation procedure. ODN components are precipitated from the DMSO reaction mixture by adding excess 2% (w/v) sodium perchlorate/acetone. Small molecular weight impurities generally remain in solution. The ODN pellet is collected by centrifugation, washed with acetone, and dried to give the sodium salt of the ODN. Reanalysis of this mixture by HPLC (Fig. 3C) shows acetone-insoluble (ODN) products.

Step 3. Finally, the desired ODN conjugate is isolated from side products by HPLC. The fraction-containing product is concentrated with butanol as described earlier to a volume of ~0.1 ml and is then precipitated with 2% sodium perchlorate/acetone to give the sodium salt of the ODN conjugate. Using combustion analysis, we have shown that cholesterol-modified ODNs isolated using this precipitation procedure have no residual sodium perchlorate.[9] Simply drying the HPLC-purified conjugate is feasible for aqueous stable conjugate groups. Traces of TEAA can be pumped off more easily if excess sodium bicarbonate is added to the fraction before drying (unpublished results).

After isolation, the ODN conjugates are dissolved in water and further characterized for purity by HPLC (Fig. 3D). Concentration is determined from the UV absorbance at 260 nm. Extinction coefficients for ODNs are calculated using the nearest neighbor model.[10] If the conjugate group absorbs strongly at 260 nm, its extinction coefficient must be added to that of the ODN. Usually the identity of the conjugate is clear from the "peak to peak" conversion shown in Fig. 3 and its UV spectrum. Sometimes further confirmation is required using mass spectroscopy or physical properties such as melting temperature (T_m) or fluorescence. After identity is confirmed, the ODN conjugate can be used in the functional assay for which it was designed.

Method B

Because the MGB conjugate group can survive ODN deprotection conditions, acylation of the immobilized ODN-NH$_2$ with MGB–TFP 2 is an option (Fig. 1). After addition of the hexylamine phosphoramidite, the MMT group is removed on the DNA synthesizer. The dried controlled pore glass (CPG) support is removed from the synthesis column and treated with a solution of 2 in anhydrous DMSO and excess triethylamine. Because effcient utilization of 2 is an important consideration, only 2.5 molar equivalents are used. Generally, 20 equivalents of activated ester are used. Unlike Method A, the kinetics of the reaction is difficult to follow. Generally the reaction is kept at room temperature overnight, the CPG is washed well, and the ODN conjugate is cleaved from the CPG and deprotected with ammonia as usual. HPLC analysis of the crude reaction mixture (Fig. 4A) shows some unmodified ODN-NH$_2$ (8 min), but the major peak corresponding to MGB–ODN 3 (16 min) is easily identified. HPLC purification gave the desired product as usual in 95% purity (Fig. 4B).

Yield data for Method A and Method B are compared in Table I. For preparation of 3, Method B is less labor intensive than Method A and isolated yields (based on 1 μmol scale DNA synthesis) are slightly better. However, Method B was much more consumptive of TFP ester 2 and required more careful chromatography. For that reason, Method A was chosen for further scaleup to provide 3 for NMR studies.[3] Because Method A requires several HPLC and precipitation steps, yields of conjugates can be low. Higher yields usually result if the ODNs are isolated from the HPLC eluent by drying on a centrifugal evaporator. The butanol concentration method is more rapid, but a steady hand and keen eye are required during the precipitation process. ODN conjugates bearing chemically reac-

[10] C. R. Cantor, M. M. Warshaw, and H. Shapiro, *Biopolymers* 9, 1059 (1970).

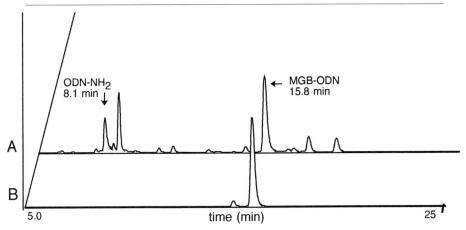

FIG. 4. HPLC analysis showing preparation of MGB–ODN **3** (Method B).

tive groups are unable to survive the aqueous dry down conditions and must be isolated by the butanol method.[7]

Experimental Protocols

Method A. Preparation of ODN Conjugates from HPLC-Purified ODN-NH₂

Equipment

Gradient HPLC system with appropriate column and UV detector
Vortexer, sonic bath
Microcentrifuge

TABLE I
YIELD DATA FOR SYNTHESIS OF COMPOUND **3**

Method	ODN-NH$_2$ (μmol)	Amount of **2** used (μmol)	Yield of **3** (μmol)	Yield based on ODN-NH$_2$ (%)	Yield based on **2** (%)
A	0.22[a]	0.55	0.082	37	15
B	1[b]	2.5	0.108	11	4.3

[a] For Method A, HPLC-purified ODN-NH$_2$ (**1**, TEA salt) from a 1-μmol scale DNA synthesis was used.
[b] For Method B, a 1-μmol column of dG-CPG was used.

Centrifugal evaporator with high vacuum pump
Pipettors (20, 200 μl), Pasteur pipettes
14- and 1.7-ml polypropylene tubes
UV spectrometer

Reagents

HPLC-purified ODN-NH$_2$ (detritylated) from 1 μmol scale DNA synthesis
Activated ester derivative of conjugate group
Triethylamine (stored over crushed calcium hydride)
DMSO (stored over crushed calcium hydride)
0.1 M triethylammonium acetate (pH 7.5)
Acetonitrile
n-Butanol
2% sodium perchlorate in acetone
Acetone

Procedure. Dried, detritylated ODN-NH$_2$ from a 1-μmol scale synthesis is dissolved in 0.5 ml of water and injected on a Hamilton PRP-1 column (Reno, NV) that is equilibrated with 0.1 M TEAA (pH 7.5). The TEA salt of the ODN is eluted from the column using a gradient of 0–60% acetonitrile/30 min and a flow rate of 2 ml/min. The desired peak (~13 min) is collected and dried *in vacuo* on a centrifugal evaporator. Alternatively, the TEA salt is precipitated from the HPLC eluent with *n*-butanol. The residue is dissolved in 0.2 ml of water and the concentration is determined. The TEA salt (0.22 μmol) is redried in a 1.7-ml Eppendorf tube, and the residue is dissolved in 0.1 ml DMSO with 15 μl of triethylamine. A 10-mg/ml solution of the desired activated ester in DMSO is prepared, and the desired number of molar equivalents is added to the ODN. Warning! DMSO solutions of activated esters can damage skin. Disposable nitrile gloves are highly recommended for this step. The mixture is vortexed and incubated until deemed complete by HPLC analysis. The crude ODN conjugate is then precipitated by adding 1.5 ml of 2% NaClO$_4$/acetone and vortexing. The mixture is centrifuged at 3000 rpm for 5 min (ambient temperature), and the pellet is sonicated with 1.5 ml of acetone and recentrifuged. The pellet is dried *in vacuo* for 15 min, dissolved in 0.2 ml of water, and analyzed by HPLC. After identifying the conjugate peak, the ODN conjugate is isolated by HPLC using a 4.6 × 250-mm column and the gradient as specified later. The peak is collected in ~1 ml of TEAA/acetonitrile, concentrated with butanol to ~0.1 ml, and precipitated with 1.5 ml of 2% NaClO$_4$/acetone. The pellet is washed with acetone and dried *in vacuo* for 15 min and the purified product is dissolved in 0.2 ml of water. Aliquots are removed for C$_{18}$ HPLC analysis and for concentration determination. The bulk solution of the ODN conjugate is stored at $-20°$ for future use.

Method B Preparation of ODN Conjugates from Immobilized ODN-NH$_2$

Reagents

Protected ODN-NH$_2$ (detritylated) on controlled pore glass support (1 μmol)Concentrated ammonium hydroxide

5 ml Reactivial (Pierce Chemical Co., Rockford, IL) with Teflon septum

Other equipment and reagents are as described for Method A

Procedure. The trityl-off solid support is washed and dried on the DNA synthesizer and then emptied into a 1.7-ml Eppendorf tube. A 20-mg/ml solution of the desired activated ester in DMSO is prepared, and the desired number of molar equivalents is added to the solid support DMSO with 15 μl of triethylamine. Enough dry DMSO (~0.1 ml) is added to cover the support. The mixture is capped, vortexed, and incubated overnight. DMSO is added to the mixture and the supernatant is removed (warning! disposable nitrile gloves are highly recommended for this step). The CPG is washed with another 1.5-ml portion of DMSO and then with 1.5 ml of acetone. The CPG is dried *in vacuo* for 15 min and then transferred to a Reactivial with 1 ml of concentrated ammonium hydroxide. The ODN is deprotected as usual (5–15 hr at 55°) and cooled before opening. The ammonia solution is injected directly onto a Hamilton PRP-1 column (Reno, NV) that is equilibrated with 0.1 M TEAA (pH 7.5). The ODN conjugate is eluted from the column using a gradient of 0–60% acetonitrile/30 min and a flow rate of 2 ml/min. The desired peak is collected and dried *in vacuo* on a centrifugal evaporator. Alternatively, the conjugate is isolated from the HPLC eluent by *n*-butanol concentration and 2% NaClO$_4$/acetone precipitation. The residue is dissolved in 0.2 ml of water and the concentration is determined. Aliquots are removed for C$_{18}$ HPLC analysis and for concentration determination. The bulk solution of the ODN conjugate is stored at $-20°$ for future use.

HPLC Analysis Conditions

All modified ODNs are analyzed by reversed-phase HPLC (C$_{18}$ HPLC) using a 4.6 × 150-mm C$_{18}$ column (3-μm particle size, 300-Å pore size) equipped with a guard column (Varian Dynamax). A gradient of 5–65% acetonitrile over 30 min is used (flow rate of 1 ml/min) where the aqueous buffer is 0.1 M TEAA (pH 7.5); detection is by UV absorbance at 260 nm.

[10] Gene Switching: Analyzing a Broad Range of Mutations Using Steric Block Antisense Oligonucleotides

By PAUL A. MORCOS

Introduction

The basic idea of turning off one gene and replacing it with an altered or altogether different gene is not a new one. Yeast geneticists have, for many years, used a variety of techniques to perform similar tasks. For example, yeast can be transformed easily with a plasmid expressing a mutant or human form of an endogenous gene, and the endogenous gene can be disrupted by targeted integration via homologous recombination. This allows the yeast geneticist to perform complementation studies as well as determine the effects of engineered mutations. This technique, however, is time-consuming, expensive, and not well suited for studies in mammalian tissue culture.

This article describes gene switching, a simple but powerful method that employs antisense technology to specifically shut down an endogenous gene and replace it with an engineered construct transfected into the cells under study (Fig. 1). One or more functions have been determined for many human genes, and additional interest lies in characterizing functional domains and assessing the effects of truncation and point mutations on the function of a gene product. The gene switching method allows one to utilize antisense technology as a powerful tool to carry out a variety of genetic studies, including determination of functional domains, analysis of mutational or expression level effects, and even genetic complementation studies with minimal cost, time, and additional variables. This method is easily adaptable to a variety of systems; however, there are requirements that must be met for a successful outcome. First, the chosen tissue culture cells must be amenable to standard delivery methods such as streptolysin O treatment[1] or the scrape-load procedure,[2] which we describe for introducing antisense molecules into adherent cells in culture. Second, the transfected gene must be introduced in an acceptable expression plasmid for the chosen cell type and cloned so that the endogenous (chromosomal copy) gene and the cloned gene have different leader sequences. The leader sequence of the endogenous gene serves as the antisense target in the gene switching

---

[1] D. Spiller and D. Tidd, *Antisense Res. Dev.* **5,** 13 (1995).

[2] M. Partridge, A. Vincent, P. Mathews, J. Puma, D. Stein, and J. Summerton, *Antisense Nucleic Acid Drug Dev.* **6,** 169 (1996).

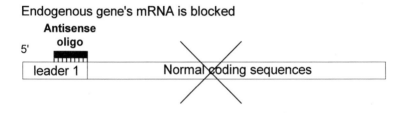

Endogenous gene's mRNA

5'

| leader 1 | Normal coding sequences |

Antisense target

Transfect exogenous gene

Introduce antisense oligo

Endogenous gene's mRNA is blocked

Antisense oligo

5'

| leader 1 | Normal coding sequences |

Exogenous gene's mRNA is translated

5'

| leader 2 | Mutant coding sequences |

FIG. 1. A schematic illustrating the gene switching method.

method. Finally, an antisense structural type must be used that specifically shuts down the endogenous gene without affecting expression of the cloned gene or any other gene in the cell. Not unexpectedly, this last requirement proves to be the most difficult to satisfy.

Readily available antisense structural types, including phosphorothioate-linked DNAs[3] (thiophosphates), chimeric oligonucleotides,[4] 2'-O-

[3] C. Stein and J. Cohen, in "Oligodeoxynucleotides: Antisense Inhibitors of Gene Expression" (J. Cohen, ed.), p. 97. CRC Press, Boca Raton, FL, 1989.
[4] B. Monia, E. Lesnik, C. Gonzalez, W. Lima, D. McGee, C. Guinosso, A. Kawasaki, P. Cook, and S. Freier, J. Biol. Chem. **268**, 14514 (1993).

methyl-RNAs,[5] peptide nucleic acids[6] (PNAs), and morpholino oligonucle-otides,[7] all provide a means of shutting down a selected gene. However, only thiophosphates and chimeric oligonucleotides (RNase H-competent structural types) are touted as having the ability to target the coding sequence of a gene. The most readily available RNase H-competent antisense molecules, the thiophosphates, lack the specificity necessary to differentiate point mutations and, in most cases, lack even the specificity to target a selected gene. Thiophosphates suffer a number of fundamental problems.[8] First, the secondary structure of a typical target RNA can effectively block thiophosphates from gaining access to a desired target sequence. Second, thiophosphates exhibit a variety of nonantisense effects that can lead to erroneous conclusions about gene function. In addition, thiophosphates are sensitive to both exonucleases and endonucleases. While more resistant to nucleases than thiophosphates, chimeric oligonucleotides still exhibit less than optimal specificity and generate significant nonantisense effects. PNAs do not suffer from undue nonantisense effects, but they lack the specificity of morpholino oligonucleotides, and PNAs are replete with target sequence constraints and composition limitations. Morpholino oligonucleotides provide better specificity than PNAs and thiophosphates and are free of the multiplicity of nonantisense effects observed with thiophosphates.[7,9] Morpholino oligonucleotides, which work via a steric block mechanism like PNAs, are extremely effective in specifically inhibiting mRNA sequences from the 5'-cap through the first 25 bases of the protein-coding region. The extremely high specificity of morpholinos is demonstrated by a 4-bp mismatch in a 25-mer oligonucleotide that virtually abolishes the affects on that mRNA.[9]

This article describes the gene switching method, including general considerations in planning a gene switching experiment. In our experimental gene switching example, we employ a dual reporter system to compare the effectiveness of an RNase H-independent structural type, a morpholino oligonucleotide, to that of a standard thiophosphate and an advanced chimeric oligonucleotide. Morpholino oligonucleotides were chosen over the other readily available RNase H-independent steric blockers, 2'-O-methyl phosphorothioates, and PNAs based on an experiment comparing their

[5] S. Shibahara, S. Mukai, H. Morisawa, H. Nakashima, S. Kobayashi, and N. Yamamoto, *Nucleic Acids Res.* **17**, 239 (1989).
[6] M. Egholm, O. Buchardt, P. Nielsen, and R. Burg, *J. Am. Chem. Soc.* **114**, 1895 (1992).
[7] J. Summerton and D. Weller, *Antisense Nucleic Acid Drug Dev.* **7**, 187 (1997).
[8] C. Stein, *Nature Med.* **1**, 1119 (1995).
[9] J. Summerton, D. Stein, S. Huang, P. Mathews, D. Weller, and M. Partridge, *Nucleic Acid Drug Dev.* **7**, 63 (1997).

Endogenous gene

Exogenous gene

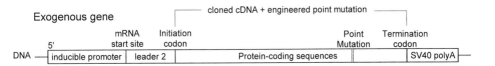

Endogenous gene's mRNA

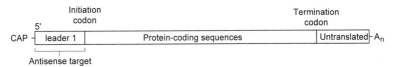

Exogenous gene's mRNA

Fig. 2. The typical structure of an endogenous gene and a transfected gene in a gene switching experiment. In this example, the transfected gene is derived from cloning the cDNA of the endogenous gene into a plasmid and engineering the mutation of interest. The transfected gene is under the control of an inducible promoter, and leader 2 is derived from the plasmid.

activity in a new "positive" test system for RNase H-independent antisense structural types.[10]

Gene Switching: Principle and Advantages

The gene switching method makes use of antisense technology to shut down expression of an endogenous gene in tissue culture and replace it with plasmid-mediated expression of a desired gene construct. The basis for the gene switching method lies in selectively inhibiting expression of the endogenous gene by targeting its mRNA leader sequence and providing a replacement gene on a plasmid that generates mRNA with a different leader sequence not affected by the antisense oligonucleotide (Fig. 2).

[10] S. Kang, M. Cho, and R. Kole, *Biochemistry* **37**, 6235 (1998).

Colige *et al.*[11] used antisense oligonucleotides for a somewhat similar purpose. They transfected an internally deleted human proα1(I) procollagen gene into NIH 3T3 cells and then used a thiophosphate antisense oligonucleotide to inhibit expression of the transfected gene. Their most effective target, which was a 20-base sequence of the human gene, which differed by nine nucleotides from the endogenous mouse gene, yielded up to 80% inhibition of the transfected gene and 10% inhibition of the endogenous gene.

The gene switching method provides several distinct advantages over this earlier antisense strategy. First, in its simplest form, a single antisense oligonucleotide can be used for shut down of the endogenous gene while analyzing many different plasmid constructs, including unchanged or even subtly altered forms of the endogenous gene. Second, targeting the leader sequence of the gene allows the researcher to utilize a high-specificity RNase H-independent antisense type instead of having to use an inherently low-specificity RNase H-competent type such as thiophosphates. Targeting the noncoding sequence allows the researcher to analyze the effects of point mutations, insertion mutations, deletion mutations, and even expression level effects associated with increased or decreased expression of the normal gene. Third, in many cases the researcher may already have cDNA constructs that are ready for use in this system. Additional desired constructs including point mutations, are created easily using simple polymerase chain reaction (PCR) techniques.

Requirements

Several criteria must be met in order to utilize the gene switching method. First, the tissue culture cell type must be amenable to the delivery of antisense oligonucleotides. Unassisted delivery of antisense oligonucleotides has not yet been achieved, and many of the current methods for delivery are detrimental to the cell. Lipofectin and other lipid-derived cationic transfection compounds are toxic and only work with charged antisense structural types.[12] Streptolysin O can be used with charged or uncharged antisense structural types, but streptolysin O is also toxic and often results in suboptimal delivery.[1] If nonadherent cells are required, then streptolysin O is the delivery method of choice because it affords the researcher the ability to choose among all of the readily available antisense structural types. However, the consistent and efficient delivery of oligonu-

[11] A. Colige, B. Sokolov, P. Nugent, R. Baserga, and D. Prockop, *Biochemistry* **32,** 7 (1993).
[12] J. Lewis, K. Lin, A. Kothavale, W. Flanagan, M. Matteucci, R. DePrince, R. Mook, R. Hendren, and R. Wagner, *Proc. Natl. Acad. Sci. U.S.A.* **93,** 3176 (1996).

cleotides with streptolysin O may require substantial optimization. For adherent cells, scrape loading is the method of choice for delivering antisense oligonucleotides.[2] Described in detail later, scrape loading results in the efficient delivery of charged and uncharged antisense structural types with minimal detrimental effect on the cells.

Second, the antisense target sequence must be within the leader sequence of the endogenous gene, and the transfected gene must be cloned in a manner such that it does not share the leader sequence of the endogenous gene. At minimum, a portion of the endogenous leader sequence must be known so that a precise antisense target can be defined. This is the crux of the gene switching method. Cloned sequences that are expressed in mammalian cells almost invariably include the initiation codon of the endogenous gene, but typically do not include all of the leader sequence and ribosomal assembly site. Many of today's expression plasmids include the critical sequences necessary for efficient expression, including the transcriptional and translational start sites, requiring the researcher to clone only the coding sequence beginning with the initiation codon. Use of such expression plasmids is the optimal scenario for using the gene switching method as it allows the researcher to utilize the entire endogenous leader sequence in selecting a target. In situations where the leader of the endogenous gene is poorly defined and partially (at least 20 bases) or entirely replicated in the transfected gene, the engineered removal of 20 or more bases immediately upstream of the initiation codon in the transfected gene is a potential solution and is accomplished relatively easily using PCR. In most cases, this deletion will not affect expression of the gene, but will allow antisense targeting encompassing the same bases not deleted in the endogenous gene.

General Considerations

Once a tissue culture cell type and antisense target are selected, the remaining tasks include choosing a plasmid, optimizing expression of the transfected gene, and delivering antisense oligonucleotides.

There are many options when it comes to choosing a plasmid, particularly in terms of regulating and monitoring gene expression. While the choice of plasmid will likely be determined by a particular study, the following general guidelines can help in making the right choice. Inducible or repressible plasmid expression is desired in a typical study using gene switching. This level of control allows a researcher to induce [e.g., with isopropylthiogalactoide (IPTG)[13]] or derepress (e.g., removal of tetracy-

---

[13] D. Wyborski and J. Short, *Nucleic Acids Res.* **19,** 4647 (1991).

cline[14]) expression of the transfected gene while concomitantly shutting down the endogenous gene by an antisense mechanism. There are several advantages to having control over plasmid expression. First, many genes give rise to distinct phenotypes when overexpressed. Having both plasmid-derived and endogenous expression prior to the addition of antisense oligonucleotides may yield results that have little or nothing to do with the specifics of the transfected gene, but rather due to a combined effect. Second, using a repressible or inducible expression system allows the researcher to have a defined "switch," a time at which the endogenous gene is switched "off" and a time at which the transfected gene is switched "on." Third, inducible/repressible systems are also better suited for experimental controls, which include not only noninduction or maintenance of repression, but also performing a reverse gene switching experiment in which expression of the transfected gene is inhibited with antisense. Fourth, maintaining plasmid control should result in a reduction of background protein levels expressed by the transfected gene in cases where the transfected gene serves as the antisense target. Transfection with constitutively expressing plasmids may give rise to such high levels of protein expression that proteins with long half-lives could remain for days. In such a system, antisense activity measured in the short term could be misconstrued as poor or nonactive if one is measuring gene product or reporter levels.

Monitoring protein expression from the transfected gene can be performed with protein specific tools (i.e., antibodies) or by utilizing one of the many available plasmids that generate fusion proteins of the cloned gene product with a reporter. In general, the latter is much more effective and can be more accurately measured. Reporters are also less susceptible to experimental variation, and the potential background generated by the endogenous gene is nonexistent. In the gene switching example described in the next section, the firefly[15] and the *Renilla*[16] luciferases are used as reporters. In this case, very accurate measurements of luciferase activity can be obtained with a luminometer. Other common reporters that allow for accurate expression measurements using commercially available kits are β-galactosidase and chloramphenicol acetyltransferase (CAT). If microscopy and subcellular localization studies are important, then a vector that generates a fusion to a fluorescent reporter such as green fluorescent

[14] F. Sturtz, L. Cioffi, S. Wittmer, M. Sonk, A. Shafer, Y. Li, N. Leeper, J. Smith-Gbur, J. Shulok, and D. Platika, *Gene* **221,** 279 (1998).

[15] J. de Wet, K. Wood, D. Helinski, and M. Deluca, *Proc. Natl. Acad. Sci. U.S.A.* **82,** 7870 (1985).

[16] W. Lorenz, R. McCann, M. Longiaru, and M. Cormier, *Proc. Natl. Acad. Sci. U.S.A.* **88,** 4438 (1991).

protein[17] (GFP) may be a good choice; however, accurate expression measurements are typically sacrificed.

Generating the desired level of protein expression from the transfected gene is best achieved with a combination of optimal transfection and induction/derepression. Optimizing transfection is entirely DNA and tissue type specific and will be unique for each study, and the ultimate level of protein expression is gene specific. It may take several pilot experiments to optimize transfection and protein expression before delivering antisense oligonucleotides.

Gene Switching Method in Practice

A dual reporter system was employed in order to demonstrate the gene switching method. In our example, the endogenous gene is derived from a stably integrated plasmid in HeLa cells carrying a dexamethasone-inducible firefly luciferase gene with a leader sequence derived from rabbit globin.[9] The transfected gene is a CMV promoter-driven *Renilla* luciferase in an unmodified, commercially available plasmid called pRLCMV.[18] The stably transfected HeLa cell was chosen because of its use in previously published antisense experiments.[9] The pRLCMV plasmid was chosen because of the similar nature of *Renilla* luciferase to that of firefly luciferase and additionally allowing the use of a very sensitive and commercially available dual luciferase assay. In order to demonstrate the effectiveness of antisense oligonucleotides to specifically shut down expression of the endogenous gene without affecting expression of the transfected gene, we chose to compare a steric blocker to that of two RNase H-dependent structural types, a classic thiophosphate oligonucleotide and a new generation chimera.

In order to determine the best steric blocker for use in the gene switching method, we employed a new "positive" antisense assay system for RNase H-independent structural types.[10] This commercially available positive assay, developed by Ryszard Kole, incorporates a globin splice site mutation fused upstream to the firefly luciferase coding sequence. This mutation generates a favorable splice site, resulting in a premature stop codon and failed expression of firefly luciferase. Kole has shown that steric blocking antisense structural types can act as a "splice corrector" by blocking the splice site generated by the mutation, restoring normal splicing and inducing firefly luciferase expression. This "positive" assay imparts an important advantage over the typical "negative" assays where one looks for downregu-

---

[17] M. Chalfie, Y. Tu, G. Euskirchen, W. Ward, and D Prasher, *Science* **263,** 802 (1994).
[18] E. Schmidt, G. Christoph, R. Zeller, and P. Leder, *Mol. Cell Biol.* **10,** 4406 (1990).

lation in that there is little or no possibility of nonantisense effects generating a positive result in the splice-correcting system. In a negative assay, a variety of nonantisense effects can easily be misconstrued as antisense derived. In fairness to PNAs, which have a synthesis length barrier of roughly 18 bases, we compared identical 18-mer PNA, 2'-O-methyl phosphorothioate, and morpholino oligonucleotides targeted against the globin splice site mutation in the positive assay. An 18-mer PNA is the longest available commercially and thus represents a "best case" for the PNA in our system. Morpholino oligonucleotides do not have similar synthesis constraints, and as such we chose to add a 28-mer morpholino oligonucleotide targeted against the same splice site mutation to represent the "best case" scenario for morpholino oligonucleotides. In comparing the 18-mers, morpholino oligonucleotides and PNAs were equally capable of restoring normal splicing (Fig. 3). PNAs appear to have 25–100% greater activity than the 18-mer morpholino in the 250 to 750 nM range, but the 18-mer morpholino and PNA exhibit equivalent activity at 1000 nM. The 2'-O-methyl phosphorothioate, however, exhibited approximately 16% of the activity of the corresponding morpholino and PNA oligonucleotides at 1000 nM. Interestingly, the 28-mer morpholino was in a league by itself in comparison to the 18-mer oligos, with 2.3 times the activity of the 18-mer PNA and 84% of maximal splice correction at 1000 nM final concentration. The maximal splice correction was determined as the relative luciferase units achieved with the 28-mer morpholino oligonucleotide at 3000 nM (data not shown). In summary, these results indicated that the morpholino oligonucleotide was the best choice to represent the RNase H-independent structural types.

In our demonstration of the gene switching method, a 25-mer morpholino oligonucleotide directed against the globin leader sequence was compared to a standard thiophosphate and an optimized chimera directed against the same sequence. The optimized chimera was composed of 9 thiophosphate subunits at the 5' end coupled to 16 2'-O-methyl-linked thiophosphate subunits ending with a 3' inverted base. The scrape-load procedure described later was used to deliver oligonucleotides at final concentrations of 250, 500, and 1000 nM into HeLa cells transfected stably with the globin leader–firefly luciferase fusion gene and transfected transiently with pRLCMV. It is important to note that in pilot experiments, the thiophosphate oligonucleotides and, to a lesser extent, the chimeric oligonucleotides interfered with reattachment of the scrape-loaded cells (unpublished observation). Thus, in this experiment, to allow reattachment of cells treated with thiophosphates, all cells were transferred to fresh media free of oligonucleotides after scrape loading. As seen in Fig. 4, the thiophosphate showed no ability to shut down endogenous expression of

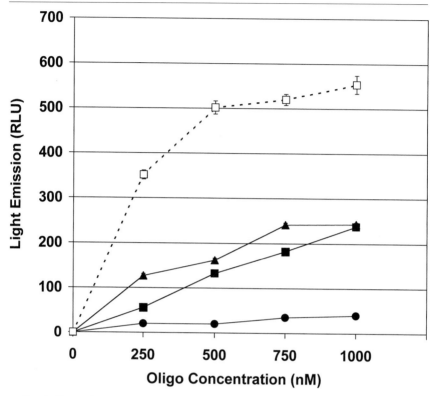

Fig. 3. Comparing morpholino, PNA, and 2'-*O*-methyl phosphorothioate oligonucleotides in a positive test system. An 18-mer morpholino (■), an 18-mer PNA (▲), an 18-mer 2'-*O*-methyl phosphorothioate (●), and a 28-mer morpholino (□) oligonucleotide were tested for the ability to correct a globin splice site mutation. Oligonucleotides were tested in duplicate at the indicated concentrations in HeLa cells transfected stably with pLUC/705,[10] a plasmid composed of the luciferase gene interrupted with human β-globin intron 2 containing a favored splice site mutation (IVS2-705).

firefly luciferase, and the chimera managed only a 25% reduction of expression at 600 n*M*. Conversely, the morpholino oligonucleotide effectively reduced endogenous firefly luciferase expression by 75% at 600 n*M* without significantly affecting expression of the transfected *Renilla* luciferase. The chimera, however, appeared to stimulate pRLCMV expression by 50% at 600 n*M*.

Discussion

In our test system, only the morpholino oligonucleotides were capable of efficiently and specifically shutting down endogenous firefly luciferase

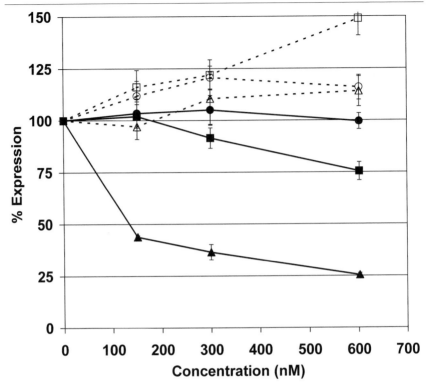

FIG. 4. Comparing morpholino, standard thiophosphate, and optimized chimera in a gene switching experiment. A 25-mer morpholino (triangles), a 25-mer standard thiophosphate (circles), and a 25-mer optimized chimera (squares) were tested for effects on expression of the targeted firefly luciferase (closed symbols) and nontargeted *Renilla* luciferase (open symbols). Data are displayed as a percentage of the expression achieved without addition of antisense oligonucleotides. Antisense oligonucleotides were tested in triplicate at the indicated concentrations in HeLa cells transfected stably with a plasmid composed of a dexamethasone-inducible MMTV promoter and the rabbit α-globin leader fused to firefly luciferase[2] and transfected transiently with pRLCMV expressing *Renilla* luciferase.

expression without affecting expression of the transfected *Renilla* luciferase gene. The classic thiophosphate oligonucleotide had little effect on either firefly or *Renilla* luciferase expression within cells. The new generation chimera achieved only 25% shutdown of the targeted firefly luciferase and caused a 50% stimulation of the nontargeted *Renilla* luciferase.

The comparison of morpholino, PNA, and 2′-*O*-methyl phosphorothioate oligonucleotides in the positive test system indicated that a "best case" 28-mer morpholino oligonucleotide has more than 2-fold greater activity than a "best case" PNA and roughly 10-fold greater activity than the 2′-

O-methyl phosphorothioate directed against the same target. These data suggest that the RNase H-independent structural types, including morpholinos and, to a lesser extent, PNA and $2'$-O-methyl phosphorothioates, should provide useful tools in implementing the gene switching method.

In principle, the gene switching requirement of targeting the endogenous leader sequence should be achievable with any antisense type; however, the RNase H-competent structural types showed no antisense effect in our system. This lack of in-cell activity by phosphorothioates is not unexpected based on several reports.[7-9] Specifically, it has been reported that thiophosphates which are quite active in cell-free translation assays, can require more than a thousandfold higher concentration in scrape-load experiments in order to achieve a similar level of activity.[9] Experiments carried out by microinjecting thiophosphates into cells also indicate that most thiophosphates are largely inactive within cells.[19] It is suspected that the ionic nature of thiophosphates may afford interaction with cellular strand-separating factors and other cellular components, thus reducing activity.[8,20]

The high degree of variability encountered with endogenous gene expression and plasmid-mediated expression of transfected genes may require optimization of methods for particular studies to yield good results with the gene switching method. The factors that will likely require optimization are type and concentration of transfection agent and plasmid DNA, type and concentration of the oligonucleotide delivery agent, and the amount of time after oligonucleotide delivery before analyzing cells for phenotype or reporter activity. Determining the type of the transfection agent is primarily a function of the cell type, and typically a particular cationic lipid-based agent favors certain cell types over others. Pilot experiments will be required to define the optimal concentration and ratio of transfection agent to DNA for a particular level of expression of the transfected gene. The right plasmid for a particular gene switching experiment depends on the desired level of regulation, the desired level of expression, and a chosen means of monitoring expression. As mentioned previously, the scrape-load procedure is the oligonucleotide delivery method of choice for adherent cells, and cationic lipid-based transfection agents are not a viable alternative for delivering morpholino or PNA oligonucleotides due to their nonionic nature. Prior to publication, GENE TOOLS introduced an optimized Osmotic Delivery System that provides uniform delivery of oligos in adherent and suspension cells that surpasses the efficacy and viability obtained with streptolysin O. In our study, cells were analyzed for antisense activity 16

---

[19] C. Moulds, J. Lewis, B. Froehler, D. Grant, T. Huang, J. Milligan, M. Matteucci, and R. Wagner, *Biochemistry* **34,** 5044 (1995).

[20] A. Krieg and C. Stein, *Antisense Res. Dev.* **5,** 241 (1995).

hr after scrape loading, and in our system endogenous expression was induced with dexamethasone after the antisense oligonucleotides were delivered. It is quite possible that a given gene switching experiment may require increasing the time between oligonucleotide delivery and measurement of antisense activity, particularly if the endogenous gene exhibits high levels of expression and/or the proteins expressed have long half-lives.

The gene switching method provides a powerful means of studying gene function. By targeting the leader sequence of an endogenous gene, the researcher can utilize high-specificity RNase H-independent antisense structural types. Further, the affects of replacing an endogenous gene with virtually any desired construct, from different genes to subtle point mutations, can be studied. It is important to realize that all of these studies can be accomplished with a single effectively targeted antisense oligonucleotide. We have shown that morpholino oligonucleotides work well in this system, and for researchers working with adherent cells, particularly HeLa or NIH 3T3 cells, gene switching is easily tailored for a variety of studies. Experiments with nonadherent cells will be more difficult due to the need for using less reliable delivery methods. Such experiments will likely require more preparation before viable data are generated.

Materials and Reagents

Materials

AVI Biopharm Inc. provided HeLa cells transfected stably with a plasmid composed of a dexamethasone-inducible mouse mammary tumor virus (MMTV) promoter and the rabbit α-globin leader fused to firefly luciferase.[2] HeLa cells transfected stably with pLUC/705,[10] a plasmid comprising the firefly luciferase gene interupted with human β-globin intron 2 containing a favored splice site mutation (IVS2-705), were from Dr. Ryszard Kole. Morpholino oligonucleotides were synthesized by GENE TOOLS, LLC (www.gene-tools.com). Standard and high-performance liquid chromatography (HPLC)-purified chimeric phosphorothioate oligonucleotides were purchased from Oligos Etc. Inc (www.oligosetc.com). PNA oligonucleotides were purchased from Perceptive Inc. (www.pbio.com). 2'-O-Methyl phosphorothioate oligonucleotides were a gift from Dr. Ryszard Kole. The plasmid pRLCMV, which is composed of a CMV promoter driven-*Renilla* luciferase, was purchased from Promega (Madison, WI).

Reagents

The transfection compound Tfx20 was purchased from Promega and prepared as described in Promega note TB216. Dexamethasone (Sigma,

St. Louis, MO) was resuspended in sterile water to a final concentration of 1 mM. Tissue culture cells were grown in D-MEM/F12 (GIBCO-BRL, Gaithersburg, MD) supplemented with 10% fetal bovine serum (GIBCO-BRL). Cells were trypsinized with 0.25% trypsin–EDTA (Sigma). Reduced serum media for transfection were purchased from GIBCO-BRL. Dulbecco's phosphate-buffered saline (PBS) was purchased from GIBCO-BRL. The firefly luciferase assay system and dual luciferase reporter assay system, including the firefly and *Renilla* luciferase substrates and passive cell lysis buffer were purchased from Promega. The protein assay reagent was purchased from Bio-Rad (Richmond, CA).

Methods

Scrape-Loading Antisense Oligonucleotides

Approximately 10^6 trypsinized cells in 1 or 2 ml of growth medium are seeded into a well of a six-well tissue culture plate. After incubating at 37° for 16–24 hr, antisense or control oligonucleotide is added to the medium at a desired final concentration, swirled briefly, placed on a flat surface, and scraped gently with a disposable, sterile cell scraper (Sarstedt). Best results are achieved with a low-force sweeping motion using cell scrapers that have a rubber blade. Using a disposable pipette, gently transfer medium to the well of another culture plate and restore to incubator. Induction reagents are added 6 hr after scraping. Cells should be incubated at least 16 hr prior to measuring antisense activity.

Testing RNase H-Independent Antisense Oligonucleotides in the Positive Test System

Approximately 10^6 HeLa cells transfected stably with pLUC/705 are seeded into six-well tissue culture plates. After incubation at 37° for 16 hr, morpholino, PNA, 2'-*O*-methyl phosphorothioate 18-mer oligonucleotides (each with the sequence 5' CCTCTTACCTCAGTTACA), and 28-mer morpholino oligonucleotides (5' CCTCTTTACCTCAGTTACAATTTTATATGC) are scrape loaded in duplicate as described earlier at final concentrations of 0, 250, 500, 750, and 1000 nM. Additionally, 28-mer morpholino oligonucleotides are scrape loaded at 3000 nM to serve as the maximum achievable splice correction. After 24 hr incubation at 37°, cells are washed 1× with PBS, and then 250 μl passive lysis buffer is added to each well. Flasks are swirled to distribute lysis buffer evenly and then cells are scraped with a disposable cell lifter (Fisher). Cells are transferred to microfuge tubes, vortexed for 10 sec, and centrifuged at 14,000 rpm for 1 min. Twenty microliters of each lysate is assayed for firefly luciferase activity and protein concentration as described later.

Firefly Luciferase Assay and Calculations for the Positive Test System

In a clean test tube, 20 μl of cell lysate is mixed with 100 μl of Promega luciferase assay reagent at ambient temperature and vortexed for 5 sec. Light emission is measured for 15 sec in a Model TD-20e luminometer (Turner Designs, Inc.).

Following assessment of light emission, protein in the lysate is quantitated by adding 20 μl of cell lysate to 780 μl sterile water and 200 μl protein assay reagent, vortexing for 1 min, and then reading the optical absorbance in a spectrophotometer at 595 n*M*. Values are normalized by dividing light emission by protein determinations to give relative light units (RLUs) that are proportional to cell number. Specifically, RLUs are calculated by dividing the measured light units by optical absorbance measured in the protein assay. The positive test system has a slight background level of splice correction. For graphing results, background RLUs generated without the addition of antisense oligonucleotides (15.55 RLUs) are subtracted from RLUs generated from the addition of antisense oligonucleotides to reveal splice correction that is exclusively due to antisense oligonucleotide activity.

Gene Switching Experiment

HeLa cells transfected stably with a plasmid composed of the rabbit α-globin leader fused to firefly luciferase are grown to 80% confluency in a 75-cm^2 tissue culture flask prior to transfection with *Renilla* luciferase plasmid pRLCMV. In a sterile 15-ml screw-cap test tube, 9 ml serum-reduced media, 40 μl Tfx20 transfection reagent, and 12 μg pRLCMV plasmid DNA are mixed gently, incubated at ambient temperature for 15 min, and added to the cells that were first washed once with PBS. The cells are allowed to incubate for 1 hr at 37°, and then 36 ml growth medium is added to the flask and allowed to incubate at 37° for another 16 hr. The medium/transfection mixture is removed from the confluent monolayer, and the cells are trypsinized, resuspended in 33 ml growth medium, and seeded in 1-ml aliquots into 33 wells of 6-well tissue culture plates. After incubation at 37° for 24 hr, the medium in each well is replaced with 1 ml fresh growth media. Morpholino, phosphorothioate, and chimeric 25-mer oligonucleotides (5' GGTGGGTTCCTTCTCAGTCGGACTGG) are scrape loaded in triplicate as described earlier at final concentrations of 0, 150, 300, and 600 n*M*. Six hours after scrape-loading oligonucleotides, endogenous globin–luciferase expression is induced with dexamethasone at 3 μM final concentration. After another 16 hr of incubation, lysates are prepared as described for the positive test system, except 150 μl of passive lysis buffer is used for each well. Dual luciferase assays are performed as described next.

Dual Luciferase Assay and Calculations for Gene Switching Experiment

Firefly (LARII) and *Renilla* luciferase substrate (Stop & Glo) reagents are prepared as described in Promega technical manual TM040. In a clean test tube, 10 μl of cell lysate is mixed with 50 μl of LARII at ambient temperature and vortexed for 5 sec. Light emission is measured for 15 sec in a luminometer. To the same tube, 50 μl of Stop & Glo buffer is added to quench the firefly luciferase reaction and simultaneously activate the *Renilla* luciferase reaction. The tube is vortexed for 5 sec and light emission is again measured for 15 sec in a luminometer.

Following the assessment of light emission, protein in the lysate is quantitated as described for the firefly luciferase assay. RLUs are also determined as described for the firefly luciferase assay. In the gene switching experiment, 100% expression is equivalent to the RLUs generated from induced cells without addition of an antisense oligonucleotide. For graphing results, the expression level is determined for each antisense oligonucleotide by dividing the RLUs generated with the addition of oligonucleotides by the RLUs generated from induced cells without the addition of an oligonucleotide.

[11] Vesicular Stomatitis Virus as Model System for Studies of Antisense Oligonucleotide Translation Arrest

By IAN ROBBINS and BERNARD LEBLEU

Vesicular Stomatitis Virus

The Rhabdoviridae, of which vesicular stomatitis virus (VSV) is a member, are cosmopolitan viruses that infect vertebrates and invertebrates, as well as many species of plants. They characteristically have a wide host range and include rabies and rabies-like viruses. VSV can be divided into two genetically distinct serotypes referred to as VSV-Indiana and VSV-New Jersey. They cause nonfatal diseases that are of significant economic importance in cattle and swine. Infections in humans are rare and result in influenza-like symptoms. Most human infections are associated with exposure to infected animal carcasses, but infections have occurred in the laboratory[1] and this virus must be considered as a pathogen. Suitable precautions must be taken in the handling and containment of these viruses in the laboratory. In certain countries, permission may be required from the public health authorities.[2]

The goal of this article is neither to describe the morphology and ecology of VSV or the Rhabdoviridae (reviews may be found in Refs. 1, 3, and 4) nor to claim that VSV is a suitable target for antisense drug development. Rather, our aim is to outline some of the characteristics of the molecular physiology of this virus and how they can be exploited to create a model system in which fundamental studies of antisense action may be studied.

Vesicular Stomatitis Virus as Model of Translation Arrest

As with most other rhabdoviruses, VSV comprises five proteins (Fig. 1) that are designated G (glycoprotein), N (nucleoprotein), P (phosphoprotein; formerly designated NS for nominal phosphoprotein), M (matrix protein), and L (RNA-dependent RNA polymerase).[4]

Following penetration of a host cell and uncoating, the negative-strand genomic RNA is transcribed by the viral RNA-dependent RNA polymerase, protein L, with which it had been encapsidated.[5] Transcription begins at the 3' terminus of the genomic RNA, producing a 48 nucleotide leader RNA followed by the sequential synthesis of individual mRNA encoding, respectively, the N, P, M, G, and L proteins.[6–8] The transcriptase pauses and transcription is attenuated at each intergenic region,[9] resulting in a gradient of mRNA production with N > P > M > G > L. The mRNA transcripts are capped, methylated, and polyadenylated. These, as well as the leader sequence and the intergenic regions, are potential sites for

[1] R. R. Wagner and J. K. Rose, *in* "Fields Virology" (B. N. Fields, D. M. Knipe, and P. M. Howley, eds.), 3rd ed., p. 1121. Lippincott-Raven, Philadelphia, 1996.

[2] W. H. Wunner, *in* "Virology: A Practical Approach" (B. W. J. Mahy, ed.), p. 79. IRL Press, Oxford, 1985.

[3] R. E. Shope and R. B. Tesh, *in* "The Rhabdoviruses" (R. R. Wagner, ed.), p. 509. Plenum, New York, 1987.

[4] B. Dietzschold, C. E. Rupprecht, Z. Fang Fu, and H. Koprowski, *in* "Fields Virology" (B. N. Fields, D. M. Knipe, and P. M. Howley, eds.), 3rd ed., p. 1137. Lippincott-Raven, Philadelphia, 1996.

[5] D. Baltimore, A. S. Huang, and M. Stampfer, *Proc. Natl. Acad. Sci. U.S.A.* **66,** 572 (1970).

[6] G. Abraham and A. K. Banerjee, *Proc. Natl. Acad. Sci. U.S.A.* **73,** 1504 (1976).

[7] L. A. Ball and C. N. White, *Proc. Natl. Acad. Sci. U.S.A.* **73,** 442 (1976).

[8] L. E. Iverson and J. K. Rose, *J. Virol.* **44,** 356 (1982).

[9] L. E. Iverson and J. K. Rose, *Cell* **23,** 477 (1981).

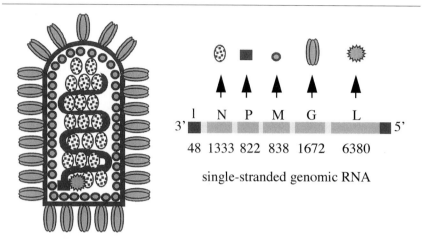

FIG. 1. Schematic representation of the organization of vesicular stomatitis virus and its single-stranded, negative polarity genomic RNA showing regions coding for the five proteins and for the leader sequence (1). N, nucleocapsid; P, phosphoprotein. M, matrix protein; G, glycoprotein; L, RNA-dependent RNA polymerase. The numbers correspond to the length in nucleotides of each coding region.

transcription inhibition. Studies along these lines have been performed in our group but will not be detailed here.[10]

Shortly after infection (within 1 hr, depending on the multiplicity of infection) there is a progressive shutdown of host macromolecular synthesis.[11] This inhibition of cellular RNA transcription can be complemented by the addition of actinomycin D that binds DNA, preventing the progression of cellular RNA polymerases. VSV mRNA production escapes this blockade as it is transcribed from an RNA template. The use of actinomycin D, therefore, allows the production of an mRNA pool, in infected cell cultures, that is comprised almost exclusively of viral transcripts. This, coupled to the distinct gradient of viral mRNA transcription, can be exploited to isolate an mRNA population that, when translated *in vitro,* will produce only three major protein products (N, P, and M). This is an ideal situation for *in vitro* translation arrest assays. Indeed, it is one of the rare systems in which natural mRNAs (as opposed to RNA transcribed *in vitro* from recombinant plasmid templates) can be used, there being a limited number of transcripts at high concentration. In addition, if one of the three

[10] G. Degols, J. P. Leonetti, C. Gagnor, M. Lemaitre, and B. Lebleu, *Nucleic Acids Res.* **17,** 9341 (1989).
[11] P. K. Weck and R. R. Wagner, *J. Virol.* **30,** 410 (1979).

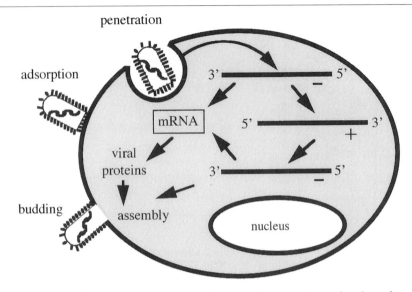

FIG. 2. An outline of the vesicular stomatitis virus life cycle. A negative sign refers to negative-stranded genomic RNA and a plus sign to the replicative intermediate. As shown, all steps take place in the cytoplasm.

mRNAs is targeted with an antisense oligonucleotide, the two others will serve as internal controls.

In addition, it is possible to compare the antisense effects observed *in vitro* with the results obtained in infected cell cultures.[10,12–14] Indeed this is one of the rare models in which the translational compartment can be specifically targeted as VSV has an entirely cytoplasmic life cycle (Fig. 2). In many other cases it is difficult to establish whether nuclear or cytoplasmic stages of RNA processing are involved in the observed antisense effect.

Preparation and Titering of Viral Stock Suspensions

Preparation of Viral Stock Suspensions

Vesicular stomatitis virus replicates efficiently in virtually all mammalian and avian cells, whether they are maintained in suspension culture or in

[12] M. Lemaitre, B. Bayard, and B. Lebleu, *Proc. Natl. Acad. Sci. U.S.A.* **84,** 648 (1987).
[13] J. P. Leonetti, B. Rayner, M. Lemaitre, C. Gagnor, P. G. Milhaud, J. L. Imbach, and B. Lebleu, *Gene* **72,** 323 (1988).
[14] C. Gagnor, J. R. Bertrand, S. Thenet, M. Lemaitre, F. Morvan, B. Rayner, C. Malvy, B. Lebleu, J. L. Imbach, and C. Paoletti, *Nucleic Acids Res.* **15,** 10419 (1987).

monolayer cultures. We routinely use the mouse fibroblastic cell line L929, but we have also used the baby hamster kidney cell line, BHK/21, to good effect.

The L929 cell line that we use routinely is grown in minimal essential medium (MEM) supplemented with 5% (v/v) fetal calf serum. Released progeny virus is detectable within 2 hr after infection and the virus yield rises exponentially for 6–8 hr, reaching maximum yields of 1000 infectious viral particles (IVP) at 10–12 hr postinfection.[2]

VSV stock suspensions can be produced by infecting subconfluent L929 or BHK/21 cells at an m.o.i (multiplicity of infection is the number of viral particles per milliliter) of 5. The supernatants are collected 12 hr later and centrifuged at 1200g for 5 min to remove detached cells and cellular debris. The culture supernatants are stored at −80°.

Titration of Viral Suspensions

Limit Dilution. Although less accurate than the plaque assay, the limit dilution method maintains acceptable accuracy while offering great ease and flexibility of utilization. It is especially applicable to situations where large numbers of titrations must be made, as when assessing the effects of antisense oligonucleotides in infected cell cultures.

The method consists of the inoculation of cells with serial dilutions of viral suspensions and assessing the presence or absence of viral growth at each concentration using cell lysis as a criterium.

Experimental conditions necessary for the correct statistical analysis of data have been shown to be fulfilled with VSV infections, in the conditions described later, when L929 mouse fibroblasts are used.[15]

A subconfluent culture of L929 in a 10-cm-diameter petri dish is detached with trypsin. The adherent cells are washed two times with phosphate-buffered saline (PBS) and then 2 ml of a 0.025% (w/v) solution of trypsin in PBS is added to the plate. The plate is swirled to distribute the trypsin evenly and the trypsin solution is aspirated immediately. This leaves a film of trypsin on the cells, sufficient to detach them when the plate is incubated at 37° for 5 min. Ten milliliters of culture medium (MEM supplemented with 5% fetal calf serum) is added to the plate and the cells are detached from the plate by repeated gentle pipetting. The concentration of the cell suspension is estimated by a hemocytometer, or with an electronic particle counter, and adjusted, by adding medium, to 2×10^5 cells/ml. Each well of a 96-well cell culture plate is seeded with 100 μl of this cell suspension. Agitation of the cell suspension from time to time is necessary if all

---

[15] P. G. Milhaud, M. Silhol, T. Faure, and X. Milhaud, *Ann. Virol.* (*Inst. Pasteur*) **134**, 405 (1983).

wells of the culture plate are to receive approximately the same number of cells (the nature of the assay does not necessitate that all wells receive exactly the same number of cells). The cells are incubated overnight at 37° in a 5% (v/v) CO_2 incubator and each well is infected with 100 μl of viral suspension the following day. Serial, 10-fold dilutions of the viral stock are made in culture medium. It is important to change the pipette tip between each dilution. The 8 wells of each column are infected with virus at the same dilution (i.e., 12 dilutions can be assayed on a single plate). A fixed volume pipette is recommended for this purpose as the degree of precision over the entire plate is greater. It is very easy to cross-contaminate adjacent wells via aerosols created during pipetting. It is therefore recommended to start with the lowest dilution and to cover the columns adjacent to that being pipetted. The dilution range should be sufficient that at least one column comprises only infected wells and that at least one column comprises no infected wells. The plates are incubated for a further 48 hr (this time can be extended to 3 or 4 days if necessary, unlike the plaque assay in which the incubation time is fairly strict). The plates are then scored for cell lysis using an inverted microscope. The viral titer is calculated from Table I, which is modified from Milhaud *et al.*[15]

The greatest dilution at which 100% of the cells is lysed is taken as 10^{-L}. The following two dilutions are taken as $10^{-(L+1)}$ and $10^{-(L+2)}$, respectively. The number of wells in which the cells are lysed at the latter two dilutions are entered into Table I, with the value derived at their intersection being the number of IVP per ml at dilution 10^{-L}. Occasionally a lysed well may

TABLE I
LIMIT DILUTION CALCULATION TABLE[a]

$10^{-(L+1)}$ $10^{-(L+2)}$	0	1	2	3	4	5	6	7
0	24	28.7	36.2	46.7	62.2	84.2	114.5	158.7
1		35.0	44.3	57.7	76.5	101.5	135.5	186.0
2			54.5	71.0	92.7	121.0	159.5	220.3
3				85.7	110.0	141.5	185.8	259.2
4					127.7	162.9	213.5	303.7
5						185.5	244.0	355.0
6							277.0	414.7
7								484.0

[a] Columns represent the number of infected wells at dilution $10^{-(L+1)}$ and the rows those at dilution $10^{-(L+2)}$. The value read at the intersection is the number of infectious viral particles (IVP) at dilution 10^{-L}. Modified from P. G. Milhaud, M. Silhol, T. Faure, and X. Milhaud, *Ann. Virol. (Inst. Pasteur)* **134**, 405 (1983).

be encountered at dilution $10^{-(L+3)}$. In this case it should be scored along with the wells at dilution $10^{-(L+2)}$. An example of the use of Table I is given in Fig. 3.

Plaque Assay. As VSV provokes the lysis of the host cells, it may be titered by its capacity to form plaques in monolayer cultures. This is proba-

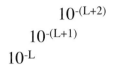

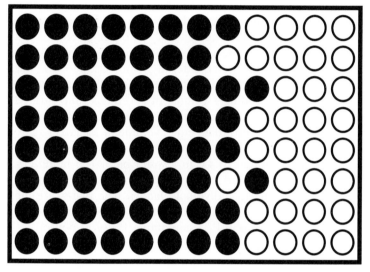

$$10^{-1}\ \ 10^{-2}\ \ 10^{-3}\ \ 10^{-4}\ \ 10^{-5}\ \ 10^{-6}\ \ 10^{-7}\ \ 10^{-8}\ \ 10^{-9}\ \ 10^{-10}\, 10^{-11}\, 10^{-12}$$

dilution

FIG. 3. An example illustrating the use of Table I for the titration of vesicular stomatitis virus by limit dilution. The diagram represents a 96-well cell culture plate seeded with L929 cells and infected with 100 μl of serial dilutions (from 10^{-1}– 10^{-12}) of a viral suspension. The filled circles correspond to wells in which cell lysis has occurred and open circles to wells containing unlysed cells. The greatest dilution at which 100% of the wells contain lysed cells is taken as 10^{-L} (in this example L = 7). The number of wells containing lysed cells at the following dilution (6 in this example) is entered into Table I as row $10^{-(L+1)}$. The number of wells containing lysed cells at dilution $10^{-(L+2)}$ (2 in this example) is entered into Table I. The intersection gives the viral titer in infectious viral particles per milliliter at dilution 10^{-L}. In this example, it is 159.5 at dilution 10^{-7}, equivalent to 1.6×10^9 IVP/ml for the undiluted viral suspension.

bly the most accurate titration method, but its application is more difficult than the limit dilution method. The number of plaques, x, produced in a monolayer of cells by plating a volume, v, of a dilution, y, of virus can be used to determine the titer, n, in plaque-forming units (pfu) per ml, according to the equation $n = x/vy$ (strictly speaking, pfu is not equivalent to IVP, but for fast-growing cytopathic viruses such as VSV, the difference is negligible).

L929 cells (6×10^6) are plated on 10-cm-diameter tissue culture dishes and incubated overnight to produce confluent monolayers. Serial 10-fold dilutions of the viral stock are made in culture medium. The culture medium is aspirated from the culture dish, and 1 ml of diluted viral suspension is pipetted carefully (such that the monolayer is not disrupted) over the cell monolayer. The cells are incubated at 37°, and the viral supension is redistributed over the surface occasionally in order to ensure a uniform infection rate. After 1 hr, the viral suspension is removed by aspiration and the cell monolayer is washed carefully with warmed PBS (washing is important to avoid secondary infections). The PBS is aspirated. MEM agarose is prepared by adding 1 volume of two times concentrated MEM without phenol red [supplemented with 4% (v/v) fetal calf serum and equilibrated to 37° in a water bath] to an equal volume of 1% (w/v) agarose that has been boiled first (to dissolve and sterilize the agarose). The MEM agarose is then equilibrated to 42° in a water bath. Ten milliliters of this solution is immediately spread over the cell monolayer. The agarose serves to restrain viral progeny to individual plaques. The plates are incubated for 18 hr at 37° in a 5% CO_2 incubator. The plaques are revealed by adding 2 ml of 0.5% (w/v) agarose in PBS containing 1% (v/v) neutral red.

Preparation of VSV Messenger RNA

Viral Infection

Confluent cells from two 15-cm-diameter culture dishes are detached with trypsin and seeded to five new 15-cm-diameter dishes. The cells are cultured overnight to near confluence. The cell monolayers are washed twice with warmed PBS. Ten milliliters of prewarmed MEM complemented with 5% (v/v) fetal calf serum containing 5 μg/ml actinomycin D is added to each plate, and the plates are returned to the 37° CO_2 incubator for 20 min. The actinomycin D-containing medium is collected by pipette into a 50-ml sterile Falcon tube, and the VSV stock suspension is added to give a final m.o.i. of 5 IVP/ml. The diluted viral suspension is mixed by repeated inversion of the Falcon tube, and 10 ml is redispatched to each tissue-culture dish. The infected cells are incubated a further 5 hr prior to the extraction of RNA.

Total RNA Extraction

The culture medium is aspirated into a vessel that contains bleach (it contains a high titer of infectious virus). The cell monolayers are washed twice with ice-cold PBS (aspirate the last drops thoroughly by inclining the plates at 45° for a few seconds).

All reagents used in the extraction of RNA should be RNase-free, molecular biology grade, dissolved in RNase-free water, and autoclaved. Freshly prepared 18 MΩ water is usually RNase free, as are many brands of injectable water; otherwise the water must be treated with diethyl pyrocarbonate (see Sambrook *et al.*[16] for details).

Six milliliters of guanidine isothiocyanate (5 *M*) is added to one of the dishes, and the cells are scraped from the dish with a sterile plastic scraper or with a rubber policeman that has been treated for 20 min with 0.1 *N* NaOH and rinsed abundantly with RNase-free water. The viscous solution is scraped to the second plate and the cells from the second plate are scraped into it. The process is repeated for the five plates, and the combined lysate is poured into a sterile 50-ml Falcon tube. Six hundered microliters of 2-mercaptoethanol (14.3 *M*) is added, followed by 30 ml of lithium chloride (3 *M*). The solution is mixed thoroughly with a vortex and stored overnight at 4°.

The tube is then centrifuged for 60 min at 10,000*g* (4°), and the supernatant is carefully poured off. The tube is left inverted for a few minutes on an absorbant tissue to eliminate the residual supernatant. The pellet is rehydrated (by repeated pipetting) in 1 ml TES (10 m*M* Tris–Cl, pH 8, 1 m*M* disodium ethylenediaminetetraacetate (EDTA), 0.1% (w/v) sodium dodecyl sulfate (SDS); because the SDS will precipitate if autoclaved, the dry powder should be dissolved in autoclaved Tris–EDTA solution).

The dissolved RNA is transferred to a 2-ml Eppendorf tube and extracted twice with water saturated phenol/dichloromethane/isoamyl alcohol (25:24:1) and once with dichloromethane/isoamyl alcohol (24:1). The final aqueous phase is split equally between two 1.5-ml Eppendorf tubes. RNA is precipitated with a 1/20 volume (≈25 μl) of NaCl (5 *M*) and 2 volumes of ethanol (≈1 ml). The tubes are left at −80° for 30 min (or longer if desired) and then centrifuged for 10 min at 15,000*g* at 4°. The supernatant is aspirated, and the pellet of RNA is dispersed by vortexing

---

[16] J. Sambrook, E. F. Fritsch, and T. Maniatis, "Molecular Cloning: A Laboratory Manual," 2nd ed. Cold Spring Harbor Laboratory Press, Cold Spring Harbor, NY, 1989.

in 200 μl of 70% (v/v) ethanol. The tubes are recentrifuged for 10 min at 15,000g at 4° and the RNA pellets are dried under vacuum.

Each pellet is dissolved in 200 μl of TES, and the solutions are pooled. The optical density (OD) of 5 μl of this solution, diluted to 1 ml in H_2O, is read at wavelengths of 260 and 280 nm (an OD_{260} of 1.0 = 40 μg/ml RNA; the ratio OD_{260}/OD_{280} should be between 1.8 and 2.0 if the RNA is pure). A total quantity of RNA in excess of 1 mg is obtained routinely.

Purification of Poly(A)$^+$mRNA

Preparation of Oligo(dT) Columns. A small wad of autoclaved glass wool is pushed to the bottom of a 2-ml syringe barrel and several column volumes of 0.1 M NaOH are passed through the column. The column is rinsed thoroughly with RNase-free water. Oligo(dT) cellulose beads (Boehringer) are weighed according to the amount of total RNA extracted (100 mg per mg RNA extracted) and suspended in TES–0.5 M NaCl. The slurry of oligo(dT) cellulose is added to the column and allowed to settle to a homogeneous bed. Because it is important that the gel never dries out, TES–0.5 M NaCl must be added continually. The gel is equilibrated with 10 column volumes of TES–0.5 M NaCl.

Preparation of Sample. The total RNA sample (now in 395 μl) is incubated at 65° for 10 min and is then plunged immediately into half-melted ice. Three hundred microliters of 5 M NaCl and 2.3 ml of TES are added, and the tube is agitated on a vortex.

Affinity Chromatography. The 3-ml sample is passed over the oligo(dT) column, and the column outflow is collected and passed two more times over the column. The gel is washed with 10 column volumes of TES–0.1 M NaCl and is eluted four times with 500 μl of TES preheated to 65°, with the eluate being collected in four separate 1.5-ml Eppendorf tubes. One hundred microliters of 5 M NaCl is added to each tube, and the poly(A)$^+$ RNA is precipitated with 1 ml ethanol at −80° for at least 1 hr. The tubes are centrifuged at 15,000 g for 10 min at 4°, and the pellets are resuspended by vortexing in 200 μl of 70% (v/v) ethanol and recentrifuged. The RNA pellets are dried under vacuum. Each pellet is dissolved in 100 μl of TES. The solutions are pooled, and the RNA is reprecipitated with 100 μl of 5 M NaCl and 1 ml of ethanol as described earlier. The final RNA pellet is dissolved in 40 μl of RNase-free water. Five microliters is used to estimate the concentration by spectrophotometry, as described earlier, and the final concentration is adjusted to 1 μg/μl. The total amount of poly(A)$^+$ RNA should be equal to 4–5% of the total RNA loaded on the column. Serial dilutions of the mRNA preparation are tested in the *in vitro* translation assay, described in the following section, and appropriately diluted aliquots of the mRNA are stocked at $-80°$. We have generally found optimal concentration to lie between 100 and 300 ng per 10-μl translation assay.

In Vitro Translation Arrest Assay

Rabbit reticulocyte lysates (Promega, Madison, WI) are fragile, should be stored at $-80°$, and should not be thawed and refrozen repeatedly. They are viscous and great care should be taken in pipetting them. We allow a new batch to thaw on ice (about 15 min) and carefully pipette 100-μl aliquots into Eppendorf tubes that are immediately stocked at $-80°$ until use. Results vary somewhat from batch to batch and it is therefore advisable to note the batch number as it is often possible to order a specific batch from the supplier.

We also dispatch L-[^{35}S]methionine (*in vivo* cell labeling grade; 1000 Ci/mmol; 10 mCi/ml) into 10-μl aliquots and store it at $-80°$.

In some cases, it will be necessary to prehybridize oligonucleotides to the RNA before adding the reticulocyte lysate. For this purpose we add 0.5 μl of a 20$\times$ stock solution of oligonucleotide and 0.5 μl of a 4$\times$ prehybridization buffer (40 mM Tris, 20 mM $MgCl_2$, 200 mM KCl, 200 nM dithiothreitol, 0.5 M potassium acetate, pH 7.5) to 1 μl of mRNA (at an appropriate dilution, as established by serial dilution of the stock). This mixture is then incubated in a thermocycler for 1 min at 65°, for 10 min at 45°, and for 10 min at 37°. The tubes are pulse centrifuged at 4° to collect the full 2 μl at the bottom of the tube. Eight microliters of the following lysate premix is added to each tube: 100 μl rabbit reticulocyte lysate, 5 μl 1 mM amino-acid mixture (minus methionine), 10 μl [^{35}S]methionine, 1 μl RNasin (Promega), and 0.5 μl of RNase H (*Escherichia coli;* Boehringer) (or water). This premix is sufficient for 14 translation reactions that are incubated at 30° for 1 hr. Four microliters of 4$\times$ gel loading buffer (125 mM Tris–Cl, pH 6.8, 8% (w/v) SDS, 30% (v/v) glycerol, 1.43 M 2-mercaptoethanol, 0.03% (w/v) bromphenol blue) is added to each tube, and the tubes are stored at $-20°$ until they are run on a 12.5% (w/v) polyacrylamide–SDS gel (PAGE–SDS; 37.5:1 acrylamide:bis acrylamide) with a 4% (w/v) stacking gel.

Gel-, or high-performance liquid chromatography-, purified oligonucleotides obtained from most suppliers are sufficiently pure to be used directly in the assay. The translation rate, however, is sensitive to the ionic conditions of the reaction cocktail and it is advised to dilute the oligonucleotides in RNase-free water rather than in buffered solutions. This decreases their stability during long-term storage and they should preferably be stocked at $-80°$.

Three major radioactive protein bands are separated (N, P, and M), as discussed earlier (Fig. 5). The M protein is a 229 amino acid protein[17] with a predicted molecular mass of 26 kDa. The N protein is a 422 amino acid protein with a predicted molecular mass of 47 kDa. The P (NS) protein is a 265 amino acid protein (for the Indiana strain) with a predicted molecular mass of 30 kDa. The latter protein runs anomalously on PAGE–SDS, however, due to heavy phosphorylation and the presence of 18 negatively charged, and no positively charged, residues within a 44 amino acid stretch in the N-terminal half of the protein.[18] The apparent molecular mass on PAGE–SDS varies from 40 to 50 kDa according to experimental conditions. In the gel described earlier, the P protein migrates slightly further than the N protein. However, if glycerol is included in the gel formulation, the N protein migrates further than the P protein. In any case, the two proteins migrate in close proximity. If the conditions are optimized, the translation mixtures can be run 1 hr on minigels. It is, however, recommended that larger gels be used with a slow migration rate to increase resolution. The gels are dried and exposed to autoradiographic film. Free [^{35}S]methionine can cause background, which can be reduced significantly by transferring the proteins to nitrocellulose membranes. In this case, the gel should be dried after transfer and exposed as well in order to verify that the three bands have all been completely transferred. Sensitivity can be increased greatly by using a phosphor screen (Kodak, Rochester, NY) and appropriate autoradiographic film. An outline of the assay is presented in Fig. 4.

In many cases, when an antisense oligonucleotide and RNase H are added to the translation mix, the expected, truncated N-terminal portion of the protein can be observed (Fig 5). Reticulocyte lysates display low endogenous RNase H activity in the conditions of the translation assay.[19] This unique situation allows the discrimination of RNase H-dependent from RNase H-independent mechanisms for antisense oligonucleotides. Certain batches may contain sufficient endogenous RNase H to promote a slight decrease of the target protein product even in the absence of exogenous enzyme. In this case, 0.4 μg poly(rA-dT) (Pharmacia) can be added to each translation reaction in order to compete out endogenous RNase H.[20]

This type of assay has a number of applications in the antisense field. First, it is a simple and sensitive *in vitro* assay for evaluating the accessibility to hybridization of RNase H-competent antisense oligonucleotides comple-

[17] J. K. Rose and C. J. Gallione, *J. Virol.* **39**, 519 (1981).
[18] C. J. Gallione, J. R. Greene, L. E. Iverson, and J. K. Rose, *J. Virol.* **39**, 529 (1981).
[19] C. Cazenave, P. Frank, and W. Busen, *Biochimie* **75**, 113 (1993).
[20] R. Y. Walder and J. A. Walder, *Proc. Natl. Acad. Sci. U.S.A.* **85**, 5011 (1988).

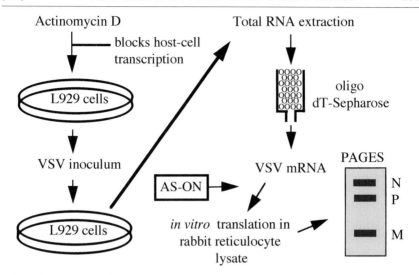

FIG. 4. An outline of vesicular stomatitis virus mRNA production and use in the *in vitro* translation assay. AS-ON, antisense oligonucleotide; PAGES, polyacrylamide gel electrophoresis–sodium dodecyl sulfate.

mentary to various portions of the mRNA target.[14] This assay can be adapted to monitor the binding of RNase H-incompetent oligonucleotides by their capacity to compete with an isosequential RNaseH-competent oligonucleotide.[21] This assay has allowed us to establish that peptide nucleic acid and *N*-3′-phosphoramidate analogs are unable to arrest translation despite their very high affinity binding to the target site.[21] This is an important issue as it is often difficult to discriminate true translation arrest from the RNase H-mediated degradation of targeted mRNA. In this respect, it is a convenient system for the characterization of new antisense analogs. As an example, we have established that a *cis*-platinated 2′-*O*-methyl antisense derivative was capable of effectively arresting translation in an RNase H-independent fashion at a virtually stoichiometric ratio.[21] In addition, novel antisense strategies can be evaluated. As an example, we have demonstrated with this assay that an antisense oligonucleotide coupled to 2,5-A (a specific activator of the ubiquitous cellular RNase L) also exhibited RNase H-dependent activity.[22]

[21] J. E. Gee, I. Robbins, A. C. van der Laan, J. H. van Boom, C. Colombier, M. Leng, A. M. Raible, J. S. Nelson, and B. Lebleu, *Antisense Nucleic Acid Drug Dev.* **8,** 103 (1998).
[22] I. Robbins, G. Mitta, S. Vichier-Guerre, R. Sobol, A. Ubysz, B. Rayner, and B. Lebleu, *Biochimie* **80,** 711 (1998).

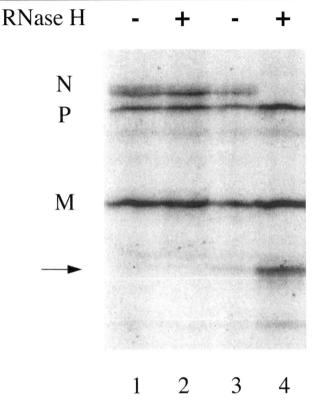

Fɪɢ. 5. PAGE–SDS analysis of *in vitro*-translated vesicular stomatitis virus mRNA. Three major bands are observed, corresponding to the N, P, and M protein products as indicated. The addition of exogenous (*E. coli*) RNase H has no effect on translation (compare lane 2 to lane 1) in the absence of an antisense oligonucleotide. Incubation with an oligonucleotide complementary to nucleotides 637–651 of the N protein mRNA decreases translation slightly in the absence of exogenous RNase H (lane 3) and inhibits translation completely in the presence of exogenous RNase H (lane 4). In this case, a truncated polypeptide (indicated by an arrow), of the expected molecular weight, accumulates.

Conclusions

In vitro RNase H-mediated cleavage of an mRNA target has been demonstrated to reflect efficiency in cell culture systems in many cases.[23–27] Most of the new chemistries, however, do not elicit RNase H activation.

[23] S. P. Ho, D. H. Britton, B. A. Stone, D. L. Behrens, L. M. Leffet, F. W. Hobbs, J. A. Miller, and G. L. Trainor, *Nucleic Acids Res.* **24,** 1901 (1996).

[24] S. P. Ho, Y. Bao, T. Lesher, R. Malhotra, L. Y. Ma, S. J. Fluharty, and R. R. Sakai, *Nature Biotechnol.* **16,** 59 (1998).

This major hurdle can be circumvented by two strategies: (i) restoring nucleolytic activity by the conjugation of suitable pendant groups or by incorporating an RNase H-competent window (mixed backbone oligonucleotide) or (ii) increasing the efficiency of steric blockade. In this respect, the simple assay described in this article has the advantage of discriminating true translation arrest from target RNA destruction. Because natural mRNA are used in this assay, both untranslated and coding regions can be examined. Finally, the stimultaneous translation of three mRNA provides a convenient, and important, internal control.

[25] W. F. Lima, V. Driver-Brown, M. Fox, R. Hanecak, and T. W. Bruice, *J. Biol. Chem.* **272,** 626 (1997).

[26] O. Matveeva, B. Felden, S. Audlin, R. F. Gesteland, and J. F. Atkins, *Nucleic Acids Res.* **25,** 5010 (1997).

[27] O. Matveeva, B. Felden, A. Tsodikov, J. Johnston, B. P. Monia, J. F. Atkins, R. F. Gesteland, and S. M. Freier, *Nature Biotechnol.* **16,** 1374 (1998).

[12] Purification of Antisense Oligonucleotides

By Ranjit R. Deshmukh, Douglas L. Cole, and Yogesh S. Sanghvi

Introduction

Despite ongoing and significant advances in synthesis chemistry and reactor/reagent delivery systems design,[1] crude products of antisense phosphorothioate oligonucleotide (AO) solid-phase syntheses still typically contain only about 75% full-length oligomer. While this could be seen as an excellent overall yield for a 19-step synthesis, the products must be purified to a much higher full-length oligomer content prior to use as antisense drugs. Oligonucleotide purification research has thus been a key part of successful antisense drug development.[2]

Selective separation methods are required for the preparation of high-purity AO for antisense discovery research and clinical development and for the development of meaningful analytical control procedures. Finally, preparative separation methods must be created and scaled up to support AO manufacture for therapeutic use.

[1] M. Andrade, A. S. Scozzari, D. L. Cole, and V. T. Ravikumar, *Nucleosides Nucleotides* **16,** 1617 (1997).

[2] S. T. Crooke, *in* "Antisense Research and Application" (S. T. Crooke, ed.), p. 1. Springer-Verlag, Berlin, 1998.

Concern for synthetic oligonucleotide impurity types and quantities can vary, depending on the target application. High-purity oligonucleotides are required for diagnostic probe and antisense drug applications, whereas lower purity oligomers may be acceptable as oligonucleotide primers. Microgram to gram AO quantities are often sufficient for *in vitro* and pharmacological screening purposes, whereas kilograms are required for AO drug clinical trials. Actual and potential AO drug markets range from kilogram to metric ton levels. A single broad purification strategy could have utility over this range of scale and purity requirements, but the optimal systems differ for specific AO compounds. This article first broadly reviews the available methods for AO purification and then stresses techniques suitable for milligram, gram, and kilogram scale purification.

This article discusses methods for purifying crude AO by two key chromatographic techniques, reversed phase (RP) and anion exchange (AX). Other currently available techniques are also summarized briefly. For RP and AX techniques, small-scale results are presented first, followed by a discussion of large-scale methods. Examples presented are directed exclusively to the phosphorothioate oligonucleotides, first-generation antisense oligonucleotide drugs.

General Purification Strategies for Oligonucleotides

Characteristic properties of synthetic oligonucleotides and their process-related impurities, such as polarity, charge, and size, may be exploited for their purification.[3–5] For example, a hydrophobic protecting group left attached at the 5'-*O*-oligonucleotide terminus, such as 4,4'-dimethoxytrityl (DMT), allows effective use of RP-high-performance liquid chromatography (HPLC)[6,7] for purification. Hydrophobicity of the heterocyclic bases at mild pH also aids RP-HPLC purification, although selectivity of this effect is most significant for short oligonucleotides. Both hydrophobic effects are

[3] W. Haupt and A. Pingoud, *J. Chromatogr.* **260**, 419 (1983).
[4] G. Zon, *in* "High-Performance Liquid Chromatography in Biotechnology" (W. Hancock, ed.), p. 301. Wiley, New York, 1990.
[5] G. Zon and J. A. Thompson, *BioChromatography* **1**, 22 (1986).
[6] Y. S. Sanghvi, M. Andrade, R. R. Deshmukh, L. Holmberg, A. N. Scozzari, and D. L. Cole, *in* "Manual of Antisense Methodology" (G. Hartman and S. Endres, eds.), p. 3. Kluwer Academic Publishers, Norwell, 1999.
[7] W. J. Warren and G. Vella, *in* "Protocols for Oligonucleotide Conjugates" (S. Agrawal, ed.), p. 233. Humana Press, Totowa, NJ, 1994.

useful in the preparative hydrophobic interaction chromatographic (HIC) purification of oligonucleotides.[8]

Oligonucleotide backbone phosphates impart a strong negative charge to the molecule, with the mass to charge ratio related (but not linearly so) to the number of nucleotides in the oligomer. Anion-exchange chromatography is thus a selective separation method for synthetic oligonucleotide separations and is widely used in both analysis and purification.[7,9–16]

To produce very pure oligonucleotides, highly specific affinity Watson–Crick base pairing can be exploited for purification under nonequilibrium conditions in the affinity chromatography mode.[17–19]

Gel permeation chromatography (GPC)[20] is useful for the size-based separation of oligonucleotides from one another or for gross size-class chemical separations such as the isolation of total isolated oligonucleotide from contaminating protein. This technique is not particularly useful for the purification of synthetic oligonucleotides less than 50 bases in length.

Two or more stationary phase–solute interaction mechanisms may be combined in mixed-mode chromatographies in order to enhance separation selectivity. For example, by the addition of ion-pairing agents to RP-HPLC mobile phases,[5,21–24] hydrophobic and charge–charge interactions may be combined in a single separation. Direct interaction of the stationary phase with oligonucleotide phosphate diester groups and other stationary phase–

[8] P. Puma, in "HPLC: Practical and Industrial Applications" (J. Swadesh, ed.), p. 81. CRC Press, Boca Raton, FL, 1997.

[9] W. A. Ausserer and M. L. Biros, *BioTechniques* **19**, 136 (1995).

[10] B. J. Bergot and W. Egan, *J. Chromatogr.* **599**, 35 (1992).

[11] B. J. Bergot, U.S. Patent 5,183,885 (1993).

[12] R. R. Drager and F. E. Regnier, *Anal. Biochem.* **145**, 47 (1985).

[13] V. Metelev and S. Agrawal, *Anal. Biochem.* **200**, 342 (1992).

[14] J. R. Thayer, R. M. McCormick, and N. Avdalovic, *Methods Enzymol.* **271**, 147 (1996).

[15] R. R. Deshmukh, W. E. Leitch II, and D. L. Cole, *J. Chromatogr. A* **806**, 77 (1998).

[16] J. Liautard, C. Ferraz, J. S. Widada, J. P. Capony, and J. P. Liautard, *J. Chromatogr.* **476**, 439 (1989).

[17] S. Agrawal and P. C. Zamecnik, PCT International WO 90/09393 (1990).

[18] H. Orum, P. E. Nielsen, M. Jorgensen, C. Larsson, C. Stanley, and T. Koch, *BioTechniques* **19**, 472 (1995).

[19] H. Schott, H. Schrade, and H. Watzlawick, *J. Chromatogr.* **285**, 343 (1984).

[20] K.-I. Kasai, *J. Chromatogr.* **618**, 203 (1993).

[21] P. J. Oefner, C. G. Huber, F. Umlauft, G.-N. Berti, E. Stimpfl, and G. K. Bonn, *Anal. Biochem.* **223**, 39 (1994).

[22] A. P. Green, J. Burzynski, N. M. Helveston, G. M. Prior, W. H. Wunner, and J. A. Thompson, *BioTechniques* **19**, 836 (1995).

[23] C. Huber, P. Oefner, and G. Bonn, *Anal. Biochem.* **212**, 351 (1992).

[24] C. G. Huber, *J. Chromatogr. A* **806**, 3 (1998).

solute charge interactions are available simultaneously on hydroxyapatite columns.[25,26] A nucleic acid-specific chromatographic media, RPC-5,[27,28] was used to simultaneously exploit charge interaction and hydrophobicity for the purification of nucleic acids, oligonucleotides, and tRNAs. RPC-5 solid support is not currently available commercially, however.

Most of these separation techniques have been used at microgram to gram scales. RP, AX, and HIC are the only approaches that have been demonstrated at scales much greater than 100 g loading of AO. These three techniques are therefore the focus of the remainder of this article.

Experimental

Oligonucleotides

All AO used in this study were synthesized in-house at various scales. The Milligen 8800 (PE Biosystems, Framingham, MA) oligonucleotide synthesizer was used for synthesis of ISIS 2105 and ISIS 5132/CGP69846A. OligoPilot II and OligoProcess synthesizers (both from Amersham Pharmacia Biotech, Piscataway, NJ) were used to manufacture ISIS 2302. All these AO are 20-mer phosphorothioate oligodeoxyribonucleotides. Leaving the final DMT group on the 5'-nucleotide after completing oligomerization generates DMT-on crude oligomers. DMT-off crude oligomers are obtained by removing the final DMT as the final oligomerization step. All crude products are cleaved from solid support and base- and phosphate-deblocked in concentrated aqueous ammonia. Ammonia is removed by rotary evaporation *in vacuo,* and the resulting aqueous oligonucleotide solution is used for chromatography. DMT-on crude products are stored in a 1% triethylamine (v/v) solution to maintain basic pH.

Chemicals and Buffers

All general chemicals and buffers are ACS grade unless stated otherwise. Buffers are prepared from dry chemicals in most cases. A stock solution of 1 M NaOH is made and then used to make dilute NaOH solutions. The list of buffers used is given in Table I.

[25] Y. Yamasaki, A. Yokoyama, A. Ohnaka, Y. Kato, T. Murotsu, and K.-I. Matsubara, *J. Chromatogr.* **467,** 299 (1989).
[26] G. Bernardi, *Methods Enzymol.* **21,** 95 (1971).
[27] R. L. Pearson, J. F. Weiss, and A. D. Kelmers, *Biochem. Biophys. Acta* **228,** 770 (1971).
[28] R. D. Wells *et al., Methods Enzymol.* **65,** 327 (1980).

TABLE I
BUFFERS

Buffer system	Composition	pH
AX buffer I		
A	20 mM NaOH	12.0
B	20 mM NaOH + 2.5 M NaCl	12.0
AX buffer II		
A	50 mM KOH	12.6
B	50 mM KOH + 2.5 M KCl	12.6
RP buffer I		
A	2.5 M sodium acetate	7.2
B	Methanol	
C	Deionized water	
RP buffer II		
A	2.5 M sodium acetate	7.2
B	Methanol	
C	Deionized water	
D	1% trifluoroacetic acid in water	

Chromatography Instrumentation

Preparative liquid chromatography experiments are conducted on Bio-Cad workstations (PE Biosystems, Framingham, MA). The BioCad 20 provides a 20-ml/min maximum flow rate. The BioCad 60 and BioCad 250 have maximum flow rates of 60 and 250 ml/min, respectively. These chromatographic workstations have six buffer ports and automated method programming. Columns up to 600 ml volume are eluted on these chromatographs. At production scale, the KiloPrep KP100 (Biotage Division of Dyax Corp., Cambridge, MA) is used for RP-HPLC. Analytical AX analysis is conducted on a Waters Associates (Bedford, MA) system comprising a Waters 600E chromatograph with Waters 717 autosampler, 991 photodiode array detection system, and the Millennium 2.01 operating system. A thermostatted heating block is used to heat analytical columns.

Chromatographic Columns and Media

Unless otherwise stated, columns smaller than 5 ml packed volume are purchased prepacked from vendors. Larger volume anion-exchange columns from 5 to 200 ml packed volume are slurry packed in-house. A 200-ml AP-5 glass column (Waters Corp., Milford, MA, 50 mm i.d. × 100 mm length) is also used. All RP columns are packed at high pressure by the vendors. Production-scale radial compression cartridges are used in the Biotage KP100 chromatograph. A list of oligonucleotide separation chromatographic media typically available in our laboratory and discussed in this article is given in Table II.

Analytical Methods to Support Purification Development

From each purification, column fractions or pools are assayed for AO content by UV absorbance. Chromatographic impurity profiles are determined by analytical AX HPLC to quantify sulfurization-related phosphorothioate oligomer impurities, and length-altered impurities are determined by quantitative capillary gel electrophoresis (QCGE). These methods are described later in greater detail. Identity of the oligonucleotide product in crude material is determined by electrospray-mass spectrometry (ES-MS). In addition to these routine methods, additional analytical techniques may be used to completely characterize oligonucleotide products and their impurities, as summarized previously.[6]

Quantitative UV Spectroscopy. A solution concentration of oligonucleotides is determined by measuring absorbance at 266 nm. Fractions are diluted appropriately to bring absorbance into the linear response range. The measured OD/ml can be suitably converted into mg/ml values by using the factor 25.7 OD = 1 mg for most 20-mer phosphorothioate deoxyoligonucleotides.

TABLE II
CHROMATOGRAPHIC COLUMNS AND MEDIA

		Media		
Trade name	Manufacturer	Ligand chemistry	Bead chemistry	Nominal particle size (μm)
BondaPak HC$_{18}$HA	Waters Corp. (Milford, MA)	C$_{18}$	Silica	37–55
Oligo R3	PE BioSystems (Framingham, MA)	C$_{18}$	PSDVB[a]	40
Q HyperD F	BioSepra, Inc. (Marlborough, MA)	AX (Q)	Ceramic-coated silica with gel inside pores	35
Poros HQ/50	PE Biosystems	AX (Q)	PSDVB	50
Poros HQH	PE Biosystems	AX (Q)	PSDVB	10
Resource Q	Amersham Pharmacia Biotech (Piscataway, NJ)	AX (Q)	PSDVB	15

[a] Polystyrene divinylbenzene.

Analytical AX Chromatography. Analytical AX HPLC is used to quantify partial phosphodiester oligonucleotide impurities in the phosphorothioate AO used in this work. In such phosphorothioate oligonucleotides, each backbone phosphorous carries a nonbridging sulfur substituent. Synthesis-related partial phosphodiester impurities have one or more backbone phosphorous atoms substituted only by oxygen. These impurities are commonly denoted as $(P{=}O)_1$, $(P{=}O)_2$, and so on, where the subscript gives the number of phosphate diesters present in the backbone. A strong anion-exchange (SAX) column, POROS HQ/H (4.6 mm i.d. $\times$ 100 mm length) is used for most analyses. Preparative chromatography fractions are loaded directly onto the analytical column without a sample preparation step. The buffer system used is buffer A (20 mM NaOH) and buffer B (2.5 M NaCl in buffer A). A linear gradient from 0 to 100% B over 20 min is used at 70°. The Resource Q 1-ml column is used for some SAX analyses, using a similar elution profile. Other detailed protocols for analytical SAX HPLC of synthetic oligonucleotides are given by Srivatsa *et al.*[29] and Thayer *et al.*[14]

Quantitative Capillary Gel Electrophoresis (QCGE). QCGE is the most effective separation-based method for determining length-variant impurities in sequentially assembled synthetic oligonucleotides, such as (*n*-1), (*n*-2), and so on deletions, where *n*-mer is the desired full-length product. The QCGE method has been discussed in detail by Srivatsa *et al.*[30] A P/ACE 5000 (Beckman Coulter, Fullerton, CA) instrument is used for QCGE analysis. A gel-filled eCap ssDNA 100 capillary (Beckman Coulter) with a separation length of 40 cm is used at 40°. A 100 mM Tris–borate/7 M urea electrolyte is used as supplied by the vendor. The sample is injected electrokinetically at 10 kV for 5 to 20 sec, followed by separation at 14.1 kV (300 V/cm).

Samples for QCGE are prepared according to the following procedure. Fractions from AX eluates are desalted using Centricon (Amicon/Millipore) microcentrifuge cartridge 3SR modules. These cartridges have a molecular weight cutoff of 3000. Of the sample to be desalted, 1.5 ml is put in the holder and deionized water is added until the level reaches the fill-up mark (about 3 ml). This is centrifuged at 6000 rpm for 60 min. The permeate is then discarded and the retentate is again refilled with water to the fill-up mark. This is repeated three times. After the final spin, the permeate is discarded and the cartridge is inverted. It is centrifuged again at 300 rpm for 5 min to push the retentate into a collection vial. Finally,

[29] G. S. Srivatsa, P. Klopchin, M. Batt, M. Feldman, R. H. Carlson, and D. L. Cole, *J. Pharm. Biomed. Anal.* **16,** 619 (1997).

[30] G. S. Srivatsa, M. Batt, J. Schuette, R. Carlson, J. Fitchett, C. Lee, and D. L. Cole, *J. Chromatogr. A* **680,** 469 (1994).

the desalted solution is diluted to a concentration of 0.2 OD/ml and loaded on the QCGE autosampler.

Purification of Crude DMT-On Phosphorothioate Oligodeoxyribonucleotide AO

Phosphorothioate AO synthesized by the phosphoramidite approach on solid supports are usually cleaved from support with the DMT-protecting group from the final synthon intact at the 5' terminus. The acid-labile, hydrophobic DMT group quite effectively permits the purification of synthetic oligonucleotides by preparative reversed-phase chromatography. RP-HPLC has thus become the most frequently used technique for purifying crude AO products. HIC and, to a lesser degree, AX can also make use of the DMT group as an aid in purifying the full-length product.

RP-HPLC has been used for synthetic oligonucleotide purification over a wide range of scales. The reversed-phase separation of formerly capped failure sequences without the DMT function ("DMT-off") from the DMT-on product pool is straightforward. This procedure is convenient for small-scale syntheses[4] and has been shown very effective at a 100-g column loading scale for the purification of therapeutic antisense oligonucleotides.[6] A distinct advantage of RP methods is that similar protocols may be used for a variety of structurally modified synthetic oligonucleotide crude products. The RP method used for the purification of DMT-on synthetic phosphodiester DNA oligomers, for example, is very similar to the method typically used for DMT-on O,O-linked phosphorothioate 2'-O-alkyl-RNA/DNA hybrid oligomers. The same method, with minor modifications, can be used for crude oligonucleotide products protected with other 5'-O-ether groups, such as MMT. Examples of such purification are discussed next, with protocols.

RP Purification of a 20-mer Phosphorothioate Oligodeoxyribonucleotide AO

DMT-on crude ISIS 5132/CGP 69846A is purified on a polymeric RP column, Oligo R3 (Fig. 1), using buffer RP-I (Table I). A crude product solution (200 μl) is diluted in 200 mM sodium acetate solution and is injected on the column. The separation gradient used is shown in Fig. 1. At low loading, such as used in this experiment, two distinct peaks are seen. The first two are failure sequence peaks containing oligomers without the DMT group and the third peak is the DMT-on oligomer-containing peak. Typically, benzamide generated by the deprotection of adenine or cytosine base is seen as a distinct peak eluting between the main peaks.[4]

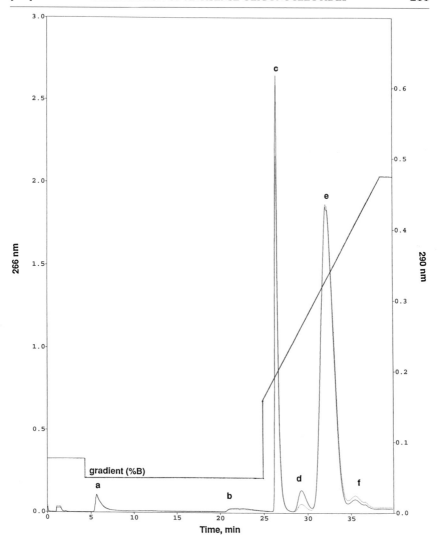

FIG. 1. RP-HPLC purification of DMT-on crude ISIS 5132/CGP 69846A (20-mer phospho-rothioate oligodeoxyribonucleotide). Peaks a, b, and c are failure sequences without the DMT group. Peak d is the benzamide peak. Peak e is the main product peak with DMT group and peaks f are failure sequences with DMT group. Sample: DMT-on crude ISIS 5132/CGP 69846A, 1 mg/ml, 4-ml injection; Column: Oligo R3 (PE BioSystems), 10 mm i.d. × 100 mm length, column volume (CV) = 7.85 ml. Buffer A: 50 mM NaOH, Buffer B: Methanol. Flow rate: 10 ml/min. Gradient program: Equilibration: 93% A + 7% B, for 4 CV, Load 4 ml sample, Wash* 97% A + 3% B for 20 CV, Elute: 20% B to 70% B in 17 CV, Hold 2 CV. (*The long wash used for this example is not necessary, 4 CV wash is sufficient).

In Fig. 1, absorption is monitored at 266 and 290 nm. The absorbance ratio is quite distinct for benzamide, allowing its distinction from the 1:3 absorbance ratio in oligonucleotide-containing peaks. It should be noted that even in the absence of separation of the benzamide peak from the DMT-off oligomer peak, good separation of the DMT-on peak from the DMT-off peak is readily achieved. The purity of a broad DMT-on heart cut is typically between 93 and 97% full-length by QCGE area-% analysis. Even higher purity material may be obtained by narrower fractionation. At higher loadings (>20 mg/ml CV), the chromatogram is more complex, but good purification is still possible with careful subfractionation of the DMT-on peak.

Silica-based C_{18} reversed-phase columns give good capacity and excellent selectivity for these crude synthetic oligonucleotide separations. The only disadvantage is that the eluent pH must be near neutrality for column stability so high pH cannot be used to denature occasionally significant secondary or tertiary structures to minimize chromatographic band widths. There are a large number of protocols for the elution of silica-based RP columns with near-neutral buffers.[4,7,31] Three commonly used buffers for RP-HPLC of AO include two volatile buffer systems: (1) 100 mM TEAA/ACN and (2) 100 mM ammonium acetate/methanol. A typical nonvolatile buffer system is (3) 200 mM sodium acetate/methanol. At large scale, sodium acetate buffer systems provide a highly useful combination of selectivity and capacity. In addition, this buffer system yields the sodium salt of the product oligonucleotide. Sodium is a preferred countercation for therapeutic antisense oligonucleotides, for physiological reasons, and gives products of manageable hygroscopicity and excellent photostability. For mass spectrometric characterization of product oligomers, volatile buffers may yield more useful salts.[32,33] Other polymeric columns with surface chemistry similar to that of the Oligo R3 column include PRP-1 (Hamilton, Reno, NV) and Source 30 RPC (Amersham Pharmacia Biotech, Piscataway, NJ). These may be used under appropriately but slightly modified elution conditions and loading capacities.

RP-HPLC with On-Column Detritylation

A key feature of polymeric RP stationary phases is that eluent buffers may have extreme pH, as in the previous example wherein a high pH buffer

[31] L. W. McLaughlin and N. Piel, *in* "Oligonucleotide Synthesis: A Practical Approach" (M. J. Gait, ed.), p. 117. Oxford Univ. Press, New York, 1990.
[32] A. Apffel, J. A. Chakel, S. Fisher, K. Lichtenwalter, and W. S. Hancock, *Anal. Chem.* **69,** 1320 (1997).
[33] L. L. Cummins, M. Winniman, and H. J. Gaus, *Bioorg. Med. Chem. Lett.* **7,** 1225 (1997).

was used to eliminate reversible secondary structure contributions to band width to improve resolution. Similarly, polymeric RP columns can tolerate contact with strongly acidic solutions for on-column product detritylation preparatory to anion-exchange LC purification. Removal of DMT postpurification requires acid treatment, cold ethanol precipitation, and physical isolation, e.g., by centrifugation. Little net processing time is typically saved, and final product purities are no greater than by reversed-phase DMT-on purification, but any potential low-level contamination by DMT-on material in detritylated product can be avoided. An example of such an alternative protocol is shown in Fig. 2. In this strategy, the DMT-off failure peak is first eluted while retaining the DMT-on peak on column. Absorbed material is then exposed to a dilute TFA solution for a well-controlled contact time, converting most absorbed oligomer to DMT-off product, which is then recovered by continued organic elution. In typical cases, 1–2% (v/v) trifluoroacetic acid (TFA) in water is satisfactory, with contact times of 5 to 10 min. This approach can reduce depurination to levels detectable, if at all, only by mass spectrometry. An advantage is that the procedure can be automated at small scale on a programmable chromatographic workstation. We have scaled up the procedure for purification of up to 10 g of crude oligonucleotide per run, but sample loading must be reduced to maintain robust separation on larger columns.

Gram Scale (1–10 g) Purification of Phosphorothioates on RP-HPLC

Phosphorothioate DNA oligonucleotide ISIS 2105 DMT-on crude product was purified on a 600-ml Oligo R3 column (50.4 mm i.d. × 300 mm length) in a sodium acetate/methanol buffer system. The chromatogram is shown in Fig. 3. The elution profile is very similar to that seen at smaller scales. In this case, 4.5 g of crude oligonucleotide is loaded on the column at 7.5 mg/ml of CV. Purification of the full-length product is good for this relatively heavy sample loading, with the benzamide peak clearly evident. Because of the high load, however, the DMT-on peak is split into multiple apparent peaks. These are analyzed by analytical SAX HPLC to determine pooling strategy. The selected product pool contains approximately 1.6 g of oligonucleotide at >95 area-% DMT-on purity.

Production Scale RP-HPLC Purification of 10–100 g Oligonucleotide

Methodology used under GMP conditions in our facility for 100-g preparative HPLC runs is similar to the method described in the previous section, except that gradient conditions are optimized to maximize sample loading and product purity. Sample loading and flow rates are linearly scaled from small column conditions. To ensure comparable physical bed

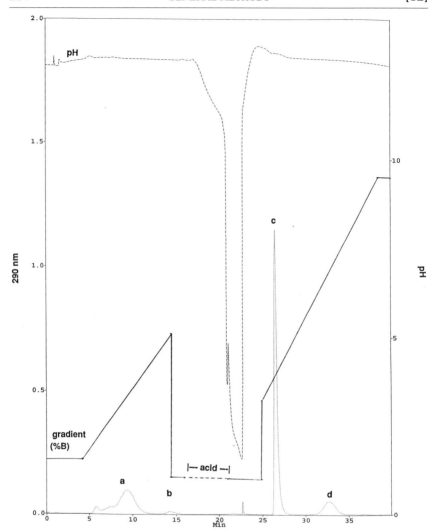

FIG. 2. On-column detritylation of 20-mer phosphorothioate AO. Peak a is the failure peak and peak b is the benzamide peak. Both these peaks elute before acid is introduced. Peaks c and d elute after acid treatment. Peak c is the detritylated product peak and peak d is the undetritylated portion of the DMT-on starting material. The solid line indicates the %B gradient profile and the dashed line indicates the pH profile. Sample: DMT-on ISIS 5132/ CGP 69846A, 1 mg/ml, 4 ml injection; Column: Oligo R3 (PE BioSystems), 10 mm i.d. × 100 mm length, column volume (CV) = 7.85 ml. Buffer A: 50 mM NaOH, Buffer B: Methanol, and Buffer C: 2% TFA in water. Flow rate: 10 ml/min. Gradient program: Equilibration: 93% A + 7% B, for 4 CV, Load 4 ml sample, Elution 1: 7% B to 35% B in 10 CV, wash: 3% B for 2 CV, Detritylation: 100% C for 5 min (flow rate 1 ml/min for last 4 min), neutralization: wash 3% B for 5 CV, Elution 2: 20% B to 70% B in 17 CV, hold 2 CV.

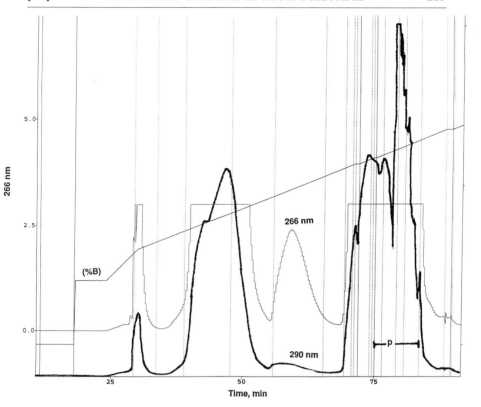

FIG. 3. RP purification of phosphorothioate AO at gram scale. The elution was monitored at 290 nm (thick trace) and 265 nm (thin trace). Vertical lines indicate the manual fraction points. p indicates the pooling for the product. Sample: DMT-on crude ISIS 2105 4.5 g injection; Column: Oligo R3 (PE BioSystems), 50.4 mm i.d. × 600 mm length, column volume (CV) = 600 ml. Buffer A: 2.5 M sodium acetate, pH 7.2, Buffer B: Methanol, and Buffer C: DI water. Flow rate: 50 ml/min. Elution 20 to 30% in 0.5 CV, 30 to 70% in CV, wash 90% B in 2 CV. 4% A was maintained through the process, diluted with Buffer C.

integrity among small-scale and large-scale columns, radial compression columns are used. The KP 100 (Biotage Division, Dyax Corp., Cambridge, MA) high-pressure automated chromatography equipment is used. The radial compression columns are prepacked with BondaPak $HC_{18}HA$ silica-based media (Waters Corp.). Figure 4 shows an example of RP purification of a 35-g crude 21-mer phosphorothioate oligonucleotide. Note that separation of DMT-on and DMT-off peaks is excellent and not compromised when a large column is used. Product purity and yield are therefore high and similar to small-column performances. A further increase in column scale successfully allows 100-g scale purifications per run.

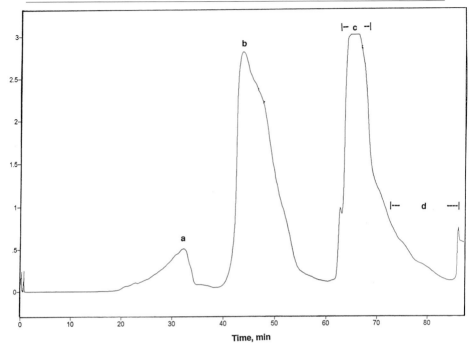

Fig. 4. Production scale (10–100 g) RP-HPLC purification of phosphorothioate AO. The *x* axis indicates the absorbance in arbitrary units. Peaks a, b indicate DMT-off failure containing peaks, c denotes the product cut and d indicates DMT-on containing failures. Sample: DMT-on crude 21-mer phosphorothioate; Column: Radial compression module, containing BondaPak $HC_{18}HA$ media (Waters Corp.), buffer system same as in Fig. 3.

Use of AX Chromatography for Purification of DMT-On Oligonucleotides

The analytical HPLC trace of DMT-on oligonucleotides (Fig. 5) shows that in this chromatographic mode the DMT-on product peak is well separated from the DMT-off failure sequences peak. It is therefore feasible to isolate DMT-on material from crude product mixtures by this method. However, the affinity of the DMT-on peak for anion exchanger media is very high, with up to 3 *M* salt being reported necessary for the complete elution of DMT-on material,[34] so greater care is needed to clean and regenerate the column after each separation.

It is also possible to load the entire DMT-on peak onto the anion-exchange resin and detritylate the material-on column. For this procedure,

---

[34] H. J. Johansson and M. A. Svensson, "Nucleic Acid-Based Therapeutics." IBC Conference, San Diego, 1995.

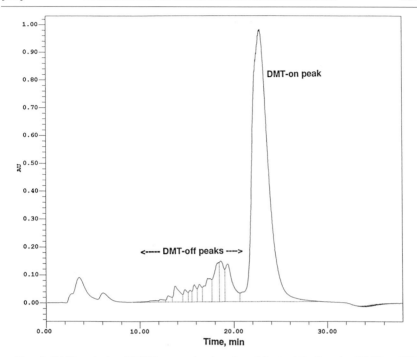

Fig. 5. SAX analysis of DMT-on crude phosphorothioate AO. Sample: DMT-on ISIS 2105. Column: POROS HQH (PE BioSystems), 4.6 × 100 mm length. Buffer A: 20 mM NaOH, Buffer B: 2.5 M NaCl in A. Flow rate: 1 ml/min. Temperature: 70°. Gradient program: Elution 0–100% B in 20 min.

the DMT-on peak is first isolated from a reversed-phase column, then loaded directly to an anion-exchange column. The bound material is detritylated by a procedure similar to that in Fig. 2. While removal of DMT-cation and residual DMT-on material is difficult, the method can be used for reducing the partial phosphodiester content of synthetic phosphorothioate oligonucleotides.

Purification of DMT-On Phosphorothioates on Hydrophobic Interaction Media

Hydrophobic interaction chromatography (HIC) offers another way to purify crude DMT-on phosphorothioate oligonucleotides.[8,35] HIC has the advantage that the ammoniacal oligonucleotide cleavage solution can be loaded directly to the column and large organic solvent volumes are unnec-

[35] Z. Zhang and J.-Y. Tang, Curr. Opin. Drug Disc. Dev. 1, 304 (1998).

essary. The method has been used at a 100-g scale, but because considerable effort is necessary to optimize the method for each sequence, it is most practical for large-scale use. HIC purification product pooling usually follows protocols similar to those used for RP-purified oligonucleotides.

Purification of Crude DMT-Off Phosphorothioate Oligodeoxyribonucleotide AO

Starting the purification process with DMT-off crude oligonucleotide has the advantage that detritylation need not be carried out after purification. At large scale, detritylation and subsequent precipitation are more difficult than at small scales. As the examples in this section indicate, high purity can be obtained without recourse to the DMT group to facilitate hydrophobic purification. Anion-exchange (AX) chromatography is convenient for the purification of DMT-off crude products. RP does have application for the purification of DMT-off crude products, but has been mainly useful for short oligonucleotides. For this reason, only AX chromatography is discussed in this section for the purification of DMT-off oligomers. At larger scales, AX processes avoid the use of large quantities of organic solvent, can use widely available low-pressure chromatography equipment, and can give product yields at least equivalent to those from RP chromatography of DMT-on oligomers.

AX can also be used as an orthogonal purification step following the RP-HPLC purification of DMT-on AO crudes. The product pool post-RP purification or post-HIC purification is detritylated and precipitated and the oligonucleotide is then redissolved in water and purified by AX HPLC. This method increases purity of the RP purified at the expense of full-length oligomer yield. The method used for the orthogonal step is, in most cases, similar to the AX method for the purification of DMT-off crudes. Therefore, only the purification of crude AO is discussed in this section.

AX Method to Purify 1–10 g DMT-Off Crude Phosphorothioate AO

ISIS 2302, an antisense oligonucleotide currently in phase IIb human clinical trials for the treatment of Crohn's disease and a range of other inflammatory conditions, is purified from crude by AX HPLC. DMT-off crude at 45 mg/ml is diluted 1:1 with equilibration buffer (20 mM NaOH). A 200-ml column (AP-5, 50.4 × 100 mm) is packed with POROS HQ/50 media followed by loading of approximately 3.9 g of crude synthesis product, which is close to 20 mg/ml of packed column volume. The chromatogram is shown in Fig. 6. The elution profile is made up of step and linear gradients

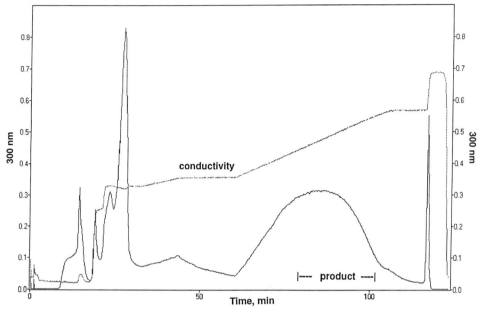

Fig. 6. AX purification of 20-mer DMT-off phosphorothioate crude. Sample: DMT-off ISIS 2302 25 mg/ml in 20 mM NaOH, 3.9 g total injection. Column: POROS HQ/50 (PE BioSystems), 50.5 × 100 mm length, column volume (CV) = 200 ml. Fractions pooled for product as indicated. Buffer A: 20 mM NaOH, Buffer B: 2.5 M NaCl in A. Flow rate: 35 ml/ min. Gradient program: Equilibration: 100% A for 0.5 CV, load 220 ml sample, wash 100% A in 0.5 CV, elution: step 0.6 M for 0.5 CV, step 0.85 M for 2 CV, gradient 0.85 to 0.95 M in 2 CV, hold 0.95 M for 3 CV, gradient 0.95 M to 1.85 in 8 CV, hold 2 CV.

to optimize purification. The main peak is fractionated, and the fractions are analyzed by QCGE and analytical SAX HPLC. Fractions meeting specified purity criteria are pooled. Analyses of pooled fractions are shown in Table III. The yield of the full-length product is 83% of that in the

TABLE III
ANALYSIS OF PURIFICATION

	Purity (%)					
	QCGE analysis			SAX analysis		
Pool	Other $(n - x, n + x)$	$(n - 1)$	n-mer	Other	$(P{=}O)_1$	$(P{=}S)$
Feed	25.1	5.5	69.4	15.7	10.4	73.9
Product	4.6	4.9	90.5	0.0	1.0	99.0

column feed, per QCGE analysis. Impurity distribution is determined by calculating the quantity of *n*-mer in each product pool. The initial waste fractions are designated as pool *w*, peak tailings at the end of the elution are pooled as *y*, *p* is the main product pool, and *r* is the recycle pool. SAX analyses for these pools are presented in Fig. 7, showing clearly that only a small fraction of the initial impurities remains in the product pool. Yield loss is mainly due to the presence of product in the recycle pool, material that may be combined with recycle pools from other runs and reprocessed on the same chromatographic column to maximize yield. This same purification can be scaled linearly by maintaining the linear flow rate, sample loading, and elution profiles. The gradient may be fine-tuned for different oligonucleotide base compositions.

It should be noted that because step and linear segments are used in optimizing the elution gradient, this method is sensitive to significant deviations from specified loading. If a lower load is used, gradient strength would require optimization. Once optimized, this procedure is effective for achieving good oligonucleotide purity at good yields.

Column loading capacity is important for large-scale purification, as in most cases an increased loading capacity directly translates to reduced equipment cost and higher throughput. Anion exchangers in general have higher dynamic binding capacities for oligonucleotides than reversed-phase media. The ability of SAX HPLC to provide high loading and throughput while maintaining moderate chromatographic yield is demonstrated in the purification of ISIS 2105, a 20-mer phosphorothioate oligonucleotide antisense drug. The crude oligomer is purified on POROS HQ/50 media under high loading (approximately 43 mg/ml of packed column bed) with optimized step gradient elution. The chromatogram is shown in Fig. 8, illustrating the effectiveness of the purification. Analysis of the product pool shows full-length purity of 95.8% by QCGE at a yield of 72% of the loaded full-length material.

Displacement and Sample Self-displacement Chromatography

Chromatographic displacement effects can be very useful in oligonucleotide purification. Applications of displacement and sample self-displacement techniques have been reviewed.[15] In displacement chromatography, a component with higher binding strength than the sample is used to displace the sample from the column in a characteristic square-wave elution profile. Dextran sulfate has been shown to be an effective displacer of oligonucleotides.[36] In sample self-displacement, the sample components themselves act

---

[36] J. A. Gerstner, P. Pedroso, J. Morris, and B. J. Bergot, *Nucleic Acids Res.* **23**, 2292 (1995).

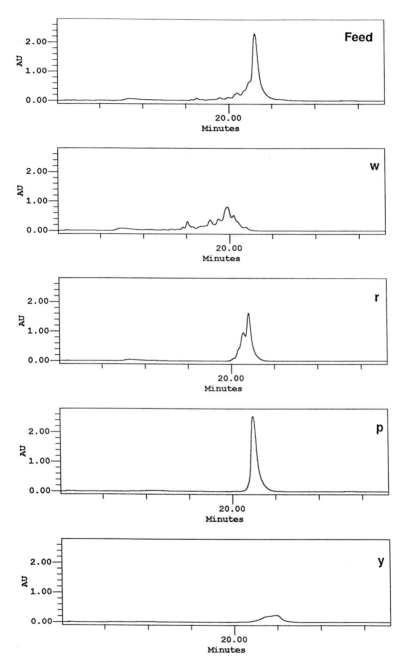

FIG. 7. Analysis of pools for SAX purification of 20-mer DMT-off phosphorothioate crude. Experiment conditions as described in Fig. 6. (Feed) Analysis for the starting material, (w) analysis of waste pool containing early eluting fractions, (r) recycle fractions, (p) product fractions, and (y) pool of late eluting impurities. This analysis shows that the purification effectively removes most of the early and late eluting impurities from the product pool.

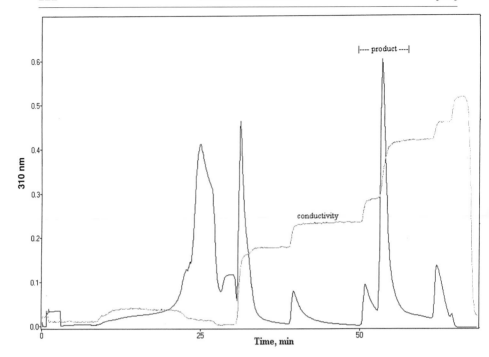

FIG. 8. AX purification of 20-mer DMT-off phosphorothioate crude. Sample: DMT-off ISIS 2105 8.6 g in 20 m*M* NaOH. Column: POROS HQ/50 (PE BioSystems), 50.5 × 100 mm length, column volume (CV) = 200 ml. Buffer A: 20 m*M* NaOH, Buffer B: 2.5 *M* NaCl in A. Flow rate: 35 ml/min. Gradient: Equilibration: 100% A for 0.5 CV, load 600 ml sample, wash 100% A in 0.5 CV, elution: step 20% B for 1.5 CV, step 28% for 2.0 CV, 40% for 0.5 CV, 68% for 1.5 CV, 80% for 0.5 CV, step 100% B for 0.5 CV.

as displacers, with stronger binding components displacing weaker binding components. Self-displacement has an advantage over displacement chromatography in that no additional compounds need to be added to the sample, thus avoiding contamination issues. Sample displacement is effective for AO purification in part because major deletion sequence impurities are chemically similar to the product but have a lower binding affinity for anion-exchange chromatographic media. Sample displacement has been applied successfully to oligonucleotide purification in the cases of phosphodiesters,[14] phosphorothioates at small and large scales,[15] and larger nucleic acids.[37] The sample load is an important variable in optimizing displacement AO purifications, and the sample self-displacement effect must be optimized during method development for either displacement mode.

[37] J. H. Waterborg and A. J. Robertson, *Nucleic Acids Res.* **21,** 2913 (1993).

Purification of Oligonucleotides with Strong Intra- and Intermolecular Interactions

Oligonucleotides rich in G residues have a tendency to form metastable secondary structures in solution. Some appropriate short sequences may also have multimeric forms. One such example is ISIS 10852, an 8-mer phosphorothioate oligonucleotide (TTGGGGTT) that can form an antiviral tetrad of four parallel strands. The tetrad formation is affected by variables, including pH, oligonucleotide concentration, salt concentration, and organics, thus complicating purification scale-up. A strongly denaturing buffer is required to simplify purification of this compound. Potassium or sodium hydroxide at a high concentration (50 mM) is effective for this purification by shifting the association equilibrium strongly to the monomeric species. High pH NaOH eluent is an effective denaturant for analytical and small-scale preparative separations when such structural motifs are present.[9,38]

Analytical scale runs are performed using two 100-mm-long POROS HQ/H columns connected in series and eluted at room temperature using the same buffers as for preparative runs (AX buffer II, Table I). The chromatogram of a DMT-off crude product is shown in Fig. 9. For short synthetic oligonucleotides, deletion sequence impurities have a significantly greater charge-to-mass than the full-length product and are thus removed more readily than failure sequences in 20-mer phosphorothioate oligonucleotides. Five grams of DMT-off crude ISIS 10852 AO is purified on Q HyperD F media (50 mm i.d. × 100 mm length, AP-5 column) under this denaturing condition. Figure 10 shows the chromatogram. Resolution is quite good, producing full-length purity greater than 92% by QCGE. Higher purities (98% by QCGE) are obtained by reprocessing the product pool under conditions enhancing the sample self-displacement effect.

Postpurification Processing

The typical product pool from the reversed-phase chromatographic purification of DMT-on synthetic oligonucleotides typically contains alcohol and low-level residual buffer salt. This material is detritylated in organic acid, and the detritylated oligonucleotide is precipitated from solution by the addition of cold ethanol, a procedure optimized for large-scale use.[39] An advantage of this overall procedure is the removal of the remaining buffer salt in the precipitation step.

---

[38] M. W. Germann, R. T. Pon, and J. H. van de Sande, *Anal. Biochem.* **165,** 399 (1987).
[39] A. Krotz, personal communication (1999).

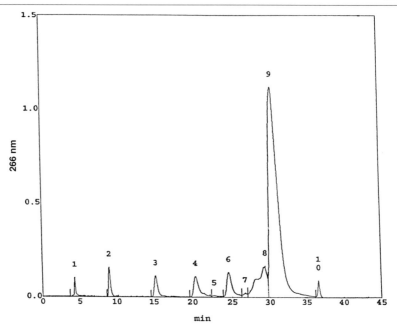

FIG. 9. SAX analysis of a self-associating phosphorothioate. Sample: DMT-off crude ISIS 10852; Column: POROS HQ/H (PE BioSystems), 4.6 mm i.d. × 200-mm length (2 100 mm columns linked in series). Buffer A: 50 mM KOH, Buffer B: 50 mM KOH + 2.5 M KCl. Flow rate: 8 ml/min. Temperature: Ambient. Gradient program: 0% A to 80% B in 10 min as shown. System: BioCad 20.

In contrast, the oligonucleotide eluate from preparative anion-exchange chromatography is DMT-off material containing very significant levels of HPLC buffer salts. AX eluates may be desalted by a variety of methods. At small scales, an effective method is to adsorb the purified oligonucleotide on a reversed-phase cartridge such as the Sep-Pak (Waters Corp., Milford, MA), wash with water to remove excess salts, and elute the oligonucleotide in an alcohol solution. Methanol is removed by drying, and the oligonucleotide is resuspended in water for isolation. This procedure may also be carried out at large scale by using an RP column, but the removal of alcohol by evaporation adds a unit operation to the process. Gel filtration and membrane-based desalting methods are alternative methods for desalting AX oligonucleotide eluates at large scale.

Summary

Chromatography is an effective tool for obtaining high-purity synthetic oligonucleotides for a variety of end uses, including antisense drug therapy.

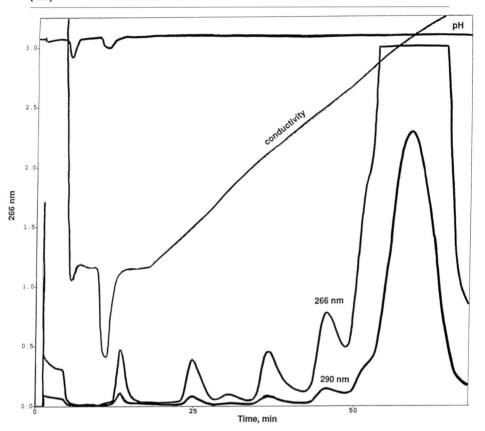

Fig. 10. Large-scale AX purification of a self-associating phosphorothioate. Chromatographic traces at 266 and 290 nm are shown. The top linear trace is the pH profile indicating effective buffering at the high pH (12.6). Sample: DMT-off crude ISIS 5320, 5 g; Column: Q HyperD F (BioSepra, Inc.), 50.4 mm i.d. × 100 mm length, column volume (CV) = 200 ml. Buffers as in Fig. 9. Flow rate: 35 ml/min. Gradient program: Equilibration: 100% A, for 1 CV, Load 100 ml sample, Wash 100% A for 0.5 CV, Elute: 20% B for 1 CV, 20% B to 80% B in 10 CV, Hold 0.5 CV.

Reversed-phase and anion-exchange chromatographies are widely used techniques for this application. While selectivity of these techniques can be modified by methods such as ion-pair RP-HPLC or affinity chromatography, these are presently used only at small scales. RP chromatography makes use of terminal hydrophobic-protecting groups to increase retention and selectivity. The main advantages of the RP method are its utility for the purification of a wide variety of modified oligonucleotide structures, its applicability across a range of terminal hydrophobic groups, such as fluorescein, and its ready use from small scale to very large scale with a minimal requirement for process development. AX-HPLC can also give high-purity products at generally higher media capacities. A more extensive method development effort is typically required for the AX-HPLC purification of AO. The AX yield per unit operation can be lower, but the isolated yield of DMT-off desalted oligonucleotide can be equal to or higher than that from RP-HPLC. As additional AO drugs enter and mature in the market, there will be a potential need for ton-scale purification processes. AX provides a way to scale up production on somewhat less expensive equipment with reduced organic solvent requirements.

[13] Boranophosphate Backbone: a Mimic of Phosphodiesters, Phosphorothioates, and Methyl Phosphonates

By Barbara Ramsay Shaw, Dmitri Sergueev, Kaizhang He, Ken Porter, Jack Summers, Zinaida Sergueeva, and Vladimir Rait

Introduction

A new class of boron-modified oligonucleotides designed for use as potential therapeutic and diagnostic agents were first reported in 1990 by this laboratory.[1] These compounds, nucleoside *boranophosphates* or *boranephosphonates,* are distinctive in that one of the nonbridging oxygens in the phosphate diester is replaced by borane (BH_3) (Fig. 1). The BH_3 group maintains the negative charge of a phosphate, but does not form classical H bonds and cannot coordinate metals well.[2] This modification imparts unique characteristics to boranophosphate nucleosides and nucleic acids.[3]

Boranophosphate **2** can be considered as a "hybrid" of three types of well-studied modified phosphates, i.e., the normal phosphate **1**, the phosphorothioate[4] **3**, and the methyl phosphonate[5] **4** shown in Fig. 2.[3] The BH_3 group in boranophosphates is isoelectronic with oxygen (O) in normal phosphates and is isolobal (pseudoisoelectronic) with sulfur (S) in phospho-

[1] A. Sood, B. R. Shaw, and B. Spielvogel, *J. Am. Chem. Soc.* **112,** 9000 (1990).
[2] J. Summers and B. R. Shaw, in preparation (1998).
[3] B. R. Shaw, J. Madison, A. Sood, and B. F. Spielvogel, *Methods Mol. Biol.* **20,** 225 (1993).
[4] F. Eckstein, *Annu. Rev. Biochem.* **54,** 367 (1985).
[5] P. S. Miller, *in* "Antisense RNA and DNA" (J. A. H. Murray, ed.), p. 241. Wiley-Liss, New York, 1992.

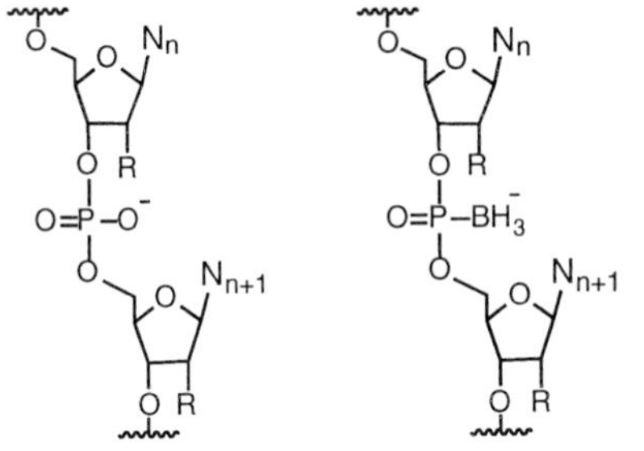

1 2

FIG. 1. Fragments of natural phosphodiester nucleic acids[1] and their boranophosphate counterpart **2**. R=H in deoxyribonucleic acids and R=OH in ribonucleic acids. Dithymidine phosphate **1** and dithymidine boranophosphate **2**. N is a purine or gyrimidine.

rothioates. The BH_3 group is isosteric with the CH_3 group in methyl phosphonates **4**. Boranophosphates would be expected to share a number of chemical and biochemical properties with both phosphorothioate and methyl phosphonate analogs.

Unique Properties of Oligodeoxynucleoside Boranophosphates (BH_3^--ODN)

The boranophosphate carries the same negative charge as the normal phosphate and the phosphorothioate, and therefore oligodeoxynucleoside boranophosphates (BH_3^--ODN) are soluble in aqueous solutions such as natural oligodeoxynucleotides O^--ODN and oligodeoxynucleoside phosphorothioates S^--ODN. The charge distribution differs among the analogs in Fig. 2, thereby changing the polarity of the compounds. In a phosphate anion **1**, the negative charge is distributed equally on the two nonbridging oxygens, whereas the charge is localized primarily on the sulfur in a phosphorothioate anion **3** or on the one nonbridging oxygen in a boranophos-

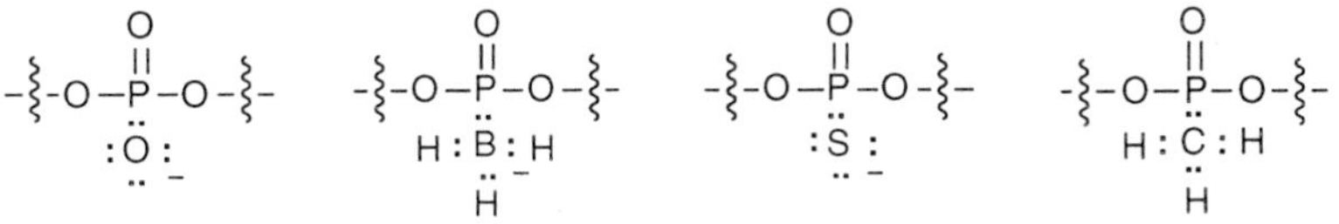

1 Normal Phosphate 2 Boranophosphate 3 Phosphorothioate 4 Methyl phosphonate

FIG. 2. Isoelectronic analogs of natural phosphodiesters. Only the first three retain a negative charge. The position of the negative charge does not necessarily reflect the actual spatial location in the molecule. Adapted from Shaw *et al.*[3]

phate anion **2**, rather than on the borane group.[6] Thus, there is a transition from a symmetric charge delocalization (between nonbridging oxygens) in a phosphate diester to an asymmetric delocalization of charge that is primarily on sulfur in a phosphorothioate ester and primarily on the nonbridging oxygen in a boranophosphate diester. The BH_3^--ODN should exhibit different hydration properties, interact differently with proteins, and cross cellular membranes more readily than normal phosphates or thiophosphates.

In a study of the partitioning of the dinucleosides **1** and **2** between octanol and water, the boranophosphate **2** distributed into the nonaqueous phase (octanol) approximately 18-fold more than the normal phosphate **1** making it more lipophilic than anionic normal phosphates but less lipophilic than nonionic methyl phosphonates.[7] Boranophosphates may be useful in transporting boron to tumor tissue selectively for radiation therapy. Boron-10, which accounts for 19% abundance among the two stable isotopes of boron, has a high cross section for thermal neutrons. On capture of a neutron as in Eq. (1), the excited boron nucleus releases α particles ($_2He^4$) that travel about 10 μm. If boron can be localized specifically in targeted cells, then these cells may be destroyed readily using boron neutron capture therapy (BNCT).[8]

$$_5B^{10} + _0n^1 \rightarrow _3Li^7 + _2He^4 + 2.4 \text{ MeV} \tag{1}$$

This article discusses methodology that makes it possible to place borane (BH_3) groups in the internucleoside phosphate linkage of DNA or RNA oligonucleotides or at the α-P of nucleoside triphosphates (as in ATP and dATP).[1,3,9–13] We briefly describe a general method used to demonstrate that the deoxyribo- and ribonucleoside 5'-(α-P-borano)triphosphates (dNTPαB or NTPαB) are good substrates for DNA and RNA polymerases.[10,14,15] We also discuss strategies employed in applications of the method. Boranophosphate linkages in DNA, like phosphorothioate linkages, are more resistant to degradation by nucleases than normal phosphate esters.[1,10,16] The borane modification therefore can be incorporated into

[6] J. S. Summers, D. Roe, P. D. Boyle, M. Colvin, and B. R. Shaw, *Inorg. Chem.* **37**, 4158 (1998).
[7] F. Huang, Ph.D. dissertation, Duke University (1994).
[8] J. G. Wilson, *Chem. Rev.* **98**, 2389 (1998).
[9] J. Tomasz, B. R. Shaw, K. Porter, B. Spielvogel, and A. Sood, *Angew. Chem. Int. Ed. Eng.* **31**, 1373 (1992).
[10] K. W. Porter, J. D. Briley, and B. R. Shaw, *Nucleic Acids Res.* **25**, 1611 (1997).
[11] K. He, A. Hasan, B. K. Krzyzanowska, and B. R. Shaw, *J. Org. Chem.* **63**, 5769 (1998).
[12] B. K. Krzyzanowska, K. He, A. Hasan, and B. R. Shaw, *Tetrahedron* **54**, 5119 (1998).
[13] D. Sergueev and B. R. Shaw, *J. Am. Chem. Soc.* **120**, 9417 (1998).
[14] H. Li, K. Porter, F. Huang, and B. R. Shaw, *Nucleic Acids Res.* **21**, 4495 (1995).
[15] K. He and B. R. Shaw, in preparation (1999).
[16] F. Huang, A. Sood, B. F. Spielvogel, and B. R. Shaw, *J. Biol. Struct. Dyn.* **10**, a078 (1993).

nucleic acids by enzymatic reactions, notably in the synthesis of boranophosphate oligonucleotides (DNA or RNA) from dNTPαB analogs[10,14] or NTPαB analogs,[15] resulting in nuclease-resistant phosphodiester linkages.

Finally, in recognition that specific targeting of an mRNA for cleavage should be an efficient way to affect expression of a protein, we describe the boranophosphate-induced cleavage of an RNA by RNase H.[17] Only three nucleic acid analogs are known to facilitate RNase H activity: S^--ODN, S_2^--ODN (dithio), and now BH_3^--ODN. The well-characterized phosphorothioate analogs, however, exhibit some undesirable nonspecific binding properties and side effects that may ultimately limit their use.[18] The phosphorodithioates analogs form less stable complementary complexes[19] and tend to form intramolecular complexes.[20] The only other antisense analog that currently offers the advantage of inducing RNase H cleavage of RNA is BH_3^--ODN.

Synthetic Methods

Chemical Synthesis of Oligonucleoside Boranophosphates (BH_3^--ODN)

One nice feature of the boranophosphate backbone modification is that the synthesis of BH_3^--ODN requires only minor modifications of either the standard phosphoramidite or H-phosphonate methodologies. The primary modification involves replacement of an iodine oxidation step by a borane exchange reaction between two Lewis bases, one of which is trivalent phosphorus. The resulting $P-BH_3$-bond is stable toward most chemicals involved in synthetic cycles and postsynthetic ammonia deprotection.

Phosphoramidite Approach (Solution Synthesis). The first synthesis of dithymidine boranophosphate was carried out in our laboratory in solution by the phosphoramidite method (Scheme 1).[1,3] The phosphite triester was boronated rapidly and quantitatively by the borane–dimethyl sulfide complex ($BH_3 \cdot DMS$). The treatment produced the stable ester of boranophosphate and simultaneously removed the 5'-O-(4,4'-dimethoxytrityl) (DMT)-protecting group, thereby allowing omission of the detritylation step in the synthetic cycle (for more details, see Shaw *et al.*[3]). Although this method is straightforward for the dithymidine boranophosphate (52% yield), it has serious limitations for bases other than thymine. Synthesis with adenine,

[17] V. Rait and B. R. Shaw, *Antisense Nucleic Acid Drug Dev.*, in press.
[18] C. A. Stein and Y.-C. Cheng, *Science* **261**, 1004 (1993).
[19] L. Cummins, D. Graff, G. Beaton, W. S. Marshall, and M. H. Caruthers, *Biochemistry* **35**, 8734 (1996).
[20] M. K. Ghosh, K. Ghosh, O. Dahl, and J. C. Cohen, *Nucleic Acids Res.* **21**, 5761 (1993).

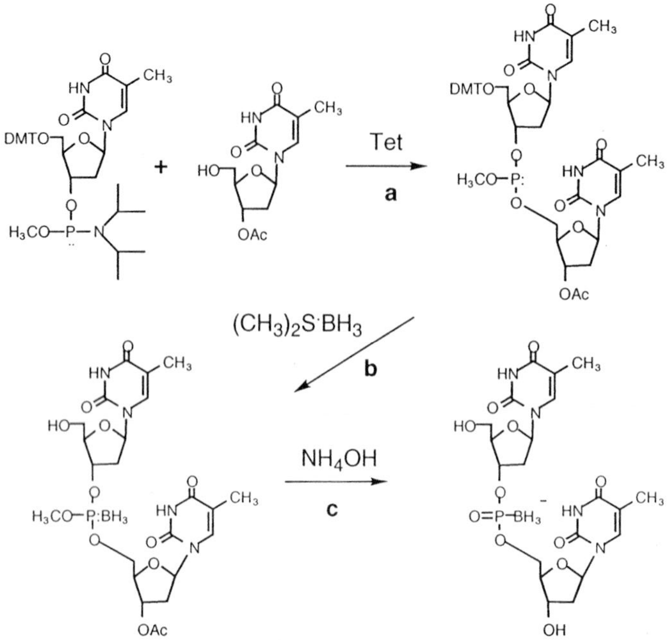

Scheme 1. Synthesis of dithymidine boranophosphate via phosphoramidite approach. (a) 0.25 M tetrazole, 15 min; (b) 0.17 M borane–dimethyl sulfide complex, 4 hr; (c) concentrated ammonia, overnight, room temperature). Adapted from A. Sood et al., J. Am. Chem. Soc. **112,** 9000 (1990).

guanine, and cytosine bases showed formation of noticeable amounts of by-products due to reaction with the borane. Substitution of $BH_3 \cdot DMS$ by milder boronating agents inevitably results in incomplete phosphite triester boronation or requires longer boronation time, neither of which is desirable for the efficient synthesis of oligomeric boranophosphates. These limitations prompted us to investigate the potential of an H-phosphonate methodology for BH_3^--ODN synthesis.

H-Phosphonate Approach. Oligonucleotide synthesis via an H-phosphonate intermediate presents a very convenient way to a variety of backbone modifications, including phosphorothioates (S⁻-ODN), phosphoramidates (NH₂-ODN), and normal phosphodiesters (O⁻-ODN).[21] The complete oligonucleotide chain can be assembled with each internucleoside linkage having trivalent P(III) phosphorous in the form of H-phosphonate. Phos-

[21] A. Kers, I. Kers, A. Kraszewski, M. Sobkowski, T. Szabo, M. Thelin, R. Zain, and J. Stawinski, *Nucleosides Nucleotides* **15,** 361 (1996).

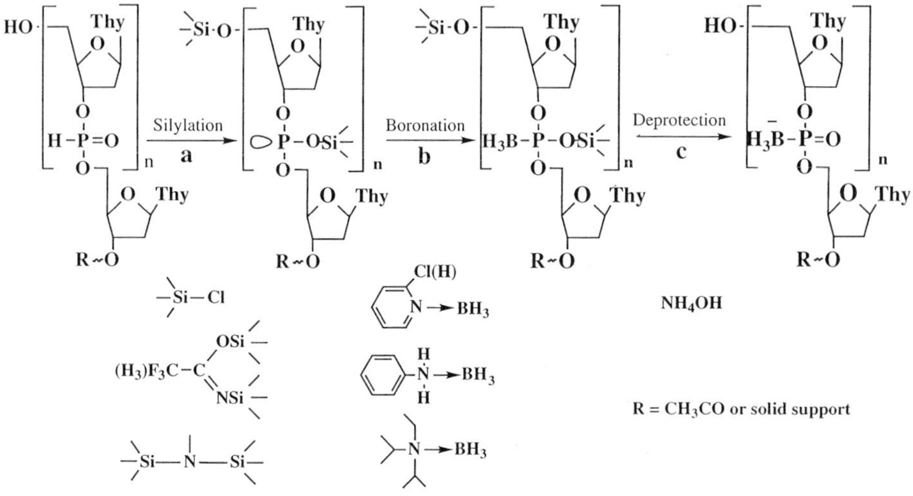

SCHEME 2. Conversion of H-phosphonate oligonucleotide backbone to boranophosphate via (a) silylation with 0.3 M agent, (b) boronation with 0.3 M agent, and (c) deprotection with concentrated ammonium hydroxide (2 hr, room temperature). Adapted from Sergueev and Shaw.[13] Reagents tested in each reaction are shown below the scheme.

phorous modification can be carried out in a single step, converting all H-phosphonate groups to the pentavalent P(V) phosphorus. This is especially important for the boronation procedure because the duration of the modification step is not as important as in the phosphoramidite approach, thereby permiting the use of milder boronating agents. Boronation of the H-phosphonate intermediate differs from the oxidation, thiolation, or amidation used to prepare O^--, S^--, or NH_2-ODN. Specifically it requires the intermediate formation of a Lewis base by conversion of the internucleoside H-phosphonate diester to a phosphite triester (Scheme 2, step a). Once the P(III) compound with its lone electron pair is formed, it exchanges with different borane–amine complexes easily, generating boranophosphate triester. The final deprotection with concentrated ammonia affords the desired BH_3^--ODN.

Procedure for H-Phosphonate Approach (Solution Synthesis). Synthesis of dinucleoside boranophosphate via the H-phosphonate approach can be done in solution (Scheme 2, $n = 1$, R $= -COCH_3$) as first reported.[22] After conventional synthesis of the protected dithymidine H-phosphonate,[23] the

[22] D. Sergueev, A. Hasan, M. Ramaswamy, and B. Ramsay Shaw, *Nucleosides Nucleotides* **16**, 1533 (1997).

[23] B. C. Froehler, *in* "Protocols for Oligonucleotides and Analogs" (S. Agrawal, ed.), p. 63. Humana Press, Totowa, 1993.

5'-DMT group is removed and the dinucleoside H-phosphonate (0.05 M) is converted smoothly to a phosphite triester by silylation in anhydrous dioxane with 0.2 M bis(trimethylsilyl)acetamide (BSA).[24] The reaction can be followed conveniently by ^{31}P nuclear magnetic resonance (NMR), showing complete disappearance of 11.5 and 12.6 ppm signals (two H-phosphonate stereoisomers) after 2 hr and formation of signals at 130.7 and 130.8 ppm that correspond to dithymidine phosphite trimethylsilyl ester; these are the only phosphorus species in the reaction mixture. Without isolation of the intermediates, the borane exchange reaction is then carried out. Instead of using $BH_3 \cdot DMS$ (which easily transfers a borane to the tricoordinated phosphorous but alternatively can transfer it to the heterocyclic base nitrogens, causing base boronation and possibly base modification by hydroboration of double bond or carbonyl group reduction[25]), we have introduced the milder boronating agent, borane–pyridine complex ($BH_3 \cdot Py$), which is also more stable and less air sensitive. At 50°, the exchange reaction with 0.5 M $BH_3 \cdot Py$ proceeds in dioxane to completion in 12 hr. The formation of dithymidine boranophosphate trimethylsilyl ester is monitored by ^{31}P NMR, where a new broad signal at 105.5 ppm appears. The only by-product observed by phosphorus NMR is the putative phosphate triester at -9.0 ppm. The amounts of this by-product depend on how strictly anaerobic conditions were observed and do not exceed 10%. Addition of an equal volume of concentrated ammonia to the reaction mixture after 2 hr gives the desired dithymidine boranophosphate (δ 93.2 ppm), which is then purified by reversed-phase (RP) HPLC. Thus, by using a combination of BSA and $BH_3 \cdot Py$ reagents, the boranophosphate dimer was synthesized from its H-phosphonate precursor in solution with 70% yield.[22]

H-Phosphonate (Solid Phase) Approach. The efficient solution synthesis encouraged us to develop a boronation procedure compatible with automated solid phase synthesis and to synthesize BH_3^--ODN longer than dimer.[13] A number of silylating and boronating agents were screened and syntheses of oligothymidine boranophosphates were performed on three types of solid supports. All tested silylating agents (see Scheme 2) proved to be effective in converting internucleoside H-phosphonate to phosphite. The time for smooth and complete conversion of dithymidine H-phosphonate attached to polystyrene resin varied from 30 min for 0.3 M heptamethyldisilazane (HMDS) to less than 5 min for 0.3 M bis(trimethylsilyl)trifluoroacetamide (BSTFA) (Table I). The subsequent borane exchange reaction was carried out with different borane–amine complexes, involving

[24] A. Kume, M. Fujii, M. Sekine, and T. Hata, *J. Org. Chem.* **49,** 2139 (1984).
[25] C. L. Lane, *Chem. Rev.* **76,** 773 (1976).

TABLE I
TIME REQUIRED FOR COMPLETE REACTION OF POLYSTYRENE RESIN-BOUND DITHYMIDINE
H-PHOSPHONATE WITH SILYLATING BORONATING AGENTS[a]

Silylating agent (0.3 M)	Time	Boronating agent (0.5 M)	Time
Bis(trimethylsilyl)trifluoroacetamide	5 min	Borane-2-chloropyridine	10 min
Chlorotrimethylsilane	10 min	Borane-aniline	15 min
Bis(trimethylsilyl)acetamide	30 min	Borane-diisopropylethylamine	15 min
Heptamethyldisilazane	30 min	Borane-pyridine	12 hr[b]

[a] Molar ratios of the dithymidine H-phosphonate to the silylating and boronating agents
were 1 : 12 and 1 : 20, respectively.
[b] At 50°.

aliphatic, aromatic, and heterocyclic amines (Scheme 2). For all complexes, a quantitative conversion of phosphite to boranophosphate was registered without any noticeable equilibrium that theoretically could be observed for a borane exchange reaction between two Lewis bases.[26] The rate of the reaction with phosphite, however, differed for different amines (Table I). The borane was transferred most easily to phosphite by borane-2-chloropyridine complex ($BH_3 \cdot CPy$), whereas the $BH_3 \cdot Py$ complex turned out to be the most inert. The significant difference in their reactivity might be explained by an inductive and, more important, steric effect of the chlorine atom on stability of the B–N bond of the amine.[27] Despite its reactivity, no modification or boronation of the thymine base by the $BH_3 \cdot CPy$ complex was detected in model experiments. ^{31}P NMR spectroscopy was used to follow conversions of phosphorus species on low cross-linked highly loaded polystyrene resin (Sigma, St. Louis, MO, 0.28 mmol nucleoside per gram of support) (Fig. 3). All conversions proceeded quantitatively; the only by-product observed was the phosphate triester. Once again the amount (<5%) depended on how strictly anaerobic conditions were observed.

Procedure for H-Phosphonate Approach (Solid-Phase Synthesis). For the automated solid-phase synthesis of oligothymidine boranophosphates, BSTFA and $BH_3 \cdot CPy$ are used as silylating and boronating agents, based on data obtained for the dimer. After conventional H-phosphonate chain elongation on a Cyclone Plus (Milligen/Biosearch) and removal of the 5'-DMT group with 3% dichloroacetic acid, the solid support (see later) is treated with 0.3 M BSTFA in tetrahydrofuran (THF) for 20 min, washed

[26] T. Reetz, *J. Am. Chem. Soc.* **82,** 5039 (1960).
[27] C. J. Foret, K. R. Korzekwa, and D. R. Martin, *J. Inorg. Nucleic Chem.* **42,** 1223 (1980).

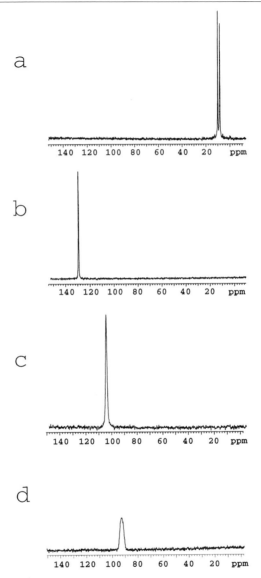

FIG. 3. ³¹P NMR spectra of dithymidine H-phosphonate: a) bound to polystyrene resin, b) after its reaction with 0.3 *M* BSTFA (10 min, room temperature, c) the subsequent reaction with 0.5 *M* BH₃·CPy, and d) after ammonia deprotection 2 hr, room temperature as in Scheme 2.

with acetonitrile for 30 sec, and allowed to react with 0.5 M BH$_3$ · CPy in anhydrous THF for 30 min. For comparison, the boronation procedure, i.e., silylation and borane exchange reaction, can likewise be performed manually with a combination of BSA or BSTFA and each of the boronating agents. On completion of the boronation procedure, the solid support is washed and the boranophosphate oligomer is cleaved and deprotected by concentrated ammonia treatment (2 hr, room temperature). ^{31}P NMR analysis of the crude reaction mixtures reveals that automated synthesis is more efficient than manual synthesis in converting the H-phosphonate group to boranophosphate due to more stringent anhydrous and anaerobic conditions. The ratio of boranophosphates to H-phosphonates and phosphates in manual mode was 91 : 5 : 4, whereas on the DNA synthesizer the ratio was 97 : 2 : 1.[13] With newer generation solid-phase synthesizers, even better yields of boranophosphates would be expected.

Oligothymidine boranophosphates up to 12 mers in length have been synthesized by the described procedure and purified by consecutive ion-exchange and reversed-phase HPLC.[13] Yields for the dodecathymidine boranophosphate prepared by different boronation procedures on different solid supports varied between 37 and 55% according to chromatographic analysis. Primarily the difference in yields was determined by the type of solid support rather than one or another set of reagents for the boronation procedure. While controlled pore glass (CPG-500, ABI, 29 μmol/g loading) and high cross-linked polystyrene resin (PS, Glen Research, 12 μmol/g loading) demonstrated high yields, the silica gel-based support (Sigma, 280 μmol/g loading) was less efficient. For instance, yields of dodecathymidine boranophosphate synthesized with bovine serum albumin (BSA) and BH$_3$ · Py reagents were 55, 53, and 37%, for CPG, PS, and silica gel, respectively.[13]

The overall yield of BH$_3$$^-$-ODN synthesis reflects the efficiencies of both the H-phosphonate chain assembly and the boronation procedure. To estimate boronation efficiency separately, we prepared samples of both normal phosphodiester oligomer (O$^-$-ODN) and BH$_3$$^-$-ODN from a single sample of H-phosphonate oligomer. After complete chain elongation, the solid support was divided: one portion was treated with iodine to yield O$^-$-ODN[28] and another portion was boronated. Because the iodine oxidation proceeds near quantitatively, comparison of the yields of O$^-$-ODN and BH$_3$$^-$-ODN allowed us to determine that the boronation procedure was 80–100% efficient relative to iodine oxidation. The efficiency depended primarily on a type of solid support, being lowest for the silica gel matrix.[13]

---

[28] P. J. Garegg, I. Lindh, T. Regberg, J. Stawinski, and R. Stromberg, *Tetrahedron Lett.* **27,** 4051 (1986).

Comments on Synthesis of BH_3^-*-ODN.* A number of groups have showed considerable interest in the synthesis and biochemical studies of backbone-boronated DNA and RNA. The synthetic methods employed were limited to those discussed earlier, i.e., phosphoramidite and H-phosphonate. In the first reported synthesis of a boranophosphate RNA,[29] the phosphoramidite approach was used to synthesize diuridine 3′,5′-boranophosphate by the same synthetic route depicted in Scheme 1, except that an additional tetrabutylammonium fluoride (TBAF) deprotection step was introduced to remove the 2′-O-silyl group. Compared to DNA synthesis, no complications were reported, but the overall yield was lower. More recently, we have reported synthesis of the same compound by the H-phosphonate approach similar to that described in Scheme 2.[30] This is an important demonstration of the applicability of both methods for the synthesis of BH_3^--RNA molecules.

Jin and Just[31] described the first stereoselective synthesis of dithymidine boranophosphate by a modified phosphoramidite approach using hydroxyethylindole as the chiral auxiliary. The reported method affords high diastereomeric excess (90–98%) and opens the possibility for the synthesis of stereoregular BH_3^--ODN. However, in its present form, the method preserves the just-mentioned limitations of the phosphoramidite synthetic approach. It is noteworthy that the borane exchange reaction proceeds stereospecifically with retention of the configuration around the phosphorus.[31] This coincides with data obtained during the stereospecific synthesis of dinucleoside boranophosphates starting with individual H-phosphonate diastereomers.[32] This feature opens opportunities to synthesize partially stereoregular BH_3^--ODN or alternating BH_3^--P, O^--P stereoregular ODN utilizing an individual diastereomer of H-phosphonate dimers as building blocks.

Syntheses of oligothymidine boranophosphate via the H-phosphonate approach have been reported independently by two other groups.[33,34] While the methodologies of the synthesis were the same (i.e., H-phosphonate chain assembly, silylation to afford a lone pair of electrons, borane exchange reaction, and final ammonia deprotection as depicted in Scheme 2), the choice of reagents and protective groups differed. Zhang *et al.*[33] used a $BH_3 \cdot DMS$ as a boronating agent, but lowered its concentration to 0.2 *M*

[29] Y.-Q. Chen, F.-C. Qu, and Y.-B. Zhang, *Tetrahedron Lett.* **36,** 745 (1995).

[30] K. He, D. Sergueev, Z. Sergueev, and B. R. Shaw, *Tetrahedron Lett.,* in press.

[31] Y. Jin and G. Just, *Tetrahedron Lett.* **39,** 6433 (1998).

[32] Z. A. Sergueeva, D. S. Sergueev, and B. R. Shaw, *Tetrahedron Lett.* **40,** 2041 (1999).

[33] J. Zhang, T. Terhorst, and M. D. Matteucci, *Tetrahedron Lett.* **38,** 4957 (1997).

[34] A. P. Higson, A. Sierzchala, H. Brummel, Z. Zhao, and M. H. Caruthers, *Tetrahedron Lett.* **39,** 3899 (1998).

to prevent thymine modification. The authors failed to achieve satisfactory boronation with milder boronating agents such as borane–N,N-diisopropylethylamine (BH$_3$ · DIPEA) or borane–pyridine (BH$_3$ · Py). Our results indicate that at appropriate conditions these agents quantitatively convert internucleoside phosphite triester to boranophosphate.[13] It should be mentioned that the choice of solvent is of great importance for the successful transfer of the borane moiety. Polar solvents such as acetonitrile or N,N-dimethylformamide inhibit the borane exchange compared to less polar THF or 1,4-dioxane.[35,36] Higson et al.[34] reported that BH$_3$ · THF, BH$_3$ · DMS, and even borane–amine complexes caused thymine reduction. Hence they used 3-N-anisoyl-protected thymidine for oligonucleotide synthesis. In our model studies we did not observe modification or boronation of the thymine base with any of the tested borane–amine complexes (see Scheme 2) at the conditions for phosphite boronation. Although we cannot exclude the possibility of minor thymine modification, it seems to be negligible, taking into account the relatively high yields of full-size oligothymidine boranophosphates with intact bases as confirmed by mass spectrometry (MS), gel electrophoresis, enzymatic digestion, and different spectroscopic methods.[13]

Problems and Prospectives in BH$_3^-$-ODN Synthesis. The major problem for the synthesis of mixed BH$_3^-$-ODN in both phosphoramidite and H-phosphonate approaches arises from borane-induced modifications of bases other than thymine. Even with a mild boronation agent like BH$_3$ · Py we still observed noticeable amounts of by-products with conventionally protected adenosine, guanosine, and especially with cytidine. An ongoing study in our laboratory is aimed at eliminating this problem by introducing new protecting groups.

Another specific feature of the borane chemistry is reactivity toward carbonium cation compounds, such as triphenylmethyl carbonium ion.[37–39] The latter is a strong hydrogen abstractor that reacts with borane, forming triphenylmethane and boronium ion.[38,39] Electron-donating substituents at phenyl rings, such as p-dimethylamino or p-methoxy, enhance the tendency of triphenylmethyl compounds to form carbonium ions and facilitate the reaction with borane complexes.[39] In view of this, one may explain the removal of the 5'-O-(4,4'-dimethoxytrityl) protection group in the presence of tetrazole and BH$_3$ · DMS as in Scheme 1.[1,3,29,40] Tetrazole acidifies the

[35] D. S. Sergueev, Z. A. Sergueeva, and B. R. Shaw, unpublished data (1998).
[36] G. E. Ryschkewitsch and E. R. Birnbaum, *Inorg. Chem.* **4**, 575 (1965).
[37] S. Matsumura and N. Tokura, *Tetrahedron Lett.* **9**, 4703 (1968).
[38] L. E. Benjamin, D. A. Carvalho, S. F. Stafiej, and E. A. Takacs, *Inorg. Chem.* **9**, 1844 (1970).
[39] G. E. Ryschkewitsch and V. R. Miller, *J. Am. Chem. Soc.* **95**, 2836 (1973).
[40] Y. Jin and G. Just, *Tetrahedron Lett.* **39**, 6429 (1998).

media, initiating the formation of the DMT cation, whereupon the borane complex reduces the latter, producing 4,4′-dimethoxytriphenylmethane and thereby shifting the equilibrium. Notably, our data showed that the $BH_3 \cdot DMS$ complex alone without tetrazole did not cause any significant detritylation.[41] A similar observation was reported by Jin and Just.[40] These studies indicate that the presence of the weak acid promotes DMT group removal by borane complexes.

Reaction of the 4,4′-dimethoxytrityl carbonium ion with borane complexes is apparently related to the reduced synthetic yield of base- and backbone-boronated compounds after detritylation.[34,40,42,43] Acidic DMT group removal caused up to 50% simultaneous removal of cyanoborane, BH_2CN, in the N7-cyanoboronated guanosine moiety in *base*-boronated guanosine.[42] Despite a diminished reductive ability of borane in boranophosphates,[26] a number of laboratories reported that the acidic deprotection step caused a partial loss of borane from the boranophosphate group accompanied by chain cleavage and formation of expected 4,4′-dimethoxytriphenylmethane.[34,42,43] Thus, either the conditions of DMT group removal have to be modified to keep borane intact or the DMT-protecting group should be replaced, e.g., with 9-fluorenylmethyloxycarbonyl group (Fmoc). We have demonstrated the utility of the Fmoc protecting group for the synthesis of base-boronated oligonucleotides[42] and believe it will prove useful for BH_3^--ODN synthesis. Ongoing studies suggest that it will be possible to synthesize chemically mixed base BH_3^--ODN in the near future.

Synthesis of Boranophosphate Analogs of Nucleoside Triphosphates: 2′-Deoxyribonucleoside and Ribonucleoside 5′-(α-P-Borano)triphosphates

Phosphate-modified nucleotide analogs have been employed extensively by Eckstein and others to elucidate mechanisms of enzyme-catalyzed reactions involving phosphate ester bond cleavage and formation.[44,45] Phosphorothioate analogs of ATP have been used elegantly to probe the interaction of metal ions with the phosphate group in nucleotidyl transferase complexes.[4] Phosphorothioates have been introduced into ribozymes, and phosphorothioate substitution interference analysis has been applied to evaluate

[41] S. Sergueev, Z. A. Sergueeva, and B. R. Shaw, unpublished data (1998).
[42] A. Hasan, H. Li, J. Tomasz, and B. R. Shaw, *Nucleic Acids Res.* 24, 2150 (1996).
[43] A. Sood, B. R. Shaw, B. F. Spielvogel, E. S. Hall, L. K. Chi, and I. H. Hall, *Pharmazie* 47, 833 (1992).
[44] F. Eckstein, P. J. Romaniuk, and B. A. Connolly, *Methods Enzymol.* 87, 197 (1982).
[45] P. A. Frey, J. P. Richard, H. Ho, R. S. Brody, R. D. Sammons, and K. Sheu, *Methods Enzymol.* 87, 213 (1982).

the importance of phosphate groups in the self-cleaving reaction of catalytic RNA.[46] Nucleotide analogs have also been used to prepare a versatile set of labeling reagents for molecular biology.[47] These studies require suitable analogs of nucleoside 5'-triphosphates (dNTPs and NTPs). It is now possible to prepare the corresponding boronated dNTP and NTP analogs.

The deoxy analogs of 5'-(α-P–borano)triphosphates (dNTPαB) were first synthesized using the phosphoramidite approach in our laboratory.[9] Ribonucleoside 5'-(α-P–borano)triphosphates (NTPαB) also can be prepared using this method. The phosphoramidite approach, however, has certain limitations (i.e., requiring exocyclic amine base protection, two ion-exchange chromatography separations, and giving low overall yields) that are overcome in our "one-pot" synthesis of dNTPαB[12,48] and NTPαB.[11] This method (outlined in Scheme 3) is a modification of that reported by Ludwig and Eckstein for nucleoside 5'-(α-P–thio)triphosphates[49] and is generally applicable for preparing both dNTPαB and ribo-NTPαB.

Synthesis of dNTPαB and NTPαB: General Procedures

Solvents and Reagents. The synthesis requires highly purified and anhydrous solvents in order to maximize yields. Anhydrous DMF, anhydrous Py, 2-chloro-4H-1,3,2-benzodioxaphosphorin-4-one (salicyl phosphorochloridite), tributylamine, and the borane–N,N-diisopropylethylamine complex were purchased from Aldrich. Tributyl ammonium pyrophosphate was purchased from Sigma. N^6,3'-O-Dibenzoyl-2'-deoxyadenosine (**5a**), 3'-O-acetylthymidine (**5b**), N^2-isobutyryl-3'-O-benzoyl-2'-deoxyguanosine (**5c**), N^4,3'-O-dibenzoyl-2'-deoxycytidine (**5d**), 2'3'-O-diacetyladenosine (**5g**), 2',3'-O-dibenzoyluridine (**5h**), and 2,3'-O-diacetylguanosine (**5i**) were purchased from Sigma and dried before use in a desiccator over P_2O_5. N^4-Benzoyl-3'-O-acetyl-5-methyl-2'-deoxycytidine (**5e**) and N^4-benzoyl-2',3'-O-diacetylcytidine (**5j**) are prepared from acetylation of the corresponding 5'-dimethoxytrityl-N^4-benzoyl-5-methyl-2'-deoxycytidine (ChemGenes) or 5'-dimethoxytrityl-N^4-benzoylcytidine (ChemGenes) followed by detritylation of the 5'-dimethoxytrityl group with 80% acetic acid.[50] 3'-O-Acetyl-2'-deoxyuridine (**5f**) is prepared from acetylation of 5'-dimethoxytrityl-2'-deoxyuridine (Chem-Impex International) followed by detritylation of the 5'-dimethoxytrityl group with 80% acetic acid.[50] The desired compound **5e**,

[46] K. R. Birikh, P. A. Heaton, and F. Eckstein, *Eur. J. Biochem.* **245**, 1 (1997).
[47] P. A. Frey, *Adv. Enzymol.* **62**, 119 (1989).
[48] K. He, K. Porter, A. Hasan, D. Briley, and B. R. Shaw, *Nucleic Acids Res.*, in press.
[49] J. Ludwig and F. Eckstein, *J. Org. Chem.* **54**, 631 (1989).
[50] R. A. Jones, in "Oligonucleotide Synthesis: A Practical Approach" (M. J. Gait, ed.), p. 23. IRL Press, Oxford, 1984.

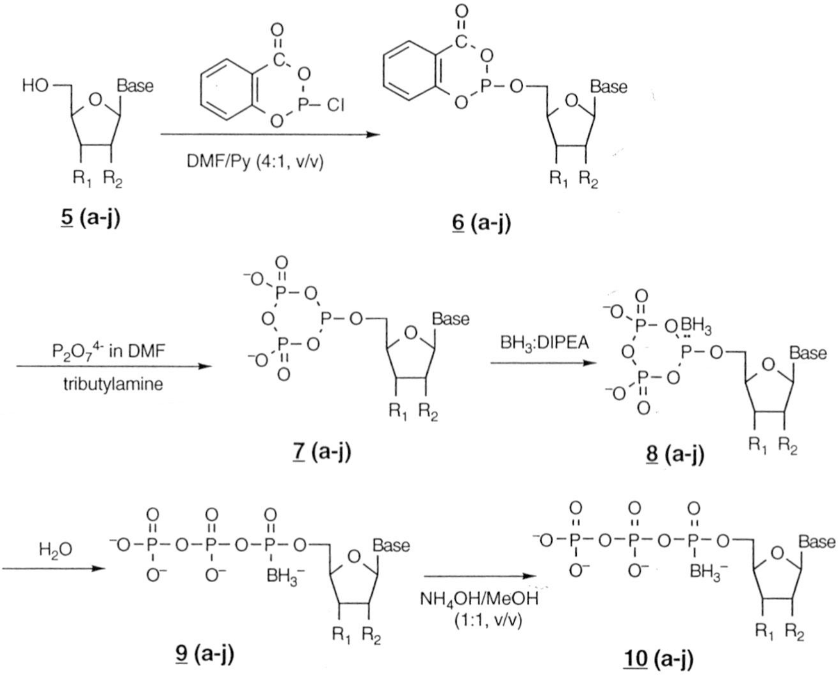

SCHEME 3. Synthesis of deoxy analogs of dNTPαB using phosphoramidite approach.

	5-9			10		
	Base (N)	R_1	R_2	Base (N)	R_1	R_2
a	A^{Bz}	OBz	H	A	OH	H
b	T	OAc	H	T	OH	H
c	G^{iBu}	OBz	H	G	OH	H
d	C^{Bz}	OBz	H	C	OH	H
e	$(5\text{-Me})C^{Bz}$	OAc	H	$(5\text{-Me})C$	OH	H
f	U	OAc	H	U	OH	H
g	A	OAc	OAc	A	OH	OH
h	U	OBz	OBz	U	OH	OH
i	G	OAc	OAc	G	OH	OH
j	C^{Bz}	OAc	OAc	C	OH	OH

5j, or **5f** is purified by chromatography on silica gel. Triethylammonium acetate is prepared by mixing equal moles of triethylamine and acetic acid and diluted with water (final 100 mM, pH 6.8).

Experimental Procedure. The base/sugar protected nucleoside (**5a–5j**) (0.25 mmol) is placed in a 25-ml flask and dissolved in anhydrous DMF (1.0 ml) under argon with magnetic stirring. Anhydrous pyridine (0.25 ml) is added and a freshly prepared solution of 2-chloro-4H-1,3,2-benzodioxaphosphorin-4-one (1.2 equivalent) in anhydrous DMF (0.5 ml) is injected dropwise. After 10 min of stirring, tributylamine (0.25 ml) is added and followed by a freshly prepared solution of tributylammonium pyrophosphate (1.2 equivalent) in anhydrous DMF (0.5 ml). An excess borane–N,N-diisopropylethylamine complex (2.0 ml) is added after 10 min of stirring. After 6 hr the reaction is stopped by adding deionized water (20.0 ml). After 1 hr of stirring at room temperature, the solvent is evaporated *in vacuo.* The residue is treated with a mixture of ammonium hydroxide and methanol (1 : 1, v/v, 20 ml) for 24 hr at room temperature. The solvent is removed under vacuum, and the residue (crude **10a–10j**) is applied to ion-exchange chromatography.

Ion-exchange chromatography is performed using a peristaltic pump and monitoring at 254 nam. Q-Sepharose FF (Pharmacia) (for purification of **10a–10d**) or QA-52 quaternary ammonium cellulose (Whatman International Ltd.) (for purification of **10e–10j**) is packed into a 1.5 × 30-cm LC column. The sample is loaded onto the column and eluted with a linear gradient of 800 ml each of 0.005 and 0.2 M ammonium bicarbonate buffer, pH 9.6, with a flow rate of 7.0 ml/min. Fractions containing the desired product (eluting between 0.1 and 0.15 mM) are collected and evaporated to dryness. Excess ammonium bicarbonate is removed by repeated lyophilization from deionized water to give the ammonium salt of 2′-deoxyribonucleoside or ribonucleoside 5′(α-P–borano)triphosphates (**10a–10j**).

^{31}P NMR Spectroscopy Monitoring of the Reaction. Phosphitylation of base/sugar protected deoxyribo- or ribonucleoside **5** with 2-chloro-4H-1,3,2-benzodioxaphosphorin-4-one (salicyl phosphorochloridite) yields the two diastereomers of nucleoside 5′(4H-1,3,2-benzodioxaphosphorin-4-one) **6**. They are identified by the appearance of two signals at δ 128.12 and 126.18 ppm observed in ^{31}P NMR spectra of the reaction mixture (Fig. 4A). Intermediate **6** is treated directly with tributylammonium pyrophosphate to form a cyclic intermediate P^2, P^3-dioxo-P^1-nucleosidylcyclotriphosphate **7**. The formation of intermediate **7** is confirmed by ^{31}P NMR spectrum (Fig. 4B) in which two signals at δ 128.12 and 126.18 ppm are shifted to a triplet at δ 106.72 (J = 40.48 Hz) and a new doublet at −18.68 ppm (J = 42.09 Hz) appears.

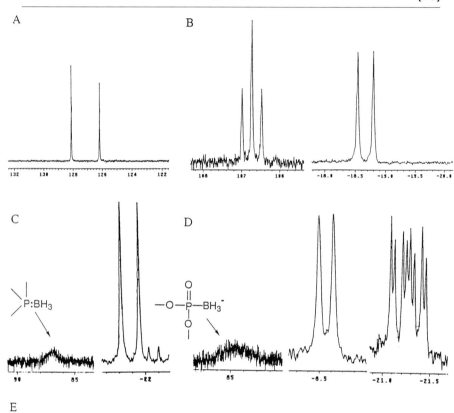

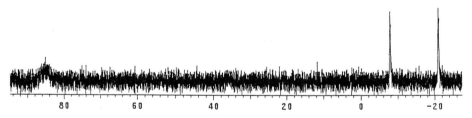

Fɪɢ. 4. Selected regions of the ^{31}P NMR spectrum (ppm) of compounds **6h**, (A) **7h**, (B) **8h**, (C) **9h**, (D) and (E) uridine 5′-(α-P–borano)triphosphate UTPαB **10h**. Reproduced in part from He *et al.*[11]

An *in situ* boronation of intermediate **7** by the exchange reaction with $BH_3 \cdot DIPEA$ is completed after 6 hr at room temperature to give nucleoside 5'-(α-P–borano)cyclotriphosphate **8**. In the reaction mixture (Fig. 4C), a ^{31}P NMR upfield shift of the triplet at δ 106.72 in compound **7** to a broad peak centered at δ 86.72 along with a slight upfield shift of the doublet to δ −21.78 ppm (J = 45.33 Hz) for P^2 and P^3 supports the formation of the P–B bond and cyclotriphosphate in compound **8**. The cyclic intermediate **8** is hydrolyzed by water to give nucleoside 5'-(α-P–borano)triphosphate **9**.[11] From ^{31}P NMR spectroscopy (Fig. 4D) of the reaction mixture, the expected intermediate **9** shows a broad peak centered at δ 85.27 (α-P), two quartets (β-P) at δ −21.25 (J = 32.87 Hz, J = 20.56 Hz) and δ −21.30 (J = 33.51 Hz, J = 19.75 Hz), and a doublet at δ −8.50 and −8.63 (J = 21.05 Hz, γ-P). Another triplet at δ −20.65 and two doublets at δ −8.04 and −3.06 result from the normal triphosphate. The singlet at δ −7.02 is due to pyrophosphate. Another singlet at δ 8.12 is attributed to H-phosphonate derived from the hydrolysis of an excess phosphitylating agent.

Treatment of intermediate **9** with ammonium hydroxide removes the base/sugar protecting groups and affords nucleoside 5'-(α-P–borano)triphosphate **10** (Fig. 4E), which is isolated in 25–46% yield by ion-exchange column chromatography. The structure, purity, and homogeneity of compound **10** are confirmed by ^{31}P NMR (Table II), ^1H NMR, UV spectroscopy, and FAB-MS. In addition to the desired product, three by-products are also isolated. One is identified as H-phosphonate (δ 10.36, about 20% yield) derived from the hydrolysis of the excess phosphitylating

TABLE II
^{31}P NMR DATA AND YIELD OF NUCLEOSIDE 5'-(α-P–BORANO)TRIPHOSPHATES[a]

| dNTPαB and NTPα B | ^{31}P NMR chemical shifts (ppm) (D$_2$O) | | | Yields (%) |
	α-P	β-P	γ-P	
dATPαB (**10a**)	82–87 (br)	−20.41 (m)	−5.39 (d)	25
dTTPαB (**10b**)	85.68 (br)	−21.21 (dd)	−8.10 (d)	43
dGTPαB (**10c**)	82–87 (br)	−20.25 (m)	−5.45 (d)	30
dCTPαB (**10d**)	82–87 (br)	−21.12 (dd)	−8.29 (m)	29
(5-Me)-dCTPαB (**10e**)	84.39 (br)	−21.21 (m)	−8.91 (m)	33
dUTPαB (**10f**)	83–87 (br)	−21.05 (dd)	−7.65 (m)	36
ATPαB (**10g**)	85.36 (br)	−21.15 (dd)	−8.85 (d)	31
UTPαB (**10h**)	85.05 (br)	−21.03 (m)	−7.87 (m)	46
GTPαB (**10i**)	85.72 (br)	−21.12 (dd)	−8.70 (d)	29
CTPαB (**10j**)	85.20 (br)	−21.26 (m)	−8.92 (d)	26

[a] ^{31}P NMR spectra are acquired at 161.9 Hz using a Varian Inova-400 spectrometer. Signals are expressed as d (doublet), m (multiplet), and br (broad). Data from Refs. 11, 12, and 48.

agent. The other two by-products are identified as nucleoside 5'-monophosphate (δ 4.75, about 30% yield) and free pyrophosphate at δ -5.12.

Separation of Two Diastereomers of dNTPαB and NTPαB. Introduction of a borane group (BH_3) to replace one of the oxygen atoms on the α-phosphate produces a pair of diastereomers for compound **10**. These diastereomers have been separated by preparative reversed-phase HPLC and named arbitrarily as isomer I and isomer II (Table III). For dNTPαB, the configuration of isomer I has been tentatively assigned[14] as R_p, as isomer I of each dNTPαB is a good substrate for DNA polymerases. For NTPαB, the R_p configuration is also putatively assigned to isomer I of NTPαB, based on the fact that only isomer I can be incorporated into RNA by T7 and SP6 RNA polymerases.[15]

Reversed-phase HPLC separation of two diastereomers of **10a–10j** is performed using a reversed-phase column (Waters Delta Pak C18, 7.8 × 300 mm, 15 μm, 300 Å) with a photodiode array UV detector. The sample of **10a–10j** (0.5 mg) is injected and eluted with isocratic elution (100 mM TEAA, pH 6.80, and methanol as buffer) at a flow rate of 3 ml/min. The fraction corresponding to each diastereomer is collected and the buffer components, TEAA and methanol, are removed by repeated lyophilization. The isomeric purity of an individual diastereomer is checked by reversed-phase HPLC under the conditions used for separation. The HPLC profiles before and after preparative HPLC separation are presented in Fig. 5. The isolated yields of isomer I and II are presented in Table III. The homogeneity and structure were confirmed by ^{31}P NMR, ^{1}H NMR, UV, spectroscopy and FAB-MS. The triethylammonium cation (TEA$^+$) can be then converted to sodium cation (Na$^+$) by passing through a sodium form cation-exchange resin Dowex 50WX8-200 (Sigma) in 6 × 100-mm glass pipettes (Baxter).

Comments. The "one-pot" method described for the synthesis of dNTPαB and NTPαB has several advantages over the previously reported phosphoramidite approach.[9] First, it does not require exocyclic amine protection of the nucleobase; introduction of the borane (BH_3) is achieved under mild conditions, eliminating the risk of the possible reduction of nucleoside. Second, this method is a time- and cost-effective approach that results in high yields. A two-step purification, ion-exchange chromatography followed by reversed-phase HPLC, gives the two diastereomers of each dNTPαB or NTPαB in high purity.

These boronated NTPs and dNTPs are available as potential substrates for the numerous studies with polymerases and nucleotide-metabolizing enzymes. It should be feasible to introduce modifications such as fluorescent labels or biotin into the α-boranotriphosphates to follow their fate in metabolic reactions or biodistribution.

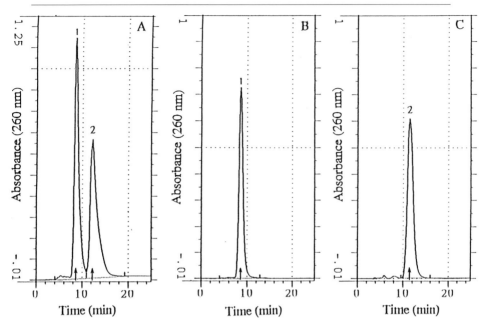

FIG. 5. Isocratic separation of diastereomers of cytidine 5'-(α-P–borano)triphosphate CTPαB (**10j**) by HPLC (A) and analysis of purity of HPLC-isolated isomer **I** (B) and isomer **II** (C). The retention times for two diastereomers of CTPαB **10j** are 8.55 min (peak 1, isomer I) and 12.24 min (peak 2, isomre II), respectively. Data from He *et al.*[11]

TABLE III
HPLC Separation of Diastereomers of Nucleoside 5'-(α-P–Borano)triphosphates[a]

dNTPαB and NTPαB	Buffer[b]	Retention time (min)	
		Isomer I (%)	Isomer II (%)
dATPαB (**10a**)	89% A and 11% B	13.54 (54%)	17.37 (46%)
dTTPαB (**10b**)	90% A and 10% B	10.29 (58%)	14.78 (42%)
dGTPαB (**10c**)	92% A and 8% B	12.64 (50%)	15.63 (50%)
dCTPαB (**10d**)	94% A and 6% B	9.61 (49%)	12.74 (51%)
(5-Me)-dCTPαB (**10e**)	92% A and 8% B	10.78 (48%)	14.80 (52%)
dUTPαB (**10f**)	94% A and 6% B	12.45 (52%)	18.20 (48%)
ATPαB (**10g**)	90% A and 10% B	11.04 (43%)	17.82 (57%)
UTPαB (**10h**)	92% A and 8% B	8.32 (46%)	11.80 (54%)
GTPαB (**10i**)	92% A and 8% B	9.31 (48%)	14.56 (52%)
CTPαB (**10j**)	94% A and 6% B	8.55 (51%)	12.24 (49%)

[a] TEAA, triethylammonium acetate. Data from Refs. 11, 12, and 48.
[b] A: 100 m*M* TEAA, pH 6.8; B: methanol.

*Enzymatic Synthesis of Oligodeoxynucleoside Boranophosphates Using
Polymerases and Boronated Precursor dNTPαB*

Enzymatic synthesis of BH_3^--ODN using boronated dNTP analogs
(dNTPαB) with polymerases[14] is a feasible and powerful alternative to their
chemical synthesis[3,13] and should provide stereoregular oligonucleotides of
any sequence. All four R_p-dNTPαB (N = A, T, G, C) (**10a–10d**) are good
substrates for a number of DNA polymerases, permitting template-directed
enzymatic synthesis of long segments of boronated DNA.[10] Because the
borane imparts chirality on a disubstituted phosphate ester and because
polymerases function in a stereospecific manner, enzymatic synthesis provides oligonucleotides of defined stereochemistry with a mixed base sequence and the potential for chimeric phosphate/boranophosphate backbones. Experiments indicate that 100% borano oligomers can be prepared
by primer extension[51] (data not shown), using the full set (**10a–10e**: N =
A, T, G, C, 5-Me-C) of the dNTPαB in Scheme 3. Additionally, it is possible
to synthesize boronated RNA,[15] substituting at least one NTP with its
boronated analog, NTPαB **10g–10j**.

Method for Primer Extension with Boronated dNTPs. In a typical reaction, a 19mer primer 5′-^{32}P-CAGGAACAGCTATGGCCTC-3′ is extended by T4 DNA polymerase in the presence of a 27mer template 5′-
GTGTAGCTGAGGCCATAGCTGTTCCTG-3′ and 100% of one R_p-
dNTPαB plus the three remaining normal dNTP, 50 μM each.[10] For each
of the four dNTPαB, the primer is extended to the full length of the 27
nucleotide template. 50mer oligonucleotides containing all-boranophos-
phate linkages have been synthesized [typical conditions include 6.5 U
modified T7 DNA polymerase (Sequenase), 37°, 15 min in the presence of
100 μM each of the four dNTPαB (**10a–10d**)]. The primer extension reactions can be carried out at low (0.1 μM) or high (to at least 25 μM) primer/
template concentrations and with primers as small as 11mers.[14] Batch-mode
syntheses for biophysical studies have been carried out by increasing the
primer/template concentrations (to 25 μM) to obtain BH_3^--ODN containing one boronated linkage.[14] Autoradiography is used to show that the
primer is extended to completion without detectable pausing.

Polymerase (PCR) with Boronated dNTPs. dNTPαB are stable at high
temperatures and are compatible with both short- and long-range (13 kb)
PCR using a variety of thermostable polymerases (*Taq*, Vent, *Taq* FS).
Amplification of templates as large as 13 kb have been carried out by
substituting one base-specific R_p-dNTPαB for the corresponding normal

[51] V. Rait and B. R. Shaw, unpublished data (1998).

dNTP.[52] Expected amounts of full-length 509-bp DNA are produced readily under normal PCR conditions (25 cycles of 95° for 1 min, 53° for 1 min, and 76° for 1 min) using three all natural dNTP (100 μM of each) and one base-specific R_p-dNTPαB (2.5 μM).[10] Following amplification, the extended product was treated with exonuclease III (Exo III), resulting in complete digestion of the natural DNA, whereas the partially boronated DNAs were nearly resistant to Exo III.[10] These results indicate that the presence of a boranophosphate in either the primer or the template has a minimal, if any, effect on chain elongation and amplification.

Because the boranophosphate linkages are quite resistant to Exo III, by carrying out for PCR reactions, with each containing only a small percentage of one base-specific boronated analog, it is possible to determine the position of the boranophosphate by treating the PCR product with Exo III and analysis on a commercially available DNA sequencer.[10] The boronated nucleotides (a) do not obstruct PCR or primer extension, (b) are more resistant to Exo III than normal nucleotides, and (c) are resistant to Exo III base specifically. These properties make boranophosphates suitable for use as the sequence delimiters in a direct PCR sequencing method.[10]

Boranophosphate Activation of RNase H-Mediated RNA Cleavage

We have found[17] that oligodeoxynucleoside boranophosphates are capable of eliciting the RNase H hydrolysis of complementary polyribonucleotides. That result was not necessarily obvious because the closest analogs of the phosphodiester linkage behave differently, *i.e.,* anionic phosphorothioates and phosphorodithioates support RNase H hydrolysis but the neutral methyl phosphonates do not.[53–56] Our data support the suggestion that the negative charge of the internucleoside linkage is necessary for RNase H activation.[57]

The RNase H assay is performed under the same conditions (buffer, polyribonucleotide, and deoxyoligonucleotide concentrations) used to determine the T_m (melting point) of the poly(rN) : oligo(dN) hybrids, e.g., the use of complementary heteroduplexes at a comparatively high concentration, 1 mM each of ribo- and deoxyribonucleoside residues. In these

[52] K. Porter and B. R. Shaw, unpublished data (1997).
[53] C. A. Stein, C. Subasinghe, K. Shinozuka, and J. S. Cohen, *Nucleic Acids Res.* **16,** 3209 (1988).
[54] L. J. Maher III and B. J. Dolnick, *Nucleic Acids Res.* **16,** 3341 (1988).
[55] P. J. Furdon, Z. Dominski, and R. Kole, *Nucleic Acids Res.* **18,** 9193 (1989).
[56] L. Cummins, D. Graff, G. Beaton, W. S. Marshal, and M. H. Caruthers, *Biochemistry* **35,** 8734 (1996).
[57] P. D. Cook, *in* "Antisense Research and Applications" (S. T. Crooke and B. Lebleu, eds.), p. 149. CRC Press, Boca Raton, FL, 1993.

experiments, the concentration of acid-soluble material in hydrolyzates was measured to trace kinetics. In addition, parallel anion-exchange HPLC of the reaction mixtures was used to identify their composition. In addition to confirming the ability of boranophosphates to mediate RNase H hydrolysis of polyribonucleotides, these experiments provide insight into the mechanism of action of the enzyme.

Method for RNase H Assay.[17] This assay (as well as the UV melting) is performed in standard RNase H buffer (20 mM Tris–HCl, pH 7.5, containing 100 mM KCl, 10 mM MgCl$_2$, 0.1 mM dithiothreitol (DTT), and 5% sucrose) at 1 mM concentrations of rA and dT nucleosides.[17] T(^{b}pT)$_{11}$, T(^{s}pT)$_{11}$, or T(pT)$_{11}$ refer to thymidine dodecamers with borano-, thio-, or normal phosphodiester internucleoside linkages respectively. In the spectrophotometric determination of concentrations, the molar extinction coefficient, ε, for poly(A), 10.4 M^{-1} cm^{-1} at 257 nm, pH 7, is taken from Blake and Fresco.[58] The T(pT)$_{11}$ ε value, 8.8 M^{-1} cm^{-1} at 267 nm, is calculated according to Cantor *et al.*[59] The same ε value is also ascribed to T(^{s}pT)$_{11}$ and T(^{b}pT)$_{11}$ at the respective λ_{max}, 268.8 and 268 nm.

A solution of poly(A) and one of the thymidine dodecamers, 100 nmol in *each* nucleoside in 88.5 μl of 20 mM Tris–HCl buffer, pH 7.5, containing 100 mM KCl, are kept overnight at ambient temperature. The samples are then equilibrated for 10 min at 20 or 30° followed by the addition of 10 μl of 100 mM MgCl$_2$, 1 mM DTT, and 50% sucrose in the Tris/KCl buffer. Hydrolysis is initiated 10 min later by the addition of 7.5 units (1.5 μl) of RNase H1 (U.S. Biochemical Corp.) and is observed by the accumulation of acid-soluble oligoadenylates. Then, 6-μl aliquots are withdrawn and treated with 114 μl of chilled 3.5% HClO$_4$. After 10 min in an ice bath, the samples are centrifuged for 5 min at 16,000g and the absorbance of the supernatant is read at 258 nm in submicro cells with a 1-cm path length.

Analytical Ion-Exchange HPLC. Ion-exchange HPLC of the hydrolysis products is done using Polysil-500 (a strong anion exchanger, Koltsovo, Russia), a 4 × 50-mm column, and HPLC with a multisolvent delivery system, photodiode array detector, and chromatography workstation; the elution rate is 1 ml/min. A 10-μl sample of a RNase H reaction mixture as described earlier (without acid treatment) is mixed with 90 μl of 30% acetonitrile, pH 6.5, loaded, washed for 2 min, and then eluted with a 60-min linear gradient (0–0.3 M) of K$_2$HPO$_4$ titrated in 30% acetonitrile to pH 6.5 by concentrated H$_3$PO$_4$. The continuous spectral monitoring of eluate is used to distinguish oligo(A) from thymidine dodecamers.

---

[58] R. Blake and J. Fresco, *Biopolymers* **12**, 775 (1973).
[59] C. R. Cantor, M. M. Warshaw, and H. Shapiro, *Biopolymers* **9**, 1059 (1970).

UV Melting. Samples of oligomer:polymer heteroduplexes, prepared as for the RNase H assay and flushed with helium, are placed in demountable closed cells with a 0.02-cm path length (Hellma Cells, Inc.); after equilibration in the cell chamber, the dependencies of absorbance, *A,* vs temperature are registered automatically with an AVIV (Lakewood, NJ) spectrophotometer (Model 17DS UV-VIS-IR) supported by AVIV 14DS V4.1L software. The temperature of the cell compartments flushed by dry nitrogen is increased in 0.5° steps with a 30-sec equilibration time and a 30-sec average time constant. The melting temperature, T_m, of a heteroduplex is taken as the temperature corresponding to the maximum in a plot of dA/dT vs *T.*

Relationship between Thermostability of ODN:RNA Hybrids and Rate of RNase H Hydrolysis. From UV melting, T_m values were 9.5, 23.5, and 44.5° for heteroduplexes formed by poly(A) and T(^{b}pT)$_{11}$, T(^{s}pT)$_{11}$, or T(pT)$_{11}$, respectively, at 1 m*M* rA:dT. In contrast, kinetic data in Fig. 6 obtained by monitoring acid-soluble oligoriboadenylates clearly show that among the thymidine dodecamers, the boranophosphate T(^{b}pT)$_{11}$ elicited the fastest rate of poly(A) hydrolysis. In the presence of the *Escherichia coli* RNase H1, 75 units/μl, the relative rates of reaction increased from 1 for T(pT)$_{11}$ to 8.2 for T(^{s}pT)$_{11}$ to 76.5 for T(^{b}pT)$_{11}$ at 20°. The same relative rate order was also found at 30°, although the ratio changed: 1:2.3:18.1.

The RNase H-stimulating effect of T(^{b}pT)$_{11}$ is remarkable by the high rates registered at temperatures that are 10 and 20° above the T_m of the T(^{b}pT)$_{11}$:poly(A) heteroduplex. This suggests that the enzyme is an active participant in the formation of a RNA:DNA hybrid and is capable of

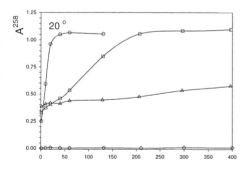
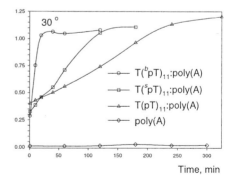

Time, min

Fig. 6. Kinetics of acid-soluble oligo(A) accumulation at 20° (left) and 30° (right) under the RNase H1 action on poly(A) in the presence of T(^{b}pT)$_{11}$ (1), T(^{s}pT)$_{11}$(2), or T(pT)$_{11}$ (3) and in the absence of dodecathymidines (4, control). Symbols mark the times of parallel HPLC analysis of hydrolysate products. Adapted from Rait and Shaw.[17]

assembling complementary strands of a potential substrate even in the thermodynamically unfavorable condition at $T > T_m$. Support for such a suggestion comes from the discovery that RNase H1 promotes the formation of a hybrid between an RNA site and a base paired strand of a stable hairprin or duplex DNA.[60]

As discussed by Tidd,[61] dissociation of the enzyme/product complex may be a rate-limiting step in the overall reaction catalyzed by RNase H, and a low T_m for the ODN : RNA hybrid would facilitate this step, allowing antisense ODN to act in a more catalytic manner. Several studies support the concept that antisense ODN with a weakened hybridization potential could provide not only an accelerated hydrolysis, but also an enhanced sequence specificity of mRNA cleavage compared to ODN with all-natural internucleoside linkages.[61–65]

Relative Content of Smallest Nondigestable Products in RNase H Hydrolysates Is Indicative of the Contribution of Enzyme Exo/Endo Mechanisms. RNase H (EC 3.1.26.4) is a hydrolase that combines both endo- and exo-nucleolyic mechanisms.[66] The exo mechanism of hydrolysis hypothetically does not contribute to an antisense effect, but is responsible for the accumulation of oligoribonucleotides that are shorter in length than the antisense ODN and which originate from the heteroduplex interior following several successive scissions of bonds in a polyribonucleotide strand. In such a mechanism, the relative content of smallest products should therefore reflect (i) the average lifetime of an enzyme/ODN : poly(rN) triple complex and (ii) the relative contribution of the endo/exo mechanisms. Oligomers that form less stable hybrids would guide RNase H to act predominantly via an endonucleolytic mechanism because any cleavage within an unstable hybrid would destabilize it drastically, causing dissociation of the whole triple complex and preventing further exonucleolytic action. In accordance with this picture, we found that the content of diadenylate, $(pA)_2$, clearly diminishes on going from the $T(pT)_{11}$: poly(A) to $T(^spT)_{11}$: poly(A) to $T(^bpT)_{11}$: poly(A) hydrolyzates in Fig. 7. This is in accordance with the

[60] J. Li and R. M. Wartell, *Biochemistry* **37**, 5154 (1998).
[61] D. M. Tidd, *in* "Antisense Therapeutics: Progress and Reports" (G. L. Trainor, ed.), p. 51. ESCOM Science Publishers B.V., Leiden, 1996.
[62] R. V. Giles and D. M. Tidd, *Anti-Cancer Drug Res.* **7**, 37 (1992).
[63] M. K. Ghosh, K. Ghosh, and J. S. Cohen, *Anti-Cancer Drug Res.* **8**, 15 (1993).
[64] R. V. Giles, C. J. Rudell, D. G. Spiller, J. A. Green, and D. M. Tidd, *Nucleic Acids Res.* **23**, 954 (1995).
[65] R. V. Giles, D. G. Spiller, J. Grzymbowski, R. E. Clark, P. Nicklin, and D. M. Tidd, *Nucleic Acids Res.* **26**, 1567 (1998).
[66] R. J. Crouch and M. L. Dirksen, *in* "Nucleases" (S. M. Linn and R. J. Roberts, eds.), p. 211. Cold Spring Harbor Press, Cold Spring Harbor, NY, 1982.

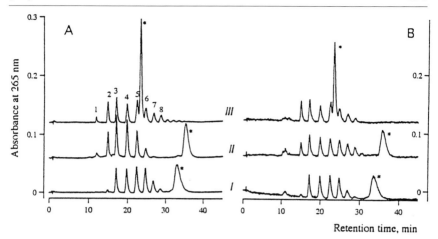

FIG. 7. Anion-exchange HPLC profiles of products of poly(A) hydrolysis by RNase H1 at 20° (A) and 30° (B) in the presence of $T(^bpT)_{11}$ (**I**), $T(^spT)_{11}$ (**II**), or $T(pT)_{11}$ (**III**). Numbers above minor peaks correspond to the extent of polymerization, n, in oligoriboadenylates, $(pA)_n$; major peaks (*) with retention times of 32.9 min (profiles **I**), 35.2 min (profile **II**), and 23.2. min (profile **III**) correspond to $T(^bpT)_{11}$, $T(^spT)_{11}$, and $T(pT)_{11}$, respectively. Adapted from Rait and Shaw.[17]

hypothesis that the boranophosphate gives rise to more internal cleavages in the hybrid.

It may also be noted that the ratio of $(pA)_2$ to $(pA)_3$ formed in each of the hydrolyzate profiles (Fig. 7) correlates with the T_m of the corresponding hybrids: the higher the T_m, the higher the ratio. Moreover, this correlation seems to be unaffected by differences in the particular extent of hydrolysis.[17] Perhaps this dimer/trimer ratio reflects a statistically averaged length of an RNase H1 step in the exonucleolytic hydrolysis and/or an intrinsic distance between the 3' end of a ribo strand and the nearest cleavable bond.

Analytical Methods

Microdetermination of Boranophosphates

Although boranophosphates have unusual hydrolytic stability,[7] $t_{1/2}$ of 40 years at pH 7, they retain some characteristics of hydroborating agents. They react with aldehydes at pH below 4 and with iminium salts, resembling cyanoborohydrides[67] in their functional group selectivity but differing from cyanoborohydrides by a markedly decreased reactivity. For instance, bora-

[67] R. F. Borch, M. D. Bernstein, and H. D. Durst, *J. Am. Chem. Soc.* **93**, 2897 (1971).

nophosphates themselves appear not to react with transition metal ions in high oxidation states; however, an ion reduction can be promoted by carbonyl-containing compounds in diluted acids. The particular color reactions of boranophosphates and Mo^{6+} of molybdic or phosphomolybdic acid in the presence of acetone and sulfuric acid were found practically useful in qualitative detection and quantitative determination of boranophosphates.[68]

Reagents

1. Standard solution, 1 mM ammonium dimethylphosphitoborohydride, DMPB. DMPB (also called dimethylboranophosphate[6]) can be synthesized easily from commercially available dimethylphosphite (Aldrich) as described.[6] Alternatively, a solution of any chromatographically pure nucleoside boranophosphate may be used after determining its concentration by UV spectrophotometry.
2. 5% ammonium molybdate in 2 N H_2SO_4
3. 5% ammonium molybdate in 12 N H_2SO_4
4. 0.1 M KH_2PO_4
5. Acetone

Determination of Boranophosphates by Reduction of Molybdic Acid. Mix 50 μl of 5% ammonium molybdate in 2 N H_2SO_4 with a sample containing up to 60–75 nmol of a boranophosphate (60–75 μl of standard solution), adjust the volume to 950 μl with water, add 50 μl of acetone, mix vigorously, and keep the mixture at 50°. In 7–10 min read the absorbance of a green-colored solution at 735 nm using an appropriate cell with a 1-cm path length. To calculate an amount (N, in nmol) of a boranophosphate in a taken sample, use the formula $N = 49.8 \times A_{735}$.

Determination of Boranophosphates by Reduction of Phosphomolybdic Acid. Mix 50 μl of 5% ammonium molybdate in 12 N H_2SO_4 and 50 μl of 0.1 M KH_2PO_4 with a sample containing up to 200 nmol of a boranophosphate (200 μl of standard solution), adjust the volume to 925 μl with water, add 75 μl of acetone, mix vigorously, and keep the mixture at 50°. In 15 min read the absorbance of a greenish solution at 835 nm using a 1-cm path length cell. To calculate an amount (N, in nmol) of a boranophosphate in a taken sample, use the formula $N = 179 \times A_{835}$.

Comments. The calibration dependencies, obtained by the reduction of molybdic and phosphomolybdic acids with DMPB and treated as linear regressions, are shown in Figs. 8A and 8B, respectively. The molar absorp-

---

[68] V. K. Rait and B. R. Shaw, in preparation (1999).

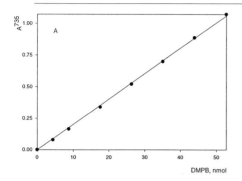
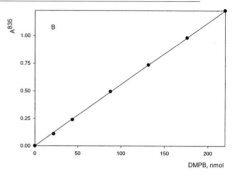

FIG. 8. Calibration dependencies obtained by the reduction of molybdic (A) and phospho-molybdic (B) acids with dimethylphosphitoborohydride.

tivity, $20.1 \times 10^3 \, M^{-1} \, cm^{-1}$, specific for the molybdate method, is twofold higher than that averaged characteristic for nucleotides at 260 nm. Thus, this method can be recommended for the precise determination of oligonucleoside boranophosphate molar absorptivities. Although fourfold less sensitive, the phosphomolybdate method has its own advantages, being insensitive to contamination such as methanol and acetonitrile (up to 10%, v/v), and to some buffer components, including phosphate. This makes the method especially valuable in the detection and quantitation of boranophosphates in complex mixtures such as chromatographic fractions or biochemical compositions.

Spectroscopic Analysis of Boronated Nucleic Acids

NMR spectroscopy is arguably the most valuable and versatile method for studies of nucleosides, nucleotides, and oligonucleotides. It can provide information about chemical composition, three-dimensional structures, and interactions with other macromolecules or ligands. Protons (^{1}H) and phosphorus (^{31}P) are NMR-active nuclei with 1/2 nuclear spin I that have been used extensively to study nucleic acids and their analogs. Nucleoside boranophosphates contain boron, which has two stable isotopes, each having an NMR-active nucleus. Naturally abundant boron consists of 80.4% ^{11}B (spin $I = 3/2$) and 19.6% ^{10}B (spin $I = 3$). ^{11}B or ^{10}B NMR spectra can provide important and easily accessible spectral information. Although ^{11}B is a quadrupolar nucleus, line broadening is not severe for the molecules studied. Although boron in the glass NMR tube and the NMR probe produce a broad background ^{11}B signal, it usually does not obstruct observations. For NMR of boronated nucleic acids, quartz NMR tubes are used for ^{11}B analysis, whereas regular tubes are used for (^{1}H) and (^{31}P) NMR.

When a ^{31}P ($I = 1/2$) atom is bonded to a single ^{11}B, as in the -P–BH$_3$ linkage, four equally spaced and equal-intensity lines (a quartet with a 1-1-1-1 pattern) are expected in ^{31}P spectra. If a ^{31}P atom is coupled to a ^{10}B atom, seven equally spaced and equal-intensity lines (a septet with a 1-1-1-1-1-1-1 pattern) are expected. Thus, a P–B bond-containing sample with naturally abundant boron can show complicated ^{31}P NMR spectra. Specifically, ^{31}P spectra usually appear as a quartet with a 1-1-1-1 pattern at ca. 80–85 ppm for boranophosphate monoester and at ca. 90–95 ppm for boranophospte diester) (see Fig. 9C and Ref. 69).

The ^1H signal from the -P–BH$_3$ moiety is located as a very broad 1-1-1-1-quartet, spanning nearly 400 Hz and centered at ca. 0.3 ppm (Figs. 9A and 9B). In ^{11}B-decoupling experiments, a sharper doublet (due to a phosphorus coupling) is observed.

UV and CD spectroscopy are useful tools for studying electronic and conformational effects in mono- and oligonucleoside analogs and for thermal denaturation studies of oligonucleotides. In UV studies of the electronic and conformational effects of the borane group, molar absorptivities and λ_{max} for dithymidyl boranophosphates were found to be the similar to normal phosphates and phosphotriester analogs.[3] However, nucleoside boranophosphates exhibit the lowest hyperchromicity compared to methylphosphonate and phosphorothioate oligomers.[3,70]

Aside from some changes in magnitudes, differences between CD spectra of dithymidine boranophosphates and unmodified dimers are minimal.[70] Signs, position, and magnitudes of the CD bands indicate that the modified dimer (both diastereomers) adopts B-type conformations, the same as the unmodified d(TpT) dimer. Increasing the temperature leads to some changes in CD spectra of dithymidine boranophosphates, suggesting that boron-modified dimers may have different intramolecular interactions than the unmodified dimer and that the configuration of boronated phosphorus exerts a small but noticeable decrease of the base-stacking interactions.

Mass spectrometry techniques combine the desirable characteristics of low-level detectability with high selectivity. Two MS techniques were used for analysis of nucleoside boranophosphate—high-resolution fast atom bombardment (HR FAB MS) and electrospray ionization (ESI). FAB MS in a negative ion mode provides reasonable analysis for substances with a molecular mass up to 1000, i.e., nucleoside boranotriphosphates (Fig. 10). Good results were obtained with polyethylene glycol-400–1000 as a matrix. Due to the approximate 4:1 ratio of $^{11}B:^{10}B$, MS spectra of all the boranophosphates have a characteristic MS signal distribution, with the main ^{11}B

[69] H. Li, C. Hardin, and B. R. Shaw, *J. Am. Chem. Soc.* **118**, 6606 (1996).
[70] H. Li, F. Huang, and B. R. Shaw, *Bioorg. Med. Chem.* **5**, 787 (1997).

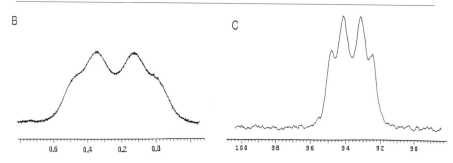

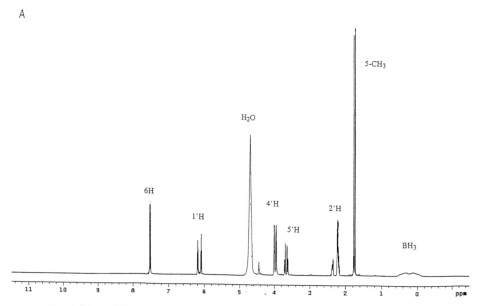

FIG. 9. ^{1}H and ^{31}P NMR spectra of individual S_p isomer Tp^{b}T (dithymidine boranophosphate): (A) ^{1}H NMR spectrum (499.9 MHz), (B) expanded region from A for -P–BH$_3$ proton signals, and (C) ^{31}P NMR spectrum (161.9 MHz).

peak having a fourfold greater signal than the neighboring lower mass ^{10}B isotope peak.

For tetra-, octa-, and dodecanucleoside boranophosphates oligomers, electrospray ionization in the negative ion mode can be applied successfully with polyethylene glycol-600 as a matrix (Fig. 11). The characteristic feature

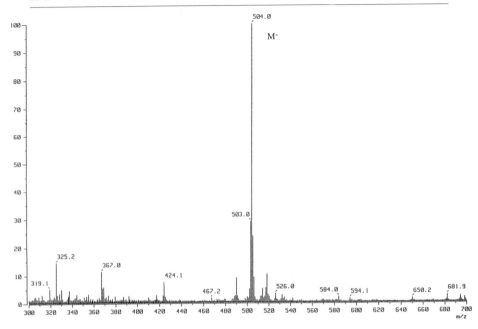

FIG. 10. FAB-MS spectra of ATPαB (α-BH₃ analog of ATP) (**10g**) in negative ion mode. An additional line with $M - 1$ mass in the spectrum is due to the presence of ^{10}B isotope (19.6% natural abundance).

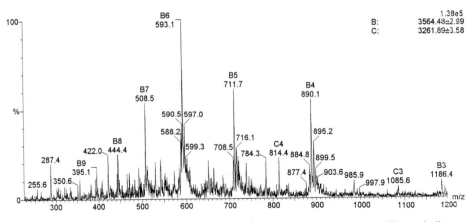

FIG. 11. Electrospray mass spectrum of T(p^{b}T)$_{11}$ in negative ion mode. Theoretically calculated mass for T(p^{b}T)$_{11}$ is 3564.9 Da.

of this method is the formation of multiple charges of the molecule being examined, resulting in the appearance of molecule ions at m/z values determined by the number of charges on each of the ions. The practical result of multiple charging is a mass accuracy of at least 0.03% in determination of the molecular weight.

Summary

Nucleoside boranophosphates are distinctive in that one of the non-bridging oxygens in the phosphate diester **1** is replaced by a borane moiety (BH_3). Although they retain the same net charge, $BH_3{}^-$-ODN have unique chemical and biochemical characteristics relative to other analogs. The change in polarity, lipophilicity, nuclease resistance, and the activation of RNase H cleavage of RNA in RNA:boranophosphate hybrids make boranophosphates very attractive for applications in enzymology and molecular biology and as potential antisense agents.

Acknowledgment

The authors are grateful to the DOE (Grant DE-FG02-97ER62376) and National Institutes of General Medical Sciences (Grant GM-57693-01) to B.R.S. for financial support.

[14] Intracellular Distribution of Digoxigenin-Labeled Phosphorothioate Oligonucleotides

By Gemma Tarrasón, David Bellido, Ramon Eritja, Senén Vilaró, and Jaume Piulats

Introduction

The great potential of antisense oligonucleotides as selective inhibitors of gene expression has led to their development and use for therapeutic purposes. Antisense technology has had to satisfy several conditions, such as stability to nucleases, enhancement of cellular permeability, and selectivity for the target sequence. Some of these requirements have been addressed by the chemical development of nucleic acid analogs. Although modifications such as phosphorothioate substitutions have been used mainly for improved stability, and C-5 propyne or morpholino modifications have been used for higher affinity,[1,2] the major concern is, at present, inefficient cellular uptake. Carrier systems such as cationic liposomes,[3,4] nanoparticles,[5] or supramolecular biovectors[6] have been proposed to enhance cellular uptake and nuclear accumulation.[7,8] Nevertheless, the characterization of intracellular distribution is important in assessing an effective antisense effect.

Studies concerning such an intracellular distribution reflect enhanced heterogenicity depending on the cell type described and the labeling technique used.[9,10] Either subfractionation techniques using ^{32}P- or ^{35}S-labeled oligonucleotides or flow cytometry and fluorescence microscopy using isotopic or fluorescent labels have been used in such studies.[10–18] Fluorescent labels normally used in conventional microscopy studies do not provide enough resolution to determine the intracellular fate of small quantities of oligonucleotide that could be responsible for antisense activity. Such resolution can only be provided by ultrastructural analysis. Isotopic labeling with ^{32}P and ^{35}S could circumvent this problem and provide accurate intracellular distribution, but this is restricted to specialized areas and requires frequent oligonucleotide preparation due to the short half-life of the iso-

[1] W. M. Flanagan, L. L. Su, and R. W. Wagner, *Nature Biotechnol.* **14**, 1139 (1996).
[2] J. Summerton, D. Stein, S. B. Huang, P. Matthews, D. Weller, and M. Partridge, *Antisense Nucleic Acid Drug Dev.* **7**, 63 (1997).
[3] D. D. Ma and A. Q. Wei, *Leuk. Res.* **20**, 925 (1996).
[4] C. F. Bennett, M. Y. Chiang, H. Chan, J. H. E. Shoemaker, and C. K. Mirabelli, *Mol. Pharmacol.* **41**, 1023 (1992).
[5] E. Fattal, C. Vauthier, Y. Aynic, Y. Nakada, G. Lambert, C. Malvy, and P. Couvreur, *J. Controlled Release* **53**, 137 (1998).
[6] C. Allal, S. Sixou, R. Kravtzoff, N. Soulet, G. Soula, and G. Favre, *Br. J. Cancer* **77**, 1448 (1998).
[7] M. Berton, S. Sixou, R. Kravtzoff, C. Dartigues, L. Imbertie, C. Allal, and G. Favre, *Biochim. Biophys. Acta* **1355**, 7 (1997).
[8] E. L. Zirbes, C. Capitini, and R. Kole, *Nucleic Acids Symp. Ser.* **36**, 112 (1997).
[9] W. Y. Gao, C. Storm, W. Egan, and Y. C. Cheng, *Mol. Pharmacol.* **43**, 45 (1993).
[10] J. Temsamani, M. Kubert, J. Tang, A. Padmapriya, and S. Agrawal, *Antisense Res. Dev.* **4**, 35 (1994).
[11] C. A. Stein and J. S. Cohen, *Cancer Res.* **48**, 2659 (1988).
[12] S. L. Loke, C. A. Stein, X. H. Zhang, K. Mori, M. Kakanishi, C. Subasinghe, J. S. Cohen, and L. M. Neckers, *Proc. Natl. Acad. Sci. U.S.A.* **86**, 3474 (1989).
[13] L. A. Yakubov, E. A. Deeva, V. F. Zarytova, E. M. Ivanova, A. S. Ryte, L. V. Yurchenko, and V. V. Vlassov, *Proc. Natl. Acad. Sci. U.S.A.* **86**, 6454 (1989).
[14] M. Ceruzzi, K. Draper, and J. Scwartz, *Nucleosides Nucleotides* **9**, 679 (1990).
[15] Y. Shoji, S. Akhtar, A. Periasamy, B. Herman, and R. J. Juliano, *Nucleic Acids Res.* **19**, 5543 (1991).
[16] P. L. Iversen, S. Zhu, A. Meyer, and G. Zon, *Antisense Res. Dev.* **2**, 211 (1992).
[17] R. Bergan, Y. Connell, B. Fahmy, and L. Neckers, *Nucleic Acids Res.* **21**, 3567 (1993).
[18] T. Iwanaga and P. C. Ferriola, *Biochem. Biophys. Res. Commun.* **191**, 1152 (1993).

topes. Moreover, the ^{32}P label is unstable in enzymatic degradation by cellular phosphatases, which would lead to misinterpretation of the results as the radioactivity detected could be an isotopic label that was not attached to the oligonucleotide.

As an alternative approach to the study of the distribution of antisense oligonucleotides, we describe the synthesis and purification of digoxigenin-labeled phosphorothioate oligonucleotides and their use in the characterization of intracellular distribution in human Epstein–Barr virus-transformed B cells using immunocytochemistry with both confocal and electron microscopy.

Synthesis of Oligonucleotides Containing Digoxigenin

Digoxigenin is used extensively as a nonradioactive labeling system for nucleic acid detection. Antibodies against digoxigenin are available either alone or conjugated to enzymes (horseradish peroxidase, alkaline phosphatase) or gold particles. Digoxigenin labeling is especially suitable for *in situ* hybridization techniques[19] because digoxigenin is not present in animal cells, so there is no need to block endogenous molecules. These reasons also support the use of digoxigenin for analyzing the cellular uptake of antisense oligonucleotides.[20] The synthesis of digoxigenin-containing oligonucleotides is performed in two steps.[21,22] First, oligonucleotides carrying an aliphatic primary amino group, usually at the 5′ end, are prepared on a DNA synthesizer (see Fig. 1). After deprotection and purification, the amino oligonucleotide is reacted with an active ester derivative of digoxigenin.[22,23] Two digoxigenin derivatives have been described (compounds A and B, Fig. 1), the first of which is available commercially. After conjugation, digoxigenin-labeled oligonucleotides are purified by reversed-phase high-performance liquid chromatography (HPLC).[22] The hydrophobicity of digoxigenin facilitates the separation from unreacted amino oligonucleotides. The presence of digoxigenin is confirmed either by mass spectrometry or by immunological detection with antidigoxigenin antibodies conjugated to colloidal gold[20] or alkaline phosphatase.[22]

[19] R. Seibl, H. Höltke, R. Rüger, A. Meindl, H. G. Zachu, R. Rasshofer, M. Roggendorf, H. Wolf, N. Arnold, J. Wienberg, and C. Kessler, *Biol. Chem. Hoppe-Seyler* **371,** 939 (1990).
[20] G. Tarrason, D. Bellido, R. Eritja, S. Vilaró, and J. Piulats, *Antisense Res. Dev.* **5,** 193 (1995).
[21] H. Zischler, I. Nanda, R. Schäfer, M. Schmid, and J. T. Epplen, *Hum. Genet.* **82,** 227 (1989).
[22] M. Escarceller, F. Rodriguez-Frias, R. Jardi, B. San Segundo, and R. Eritja, *Anal. Biochem.* **206,** 36 (1992).
[23] K. Mühlegger, E. Huber, H. von der Eltz, R. Rügger, and C. Kessler, *Biol. Chem. Hoppe-Seyler* **371,** 953 (1990).

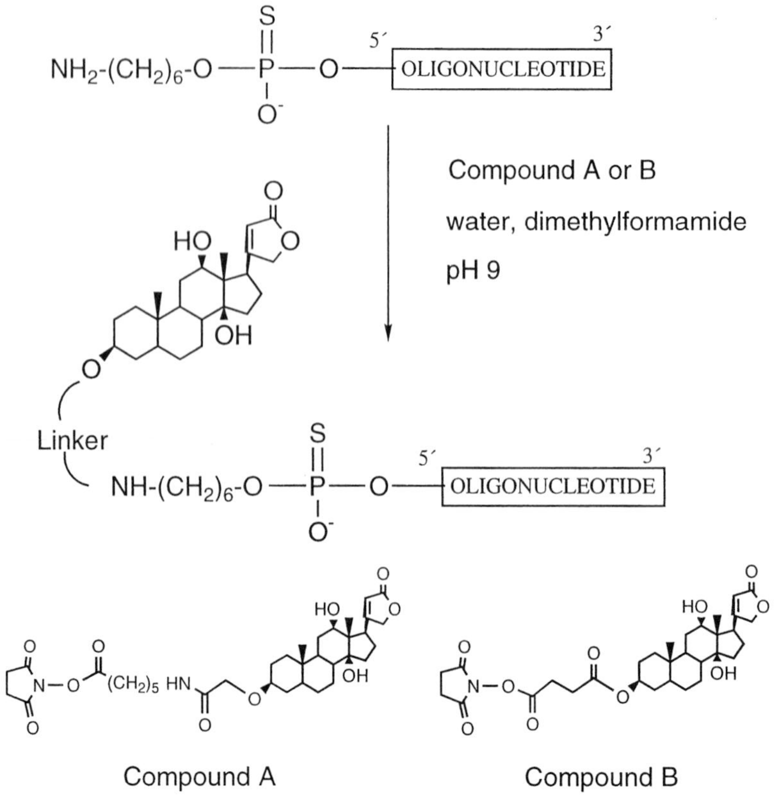

FIG. 1. Scheme for the preparation of digoxigenin-labeled oligonucleotides.

Confocal Microscopy and Colocalization Analysis

Flow cytometry and fluorescence microscopy are widely used to analyze the oligonucleotide uptake in many cell lines. The distinction between internalized and external fluorescence is difficult to make in such systems, and in the case of flow cytometry it requires enzymatic treatment of the cells to remove the membrane-bound oligonucleotide. Moreover, the complex three-dimensional nature of biological structures presents additional challenges. One goal is to observe the internal molecular distribution. Because laser-scanning microscopy combines a focused illumination spot and a detection pinhole to restrict excitation and detection to a small diffraction limited volume within the focal plane, out-of-focus planes are eliminated from the image. A full image is generated by moving either the specimen or the light beam sequentially to examine each point of the object. By obtaining optical sections through cells (Fig. 2A, see color insert) and tissues, the intracellular distribution of molecules (i.e., digoxigenin-containing oligonucleotides) can be examined, thus giving an excellent and unique approach for analyzing internalization. Colocalization experiments using double fluorescence staining detection methods (Fig. 2B, see color insert) allow the determination of the degree of intracellular codistribution between molecules. Confocal laser-scanning microscopy combined with sophisticated image processing techniques allows the cell biologist to explore the three-dimensional molecular organization of the cell and provides a better understanding of the uptake mechanisms.[24]

Previous results obtained by conventional fluorescence microscopy suggested that both digoxigenin and fluorescein-labeled oligonucleotides followed the same internalization pathway, suggesting that the label had no influence in the distribution.[20] Indirect approaches using flow cytometry or fluorescence microscopy suggest the lysosomal accumulation of phosphorothioate oligonucleotides.[25,26] However, an immunocytochemical approach based on colocalization with a lysosome marker, a lysosome-associated membrane protein[27] (Lamp-1), combined with confocal microscopy analysis, has provided direct evidence of the intracellular distribution of the oligonucleotides.[20]

Digoxigenin-labeled oligonucleotide was detected in all focal planes of the LTR228 cells (Fig. 2A) and showed a diffuse cytoplasmic fluorescence pattern and bright perinuclear vesicles. Staining with the anti-lamp 1 antibody revealed that LTR228 cells presented a prominent lysosomal compartment composed of perinuclear vesicles of several diameters (Fig. 2B). The overlap of fluorescein isothiocyanate (FITC, green fluorescence) and tetramethylrhodamine isothiocyanate (TRITC) (red fluorescence) signals in the three-dimensional reconstruction of serial optical sections revealed a clear colocalization (yellow color) between vesicles containing digoxigenin-labeled oligonucleotides and immunostained lysosomes (Fig. 2B).

Ultrastructural Analysis of Intracellular Oligonucleotide Distribution

There are several reports presenting the analysis of antisense oligonucleotides at the ultrastructural level.[20,28–31] Such analysis provides not only

[24] J. B. Pawley and V. E. Centonze, in "Cell Biology: A Laboratory Handbook" (J. E. Celis, ed.), Vol. 3, p. 149. Academic Press, San Diego, 1998.
[25] J. Tonkinson and C. A. Stein, *Nucleic Acids Res.* **22,** 4268 (1994).
[26] G. Marti, W. Egan, P. Noguchi, G. Zon, M. Matsukura, and S. Bronder, *Antisense Res. Dev.* **2,** 27 (1992).
[27] S. R. Carlsson, J. Roth, F. Piller, and M. J. Fukuda, *J. Biol. Chem.* **263,** 18911 (1988).
[28] P. Zamecnik, J. Aghajanian, M. Zamecnik, J. Goodchild, and G. Witman, *Proc. Natl. Acad. Sci. U.S.A.* **91,** 3156 (1994).

higher resolution than conventional microscopy, but also the possibility of quantitation of the oligonucleotide distribution in the different subcellular structures. In ultrastructural analysis, the use of nonisotopic labels that can be recognized by immunocytochemistry requires (1) the demonstration that the antigenic capacity of the label is not affected by the methodological processing of the cellular samples and (2) the availability of antibodies to the labeling molecules. The preparation of biological samples for immuno-cytochemistry is known to interfere with the biological properties of the components, leading to problems in labeling.[32] However, the different im-munocytochemical methods used for the localization of digoxigenin-labeled oligonucleotides demonstrate that the antigenic capacity of intracellular digoxigenin is not damaged during any of the procedures used.[20]

Two different approaches to detect digoxigenin-labeled oligonucleo-tides by electron microscopy were used: (1) immunogold labeling in heavily fixed and resin-embedded cells and (2) immunogold labeling in lightly fixed and cryoultramicrotomy-processed cells. After culturing LTR228 cells and processing for Araldite embedding or cryoultramicrotomy, staining of the sections with antidigoxigenin coupled to colloidal gold reveals that both methods allow the direct detection of intracellular-accumulated oligonucle-otides. Figures 3A and 3B show the appearance of representative sections obtained by the immunostaining of resin-embedded cells. Using this ap-proach, cellular structures are very well preserved and several compart-ments can be identified easily (Fig. 3A). Antidigoxigenin gold particles are mainly restricted to electron-dense structures with a lysosomal-like appearance (Fig. 3B). However, a few gold particles are also distributed sparsely in the cytoplasm and nucleus (Fig. 3A). The immunoreactivity of digoxigenin-labeled oligonucleotides is also preserved in cryoultrasections (Fig. 3C and 3D). Both the structural preservation and the immunogold staining of digoxigenin-labeled oligonucleotides of these sections are lower than those obtained in resin-embedded sections. Immunogold particles are restricted inside dense bodies with a lysosome-like membrane structure (Figs. 3C and 3D). Parallel negative control experiments performed in LTR228 cells incubated with nonconjugated digoxigenin and processed for either resin embedding or cryoultramicrotomy (not shown) indicated that

[29] C. Beltinger, H. U. Saragovi, R. M. Smith, L. LeSauteur, N. Shah, L. DeDionisio, L. Christensen, A. Raible, L. Jarett, and A. M. Gewirtz, *J. Clin. Invest.* **95,** 1814 (1995).
[30] K. Lappalainen, R. Miettinen, J. Kellokoski, Y. Jääskeläinen, and S. Syrjänen, *J. Histochem. Cytochem.* **45,** 265 (1997).
[31] W. Sommer, X. Cui, B. Erdmann, L. Wiklund, G. Bricca, M. Heilig, and K. Fuxe, *Antisense Nucleic Acid Drug Dev.* **8,** 75 (1998).
[32] M. Bendayan, A. Nanci, and F. W. K., *J. Histochem. Cytochem.* **35,** 983 (1987).

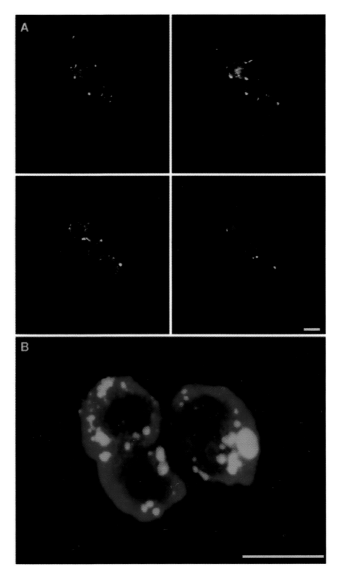

FIG. 2. Confocal immunofluorescence microscopy of digoxigenin-labeled oligonucleotides and lysosomes. Exponentially growing LTR228 cells were resuspended in fresh medium at 1.25×10^5 cells/ml in the presence of $10 \, \mu M$ digoxigenin-labeled oligonucleotide. Immunofluorescence procedures for digoxigenin and anti-Lamp 1 detection (used as a lysosomal marker) were as indicated in the text. (A) Four different optical sections (0.3 μm) of the intracellular distribution of digoxigenin-labeled oligonucleotides. Note the bright spots corresponding to the digoxigenin immunostaining, which are different on the four optical sections. (B) Merge image from the three-dimensional reconstruction of seven optical sections from LTR228 double immunostained with antidigoxigenin and anti-Lamp 1. The yellow color indicates sites where digoxigenin-labeled oligonucleotides and lysosomes colocalize. Bar: 10 μm.

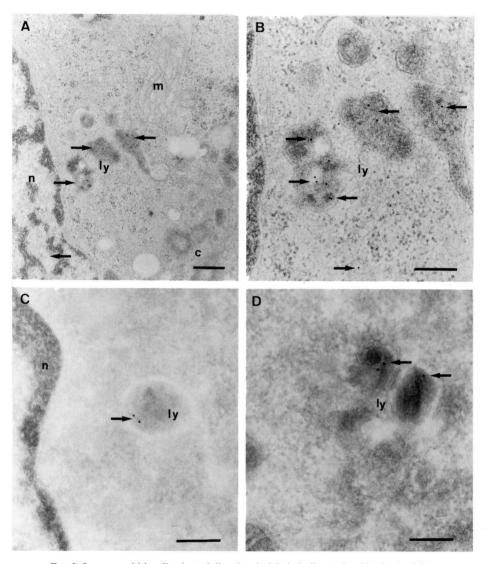

Fig. 3. Immunogold localization of digoxigenin-labeled oligonucleotides in Araldite-embedded LTR228 cells (A and B) and in cryoultrasections of LTR228 cells (C and D). LTR228 cells cultured with digoxigenin-labeled oligonucleotides were fixed with a mixture of 2% paraformaldehyde and 2.5% glutaraldehyde, processed for Araldite embedding or a mixture of 2% paraformaldehyde and 0.1% glutaraldehyde, and processed for cryoultramicrotomy. Sections of Araldite-embedded cells were etched on a saturated solution of sodium metaperiodate for 1 hr and processed for digoxigenin immunolabeling. For immunolabeling, sheep antidigoxigenin coupled to 10 nm colloidal gold was used. (A) Low magnification photomicrograph of the general appearance of Araldite-embedded LTR228 cells. (B) High magnification of the lysosome-like (ly) structures observed in (A). (C and D) Representative lysosome-like structures where colloidal gold was detected. Arrows point to colloidal gold particles. n, nucleus; m, mitochondria; c, centriole. Bar: (A) 0.5 μm; (B–D) 0.250 μm.

all the immunolabel observed in cells cultured with digoxigenin-labeled oligonucleotides was specific.

Therefore, double immunofluorescence and two different methods of immunoelectron microscopy confirm the major lysosomal accumulation of digoxigenin-labeled phosphorothioate oligonucleotides, as well as their scarcity in the nucleus, when they are delivered to the cells without using any carrier system. Other authors using digoxigenin-labeled oligonucleotides in a similar study show a slight increase in nuclear accumulation when cationic lipids are used to deliver the oligonucleotides to the cells.[30]

Understanding the intracellular transport and distribution of antisense oligonucleotides is of extreme importance in the development of oligonucleotide analogs as synthetic gene inhibitors and in their use in therapeutic applications. Digoxigenin can be considered a reliable label for the ultrastructural analysis of such processes.

Experimental Procedures

Preparation of Digoxigenin-Labeled Oligonucleotides

Oligonucleotide sequences with phosphorothioate linkages are prepared using standard solid-phase 2-cyanoethyl phosphoramidites[33,34] in a 392 DNA synthesizer (Applied Biosystems). Phosphorothioate linkages are introduced by replacing the iodine solution with a solution of either tetraethylthiuram disulfide[35] or 3H-1,2-benzodithiol-3-one 1,1-dioxide.[36] The primary amino group is introduced at the 5' end using N-MMT-6-aminohexanol phosphoramidite.[37] After ammonia deprotection for 6 hr at 55°, MMT-NH-oligonucleotides are purified by COP cartridges (Cruachem, Glasgow, Scotland) essentially as described by the manufacturer with minor modifications, as described. Removal of the MMT group on the cartridges is performed with a 5-min treatment with a 2% aqueous solution of trifluoroacetic acid, and the elution of oligonucleotides from the COP cartridge is performed with a 35% aqueous solution of acetonitrile. Oligonucleotides are converted to sodium salt form by passage of an aqueous solution of oligonucleotides through a small column (30 × 6 mm) filled with a cation-exchange resin (Dowex 50X2, sodium form). Aliquots of 10 optical density

[33] N. D. Sinha, J. Biernat, J. McManus, and H. Köster, *Nucleic Acids Res.* **12,** 4539 (1984).
[34] M. H. Caruthers, A. D. Barone, S. L. Beaucage, D. R. Dodds, E. F. Fisher, L. J. McBride, M. Matteucci, Z. Stabinsky, and J. Y. Tang, *Methods Enzymol.* **154,** 287 (1987).
[35] H. Vu and B. L. Hirsbeim, *Tetrahedron Lett.* **32,** 3005 (1991).
[36] R. P. Iyer, L. R. Phillips, W. Egan, J. B. Regan, and S. L. Beaucage, *J. Org. Chem.* **55,** 4693 (1990).
[37] N. D. Sinha and R. M. Cook, *Nucleic Acids Res.* **16,** 2659 (1988).

units (260 nm) of 5'-amino oligonucleotide are dissolved in 0.5 ml of 50 nM sodium carbonate buffer 0.1 M (pH 9.0), and 5 mg of a digoxigenin N-hydroxysuccinimide ester derivative (see Fig. 1) dissolved in 0.5 ml of dimethylformamide is added. The mixtures are incubated at 37° overnight and then concentrated to dryness. The residues are dissolved in water and desalted on a Sephadex G-25 column eluting with 50 mM triethylammonium acetate buffer, pH 7.8. Fractions containing the oligonucleotides are pooled, concentrated, and purified by reversed phase HPLC. HPLC conditions are as follows: Nucleosil 120C$_{18}$ column (250 × 4 mm) at a flow rate of 1 ml/min, using a linear gradient of 15–75% B over 20 min. The HPLC buffer system consists of solvent A [20 mM triethylammonium acetate (pH 7.8)] and solvent B (a 1:1 mixture of water and acetonitrile). Overall yields are between 20 and 30%.

Cellular Processing

Human LTR228 cells, a derivative cell line of the WIL-2NS human B-cell line, Epstein–Barr virus-transformed, are maintained in RPMI 1640 medium supplemented with 10% fetal calf serum, glutamine, and gentamicin at 37° in a humidified atmosphere of 93.5% (v/v) air and 6.5% (v/v) CO$_2$. Exponentially growing LTR228 cells are resuspended in fresh medium at 1.25 × 10^5 cells/ml in the presence 10 μM digoxigenin-labeled oligonucleotide. After 72 hr, cells are removed from the cultures, washed twice in phosphate-buffered saline (PBS) at 400g for 5 min, and fixed. The fixation process varies, depending on subsequent processing. For fluorescence microscopy studies, fixation is carried out with 4% paraformaldehyde at 4° overnight, and processing for electron microscopy requires 2% paraformaldehyde–2.5% glutraldehyde fixation in phosphate buffer 0.1 M (pH 7.4) at 4° for 2 hr.

Immunocytochemistry and Colocalization Analysis

After 4% formaldehyde fixation, LTR228 cells cultured with vehicle or digoxigenin-labeled oligonucleotide are washed in PBS at 1500g for 5 min, placed on gelatin-coated slices, and permeabilized further at −20° cold acetone for 30 sec. After blocking with PBS containing 20 mM glycine and 1% bovine serum albumin for 30 min, cells are incubated with FITC-conjugated sheep antidigoxigenin antibody (Fab fragment) (Boehringer Mannheim) at 50 μg/ml in blocking medium in a humidified chamber, protected from light, for 1 hr. After washing with PBS containing 20 mM glycine, cells are counterstained with 5 μM Hoechst 33342 for 10 min, washed again in PBS, and mounted in Immunofluore mounting medium

(ICN Biomedicals, Inc.) to be analyzed on a confocal laser-scanning micro-scope.

To analyze whether the intracellular localization of the labeled oligonu-cleotides corresponds to lysosomes, immunocytochemical analysis is per-formed using a polyclonal antibody antilysosome-associated membrane protein[27] (lamp1). After fixation, cells incubated with labeled oligonucleo-tides and their respective control are processed as indicated earlier for oligonucleotide detection. After blocking with PBS containing 20 mM gly-cine and 1% bovine serum albumin for 30 min, cells are incubated with rabbit antiserum anti-Lamp 1 at 1 : 200 in blocking medium in a humidified chamber for 1 hr and, after washing, incubated with a secondary swine antirabbit antibody (Boehringer Mannheim) labeled with TRITC at 1 : 50. Cells are finally counterstained with Hoechst 33342, mounted, and analyzed as described earlier.

Confocal laser-scanning microscopy (CLSM) is performed with a Leica TCS 4D (Leica Lasertechnik GmbH, Heidelberg, Germany) adapted to an inverted microscope (Leitz DMIRB). Photographs are taken using a 63× (NA 1.3, Ph3, oil) Leitz Plan-Apochromatic objective. FITC and TRITC are excited sequentially at 488- and 568-nm lines of a krypton–argon laser, respectively. For the reflection mode, excitation is provided by the 488-nm line, and a neutral RT30/70 excitation beam splitter is used. Three-dimensional projections are made from horizontal sections, as indicated in the legends depending on the cell thickness. Each image collected is the average of eight line scans at the standard scan rate. Images of 512 × 512 pixels are printed on a Mitsubishi CP2000E color video printer. Figure 2A shows four different optical sections corresponding to the intracellular distribution of digoxigenin-labeled oligonucleotides. For colocalization analysis (Multi Color software Leica Lasertechnik), a combined image is created using the gray values of each pixel of the two single-fluorescence images: the result was one single color-coded image representing both fluorescences. Correlation of the fluorescent signal from the two channels is represented in a two-dimensional confocal cytofluorogram on which the fluorescence intensities of the first (FITC) and the second (TRITC) channels are represented on the X axis and Y axis, respectively. Coincident labeling between the two single images is shown as a cloud in the portion of the cytofluorogram located along the diagonal of the graph ($x = y$). By drawing a random window around the colocalization area[38] (central cloud), it is directly identified as yellow pixels in the combined image (Fig. 2B).

[38] R. G. Marthinho, S. Castel, J. Ureña, M. Fernández-Borja, G. Olivecrona, M. Reina, A. Alonso, and S. Vilaró. *Mol. Biol. Cell.* **7,** 1771 (1996).

Electron Microscopy

LTR228 cells cultured in the presence of a vehicle or digoxigenin-labeled oligonucleotide as indicated earlier are fixed in 2% paraformaldehyde, 2.5% glutaraldehyde in phosphate buffer 0.1 *M* (pH 7.4) for 2 hr at 4°. After fixation, the cells are washed in the phosphate buffer 0.1 *M* and pelleted, centrifuging at 1500*g* for 5 min. Cells are then dehydrated in graded acetone (20, 50, 70, 90, and 100%) and embedded in Araldite (Durcupan ACM, Fluka) according to the usual procedure. Ultrathin sections of 50–70 nm are obtained with a Reichert-Jung ultramicrotome and mounted on a 200-mesh grid with carbon-coated Formvar film. Sections are etched by floating the grids section side down on a saturated solution of 3 mg/ml sodium metaperiodate ($NaIO_4$) for 1 hr, followed by thorough rinsing in distilled water before processing for digoxigenin immunolabeling. For immunolabeling, sections are incubated with PBS–glycin, (20 m*M*), 1% BSA for 20 min and incubated with sheep antidigoxigenin (BioCell Research Laboratories) conjugated with 10 nm colloidal gold diluted 1 : 100 (210 ng/ml) for 1 hr at room temperature. Grids are washed and stained with 4% aqueous uranyl acetate for 30 min and 2% lead citrate for 10 min.

Samples for cryoultramicrotomy are fixed in 2% paraformaldehyde/ 0.1% glutaraldehyde in phosphate buffer 0.1 M (pH 7.4) for 1 hr at 4° and processed as described by Casaroli-Marano *et al.*[39] Briefly, after fixation, cells are centrifuged at 1500*g* for 5 min, pellets are warmed at 37° for 10 min, and 50 μl of warmed 10% gelatin is added to them. After resuspension in gelatin, cells are centrifuged at 1500*g* for 10 min and placed on ice for 10 min. Once embedded in gelatin blocks, the pellet is postfixed with 0.5 ml of cold 2% paraformaldehyde in phosphate buffer 0.1 *M* (pH 7.4) overnight. Blocks are then cryoprotected for 24 hr in polyvinylpyrrolidone at 4°, mounted on sample carriers for cryoultramicrotomy, and frozen by immersion in liquid nitrogen (−196°). Cryoultrasections are obtained in a Reichert-Jung ultramicrotome equipped with a FC4 system for cryosectioning. Sections are retrieved with a copper loop containing 2.3 *M* sucrose in PBS, transferred to a 100-mesh grid with carbon-coated Formvar film, and placed on 2% gelatin until processing for immunodetection. For immunolabeling on ultrathin cryosections, grids are washed on drops of 20 m*M* glycine in PBS, blocked with PBS–glycine with 1% BSA for 20 min, and incubated with sheep antidigoxigenin conjugated to 10 nm colloidal gold for 1 hr at room temperature, as indicated earlier. After three washes of 2.5 min each in PBS and six washes of 5 min each in double distilled water,

---

[39] R. P. Casaroli-Marano, R. García, E. Vilella, G. Olivecrona, M. Reina, and S. Vilaró, *J. Lipid Res.* **39,** 789 (1988).

the sections are contrasted in 0.3% uranyl acetate in methylcellulose for 10 min on ice. The grids are retrieved using a copper loop and excess fluid is removed on a filter paper. Grids are examined with a Hitachi 600 AB electron microscope (Nissai Sanyo GmbH, Deutschland) at 75 kv.

Acknowledgments

We thank Dr. S. Carlsson for the generous gift of antibody to Lamp-1. We are also grateful to Tamara Caparrós and Susana Castel for technical assistance and to Robin Rycroft for editorial help.

[15] Use of Minimally Modified Antisense Oligonucleotides for Specific Inhibition of Gene Expression

By Eugen Uhlmann, Anusch Peyman, Antonina Ryte, Annette Schmidt, and Eckhart Buddecke

Introduction

The first generation of antisense oligonucleotides (ODN) are phosphorothothiate (PS)-modified analogs in which one nonbridging oxygen of the phosphodiester internucleoside linkage is replaced by sulfur. The PS modification is particularly popular because it provides sufficient stabilization against nucleolytic degradation, whereas the duplex of PS-ODN and RNA is still recognized by cellular RNase H, which is in contrast to most other modifications. The cleavage of target mRNA by RNase H is considered as an important factor for the activity of antisense oligonucleotides. Although a number of PS ODN are in advanced stages of clinical development, certain limitations of uniformly PS modified (all-PS) ODN have emerged. All-PS ODN have a propensity for non-antisense effects that are often caused by undesired binding to proteins.[1-4] Thus PS-ODN have been reported to interact with DNA polymerases,[5] protein kinase C-β1,[6] gp120

[1] C. A. Stein and Y. C. Cheng, Science 261, 1004 (1993).
[2] J. F. Milligan, M. D. Matteucci, and J. C. Martin, J. Med. Chem. 36, 1923 (1993).
[3] C. A. Stein, Nature Med. 1, 1119 (1995).
[4] E. Uhlmann and A. Peyman, Chem. Rev. 90, 543 (1990).
[5] W. Y. Gao, J. W. Jaroszewski, J. S. Cohen, and Y. C. Cheng, J. Biol. Chem. 265, 20172 (1990).
[6] C. A. Stein, J. L. Tonkinson, L. M. Zhang, L. Yakubov, J. Gervasoni, R. Taub, and S. A. Rotenberg, Biochemistry 32, 4855 (1993).

of human immunodeficiency virus (HIV),[7] p210[bcr-abl] tyrosine kinase,[8] basic fibroblast growth factor (bFGF),[9] laminin,[10] Mac-1,[11] and serum albumin.[12] *In vivo,* immune stimulation and prolongation of activated partial thromboplastin time (aPPT) have been observed after the systemic administration of an all-PS ODN.[13,14] As a consequence, different strategies have been employed to reduce the number of PS linkages within an ODN.

The metabolism of ODN involves cleavage by exo- and endonucleases, the major degrading activity being a 3'-exonuclease.[4,15] Although capping of the 3' end, or both the 3' and 5' ends, by PS linkages protects the ODN against 3'-exonuclease degradation,[15–19] this protection strategy has proven to be only of limited success, as these PS end-capped ODN are still subject to endonuclease degradation. Similarly, random variation or alternating PS linkages without additional 3' end capping does not efficiently prevent degradation by endonucleases.[17,20] We have reported on a new "minimal protection" strategy, which is a combination of the end-capping technique and the protection at internal pyrimidine residues, which are the major sites of endonuclease degradation.[21,22] This minimal protection strategy has proven to be particularly useful because it results in ODN that are sufficiently stable to exo- and endonuclease degradation, whereas at the same time, undesirable non-antisense effects are reduced greatly.

[7] C. A. Stein, A. M. Cleary, L. Yakubov, and S. Lederman, *Antisense Res. Dev.* **3,** 19 (1993).

[8] R. Bergan, Y. Connell, B. Fahmy, E. Kyle, and L. Neckers, *Nucleic Acids Res.* **22,** 2150 (1994).

[9] M. A. Guvakova, L. A. Yakubov, I. Vlodavsky, J. L. Tonkinson, and C. A. Stein, *J. Biol. Chem.* **270,** 2620 (1995).

[10] Z. Khaled, L. Benimetskaya, R. Zeltser, T. Khan, H. W. Sharma, R. Narayanan, and C. A. Stein, *Nucleic Acids Res.* **24,** 737 (1996).

[11] L. Benimetskaya, J. D. Loike, Z. Khaled, G. Loike, S. C. Silverstein, L. Cao, J. El Khoury, T.-Q. Cai, and C. A. Stein, *Nature Med.* **3,** 414 (1997).

[12] S. T. Crooke, M. J. Graham, J. E. Zukerman, D. Brooks, B. S. Conklin, L. L. Cummins, M. J. Greig, C. J. Guinosso, D. Kornburst *et al., J. Pharmacol. Exp. Ther.* **277,** 923 (1996).

[13] S. P. Henry, P. C. Giclas, J. Leeds, M. Pangburn, C. Auletta, A. A. Levin, and D. J. Kornbrust, *J. Pharmacol. Exp. Ther.* **281,** 810 (1997).

[14] S. T. Crooke and C. F. Bennett, *Annu. Rev. Pharmacol. Toxicol.* **36,** 107 (1996).

[15] J. P. Shaw, K. Kent, J. Bird, J. Fishback, and B. Froehler, *Nucleic Acids Res.* **19,** 747 (1991).

[16] C. A. Stein, C. Subasinghe, K. Shinozuka, and J. S. Cohen, *Nucleic Acids Res.* **16,** 3209 (1988).

[17] G. D. Hoke, K. Draper, S. M. Freier, C. Gonzalez, V. B. Driver, M. C. Zounes, and D. J. Ecker, *Nucleic Acids Res.* **19,** 5743 (1991).

[18] F. Gillardon, H. Beck, E. Uhlmann, T. Herdegen, J. Sandkühler, A. Peyman, and M. Zimmermann, *Eur. J. Neurosci.* **6,** 880 (1994).

[19] F. Gillardon, I. Moll, and E. Uhlmann, *Carcinogenesis* **16,** 1853 (1995).

[20] M. K. Ghosh, K. Ghosh, and J. S. Cohen, *Anti Cancer Drug Des.* **8,** 15 (1993).

[21] A. Peyman and E. Uhlmann, *Biol. Chem. Hoppe-Seyler* **377,** 67 (1996).

[22] E. Uhlmann and A. Peyman, *Nucleosides Nucleotides,* in press.

Principles of Design of Minimally Phosphorothothiate-Modified
Antisense Oligonucleotides

Unmodified ODN are degraded rapidly in serum by 3′-exonucleases.
If the 3′ end of ODN is blocked by several PS linkages, the slower endonuclease cleavage becomes apparent (Table II). We have found that single-
strand-specific endonucleases in serum and cell extracts cleave the ODN
preferentially at pyrimidine sites.[21] This cleavage becomes especially dominant if two or more pyrimidines are adjacent. Intermediates resulting from
the preferred cleavage at pyrimidine sites often accumulate in the incubation mixture and can be detected by polyacrylamide gel electrophoresis
(PAGE) analysis.[23] The degradation of ODN by 5′-exonucleases must also
be taken into account, although it has been observed less frequently.[24]

The design of minimally PS-modified ODN involves the following criteria: (i) The ODN is end capped by two to five PS residues at the 3′ end
to render it stable against dominant 3′-exonucleases. (ii) The ODN is end
capped by two PS residues at the 5′ end to protect it against potential
degradation by 5′-exonucleases. (iii) Additional PS linkages are placed at
internal pyrimidine nucleotides, which are the major sites of degradation
by endonucleases. Two or more consecutive pyrimidines nucleotides within
an ODN are linked by PS bridges. (iv) More than four to five PS residues
in a row are avoided in order not to run into the side effects observed for
uniformly PS-modified ODN.

A typical example for a minimally modified ODN is shown in Table I.
The three 3′-terminal nucleotides of the ODN (structure I) are linked by
two PS linkages regardless of the sequence. Because the fourth nucleotide
from the 3′ terminus is a pyrimidine, a further PS linkage is introduced in
3′ position to this pyrimidine, resulting in three consecutive PS linkages at
the 3′ end. The internal motif YYY constitutes a preferred substrate for
endonucleases and is therefore stabilized by two PS linkages. A further PS
linkage is placed at the separated pyrimidine nucleotide, and finally the 5′
end is capped by two PS linkages to take care of potentially occurring
5′-exonucleases.

The minimal protection scheme can be combined easily with secondary
modifications, such as 2′-O-alkyl modification of ribose[25,26] or C5-alkynyl

[23] E. Uhlmann, A. Ryte, and A. Peyman, *Antisense Nucleic Acid Drug Dev.* **7,** 345 (1997).
[24] A. Ryte, S. Morelli, M. Mazzei, A. Alama, P. Franco, G. F. Canti, and A. Nicolin, *Anti Cancer Drugs* **4,** 197 (1993).
[25] H. Inoue, Y. Hayase, S. Iwai, and E. Ohtsuka, *FEBS Lett.* **215,** 327 (1987).
[26] B. S. Sproat, A. Iribarren, B. Beijer, U. Pieles, and A. I. Lamond, *Nucleosides Nucleotides* **10,** 25 (1991).

TABLE I
MINIMALLY MODIFIED ODN, INCLUDING
SECONDARY MODIFICATIONS

Structure	Sequence[a]
(I)	5'-N*N*NRY*RY*Y*YRY*N*N*N-3'
(II)	5'-N*N*NRY*RY*Y*YRY*N*N*N-3'
(III)	5'-N*N*NRY*RY*Y*YRY*N*N*N-3'
(IV)	5'-N*N*NRY*Ry y y r y n n n

[a] N: G, A, C, or T; R: purine, Y: pyrimidine; *PS linkage; underlined: 2'-O-methylribonucleotides; double underlined: C5-propynylpyrimidine; lowercase letters indicate peptide nucleic acid.

modification of pyrimidine bases.[27,28] The ODN of structure **II** is a so-called gap-mer,[25,29] in which 2'-O-methylribonucleotide wings are introduced to enhance binding to mRNA, whereas a window of six deoxynucleotides allows activation of RNase H. In order to enhance binding affinity to RNA, all pyrimidines in the ODN of structure **III** are substituted by C5-propynyl or hexynyl residues.[28] The activation of RNase H is not compromised by this base modification. Depending on the sequence and assay system used, ODN of structures **II** and **III** are usually 5 to 10 times more potent than the parent ODN of structure **I**. Finally, the minimal modification can also be combined with the completely nuclease-resistant and tightly binding peptide nucleic acids (PNA),[30,31] resulting in PNA–DNA chimeras (structure **IV**)[32–34] in which the minimally modified DNA part can stimulate RNase H.

[27] B. C. Froehler, S. Wadwani, T. J. Terhorst, and S. R. Gerrard, *Tetrahedron Lett.* **33,** 5307 (1992).
[28] E. Uhlmann, L. Hornung, S. Hein, S. Augustin, A. Peyman, D. W. Will, Helsberg, J. Sagi, L. Otvos, J. O. Ojwang, S. Mustain, and R. F. Rando, *Nucleosides Nucleotides* **16,** 1717 (1997).
[29] S. T. Crooke, K. M. Lemonidis, L. Neilson, R. Griffey, E. A. Lesnik, and B. P. Monia, *Biochem. J.* **312,** 599 (1995).
[30] P. E. Nielsen, M. Egholm, R. H. Berg, and O. Buchardt, *Science* **254,** 1497 (1991).
[31] B. Hyrup and P. E. Nielsen, *Bioorg. Med. Chem.* **4,** 5 (1996).
[32] E. Uhlmann, D. W. Will, G. Breipohl, D. Langner, and A. Ryte, *Angew. Chem. Int. Ed. Engl.* **35,** 2632 (1996).
[33] E. Uhlmann, A. Peyman, G. Breipohl, and D. W. Will, *Angew. Chem. Int. Ed.* **37,** 2796 (1998).
[34] E. Uhlmann, *Biol. Chem.* **379,** 1045 (1998).

Synthesis of Minimally Modified Oligonucleotides

Minimally PS-modified ODN are synthesized using a commercially available DNA synthesizer and standard phosphoramidite chemistry.[35,36] After coupling, PS linkages are introduced by sulfurization using the Beaucage reagent[37] (0.075 M in acetonitrile) followed by capping with acetic anhydride, 2,6-lutidine in tetrahydrofuran (1:1:8; v:v:v) and N-methylimidazole (16% in tetrahydrofuran). It is important that this capping step is performed after the sulfurization reaction to minimize the formation of undesired phosphodiester (PO) linkages. After cleavage from the solid support and final deprotection by treatment with concentrated ammonia (15 hr at 50°), ODN are analyzed by HPLC on a Gen-Pak Fax column (Millipore-Waters) using a NaCl gradient [buffer A: 10 mM NaH$_2$PO$_4$, in acetonitrile/water, 1:4 (v:v), pH 6.8; buffer B: 10 mM NaH$_2$PO$_4$, 1.5 M Nacl in acetonitrile/water, 1:4 (v:v), 5–60% B in 30 min at 1 ml/min]. ODN products with incomplete sulfurization can be seen as minor peaks at slightly shorter retention times on HPLC (Fig. 1). Analysis of ODN by PAGE or capillary electrophoresis will only detect $n-1$, $n-2$, etc., failure sequences, but not products containing one or more PO linkages instead of PS linkages. The ODN can be purified by HPLC or by FPLC on a Mono QR high-performance column (10/10; Pharmacia, Piscataway, NJ) using a Pharmacia Biopilot system. In the latter case, the ODN is eluted with a 0.3–1.5 M NaCl (pH 7) gradient within 60 min. For ODN that form secondary structures or have a tendency to aggregate, FPLC is performed in buffer containing 10 mM NaOH at pH 12. Thus, the ODN 5'-G*TG*CAGC*C*TG*G*G-teg (underlined: 2'-O-methylribonucleotide; teg: triethylene glycol phosphate) after purification by FPLC did not show any $n-1$ failure sequences or incomplete sulfurization (Fig. 2). The collected alkaline fractions are neutralized immediately with acetic acid. HPLC-homogeneous fractions are combined and desalted via a C$_{18}$ column (Millipore, Bedford, MA). The ODN was analyzed by negative ion electrospray mass spectroscopy (Fisons Bio-Q) to confirm the calculated mass (Figs. 3 and 4).

[35] F. E. Eckstein, "Oligonucleotides and Analogues: A Practical Approach." IRL Press, Oxford, UK, 1991.

[36] M. D. Matteucci and M. H. Caruthers, *Tetrahedron Lett.* **21,** 719 (1980).

[37] R. P. Iyer, W. Egan, J. B. Regan, and S. L. Beaucage, *J. Am. Chem. Soc.* **112,** 1253 (1990).

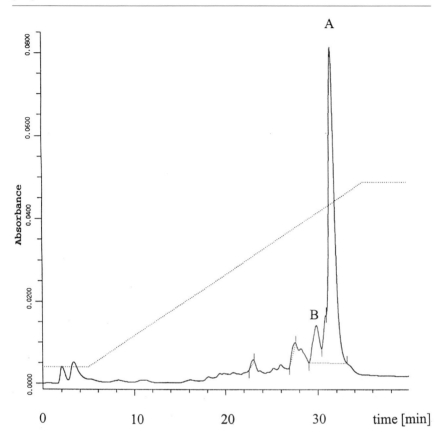

FIG. 1. HPLC of crude product from the synthesis of 5'-G*TG*CAGC*C*TG*G*G-teg (underlined: 2'-O-methylribonucleotide; teg: triethylene glycol phosphate). Analysis was performed on a Gen-Pak Fax column (Millipore-Waters, Bedford, MA) using a NaCl gradient [buffer A: 10 mM NaH$_2$PO$_4$ in acetonitrile/water, 1 : 4 (v : v), pH 6.8; buffer B: 10 mM NaH$_2$PO$_4$, 1.5 M NaCl in acetonitrile/water, 1 : 4 (v : v), 5–60% B in 30 min]. Desired main product (A) and side product with one sulfur replaced by oxygen (B).

Biological Properties of Minimally Modified Oligonucleotides

Stability against Nucleases

The stability of ODN in serum and within cells depends on their length and even more so on their sequence. We have investigated the stability of unmodified homopolymeric sequences in the presence of Vero cells and found the following order of stability $(T)_{20} < (dC)_{20} < (dA)_{20} \ll (dG)_{20}$,

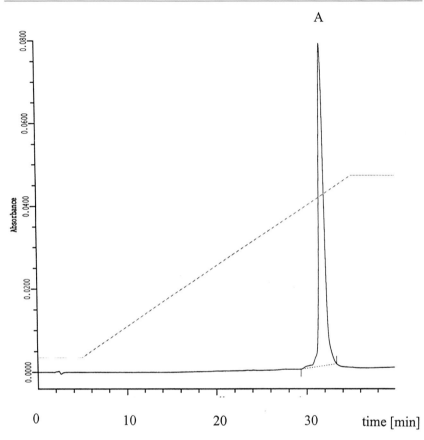

FIG. 2. HPLC of FPLC-purified 5′-G*TG*CAGC*C*TG*G*G-teg. Conditions are as described in Fig. 1.

which is in accordance with the observed preference of cleavage at pyrimidine nucleotides in heteropolymers. Guanosine-rich sequences are generally more resistant to degradation than sequences with equal base composition. The enhanced stability of G-rich ODN is most likely due to secondary structures formed by intermolecular or intramolecular hydrogen bonding. As can be seen from Table II, the stability of ODN 1 increases as the number PS linkages increases. ODN 3 with three PS linkages at either end is approximately six times more stable in serum than the unmodified oligomer. In case of pyrimidine-rich sequences, this stabilizing effect can be even more pronounced. Interestingly, the stability depends strongly on the site of PS modification at internal positions: PS linkages at purine

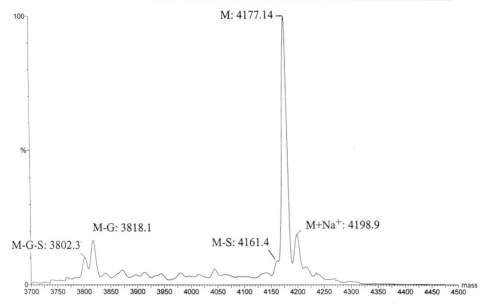

FIG. 3. ESI-MS of crude 5'-G*TG*CAGC*C*TG*G*G-teg. (M: 4177.14; calculated for $C_{128} H_{170} N_{48} O_{77} P_{12} S_6$: 4177.1; Fisons Bio-Q). The side product, in which one sulfur is replaced by oxygen, can be seen as a shoulder (M-S: 4161.4). Failure sequence side products, in which one 2'-O-methylguanosine nucleotide (M-G: 3818.1) is missing, and its incomplete sulfurization product (M-G-S: 3802.3) can also be identified in the ESI-MS. The mono sodium salt (M+NA⁺: 4198.9) of the dodecameric ODN appears as a separate peak.

positions (ODN 5) result in only a moderately increased stability, whereas PS modification at internal pyrimidines in combination with end capping yields the highly stabilized ODN 4. The uniformly PS-modified ODN 7 shows the highest stability against degradation. However, the antiviral activity of minimally modified ODN 4 containing 8 PS linkages equals that of the uniformly PS-modified ODN 7 having 19 PS linkages because of its improved binding affinity, cellular uptake, and possibly also improved activation of RNase H.

Adjacent pyrimidine nucleotides are linked preferentially by PS bridges. ODN 4 contains a pyrimidine-rich region, that is protected by four PS linkages according to the pattern C*T C*C*A*T. In order to minimize potential side effects, we prefer this noncontinuous PS pattern over the alternative C*T*C*C*A*T with five consecutive PS linkages. We have also investigated the effect of the position of the PS bridges relative to pyrimidine residues. There was basically no difference in stability if the PS linkages were in the 5' position to the internal pyrimidines (5'-G*C*AGGAG-

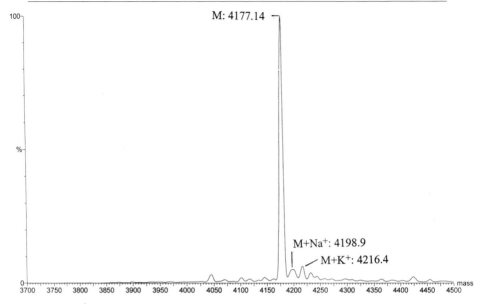

FIG. 4. ESI-MS of FPLC-purified 5'-G*TG*CAGC*C*TG*G*G-teg. (M: 4177.14). No failure sequences or incomplete sulfurization products can be recognized.

GA*TG*C*TGGGA*G*G) as compared to ODN with PS linkages at the 3' site of the pyrimidine nucleotides (5'-G*C*AGGAGGAT*GC*T*GG-GA*G*G). The corresponding ODN with PS linkages at both 5' and 3' positions of the pyrimidines (5'-G*C*AGGAGGA*T*G*C*T*GG-GA*G*G) with a total of five internal PS linkages was only marginally more stable against degradation in cell extracts than the latter two analogs having only three internal PS linkages in addition to end capping. A study of the mechanism of stabilization of minimally PS modified ODN against nucleolytic degradation revealed that ODN stability is not only caused by direct prevention of nuclease attack at the phosphate center but is supported additionally by competitive inhibition of the nucleases by the PS groups of the ODN.[23] Thus it could be shown that the stability of an unmodified ODN in serum is enhanced significantly in the presence of a second but PS-modified ODN and increases as the number of PS linkages in the second PS-modified ODN increases.

We have applied the concept of minimally modified ODN to hundreds of different sequences in various biological systems. In most cases the introduction of 40–60% of PS linkages results in sufficient nucleolytic stability of the ODN to obtain specific antisense activity. Selected examples of

TABLE II

PROPERTIES OF ANTI-HSV-1 ODN OF DIFFERENT CHEMICAL DERIVITIZATION[a]

ODN number	Sequence position of PS linkage[b]	n[c]	$t_{1/2}$[d] (hr)	Uptake level[e] (pmol/10^5 cells)	T_m[f]	MIC[g] (μM)
1	G C G G G G C T C C A T G G G G G T C G	0	3	5.6	73.5°	>80
2	G*C*G G G G C T C C A T G G G G G T*C*G	4	8	5.3	72.0°	27
3	G*C*G*G G G G C T C C C A T G G G G G*T*C*G	6	18	n.d.	71.2°	9
4	G*C*G G G G C*T C*C*A*T G G G G G T*C*G	8	30	3.2	71.0°	1
5	G*C*G G G*G C T C C A T G*G*G*G G T*C*G	8	13	n.d.	71.1°	27
6	G*C*G G*G G*C T*C C*A T*G G*G T*C*G	11	35	n.d.	69.9°	9
7	G*C*G*G*G*C*T*C*C*A*T*G*G*G*G*G*T*C*G	19	>48	2.5	65.8°	1

[a] Data are from A. Peyman, M. Helsberg, G. Kretzschmar, M. Mag, A. Ryte, and E. Uhlmann, Antiviral Res. **33**, 135 (1997).

[b] Sequences are written from 5' to 3'; an asterisk indicates the position of a PS linkage.

[c] Number of PS linkages.

[d] Half-life of nucleolytic degradation of ODN during incubation in serum in the absence of uptake enhancers.

[e] Cellular uptake by Vero cells without uptake enhancers. n.d., not determined.

[f] Melting curves as measured against the complementary DNA oligonucleotide at 1 μM each strand in 140 mM NaCl, 10 mM HEPES, pH 7.5.

[g] Minimal inhibitory concentration (in the absence of uptake enhancers); DTM values (Dosis Tolerata Maxima) were in all cases >80 μM.

effective minimally modified antisense ODN against various targets are shown in Table III.

Binding Affinity

The introduction of PS linkages of random stereochemistry into ODN as obtained from standard solid-phase synthesis[36] results in reduction of the binding affinity to RNA by 0.3–0.5° per modification. Consequently, the melting temperature of an ODN of a given sequence decreases with the increasing number of PS modifications. Thus the uniformly PS-modified ODN 7 has a melting temperature (T_m) 7.7 K lower temperature than the unmodified ODN 1, whereas for the minimally modified ODN 4, the T_m is only decreased by 2.5 K. Of course the loss in binding affinity can be compensated by secondary modifications as discussed earlier. Introduction of C5-(1-hexynyl)pyrimidine analogs at internal pyrimidine positions of PS end-capped ODN results in enhanced binding affinity and biological

TABLE III
MINIMALLY MODIFIED ANTISENSE ODN

Target	Sequence (5' to 3') and PS (*) pattern	PS (%)	Cell type	Ref.
HSV-1	G*C*GGGGC*TC*C*A*TGGGGT*C*G	42	Vero	a
c-cbl	G*C*C*GGC*CA*TGGC*CAGC*GGA*G*G	45	Osteoblasts	b
c-src	T*T*GCT*CT*TGT*TGC*TG*CC*C*A*T	53	Osteoblasts	b
c-fos	C*G*AG*AAC*AT*CAT*GG*T*C*G	56	HT-22	c
c-myc	C*A*C*GT*T*GAGGGG*C*A*T	57	SMC	d
bFGF	G*G*C*TGC*CA*TGGT*C*C*C	54	SMC	e
bax	T*G*CT*CC*C*CGGAC*C*CGT*CC*A*T	53	SCG neurons	f
Integrin αV	G*C*GGC*GGAAAAGC*CA*T*C*G	41	Osteoclasts	g

[a] A. Peyman, M. Helsberg, G. Kretzschmar, M. Mag, A. Ryte, and E. Uhlmann, *Antiviral Res.* **33,** 135 (1997).

[b] S. Tanaka, M. Amling, L. Neff, A. Peyman, E. Uhlmann, J. B. Levy, and R. Baron, *Nature* **383,** 528 (1996).

[c] F. Gillardon, T. Skutella, E. Uhlmann, F. Holsboer, M. Zimmermann, and C. Behl, *Brain Res.* **706,** 169 (1996).

[d] A. Peyman, H. Ragg, T. Ulshofer, D. W. Will, and E. Uhlmann, *Nucleosides Nucleotides* **16,** 1215 (1997).

[e] A. Schmidt, J. Sindermann, A. Peyman, E. Uhlmann, D. W. Will, J. G. Muller, G. Briethardt, and E. Buddecke, *Eur. J. Biochem.* **248,** 543 (1997).

[f] F. Gillardon, M. Zimmermann, E. Uhlmann, S. Krajewski, J. C. Reed, and L. Klimaschewski, *J. Neurosci. Res.* **43,** 726 (1996).

[g] I. Villanova, P. A. Townsend, E. Uhlmann, J. Knolle, A. Peyman, M. Amling, R. Baron, M. A. Horton, and A. Teti, *J. Bone Min. Res.,* in press.

activity.[38,39] It was interesting to find that the stability toward nucleolytic degradation is also increased by the C5-(hexynyl)pyrimidine modification without additional PS modifications at these internal pyrimidines and results in enhanced biological activity.[28]

Cellular Uptake

The cellular uptake of ODN 1 (Table II), as measured by cell association of the ^{32}P-labeled ODN on incubation with Vero cells, was found to decrease with increasing number of PS residues.[40] Thus, the minimally modified ODN 4 shows a higher cellular uptake and stronger binding affinity than the corresponding all-PS modified ODN 7. It should be noted, however, that the cellular uptake of ODN is sequence dependent, as is nuclease stability. The uptake of ^{32}P-labeled homopolymers increases in the order $(dC)_{16} < (dA)_{16} < (T)_{16} \ll (dG)_{16}$. In Vero cells, the uptake of $(dG)_{16}$ is about 40 times higher than $(dC)_{16}$. The enhanced cell association of G-rich ODN is connected with the ability to aggregate to higher molecular structures[41] as evidenced by gel electrophoresis and light-scattering measurements.

By adding three to four non-base-pairing dG residues at both 5' and 3' termini of ODN directed against HSV-1[41] or bFGF,[42] respectively, a significant increase in antiviral or antiproliferative potency was observed, respectively. In view of the reported non-antisense effects of dG-tetrad forming ODN,[43] which are most pronounced at higher ODN concentrations, special care has to be taken with respect to appropriate control experiments if this G-end-capping approach is to be used to enhance the cell association of ODN in any new experimental antisense systems.

In cultures of primary cells, we have observed antisense activity without the addition of uptake enhancers at about 1 to 10 μM ODN concentration.

[38] J. O. Ojwang, S. D. Mustain, H. B. Marshall, T. S. Rao, N. Chaudhary, D. A. Walker, M. E. Hogan, T. Akiyama, G. R. Revankar, A. Peyman, E. Uhlmann, and R. F. Rando, *Biochemistry* **36**, 6033 (1997).
[39] J. O. Ojwang, T. S. Rao, H. B. Marshall, S. D. Mustain, N. Chaudhary, D. A. Walker, A. Peyman, E. Uhlmann, G. R. Revankar, and R. F. Rando, *Nucleosides Nucleotides* **16**, 1703 (1997).
[40] A. Peyman, M. Helsberg, G. Kretzschmar, M. Mag, A. Ryte, and E. Uhlmann, *Antiviral Res.* **33**, 135 (1997).
[41] A. Peyman, A. Ryte, M. Helsberg, G. Kretzschmar, M. Mag, and E. Uhlmann, *Nucleosides Nucleotides* **14**, 1077 (1995).
[42] A. Schmidt, J. Sindermann, A. Peyman, E. Uhlmann, D. W. Will, J. G. Muller, G. Breithardt, and E. Buddecke, *Eur. J. Biochem.* **248**, 543 (1997).
[43] T. L. Burgess, E. F. Fisher, S. L. Ross, J. V. Bready, Y. X. Qian, L. A. Bayewitch, A. M. Cohen, C. J. Herrera, S. S. F. Hu, T. B. Kramer, F. D. Lott, F. H. Martin, G. F. Pierce, L. Simonet, and C. L. Farrell, *Proc. Natl. Acad. Sci. U.S.A.* **92**, 4051 (1995).

In contrast, in permanent cell lines it was usually necessary to formulate ODN with lipocationic uptake enhancers to obtain antisense activity.

Specificity

Non-antisense effects can result either from the polyanionic, polysulfate-like character of PS ODN or from certain sequence motifs (e.g., CpG, G-tetrad), which may even potentiate each other. By interrupting longer stretches of PS linkages within an ODN, non-antisense effects can be reduced by diminishing the binding of ODN to proteins. It has been reported that the minimally modified ODN G*G*G*ACC*AT*GGCA*G*C*C does not significantly inhibit high-affinity binding of bFGF to Vero cells up to 20 μM, whereas all-PS-modified ODN show inhibition levels (50% inhibition at 0.1 to 1 μM ODN) similar to polyanionic heparin.[44] Furthermore, short PS-modified ODN, such as the trinucleotide diphosphorothioate C*C*C, did not show any effect on cell proliferation up to 60 μM.[42]

The antiproliferative effect of c-*myc* all-PS ODN has been attributed to a sequence-specific non-antisense effect due to the presence of a $(dG)_4$ motif. By using the minimally modified ODN 5'-C*A*C*GT*T*GAGGG-G*C*AT, containing the G tetrad, we could specifically downregulate c-*myc* mRNA at 0.3 μM ODN concentration (5 μg/ml LipoFectamine) as evidenced by Northern blotting without affecting the expression of GAPDH. Most interestingly, a scrambled minimally modified control ODN T*A*C*-GGGGT*T*GAG*C*A*A, which also contains the $(dG)_4$ motif, as well as the sense ODN, did not reduce c-*myc* levels at 0.3 μM concentration. We have generally seen much fewer nonspecific effects using minimally modified antisense ODN as compared to all-PS ODN in cell-free translation systems and in cellular assay systems, even at ODN concentrations as high as 10 to 20 μM.

Use of Minimally Modified Antisense Oligonucleotides without Uptake Enhancers

Sequence-specific antisense effects can be achieved with all-PS ODN in cell-based assay systems when ODN are formulated at relatively low concentrations (e.g., at 0.1 to 1 μM) with uptake enhancers.[45] At a higher concentration (e.g., >10 μM), all-PS ODN often cause non-antisense effects. In contrast, minimally modified ODN have the advantage that they can often be used at higher concentrations (1 to 20 μM) without lipocationic

[44] S. M. Fennewald and R. F. Rando, *J. Biol. Chem.* **270**, 21718 (1995).
[45] C. F. Bennett, M. Y. Chiang, H. Chan, J. E. E. Shoemaker, and C. K. Mirabelli, *Mol. Pharmacol.* **41**, 1023 (1992).

uptake enhancers. This is especially advantageous when primary cells, which often suffer from treatment with lipocations, are used in the experiments.

An instructive example of this observation is the use of minimally modified antisense ODN against c-*cbl* and c-*src* to inhibit the expression of these targets in primary osteoblasts. Downregulation was successful at 10 μM ODN concentration without the use of uptake enhancers.[46] The anti c-*cbl* ODN was shown by Western blot analysis to specifically inhibit the expression of c-Cbl protein, whereas four control oligonucleotides (two different scrambled ODN, a sense ODN, and an inverted ODN) having the same number of PS linkages did not inhibit c-Cbl expression. The observed inhibition was shown to be target specific, as the expression of other proteins [actin, cortactin, and PI(3)K] remained unchanged. The inhibition of c-*cbl* expression correlated well with the inhibition of bone resorption. Similarly, a c-*src*-specific minimally modified ODN at 10 μM concentration inhibited c-Src kinase expression and the associated c-Cbl phosphorylation, but not c-Cbl protein expression, resulting in the inhibition of bone resorption, whereas the corresponding sense control oligonucleotide had no such effect.

In a further example, the expression of Bax could be specifically inhibited in primary cultures of sympathetic neurons by incubation with minimally modified antisense ODN in the absence of uptake enhancers.[47] The intensity of cytoplasmic Bax protein immunostaining was reduced significantly in neurons treated with *bax* antisense ODN at 1 μM concentration for 24 hr as compared to cultures treated with mismatch or random sequence ODN. *bax* antisense ODN-treated neurons survived when grown under suboptimal neuronal growth factor (NGF) levels, whereas control ODN-treated neurons underwent apoptosis.

Minimally modified ODN against c-*fos* were used in HT-22 cells, a subclone of the immortalized mouse hippocampal cell line HT-4, to show that c-Fos contributes to amyloid β-peptide-induced neurotoxicity.[48] ODN were added to the cells at 5 μM concentration without uptake enhancers. Because previous administration of an analogous fluorescence-labeled ODN revealed a relatively slow uptake by HT-22 cells, ODN treatment was started 12 hr before stimulation with amyloid β-peptide to cells. Pretreatment of cells with minimally modified c-*fos* antisense ODN inhibited

[46] S. Tanaka, M. Amling, L. Neff, A. Peyman, E. Uhlmann, J. B. Levy, and R. Baron, *Nature* **383,** 528 (1996).

[47] F. Gillardon, M. Zimmermann, E. Uhlmann, S. Krajewski, J. C. Reed, and L. Klimaschewski, *J. Neurosci. Res.* **43,** 726 (1996).

[48] F. Gillardon, T. Skutella, E. Uhlmann, F. Holsboer, M. Zimmermann, and C. Behl, *Brain Res.* **706,** 169 (1996).

amyloid β-peptide-induced c-Fos protein expression, whereas random sequence ODN did not show any inhibitory effect on c-Fos expression.

In another series of experiments, minimally modified antisense ODN targeted against the 5' terminus of bFGF mRNA were used in coronary smooth muscle cells.[42] In the absence of uptake enhancers at a concentration of 1.0 μM, the expression of intracellular and pericellular bFGF protein could be suppressed to about 50 and 80% of control values as judged by a highly specific enzyme immunoassay system. The inhibition was shown to be antisense specific as the corresponding minimally modified sense ODN did not affect the cellular and pericellular concentration of bFGF protein.

Autoradiographic studies showed depression of the *de novo* synthesis of bFGF by antisense ODN. [^{3}H]Leucine labeling in the presence of 0.5 μM antisense ODN resulted in a significantly lower amount of ^{3}H radioactivity incorporated into the newly synthesized bFGF than in the presence of sense ODN or under control conditions.

The specific antisense effect on the bFGF level and *de novo* synthesis of bFGF was associated with an inhibition of coronary smooth muscle cell proliferation as monitored by following the cell population growth curve over 4 days and by [^{3}H]thymidine incorporation. At about 50% inhibition at 0.5 μM, neither toxicity nor apoptosis of ODN-treated cells could be registered. The antisense ODN-mediated inhibition is time dependent. After one addition of 0.5 μM minimally modified antisense ODN the cells returned to a normal growth rate after 72–96 hr.[42]

Use of Minimally Modified Oligonucleotides in Presence of Uptake Enhancers

In order to achieve specific antisense inhibition with minimally modified ODN in permanent cell lines, the formulation of ODN with lipocationic uptake enhancers is usually essential. Minimally modified antisense ODN were applied successfully to the inhibition of a variety of targets, including c-*myc*,[49] c-*myb*, bFGF,[42] Ha-c-*ras*, c-*jun*, *junD*, integrin α V subunit,[50] VEGF, TNA-α,[39] and TNFRI.[38] The length of the minimally modified ODN was in the range of 11 to 21 nucleotides. In all investigated cases, inhibition was probably due to an antisense mechanism, as the corresponding protein levels were decreased on treatment with antisense ODN but not with the corresponding control ODN. When mRNA levels were investigated, down-

[49] A. Peyman, H. Ragg, T. Ulshofer, D. W. Will, and E. Uhlmann, *Nucleosides Nucleotides* **16,** 1215 (1997).
[50] I. Villanova, P. A. Townsend, E. Uhlmann, J. Knolle, A. Peyman, M. Amling, R. Baron, M. A. Horton, and A. Teti, *J. Bone Min. Res.,* in press.

regulation of the target protein was, in most instances, connected with a decrease in mRNA levels, suggesting a RNase H-mediated mechanism. In several experiments, the expression levels of nontargeted proteins were checked and found to be unaffected by ODN treatment.[38,46,50] Thus, expression of TNF receptor type I could be downregulated without affecting the expression of the type II receptor.[38] Similarly, minimally modified antisense ODN targeted to the αV subunit of the vitronectin receptor αVβ_3 specifically downregulated αV protein levels, whereas the β_3 protein remained unaltered.[50]

Procedure for Formulation of Minimally Modified Antisense Oligonucleotides with Lipocationic Uptake Enhancers

The antisense oligonucleotide (1.0 μmol) is dissolved in 190 μl OPTI-MEM (GIBCO-BRL, Gaithersburg, MD) by vortexing. Ten microliters of CellFectin (1 mg/ml) (GIBCO-BRL) is then added and mixed by short and gentle vortexing. For complex formation, this mixture is incubated for 20 to 30 min at ambient temperature. The resulting ODN/CellFectin complex is diluted with 800 μl OPTI-MEM to give a final concentration of 1 μM oligonucleotide formulated with 10 μg/ml CellFectin. The cells are incubated with the diluted ODN/CellFectin complex for 3–5 hr at 37° in the absence of serum. After incubation, the medium containing the ODN/CellFectin complex is replaced by serum containing medium. Depending on the cell type and target, the cells are incubated for another 24–48 hr until the inhibitory effect of antisense ODN on target gene expression is determined.

It is important that the ODN/CellFectin complex formation is performed at a relatively high ODN and CellFectin concentration as described earlier and that no serum is present during complex formation and transfection of the cells with the ODN/CellFectin complex. If the cells have to be stimulated by growth factors, the ODN/CellFectin complex has to be removed prior to this step in order to avoid nonspecific inhibition of stimulation by cell surface-bound ODN. Under certain circumstances, an additional washing step with medium may be necessary for quantitative removal of the ODN/CellFectin complex from the cell surface prior to stimulation. Similarly, cells are washed before lysis and determination of the inhibitory effect. The use of alternative uptake enhancers to CellFectin may give equivalent or superior results depending on the type of cells used. In most of our experiments, we have used a final concentration of 5–10 μg/ml CellFectin, which is not toxic to most cell lines. The final ODN concentration is usually in the range of 50 nM to 2 μM. In order to obtain optimal cellular uptake and antisense inhibition, the ODN/CellFectin ratio has to be optimized for any specific antisense ODN sequence.

Concluding Remarks

A series of studies in different cell-based assay systems have shown that minimally modified antisense ODN can be applied for the inhibition of protein expression of various targets, such as protooncogenes,[46,48,49] growth factors,[42] cytokines,[39] cytokine receptors,[38] and cell adhesion receptors.[50] The combination of PS end capping and PS protection at internal pyrimidine residues, which are the major sites of endonuclease degradation, confers sufficient nucleolytic stability to the minimally modified ODN to allow specific inhibition of gene expression. This strategy reduces the number of PS linkages required to make the ODN nuclease resistant. Concomitantly, the binding affinity to RNA and the cellular uptake of minimally modified ODN are enhanced relative to uniformly PS-modified ODN. This is why the biological activity of minimally modified antisense ODN is similar or superior to all-PS ODN, whereas non-antisense effects are reduced greatly. There are generally less non-antisense effects, such as undesired binding to proteins, when minimally PS-modified ODN are applied in cellular assay systems. This is especially true if relatively high ODN concentrations (>1 μM) are used, which are necessary for antisense inhibition without formulation of the ODN with lipocationic uptake enhancers. Initial experiments in animals show that minimally PS-modified ODN are also effective inhibitors *in vivo*.

Acknowledgments

We are grateful to Drs. R. Baron, M. Helsberg, F. Gillardon, B. Greiner, H. Ragg, R. Woessner, and A. Teti for excellent collaboration and to S. Hein, L. Hornung, S. Schluckebier, and C. Weiser for technical assistance.

Section II

Methods of Delivery

[16] Determination of Cellular Internalization of Fluoresceinated Oligonucleotides

By Lyuba Benimetskaya, John Tonkinson, and C. A. Stein

The power of antisense oligonucleotide technology lies in the ability of these compounds to bind specifically to a target mRNA via Watson–Crick base pair interactions and block the translation of that mRNA into protein. For this to be successful, the oligonucleotides must penetrate the cell membrane and achieve the appropriate concentration in the correct intracellular compartment, which may be either the cytoplasm or the nucleus, but is probably the latter.

It has long been known that this process actually can occur in intact cells. For example, Koch and Bishop[1] demonstrated that poliovirus mRNA could infect cells in tissue culture and that treatment with polycations enhanced infectivity. However, oligonucleotides unassisted by a carrier produce antisense activity only occasionally. Thus, Bergan et al.[2] showed that electroporation of U937 cells was required to increase the ability of an antisense oligonucleotide to inhibit MYC protein synthesis. Spiller et al.[3] and Giles et al.[4] permeabilized KYO1 human chronic myelogenous leukemia cells with streptolysin O to achieve sequence-specific cleavage of bcr-abl mRNA.

As described previously,[5] for antisense oligonucleotides to be effective, several criteria must be satisfied. One of the most important is that they must retain intact, both in vivo and in vitro, in the extra- and intracellular environment. Phosphodiester oligomers are highly unstable with respect to nuclease digestion. To be rendered nuclease resistant, oligonucleotides must be modified chemically. The most popular approach is the substitution of one of the nonbridging oxygen atoms at the phosphorus with sulfur or a methyl group to produce phosphorothioates[6] and methyl phosphonates,[7] respectively.

The cellular internalization of these oligonucleotides, especially phosphorothioates, is a subject of particular current interest. Several groups

---

[1] G. Koch and J. M. Bishop, Virology 35, 9 (1968).
[2] R. Bergan, Y. Connell, B. Fahmy, and L. Neckers, Nucleic Acids Res. 21, 3563 (1993).
[3] D. G. Spiller and D. M. Tidd, Antisense Res. Dev. 5, 13 (1995).
[4] R. V. Giles, D. G. Spiller, and D. M. Tidd, Antisense Res. Dev. 5, 23 (1995).
[5] C. Stein and Y. Cheng, Science 261, 1004 (1993).
[6] C. A. Stein, J. Tonkinson, and L. Yakubov, Pharmacol. Ther. 52, 365 (1991).
[7] P. Miller, Biotechnology 9, 358 (1991).

have investigated the cellular uptake of these molecules and their fate when they enter the cell.[8–17]

Phosphorothioates, which are available commercially and have entered clinical trials over the past several years, are polyanions and cannot diffuse passively across cell membranes. In fact, even uncharged methyl phosphonate oligonucleotides are too polar to partition into cell membranes. Surprisingly, however, the cellular uptake of phosphorothioate oligonucleotides occurs to a greater extent than would be expected on the basis of charge and size considerations.

A large number of experiments strongly suggest that the uptake of naked (i.e., without a carrier) oligonucleotides is an active process that requires an energy source derived from intermediary metabolism. The rate of cellular internalization can be decreased by metabolic inhibitors, such as deoxyglucose, cytochalasin B, and sodium azide.[18] The process (at least for phosphodiester oligonucleotides) is calcium[8] and temperature dependent[9,18,19,20] in many cell lines. Other parameters determining the rate of cellular internalization include the type of the cell and the medium in which uptake occurs, the class of oligonucleotide analog, its chain length and concentration, and/or the presence of linked groups.

At the present time, it is believed that the internalization of oligonucleotides depends predominantly on the active processive of adsorptive endocytosis and fluid phase endocytosis (pinocytosis).[18] The process of adsorptive endocytosis is suggested because charged oligonucleotides (i.e., phosphorodiesters and phosphorothioates) that adsorb well to the cell surface of

[8] S. Wu-Pong, T. L. Weiss, and C. A. Hunt, *Cell. Mol. Biol.* **40**, 843 (1994).

[9] R. M. Crooke, M. J. Graham, M. E. Cooke, and S. T. Crook, *J. Pharmacol. Exp. Ther.* **275**, 462 (1995).

[10] C. A. Stein, J. L. Tonkinson, L. M. Zhang, L. Yakubov, J. Gervasoni, R. Taub, and S. Rotenberg, *Biochemistry* **32**, 4855 (1993).

[11] A. M. Krieg, M. F. Gmelig, M. F. Gourley, W. J. Kisch, L. A. Chrisey, and A. D. Steinberg, *Antisense Res. Dev.* **1**, 161 (1991).

[12] W. Y. Gao, C. Storm, W. Egan, and Y. C. Cheng, *Mol. Pharmacol.* **43**, 45 (1993).

[13] W. Zhao, S. Matson, C. J. Herrara, E. Fisher, H. Yu, A. Waggoner, and A. M. Krieg, *Antisense Res. Dev.* **3**, 53 (1993).

[14] J. L. Tonkinson and C. A. Stein, *Nucleic Acids Res.* **22**, 4268 (1994).

[15] T. Iwanaga and P. C. Ferriola, *Biochem. Biophys. Res. Commun.* **191**, 1152 (1993).

[16] G. Marti, W. Egan, P. Noguchi, G. Zon, M. Matsukura, and S. Broder, *Antisense Res. Dev.* **2**, 27 (1992).

[17] R. Crooke, *Anti-Cancer Drug Design* **6**, 609 (1991).

[18] L. A. Yakubov, E. A. Deeva, V. F. Zarytova, E. I. Ivanova, A. S. Ryte, L. V. Yurchenko, and V. V. Vlassov, *Proc. Natl. Acad. Sci. U.S.A.* **8**, 6454 (1989).

[19] S. Wu-Pong, T. L. Weiss, and C. A. Hunt, *Pharm. Res.* **9**, 1010, (1992).

[20] S. L. Loke, C. A. Stein, X. H. Zhang, K. Mori, M. Nakanishi, C. Subasinghe, J. S. Cohen, and L. M. Neckers, *Proc. Natl. Acad. Sci. U.S.A.* **86**, 3474 (1989).

almost all cell types (with the exception of hematopoietic and especially T cells) are internalized to a much higher degree than uncharged species, such as methyl phosphonates or peptide nucleic acids.[10] Furthermore, other polyanions (e.g., pentosan polysulfate, suramin, and longer chain phosphorothioates such as SdC28, a 28-mer homopolymer of cytidine), which are used as competitors for the binding of charged oligomers to the cell surface, also inhibit internalization significantly. Vlassov et al.[21] have suggested that at a low oligonucleotide concentration (<0.5 μM), adsorptive endocytosis may be the primary mechanism of internalization. The critical role of adsorptive endocytosis in the internalization process is also suggested by studies of chemically modified oligonucleotides, such as 5'-cholesteryl-modified oligonucleotides.[22] This modification increases the adsorption of the oligonucleotide through a hydrophobic interaction of the cholesteryl moiety with the cell membrane; net oligomer internalization is also increased dramatically relative to nonmodified oligonucleotides. In addition, mitogen-treated lymphoid B and T cells, which have higher rates of membrane turnover than quiescent cells, also have higher rates of oligonucleotide internalization.[11]

The adsorptive process is strongly facilitated by cell surface heparin-binding proteins, which have been shown to have low nanomolar affinities for phosphorothioate oligomers. An example of such a protein is Mac-1 (CD11/CD18),[23] a lineage-restricted molecule, and others are just now beginning to be identified. The binding of a phosphorothioate oligonucleotide to such a protein is typically strongly oligomer length dependent but only weakly oligomer sequence dependent and appears to be totally independent of the sense of chirality at phosphorus.[24] An exception to the sequence dependency generality is found in oligomers containing the G-tetrad (G-quartet) motif if it is located near (within about two or three bases) either the 5' or the 3' molecular terminus. In this case, G-quartets and tetraplexes may form, which may bind to heparin-binding proteins with much higher affinity than the corresponding monomer.[25]

Fluid-phase pinocytosis is an additional mechanism of internalization that operates at a relatively high oligonucleotide concentration.[10,18] This is

[21] V. V. Vlassov, L. A. Balakireva, and L. A. Yakubov, *Biochim. Biophys. Acta* **1197**, 95 (1994).

[22] A. M. Krieg, J. Tonkinson, S. Matson, Q. Zhao, M. Saxon, L. Zhang, U. Bhanja, L. Yakubov, and C. A. Stein, *Proc. Natl. Acad. Sci. U.S.A* **90**, 1048 (1993).

[23] L. Benimetskaya, J. Loike, Z. Khaled, G. Loike, S. C. Silverstein, L. Cao, J. Khoury, T. Cai, and C. A. Stein, *Nature Med.* **3**, 414 (1997).

[24] L. Benimetskaya, J. Tonkinson, M. Koziolkiewicz, B. Karwowski, P. Guga, R. Zeltser, W. Stec, and C. A. Stein, *Nucleic Acids Res.* **23**, 4239 (1995).

[25] L. Benimetskaya, M. Berton, A. Kolbanovsky, S. Benimetsky, and C. A. Stein, *Nucleic Acids Res.* **25**, 2648 (1997).

the process by which cells constitutively engulf water and dissolved solute from the bulk or fluid phase. Pinocytosis occurs in a concentration- and time-dependent manner. Although this process is inefficient, it probably accounts for a high percentage of net oligonucleotide internalization as the fluid-phase concentration of oligomer increases. In hepatoma cells, Gao *et al.*[12] found that pinocytosis contributed 30% of the net uptake of a phosphorothioate oligomer. In all cases, however, oligonucleotides appear to enter an endosomal compartment regardless of the precise mechanism of internalization.

For internalization studies, two kinds of labeled oligomers have been employed by different researchers: 5'-^{32}P-labeled and fluorescent end labeled. The general opinion is that for the optimal determination of internalization, radiolabeled oligomers should not be used. This is because ubiquitous cell surface alkaline phosphatase activity will cleave the label, which may be internalized without the oligonucleotide. In our opinion, internalization experiments are best performed with fluorescent oligonucleotides using fluorescence-activating cell-sorting (FACS) analysis. The use of flow cytometry allows elimination of dead cells from the final analysis on the basis of propidium iodide uptake and by light-scattering parameters. This is important because nonviable cells internalize significantly more oligonucleotides than viable ones.[10,13,26] This method also allows thousands of individual cells to be analyzed rapidly and multiple populations to be distinguished. However, the attachment of a fluorescently labeled moiety to an oligonucleotide may, in theory, alter its binding properties compared to the unlabeled molecule. While this is certainly true for extremely hydrophobic molecules such as cholesterol, which will insert in the cell membrane, there is no evidence that less hydrophobic molecules such as fluorescein create similar artifacts; indeed, available data suggest otherwise.[10]

In order to determine the relative amount of internalized oligomer, it must be distinguished from cell surface, nonspecifically bound material. A simple method to do this is based on the class and length dependency of oligonucleotide binding to the cell surface.[10] It has been shown that the K_d of SdC28 bound to high-affinity-binding sites on the cell surface of HL60 cells (3 nM) is about one order of magnitude lower than that of a fluorescently labeled 15 base thymidine homopolymer phosphodiester (F-OdT15; 22 nM). Thus, by treating the washed cells briefly with a 5 μM solution of SdC28 for 2–5 min, all of the cell surface-bound oligomer can be removed. The procedure is rapid, nontoxic, and virtually 100% effective. The nonremovable fraction can be taken as being internalized material. This method thus provides an advantage over DNase digestion of cells (to remove ad-

---

[26] R. Juliano and E. Mayhew, *Exp. Cell. Res.* **73**, 3 (1972).

hered nucleic acids) or washing the cells in high salt/low pH buffer, such as glycine, pH 2.0.

Oligonucleotide Internalization and Compartmentalization

The bulk of oligonucleotide that has been internalized by either adsorptive endocytosis or pinocytosis is trapped, at least initially, in an intracellular vesicle. Tonkinson and Stein[14] studied the compartmentalization of fluorescently labeled phosphodiester and phosphorothioate oligodeoxynucleotides in HL60 cells. The authors used the phenomenon of pH quenching of fluorescein fluorescence to determine the intracellular compartmental localization and pH at which the oligomer resided. After loading the cells for 6 hr (approximating a near steady-state condition) and stripping of cell surface-bound fluorescence with SdC28, the total internalized fluorescence was determined by flow cytometry. The sodium ionophore monensin was then added to break down the pH gradient between the endosome/lysosome and the cytoplasm. As demonstrated in Fig. 1, a dramatic increase in fluorescein fluorescence was observed after monensin treatment in cells incubated with phosphorothioate and phosphorodithioate but not phosphodiester oligomers. This increase was due to the abrogation of the low pH quenching of fluorescein fluorescence in the presence of monensin. Furthermore, the site of the pH quenching was the endosome or lysosome. This was demonstrated by treatment of the cells with bafilomycin, an antibiotic that specifically inhibits the proton-pumping ability of the H^+-ATPase, the enzyme that acidifies the lumen of the endosome. In the presence of bafilomycin, the signal from FSdT15 was equivalent to that produced by the monensin treatment described earlier, indicating that no acidification took place (Fig. 1). The fact that monensin treatment did not produce a more intense signal than bafilomycin treatment alone indicated that all of the acidification was occurring in a bafilomycin-sensitive location, i.e., the endosome or lysosome.

Net uptake represents the difference between uptake and efflux. Oligonucleotides were found to undergo significant efflux as well as influx from HL60 cells (Fig. 2), which is most likely actually occurring predominantly from the vesicular structures (i.e., endosomes/lysosomes). The rate of efflux, for all classes of charged oligodeoxynucleotide studied, was best described by Eq. (1):

$$C_T = Ae^{-\alpha t} + Be^{-\beta t} \tag{1}$$

where C_T is the amount of oligomer remaining internalized in the cell at any time T, α and β are the rate constants of efflux, and $A + B = 100\%$. Each exponential component of the sum in Eq. (1) is also referred to as a

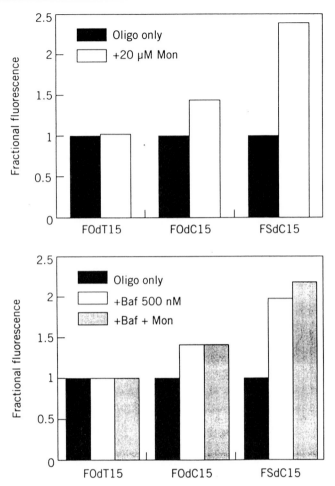

Fɪɢ. 1. Oligodeoxynucleotides are acidified depending on their class. (Top) Cells were incubated with oligomer and washed, and intracellular fluorescence was determined by flow cytometry. Monensin was then added (20 μM) to each sample and fluorescence was measured again. (Bottom) The macrolide antibiotic, bafilomycin (500 nM), was added to the incubation medium and fluorescence was determined in the presence or absence of monensin. These results indicate that acidification occurred in a bafilomycin-sensitive compartment (i.e., endosomes/lysosomes). Twelve fluoresceinated phosphodiester oligonucleotides of various sequence were studied. None were found to reside in an acidic compartment. The single exception was FOdC15, which, for uncertain reasons, behaved more like a typical phosphorothioate than a phosphodiester.

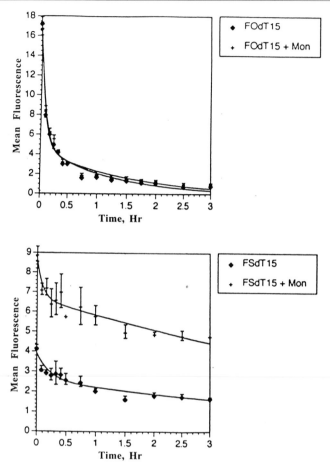

FIG. 2. Efflux of phosphodiester and phosphorothioate oligodeoxynucleotides from HL60 cells. Cells were preloaded by incubation with 5 μM FITC-OdT15 or 5 μM FITC-SdT15. Loaded cells were then placed in oligonucleotide-deficient medium and the efflux was monitored by flow cytometry. Following the last time point, cells were reloaded with tagged oligonucleotides, and efflux experiments were repeated in the presence of 20 μM monensin. Each data point is the mean $\pm$ SD of at least three experiments. Curves are a best fit to the biexponential function $C_T = Ae^{-\alpha t} + Be^{-\beta t}$.

"compartment" (although this mathematical description does not assign an actual cellular structure to each "compartment," they are in fact endosomes/lysosomes).

The efflux behavior of 12 phosphodiester oligomers in HL60 cells was evaluated.[14] These compounds predominately entered a "shallow" (relatively rapid efflux; short $t_{1/2}$) compartment. The value of $A = 61 \pm 4\%$; the value of α is 10 hr^{-1}. About 36% of the oligonucleotide enters the B, or deep compartment [relatively slow efflux, long $t_{1/2}$ ($\beta = 0.329$ hr^{-1})]. In sharp contrast, for three phosphorothioate oligonucleotides studied (15- to 28-mer), the situation was reversed. Only 18% entered the shallow (A) compartment, whereas 80% entered the deep (B) compartment ($\alpha = 3.5$ hr^{-1}; $\beta = 0.131$ hr^{-1}). Acidification of the phosphorothioate compounds occurred in the deep compartment, where the average pH were approximately 6.0 and 5.5, respectively. In contrast, phosphodiester oligonucleotides were not acidified (pH 7.2). Similar efflux data were obtained with rhodamine-labeled oligonucleotides, indicating that it was probably not the fluorescein group that was responsible for the efflux properties.

Experimental

Synthesis of 5'-Fluorescein and 5'-Rhodamine-Labeled Oligonucleotides

Oligonucleotides used for flow cytometry experiments were synthesized by standard phosphoramidite chemistry on an Applied Biosystems 380B DNA synthesizer. A 5' free amino group was added via the Aminolink 2 (Applied Biosystems) reagent. After cleavage from the controlled pore glass support by aqueous ammonia and base deblocked in concentrated ammonia at 60° for 8 hr, the sample was lyophilized. To the sample of 5'-amino-oligonucleotide, add 400 μl of sodium carbonate, pH 9.5, and 2 mg of fluorescein isothiocyanate (FITC) dissolved in 80 μl of dimethyl sulfoxide (DMSO). Allow the mixture to sit at room temperature for 16 hr in the dark. For oligomers greater than 10-mer in length, dialyze the samples (1000 MW cutoff) against 0.5 M NaCl, 1 mM EDTA for two changes, purify by HPLC (PRP-1 support, 0.1 M TEAB, from triethylamine and solid carbon dioxide, pH 8/acetonitrile, increase acetonitrile 1%/min to 25 min). Purify the oligomer by electrophoresis on 7 M urea/20% polyacrylamide gel. Excise with a scalpel the yellow, fluorescein-labeled oligomer band, which characteristically appears between the bromophenol blue and xylene cyanole markers, crush, and soak in 0.1 M TEAB overnight in the dark at 4°. Pass the eluate through a column of Sephadex G-25 to remove the last traces of free fluorescein and gel material. Dissolve the final product in a minimum volume of water and precipitate with 2% lithium perchlorate/acetone. Wash the pellet with cold acetone and reprecipitate with 3 M NaCl/ethanol. For oligomers <10-mer in length, omit the dialysis step from the purification procedure. Determine oligomer concentrations by UV spectroscopy and store in buffer (e.g., Tris–EDTA), pH 7.5–8, in the dark. Fluorescein-labeled oligonucleotides can also be synthesized using fluorescein phosphoramidite reagents available from several commercial sources. However, the amount of material obtained may be limited and costly.

Determination of Cellular Internalization by Flow Cytometric Analysis

In order to follow the internalization of fluorescent oligonucleotides, we use fluorescence-activated cell sorting (FACS).[14,23] A dual-laser flow cytometer (Becton-Dickinson FACS Calibur) with Cell Quest software is employed. Typically, flow experiments are designed as follows.

1. Plate log-phase growth cells in complete media (10^5 cells/200 μl) in 96-well microtiter plates.
2. Incubate the cells for 12–16 hr at 37°.
3. Treat the cells with oligonucleotide for the specified time. Each well represents a separate treatment and thus a distinct population. If one time point is being studied, such as a 6-hr internalization, then treat all populations at the same time. If multiple time points are being studied, such as in the determination of an internalization or an efflux function, stagger the treatments such that the longer incubation times are dosed first and the short times dosed last. Using this type of treatment, all populations can be analyzed at the same time.
4. Following the appropriate incubation times, remove media by centrifugation.
5. Add cold PBS (200 μl) containing 2.5% bovine serum albumin (PBS/BSA) with 0.1–5 μM SdC28 (purchased commercially; the concentration depends on the length and class of the fluoresceinated molecule) to each well to remove all fluoresceinated cell surface-bound oligonucleotide.
6. Allow the cells to sit for 2 min on ice, centrifuge, and remove the supernatant.
7. Wash the cells twice with cold PBS/BSA and resuspend in 300 μl PBS/BSA containing 0.3 μg/ml propidium iodide (PI).
8. Determine the relative mean fluorescence intensities for a population of 5000–10,000 live cells for each time point using a 256 log channel amplifier; subtract the background cellular autofluorescence (which is usually <10% of the measured fluorescence).
9. Following this analysis, add monensin (20 μM final concentration, in PBS) to each tube and allow to sit on ice for 10 min. Determine again the mean fluorescence intensity for a population of 5000 cells. Gate dead cells from the collection using a live gate for PI-positive cells.

10. Determine mean fluorescence from the average of triplicate samples and calculate the standard deviation among replicates. There should usually should be less than a 5% difference between the triplicates.

Efflux of Fluorescent Oligonucleotides from Cells

Plate the cells at a density of 5×10^5/ml in 96-well microtiter plates and incubate at least 16 hr. Add fluorescent oligomers to the cells at a final concentration of 5 μM. Following a cell type-dependent load time to achieve a near steady-state accumulation of oligomer, remove the media by centrifugation and add 200 μl cold PBS/BSA containing 5 μM SdC28 to each well. Allow the cells to sit 2 min on ice. After centrifugation, resuspend the cells in 200 μl prewarmed media and place in the incubator for the stated efflux times (5 min–3 hr). Following efflux, wash the cells twice in cold PBS/BSA, resuspend in 200 μl PBS/BSA, and add to 100 μl PBS/BSA with 0.3 μg/ml propidium iodide. Determine the mean fluorescence of each treatment by FACS in the absence and presence of monensin, as described earlier. The loss of fluorescent signal as a function of time can be due to a loss of internalized fluorescent oligomer or to the entry of a fluorescent oligomer into an acidic environment. Measuring the signal in the presence of monensin ensures that a true measure of the rate of exocytosis has been determined. Plot the mean fluorescent values determined for each time point as a function of time.

Fit data by DeltaGraph Professional or other commercially available curve-fitting program to the biexponential function:

$$C_T = Ae^{-\alpha t} + Be^{-\beta t}$$

Determine the parameters by multiple iterations so as to minimize the residuals, the only restriction being A, B, α $\beta > 0$.

Determination of Resident Intracellular pH of Fluorescent Oligonucleotides

This method works best with nuclease-resistant phosphorothioate oligonucleotides. Determination of the average pH of an intracellular fluorescent oligonucleotide is performed by a spectrophotometric method based on the variation in emission intensity (I_{em}) of fluorescein as a function of pH. Measurement of the I_{em} of fluorescein is made after irradiation at two wavelengths: one that produces a pH-dependent emission (490 nm) and one that produces a pH-independent emission (450 nm). The ratio of the intensities of the two emissions (I_{490}/I_{450}) can then be compared to a standard curve of I_{490}/I_{450} vs pH. The standard curve is generated by placing cells containing the 5'-fluorescein oligonucleotide in nigericin-containing, high K^+ buffers of various pH. [Nigericin is an ionophore that equilibrates the intracellular (and intraendosomal) pH with the extracellular environment.]

Incubate 2×10^6 cells with fluorescein-labeled oligonucleotide for 6 hr. Wash the cells with cold PBS. Add 5 μM SdC28 in cold PBS for 2 min and then wash twice with PBS. Place subpopulations of 5×10^4 cells in 3 ml of 132 mM K^+/20 μM nigericin-containing buffers of various pH: 10 mM MES, pH 5.5, 6.0, and 6.5; 10 mM HEPES, pH 7 and 7.5 (for the calibration curves); or in PBS or Hanks' balanced salt solution (HBSS) with HEPES buffer, pH 7.1 (for pH determination of the oligonucleotide). Determine the fluorescent emission intensity from each population by fluorescence spectrophotometry, using the excitation spectrum setting. Collect emitted light >515 nm using a cutoff filter. (Correct variations in lamp intensity with a rhodamine standard.) Determine the intensity of light emission at excitation wavelengths of 450 and 490 nm, and plot the 490/450 ratio as a function of pH. After the standard curve has been determined, perform the identical experiments with the cells treated at physiologic pH in either PBS or HBSS/HEPES. Compare the 490/450 ratio to the standard curve, as this will provide a high estimate of the resident intracellular pH of the oligonucleotides.

[17] Intrabody Tissue-Specific Delivery of Antisense Conjugates in Animals: Ligand–Linker–Antisense Oligomer Conjugates

By Robert J. Duff, Scott F. Deamond, Clinton Roby, Yuanzhong Zhou, and Paul O. P. Ts'o

Introduction

The selective inhibition of gene expression through specific oligonucleotide binding to key mRNA target sequences is the goal of antisense strategies. However, from the beginning, antisense strategies have faced several obstacles, such as *in vivo* stability of the oligonucleotide, cellular uptake, efficiency of hybridization of the antisense agent to the mRNA target, selectivity of oligonucleotide binding, and inhibition of gene expression.[1] Of these factors, the low and nonselective cellular uptake of antisense

---

[1] J. S. Cohen, in "Oligodeoxynucleotides: Antisense Inhibitors of Gene Expression" (J. S. Cohen, ed.), p. 1. CRC Press, Boca Raton, FL, 1989.

oligonucleotides has hindered its therapeutic usefulness most severely. Certainly then, a versatile method to specifically and efficiently deliver an antisense oligonucleotide to the intracellular medium of a particular cell within a single organ would have a pronounced effect on the realization of effective antisense therapies.

This article describes our approach for the ligand-directed delivery of antisense oligonucleotides. Paramount to this approach was the design and synthesis of a molecular scaffold symbolized as "A-L-P" with "A" representing an unique ligand, specific for a receptor on the surface of the target cell. The "P" represents the "payload" portion, typically an oligonucleotide, which is linked uniquely to the ligand through the linker (the "L" portion). The general construct is shown in Fig. 1. The methods here describe A-L-P neoglycoconjugate formation, the cellular uptake of these ligand–linker–oligonucleotide conjugates via receptor-mediated endocytosis, their *in vivo* biodistribution, and their effectiveness toward the inhibition of expression of an integrated hepatitis B virus (HBV).

Principle and Advantage of Method

The principle of this method is the strategy of the A-L-P delivery system through which the ligand-specific, receptor-mediated endocytosis greatly facilitates the cellular uptake and delivery of exogenous DNA and antisense oligonucleotides into the intracellular medium. Ligand-directed, receptor-mediated endocytosis has been used previously in conjunction with complex, chemically undefined and structurally heterogeneous glycoconjugates to achieve increased intracellular concentrations. Wu and Wu[2] were the first to demonstrate that exogenous genes or oligonucleotides (negatively charged) complexed electrostatically to a poly(L-lysine) (positively charged)-linked asialoorosomucoid were efficiently and specifically taken into human hepatocellular carcinoma (HepG2) cells through direct interaction with the asialoglycoprotein (ASGP) receptor. Since then, other examples of receptor-mediated delivery of DNA have appeared, including a tetra-antennary galactose neoglycopeptide–poly(L-lysine) conjugate,[3] folate conjugates,[4,5] and 6-phosphomannosylated protein linked to an antisense oligonucleotide via a disulfide bond.[6] Another study used human

[2] G. Y. Wu and C. H. Wu, *J. Biol. Chem.* **262**, 4429 (1987).
[3] C. Plank, K. Zatloukal, M. Cotten, K. Mechtler, and E. Wagner, *Bioconj. Chem.* **3**, 533 (1992).
[4] B. A. Kamen, M.-T. Wang, A. J. Streckfuss, X. Peryea, and R. G. W. Anderson, *J. Biol. Chem.* **263**, 13602 (1988).
[5] C. P. Leamon and P. S. Low, *Proc. Natl. Acad. Sci. U.S.A.* **88**, 5572 (1991).
[6] E. Bonfils, C. Dupierreux, P. Midoux, N. T. Thuong, M. Monsigny, and A. Roche, *Nucleic Acids Res.* **20**, 4621 (1992).

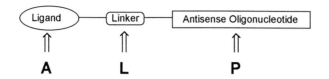

Conjugate	Ligand	Linker	Oligonucleotide	Radiolabel
1	YEE(ahGalNAc)$_3$	SMCC-AET	$^{5'}$U^m<u>TTTTTTT</u>	^{32}P
1b	-	H$_3$N$^+$CH$_2$CH$_2$NH-	$^{5'}$pU^m<u>TTTTTTT</u>	^{32}P
1c	YEE(ah)$_3$	SMCC-AET	$^{5'}$U^m<u>TTTTTTT</u>	^{32}P
1d	Y	SMCC-AET	$^{5'}$U^m<u>TTTTTTT</u>	^{32}P
1e	YEE(ahGalNAc)$_2$	SMCC-AET	$^{5'}$U^m<u>TTTTTTT</u>	^{32}P
1f	YEE(ahGalNAc)$_3$	SMCC-AET	$^{5'}$pU^m	^{32}P
1g	-	-	$^{5'}$pU^m<u>TTTTTTT</u>	^{32}P
2	YEE(ahGalNAc)$_3$	SMCC-AHT	$^{5'}$GTTCTCCATGTTCAG	^{35}S
3	YEE(ahGalNAc)$_3$	SMCC-AHT	$^{5'}$AAAGCCACCCAAGGCA	-
4	YEE(ahGalNAc)$_3$	SMCC-AHT	$^{5'}$TGAGCTATGCACATTCAGATTT	-

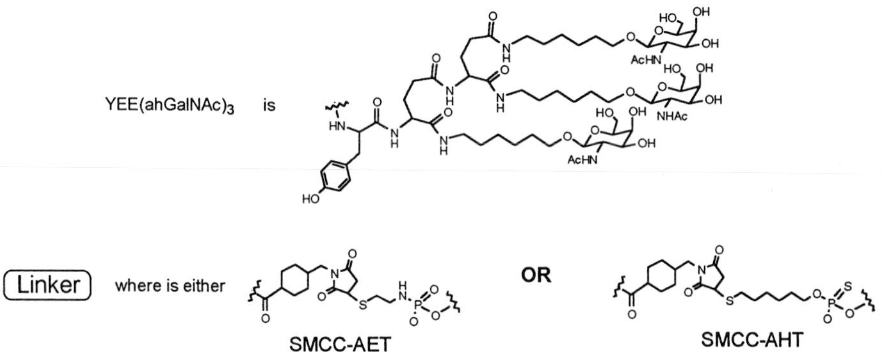

FIG. 1. Structures of neoglycoconjugates (1–4). AET and AHT are aminoethylthiol and aminohexylthiol moieties, respectively; underlined bases represent methyl phosphonate linkages. U^m represents a uridine 2'-deoxy-2'-methoxyribose unit.

serum albumin, which was linked to the triantennary, N-acetylgalactosamine neoglycopeptide, YEE(ahGalNAc)$_3$-modified poly(L-lysine), to deliver DNA into cells.[7] Hangeland et al.[8] found that the enhanced cellular uptake of the antisense oligonucleotide could be achieved if YEE(ahGalNAc)$_3$, a glycotripeptide, known to bind to Gal/GalNAc

[7] J. R. Merwin, G. S. Noell, W. C. Thomas, H. C. Chion, M. E. De Rome, T. D. McKee, G. L. Spitalny, and M. A. Findeis, Bioconj. Chem. 5, 612 (1994).
[8] J. J. Hangeland, J. T. Levis, Y. C. Lee, and P. O. P. Ts'o, Bioconj. Chem. 6, 695 (1995).

receptor sites on hepatocytes,[9] was attached covalently to a nuclease-resistant antisense oligonucleotide. This ligand has been shown to have a very high affinity for the mammalian hepatic ASGP receptor[9] ($K_d = 7$ nM) and, therefore, would presumably deliver the complexed oligomer to the hepatocytes of the liver.

The A-L-P construction has provided several advantages that do benefit cellular uptake, *in vivo* organ localization, and bioefficacy. These advantages are (1) the low toxicity of the components of the conjugates, even at elevated levels, (2) the structural homogeneity of each starting material and product, and (3) the components of the A-L-P conjugate are in defined relative proportions with each other. The first advantage of the A-L-P method is the low toxicity of the components. In order for antisense therapeutic agents to be effective, their intracellular concentration must reach concentrations up to 100 μM.[10] Polycationic compounds [e.g., poly(L-lysine) and cationic peptides] are known to be toxic at certain concentrations.[11,12]

The second advantage is that the components of the A-L-P construct are characterized readily and extensively by standard biochemical and chemical methods, i.e., mass spectrometry and ^{1}H nuclear magnetic resonance (NMR) spectroscopy, to a unique and unambiguous structure. This is in contrast to human and bovine serum albumins or poly(L-lysine)-derived molecular scaffolds. These large polymeric molecules are difficult to fully define due to their indeterminate physical and molecular properties. The consequences, therefore, are that products resulting from these materials will also be difficult to characterize fully. For example, poly(L-lysine) exists in varying degrees of polymerization (from 16 to 430 residues)[13] and so the products of poly(L-lysine) derivatization will be a family of compounds of varying degrees of functionalization. The defined properties of the A-L-P delivery system method do indeed represent an advantage.

The final advantage would be the 1 : 1 ratio of the structural components associated within the A-L-P system. To be more specific, it can be established unequivocally that there is a *single* ligand moiety attached to *one* oligonucleotide portion through a *unique* linker. The ligand is attached to the linker by an amide bond, and the oligonucleotide is joined covalently to the linker through a stable thioether bond. The chemistry used here enabled this linker to make the connection between the ligand and the

[9] R. T. Lee and Y. C. Lee, *Glycoconj. J.* **4**, 317 (1987).
[10] C. H. Agris, K. R. Blake, P. S. Miller, M. P. Reddy, and P. O. P. Ts'o, *Biochemistry* **25**, 6268.
[11] B. Mauersberger, C. U. Mickiwitz, J. Zipper, J. Axt, and G. Heder, *Exp. Pathol.* **13**, 268 (1977).
[12] C. L. Bashford, G. M. Alder, G. Menestrina, K. J. Mickelm, J. J. Murphy, and C. A. Pasternak, *J. Biol. Chem.* **261**, 9300 (1986).
[13] J. G. Tatake, M. M. Knapp, and C. Ressler, *Bioconj. Chem.* **2**, 124 (1991).

oligonucleotide in a regio- and site-specific fashion yielding one product. Therefore, the active compound is the only molecular species in solution and should lead to a more specific inhibition independent of carrier effects. As a result, there is no need to determine the empirical ratios of the components within the A-L-P system. In other conjugates, the ratio of oligonucleotides, or DNA, to cationic conjugate must be determined empirically in each case. Furthermore, these conjugates present difficulties in the formulation of therapeutic applications due to the need to perform these empirical calculations. In summary, the A-L-P neoglycoconjugates used in the present experiments are synthetic, chemically defined, structurally homogeneous, unique chemical entities.[8]

Materials and Reagents

Reagents and Buffers

YEE(ahGalNAc)$_3$ (a generous gift from Dr. Y. C. Lee, Johns Hopkins University, Baltimore, MD)

Bacteriophage T4 polynucleotide kinase (Life Technologies, Grand Island, NY)

Dulbecco's phosphate-buffered saline, pH 7.2 (Mediatech, Sterling, VA)

RPMI + 2% and +4% fetal calf serum (Mediatech)

Silicon oil (Nye Lubricants Inc., New Bedford, MA)

Trypsin/EDTA (0.05% trypsin: 0.53 mm EDTA; Mediatech)

[α-^{35}S]dATP, [γ-^{32}P]dATP, and [α-^{32}P]dCTP (Amersham, Piscataway, NJ)

1-Ethyl-3-[(dimethylamino)propyl]carbodiimide (EDAC) (Aldrich Chemical Co., St. Louis, MO)

N-Hydroxysuccinimidyl 4-(N-methylmaleimido)cyclohexane 1-carboxylate (SMCC)(Pierce, Rockford, IL)

Lysis buffer (0.5% NP-40, 100 mM NaCl, 14 mM Tris–Cl, 30% CH$_3$CN)

N-Acetylglucosamidase and chymotrypsin (Sigma, St. Louis, MO)

Solvable tissue solubilizer (Life Technologies)

Anhydrous ethyl ether (J.T. Baker, Sanford, ME)

30% acrylamide/7 M urea

1× TBE/7 M urea (89 mM Tris–HCl, 89 mM boric acid, 0.002 M EDTA)

Formamide-loading buffer [0.2% bromphenol blue, 0.2% xylene cylanol(w/v) in 90% formamide, 10% 1× TBE]

A purified 3.2 kb EcoRI HBV fragment (a generous gift of Dr. B. Korba)

G418 (Life Technologies)
Ausyme monoclonal HbsAG kit (Abbott Laboratories, Napierville, IL)
HBsAG standard (Chemicon, Temacia, CA)
Nick translation kit (Life Technologies)
Probequant microspin columns (Pharmacia Biotech, Piscataway, NJ)
1 N sodium hydroxide/10× SSC
0.4 M Tris–HCl
Hybridization solution (6× SSC, 5× Denhardt's solution, 50% formamide, 0.5% SDS, 10% dextran sulfate)
Formula 989 (Amersham)

Equipment

SepPak columns (Waters, Milford, MA)
48-well tissue culture-treated plates (Costar, Cambridge, MA)
Coulter counter Model ZBI (Coulter Electronics, Hialeah, FL)
Humidified incubator-Forma 3052 (Forma Scientific, Marietta, OH)
Polytron homogenizer Model PCU 2-110 (Brinkman Inst., Westbury, NY)
Centricon filters: 30,000 molecular weight cutoff (Millipore, Bedford, MA)
Vertical gel apparatus (Hoefer Pharmacia Biotech, San Francisco, CA)
Power supply (ISCO, Lincoln, NE)
GeneScreen plus nylon membranes (NEN, Boston, MA)
BioTek EIA plate reader; wavelength 492 nm (BioTek, Burlington, VT)
Fujix Bas 1000 phosphoimager (Fuji Medical Systems, Stamford, CT)
Packard Tricarb 1900TR scintillation counter

Animals

CD-1 male mice 22–35 g in weight (Charles River, Wilmington, MA)

Methods

All procedures described are adapted from previous works.[8,14,15]

---

[14] J. J. Hangeland, J. E. Flesher, S. F. Deamond, Y. C. Lee, and P. O. P. Ts'o, *Antisense Nucleic Acid Drug Dev.* **7,** 141 (1997).
[15] S. F. Deamond, R. J. Duff, C. Roby, Y. Zhou, and P.O.P. Ts'o, submitted for publication.

Syntheses of [³²P]-[YEE(ahGalNAc)₃-SMCC-AET pUᵐpT₇](1), [³²P]-
[EDA-pUᵐpT₇](1b), and [³²P]-[YEE(ah)₃-SMCC-AET-pUᵐpUT₇](1c)

The parent oligomer, ³²P-end-labeled pUᵐpT₇, is synthesized and purified according to established procedures.[16,17] The ligand, YEE(ahGalNAc)₃, is synthesized and purified as described previously[9,18] and stored at −20° as an aqueous solution. The radiolabeled conjugate **1** is synthesized and characterized as described.[8] Briefly, UᵐpT₇(16 nmol) is 5′ end labeled using 0.95 equivalents of unlabeled ATP, 286 μCi of [γ-³²P]ATP, and bacteriophage T4 polynucleotide kinase.[19] The reaction is lyophilized and redissolved in 0.2 *M* 1-methylimidazole, pH 7.0, and treated with 1.0 *M* cystamine hydrochloride, pH 7.2, containing 0.3 *M* 1-ethyl-3-[(dimethylamino)propyl]carbodiimide.[20] After heating at 50° for 2 hr, excess reagents are removed using a SepPak (Waters). The crude cystamine adduct is redissolved in 10 m*M* sodium phosphate, pH 8.0, containing 50 m*M* dithiothreitol (DTT) and heated to 37° for 30 min. The excess reductant is removed by SepPak (Waters), and the crude radiolabeled methyl phosphonate oligonucleoside, modified to contain a free thiol at the 5′ end, is dried *in vacuo*. In a separate vessel, the ligand, YEE(ahGalNAc)₃ (160 nmol), is dissolved in anhydrous DMSO and treated with one equivalent each of anhydrous diethylisopropylamine and SMCC. The final step of the synthesis is carried out by adding the contents of the second reaction [YEE(ahGalNAc)₃-SMCC] to the modified methyl phosphonate oligonucleoside and evaporating DMSO slowly under vacuum overnight. Following purification by polyacrylamide gel electrophoresis (PAGE), conjugate **1** is isolated in 14% yield based on the starting methyl phosphonate oligonucleoside and has a specific activity of 17.9 Ci/mmol. Conjugate **1b** is synthesized by mixing the ³²P-labeled pUᵐpT₇ with 0.1 *M* EDAC[20] in a buffer containing 0.1 *M* imidazole (pH 7.0) at 37° for 2 hr, followed by overnight incubation with an aqueous solution of 0.3 *M* ethylenediamine hydrochloride (pH 7.0). Excess reagents are removed using a SepPak (Waters). Conjugate **1c** (Fig. 1) is obtained in 61% yield following treatment

[16] P. S. Miller, C. D. Cushman, and J. T. Levis, *in* "Synthesis of Oligo-2′-deoxyribonucleoside Methylphosphonates: Oligonucleotides and Analogues. A Practical Approach" (E. Eckstein, ed.), p. 137. IRL Press, Oxford, 1991.

[17] R. I. Hogrefe, M. A. Reynolds, M. M. Vaghefi, K. M. Yang, K. M. Riley, R. E. Klem, and L. T. Arnold, Jr., *in* "Methods of Molecular Biology" (S. Agrawal, ed.), Vol. 20, p. 143. Humana Press, Totowa, New Jersey, 1993.

[18] Y. Oshumi, Y. Ichikawa, and Y. C. Lee, *Cell Technol.* **9**, 229 (1990).

[19] J. Sambrook, E. F. Fritsch, and T. Maniantis, *in* "Molecular Cloning: A Laboratory Handbook," 2nd ed. Cold Spring Harbor Laboratory Press, Cold Spring Harbor, NY, 1989.

[20] B. C. F. Chu, G. M. Wahl, and L. E. Orgel, *Nucleic Acids Res.* **11**, 6513 (1983).

of the full-length conjugate (700 pmol) with N-acetylglucosamidase (0.13 U; Sigma) in 50 mM sodium citrate, pH 5.0, at 37° for 1 hr.

Phosphorothioate Oligomer Synthesis and Ligand Conjugation

Phosphorothioate oligodeoxynucleotides are synthesized via automated methods using an ABI 392 DNA/RNA synthesizer with β-cyanoethyl phosphoramidites (Glen Research, Sterling, VA). An additional 5'-thiol modification, essential for ligand conjugation, is accomplished using the corresponding 5'-disulfide containing thiol modifier (Glen Research, Sterling, VA) in conjunction with the automated methods and reductive cleavage of this 5'-disulfide to provide thiol-modified oligonucleotide. All oligonucleotides are deprotected with concentrated ammonium hydroxide for 16–20 hr at 55°. Oligonucleotides are subsequently purified by PAGE using a 15% polyacrylamide, 7 M urea, and a Tris–borate EDTA buffer. The appropriate bands are excised, extracted with 50 mM sodium phosphate (pH 5.8), and desalted on SepPak columns (Waters).

Ligand conjugation for neoglycoconjugates **2–4** is carried out in two steps (experimental details are published elsewhere) (Fig. 2): (1) the 5'-terminal disulfide is reduced to a thiol by the action of DTT and (2) treatment of the thiol oligonucleotide with the ligand that was modified previously with SMCC. The final ligand–oligonucleotide conjugate is confirmed by PAGE by showing an altered mobility for the product and identified by mass spectrometry. The ^{35}S radiolabel on conjugates **2** and **3** and also unconjugated **2b** is installed at the 3'-terminal using the combined action of terminal deoxynucleotidyltransferase (Life Technologies) and [α-^{35}S]dATP (>1000 Ci/mmol) (Amersham).

Cell Culture

Human hepatoma cell line HepG2 [American Type Culture Collection (ATCC) Bethesda, MD] is maintained on RPMI + 10% fetal calf serum, and the HBV-transfected counterpart HepG2 2.2.15 (a gift of Dr. G. Wu) is maintained on RPMI + 4% fetal calf serum containing 4 mM glutamine (Mediatech) and incubated at 37°, 5% CO_2 in a humidified atmosphere. Cultures are refed two to three times/week. Cells are selected with G418 (Life Technologies) and reselected every two to three passages.

Cellular Uptake

Cellular uptake experiments are performed using both HepG2 cells and HepG2 2.2.15 cells as described previously.[8,15] For a typical experiment, cells are passaged into wells and grown in the appropriate medium (RMPI

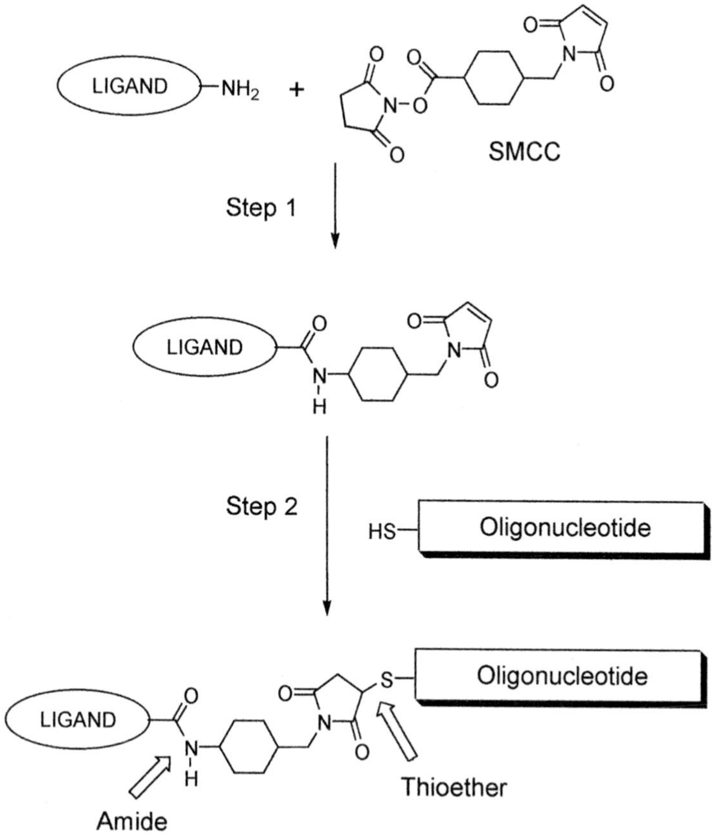

FIG. 2. Reaction scheme for the syntheses of YEE(ahGalNAc)₃ containing neoglycoconjugates (**1–4**).

1640) to a density of ca. 10^5 to 10^6 cells per well. The cells are then incubated at the appropriate temperature with medium that contains 2% fetal calf serum (FCS) and is made 1 μM containing the ^{35}S-labeled or ^{32}P-labeled conjugate. We examined the association of the radiolabeled ligand–oligonucleotides with the cell lines HepG2, HL-60, and HT 1080 (ATCC) using three different sets of conditions: (i) incubation at 37°, which allows cell surface binding and internalization of the conjugate; (ii) incubation at 4°, which prevents internalization, but permits cell surface binding; and (iii) incubation at 37°, followed by washing the cells with 50 mM EDTA in PBS, which depletes the carbohydrate receptor domain of calcium, causing the release of surface-bound ligand. In each case the cells are pelleted through silicon oil just prior to lysis in order to remove cell surface-associated radioactivity and then quantification of radioactivity by scintillation counting.

Tissue Distribution and Time Course of Clearance

Male CD-1 mice (Charles River), weighing 22–35 g, received a single injection via the tail vein of 7–30 pmol of $[^{32}P]$-[YEE(ahGalNAc)$_3$]-SMCC-AET-pU^m p$\underline{T}_7$ (1) or 7 pmol of $[^{32}P]$-[YEE(ah)$_3$]-SMCC-AET-pU^{m}p$\underline{T}_7$ (1c) contained in 0.2 ml of saline. A radioactive dose of ~0.5 μCi is administered to each mouse. The mass of dose injected varys with variation in the specific activity of the conjugate. Three mice per time point are injected. The experiment is repeated once. The mice are sacrificed by cervical dislocation at 15, 30, and 60 min and at 2, 4, 6, and 24 hr. Blood, liver, kidneys, spleen, muscle, upper and lower gastrointestinal tract, and feces are collected and weighed. Representative samples from these organs and tissues are weighed and placed in glass vials. In order to collect the urine (~2 hr postinjection), the external urethra of the mice is ligated under brief ether anesthesia and, after sacrifice, the bladders are removed and placed into glass vials. Solvable 2% NaOH (1 ml) is added to each sample. The samples are then placed on a slide warmer to be digested overnight, removed the next morning to cool, and decolorized with three to seven drops of H_2O_2 (30%, w/v). The samples are dissolved in a 10-ml Formula 989 (Amersham) scintillation cocktail, and the amount of radioactivity is determined by scintillation counting (Packard Tricarb 1900 TR; <3% error). Aliquots of the injected dose are counted to calculate the percentage accumulated dose per organ or tissue amount. Use of a standard aliquot of the injected dose during the counting procedure permits correction for radioactive decay. A quench correction is performed using the internal standard method supplied with the scintillation counter.

Analysis of Metabolites Isolated from HepG2 Cells

Approximately 10^5 HepG2 cells (ATCC) are incubated at 37° in RPMI supplemented with L-glutamine and 2% fetal calf serum (Life Technologies) containing 1 μM ^{32}P-labeled 1 for 2, 4, 8, 16 and 24 hr. Cultures are then washed twice with PBS, pelleted through silicon oil (Nye Lubricants), and lysed (0.5% NP-40, 100 mM NaCl, 14 mM Tris–HCl, pH 7.5, 30% CH_3CN). The lysate is extracted with 50% aqueous CH_3CN (v/v) twice. The extracts are lyophilized, redissolved in formamide-loading buffer, and analyzed by 15% polyacrylamide gel electrophoresis (2 V/cm, 30 min).

Gel analysis is performed on a 15% polyacrylamide gel (19:1 acrylamide:bisacrylamide). Gels are 16 cm × 18 cm × 0.75 mm in size and are run at 2 V/cm for 45 min. The buffer is 89 mM TBE (89 mM Tris–HCl, 89 mM boric acid, and 0.2 mM ethylenediamine tetraacetate) buffered at pH 8.0. The gel is subsequently autoradiographed at $-80°$ for 2 days using Kodak (Rochester, NY) X-OMAT X-ray film.

Analysis of in Vivo Conjugate Metabolism

Male CD-1 mice, weighing between 22 and 35 g, receive a single injection via the tail vein of ^{32}P-labeled [YEE(ahGalNAc)$_3$]-SMCC-AET-pU^{m}pT$_7$ (**1**). Animals are sacrificed after 15, 60, and 120 min. Livers and bladders are collected as before, placed into plastic vials, and frozen immediately ($-80°$). Samples of liver are thawed to $0°$ and weighed (average mass 0.25 g). The tissue is homogenized (Polytron PCU-2-110 tissue homogenizer) in 4 volumes of acetonitrile/water (1:1, v/v). Tissue debris is removed by centrifugation (10,000g, 20 min, $0°$; Sorval Model RC-5B refrigerated superspeed centrifuge). The supernatant is removed and the extraction procedure is repeated. Recovery is calculated as the percentage of the total radioactivity (i.e., the decolorized homogenate) contained in the supernatant. A typical recovery of radioactivity from the liver samples is ~90%. A portion of the supernatant is filtered through a Centricon filter (30 kDa molecular mass cutoff; 20 min, $0°$, 10,000g; Hermle Z 360 K refrigerated microcentrifuge) and lyophilized. The residue is redissolved in 10 μl of formamide-loading buffer (90% formamide, 10% 1× TBE, 0.2% bromphenol blue, and 0.2% xylene blue) in preparation for analysis by 15% PAGE (16 cm × 18 cm × 0.75 mm, 2 V/cm, 45 min) and subsequent autoradiography. Urine is collected from the bladder, which has been thawed to $0°$, and is deproteinized with ethanol (1:2, v/v) at $0°$ for 30 min. The precipitated proteins are removed by centrifugation (16,000g, 20 min, $0°$). Recovery of radioactivity is estimated to be ~90% by comparing the aliquots of the supernatant and the protein pellet. A portion of the supernatant is lyophilized, redissolved in formamide-loading buffer, and analyzed by 15% PAGE (16 cm × 18 cm × 0.75 mm, 2 V/cm, 45 min). Standards are produced by the incubation of full-length conjugate **1** with, in separate reactions, N-acetylglucosamidase (0.13 U) in 50 mM sodium citrate, pH 5.0, chymotrypsin in 10 mM Tris–HCl containing 200 mM KCl, pH 8.0, and 0.1 N HCl each at $37°$ for 30 min.

Antiviral Treatment

The three neoglycoconjugates **2–4** utilized in the HBV antiviral studies are synthesized by the conjugation of phosphorothioate (ps) oligomers, shown previously to inhibit HBV replication *in vitro*,[21] to the liver-specific

[21] B. E. Korba and J. L. Gerin, *Antiviral Res.* **28,** 225 (1995).

ligand YEE(ahGalNAc)$_3$: $^{5'}$GTTCTCCATGTTCAG$^{3'}$ (**2b**) targets the translation initiation site of the surface antigen gene (*sa* gene) and $^{5'}$AAAG-CCACCCAAGGCA$^{3'}$ (**3b**) targets the unpaired loop of the encapsidation site of the HBV pregenome (e site). The base sequences used to synthesize these are HBV subtype ayw,[22] the same subtype expressed *in vitro* by HepG2 2.2.15.[23] In addition, a third conjugate (**4**) using a ps oligomer, which is noncomplementary to the HBV genome, $^{5'}$TGAGCTATGCA-CATTCAGCTT$^{3'}$, is prepared as a control to assay for nonspecific effects of the neoglycoconjugates.

Antiviral activity of the oligonucleotides is assessed using confluent cultures of HepG2 2.2.15 cells. Cells are seeded into 48-well plates (Costar) at a density of 3–5 × 10^4 cells/well in RPMI + 2% fetal calf serum containing 4 mM glutamine and allowed to grow for 3–4 days until confluence is achieved. At this point, treatment is initiated with either neoglycoconjugates containing the conjugates described previously or the corresponding uncon-jugated ps oligomers alone at concentrations ranging from 1.0 to 20 μM. Cell numbers are quantitated using a Model ZB I Coulter counter (Coulter Electronics). Confluent cultures are used because HBV replication has been shown to reach stable, maximal levels only at this density in HepG2 2.2.15 cells.[24] All treatments are performed in triplicate and continued for 96 hr. Antiviral effects are assayed as detailed later and compared to untreated control cultures in order to determine the degree of inhibition. Values are reported as the average of six trials ± one standard deviation.

Analysis of HBV Nucleic Acids and Proteins

The effect of antiviral treatment on HBV surface antigen expression (HBsAG) by HepG2 2.2.15 cells is determined by semiquantitative EIA analysis[25] using the Ausyme Monoclonal kit (Abbott Laboratories). Test samples are diluted so that values are in a linear dynamic range of the assay. Standard curves using HBsAG (Chemicon) are included in each set of analyses. Values are quantitated on a Bio-Tek EIA plate reader at a fixed wavelength of 492 nm.

Extracellular HBV DNA is analyzed by quantitative dot-blot hybridiza-tion using a modification of procedures described previously.[26,27] Experi-

[22] F. Gailbert, E. Mandart, F. Fitoussi, P. Tiollais, and P. Charnay, *Nature* (*London*) **281**, 646 (1979).

[23] G. Acs, M. A. Sells, R. H. Purcell, P. Price, R. Engle, M. Shapiro, and H. Popper, *Proc. Natl. Acad. Sci. U.S.A.* **84**, 4641 (1987).

[24] M. A. Sells, A. Z. Zelant, M. Shvartsman, and G. Acs, *J. Virol.* **62**, 2836 (1988).

[25] C. Muller, K. F. Bergman, J. L. Gerin, and B. E. Korba, *J. Infect. Dis.* **165**, 929 (1992).

[26] B. E. Korba and G. Milman, *Antiviral Res.* **15**, 217 (1991).

[27] B. E. Korba and J. L. Gerin, *Antiviral Res.* **19**, 55 (1992).

mental and control media samples are centrifuged and treated with an equal volume of 1 N NaOH–10× SSC and incubated at room temperature for 30 min. Samples are then applied directly to nylon membranes (Gene Screen Plus, Amersham) presoaked with 0.4 M Tris–HCl (pH 7.5) using a dot-blot apparatus (Life Tech., Grand Island, NY). Membranes are neutralized with 0.5 M NaCl–0.5 M Tris–HCl (pH 7.5), rinsed in 2× SSC, and baked at 80° for 2 hr.

A purified 3.2-kb EcoRI HBV fragment (a gift of Dr. B. Korba)[28] is labeled with $[\alpha\text{-}^{32}P]$dCTP using a nick translation kit (Life Tech.) and purified using ProbeQuant microspin columns. Blots are prehybridized for 3–4 hr at 42° in a solution containing 6× SSC, 5× Denhardt's solution, 50% formamide, 0.5% SDS, and 125 μg/ml denatured, sheared salmon sperm DNA. Hybridization is carried out for 18–22 hr in a solution of the same composition with the addition of 10% dextran sulfate. Blots are washed sequentially at 42° and densitometric measurements are quantitated with a Fujix Bas 1000 phosphoimager (Fuji Medical Systems). Virion DNA levels are determined by comparing these measurements to known amounts of HBV DNA standards applied to each blot.

Neoglycoconjugate Stability

The *in vitro* stability of the neoglycoconjugates and unconjugated oligomers used in this study is determined by PAGE analysis. Mixtures from antiviral treatments containing the appropriate conjugate are incubated in RPMI + 2% FCS at 37° for 24, 48, and 96 hr. Aliquots containing 2 μl of either neoglycoconjugates or unconjugated oligomers are added to 10 μl of gel-loading buffer and electrophoresed for 20 min at 800 V on a 20% polyacrylamide gel containing 7 M urea. The resulting gels are analyzed with a Fujix Bas 1000 phosphoimager (Fuji Medical Systems).

Toxicity Analysis

After the treatment period, the number of viable cells is determined microscopically by trypan blue exclusion. A minimum of at least 200 cells from each well are counted. All determinations are performed on triplicate wells.

Results and Discussion

The methods described here have demonstrated the utility of chemically defined, structurally homogeneous conjugates when coupled to a receptor-

[28] B. E. Korba, F. V. Wells, B. Baldwin, P. J. Cote, B. C. Tennant, H. Popper, and J. L. Gerin, *Hepatology* **9**, 461 (1989).

mediated endocytotic process. The structures of the specific conjugates are shown in Fig. 1. By employing the A-L-P delivery system, we have shown that these unique neoglycoconjugates containing the ligand YEE(ahGalNAc)$_3$ are capable of traversing the cellular membrane, of *in vivo* localization within the liver, and able to inhibit hepatitis B virus expression in a model cell culture system.

Neoglycoconjugate Synthesis

The general conjugate syntheses is shown in Fig. 2. Initially described by Hangeland *et al.,*[8] the conjugate assembly route involves two steps. The first step, as indicated in Fig. 2, was the modification of the ligand with the heterobifunctional linker, SMCC. Second, the single-site and covalent attachment of the ligand linker to the oligonucleotide portion proceeded via a Michael addition reaction through a thioether bond. The overall yield of conjugate **1** was 24%. The ^{32}P radiolabel used with conjugate **1** and congeners was added upon the synthesis of the thiol oligonucleotide. The combined action of [γ-^{32}P]ATP and T4 polynucleotide kinase was used to install the radiolabel at the 5' end of U^mT_7. To reduce phosphatase activity and permit ligand conjugation, this terminal phosphate was then converted to the cystamine-derived phosphoramidate via reaction with EDAC.

A similar conjugation procedure was developed that provided >90% yields (experimental details of this procedure will be published elsewhere). This new route has yielded four YEE(ahGalNAc)$_3$-containing phosphoro-thioate oligodeoxyribonucleotide conjugates in nearly quantitative yield (Fig. 1). These four conjugates are YEE(ahGalNAc)$_3$-SMCC-AHT-5'GTTCTCCATGTTCAG3' (**2**), which targeted the *sa* gene [unconjugated form: 5'GTTCTCCATGTTCAG3' (**2b**)], YEE(ahGalNAc)$_3$-SMCC-AHT-5'AAAGCCACCCAAGGCA3' (**3**), which targeted the e-site [unconjugated form: 5'AAAGCCACCCAAGGCA3' (**3b**)], and the random control, YEE(ahGalNAc)$_3$-SMCC-AHT-5'TGAGCTATGCACATTCAGCTT3' (**4**) [unconjugated form: 5'TGAGCTATGCACATTCAGCTT3' (**4b**)]. The ^{35}S radiolabel of conjugate **2** and unconjugated **2b** was installed by the combined action of [α-^{35}S]dATP and terminal deoxynucleotidyltransferase (TdT) (Life Technologies) through the addition of a polyadenylated tail, which typically extends the oligomer between one and four adenosine residues. Antisense neoglycoconjugates were assayed for their biological effects on integrated HBV in human hepatoma cells.

Cellular Uptake

Data from the cellular uptake experiments are summarized graphically in Figs. 3 and 4. These cellular uptake experiments demonstrate the enhanced uptake and cell-type-specific binding of the conjugate via the ligand YEE(ahGalNAc)$_3$-directed, receptor-mediated process by human hepatocellular carcinoma cells (HepG2). The peak cellular uptake of conjugate **1** was measured at 26 pmol/10^6 cells, a 43-fold enhancement over the unconjugated, ethylenediamine (EDA)-protected oligomer (**1b**) (0.6 pmol/10^6 cells) (Fig. 3). The next experiment was to test if the process was indeed mediated by the hepatic carbohydrate receptor. This procedure involved the preincubation of the cells with 100 equivalents of the unlabeled free ligand, which was then followed by the addition of the radiolabeled conjugate (Fig. 3). Due to the competitive binding from the free, unlabeled ligand, the result was a minimal cellular uptake of the radiolabeled conjugate. Additionally, when the carbohydrate moieties were removed through enzymatic digestion through *N*-acetylglucosamidase-mediated hydrolysis, the uptake of the ligand-degraded conjugate **1c** was again at background levels (~1 pmol/10^6 cells). In summary, suppression of conjugate uptake resulted if the ligand was absent or partially disabled.

To demonstrate the cell-type specificity of the ligand, similar cellular uptake experiments were carried out using two receptor-negative cell lines: HL-60 cells and HT-1080 (Fig. 4). It is well established that the asialoglycoprotein receptor is found on the surface of hepatocytes[29] and represents an efficient means for selectively targeting this tissue for intracellular delivery for a variety of therapeutic agents. As shown in Fig. 4, the result was a negligible cellular uptake of conjugate **1** by either of these cell lines. These results have confirmed that cellular uptake of the conjugate is indeed a cell-type-specific, ligand-dependent and receptor-mediated event occurring on the surface of the hepatic cell.

In subsequent experiments, the A-L-P delivery system was modified to contain an antisense oligonucleotide targeted against the integrated hepatitis B virus. Phosphorothioate oligodeoxynucleotides used for these experiments were negatively charged, whereas the oligomer used for the first cellular uptake studies consisted of nonionic methyl phosphonate linkages. However, despite this difference, HepG2 cells displayed a rapid and linear uptake of conjugate **2** to the extent of 17.3 pmol/10^6 cells at 24 hr. As before, cellular association of the corresponding, unconjugated ps-oligomer **2b** with the hepatoma cells accumulated only to the extent of 1.0 pmol/10^6 cells at 24 hr (Fig. 3). It is important to note that when identical experiments were performed using HepG2 2.2.15 cells, comparable results were obtained (data not shown).[15] Overall, each neoglycoconjugate tested, whether using a charged or uncharged backbone, displayed linear and rapid uptake to

---

[29] M. A. Findeis, C. H. Wu, and G. Y. Wu, *Methods Enzymol.* **247**, 341 (1994).

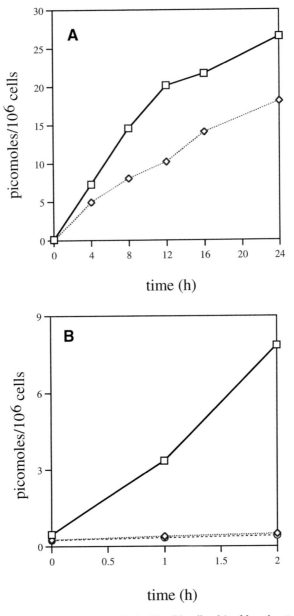

FIG. 3. (A) Time course for the uptake by HepG2 cells of 1 μM conjugates **1** (□) and **2** (◇). Cells were incubated at 37° and collected as described in the Methods section. Each data point represents the average of three experiments. (B) Time course for the uptake by HepG2 cells of 1 μM conjugate **1** (□) alone and (○) in the presence of free YEE(ahGalNAc)$_3$, and unconjugated **1b** (◇). Cells were incubated at 37° for 0, 1, and 2 hr and the samples were collected as described in the Methods section. Each data point represents the average of three experiments.

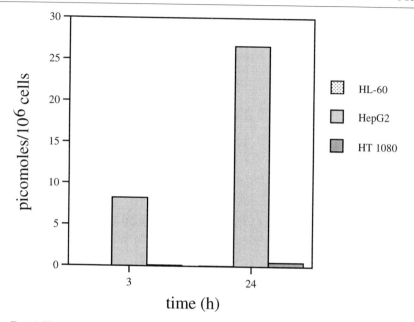

FIG. 4. Tissue-specific uptake of conjugate **1** by HepG2, HL-60, and HT 1080 cells. Cells were collected and the amount of ^{32}P was determined at 3 and 24 hr for each cell line. Experiments were done in triplicate and data expressed as the average ± one standard deviation.

the extent of ~20 pmol/10^6 cells at 24 hr, whereas the corresponding unconjugated oligomer associated poorly at ~1.0 pmol/10^6 cells at 24 hr (Fig. 3).

Tissue Distribution and Kinetics of Liver Uptake and Clearance

In vivo tissue distribution data show results similar to that obtained with both methylphosphonate and phosphorothioate oligomers in cultured human cells. Table I shows the kinetic distribution of conjugate **1** following intravenous injection. Likewise, Table II shows that the phosphorothioate-derived conjugate **3** (Fig. 1) localized primarily within the liver (46.2 ± 8.5%) in 15 min as opposed to the unconjugated form (21.7 ± 5.1%).[30] With conjugate **1**, the radiolabeled material associated to the greatest extent with the liver, reaching a peak value of 69.9 ± 9.9% of the injected dose at 15 min postinjection. The ranking of total radioactivity of the other tissues at 15 min postinjection was, in decreasing order: muscle > kidney >

[30] J. J. Hangeland, S. F. Deamond, R. J. Duff, C. Roby, Y. Zhou, Y. C. Lee, and P. O. P. Ts'o, unpublished result (1996).

TABLE I

PERCENTAGE INJECTED DOSE PER ORGAN VS TIME POSTINJECTION FOLLOWING SINGLE
INJECTION OF $[^{32}P]$-$[YEE(ahGalNAc)_3$-SMCC-AET-pU^{m}p$\underline{T}_7]$ INTO MICE via TAIL VEIN[a,b]

Time postinjection (min)	Percentage initial dose per organ				
	Blood	Liver	Spleen	Kidney	Muscle
15	2.79 ± 0.18	69.9 ± 9.9	0.08 ± 0.04	3.00 ± 1.26	7.83 ± 1.49
30	2.25 ± 0.48	41.8 ± 9.3	0.08 ± 0.03	2.12 ± 0.27	8.42 ± 1.51
60	1.42 ± 0.38	25.2 ± 2.4	0.20 ± 0.01	1.58 ± 0.15	8.46 ± 2.32
120	0.90 ± 0.26	14.2 ± 2.2	0.17 ± 0.04	1.26 ± 0.19	8.76 ± 0.92
240	1.09 ± 0.16	10.6 ± 4.2	0.24 ± 0.02	1.25 ± 0.21	13.0 ± 3.9
360	1.23 ± 0.30	8.5 ± 0.6	0.16 ± 0.08	1.80 ± 0.70	17.2 ± 4.6
1440	0.61 ± 0.11	3.2 ± 1.4	0.25 ± 0.04	0.92 ± 0.19	13.9 ± 1.3

[a] Values are reported as the average of three trials ± one standard deviation. Approximately 0.5 μCi (30 pmol) of conjugate was injected into each animal. The mass of each organ was determined separately and was used to determine the percentage dose per organ from percentage dose per gram of tissue. Typical values for the mass of each organ or tissue are blood = 0.07 × body mass; liver = 1.6 ± 0.21 g; spleen = 0.17 ± 0.05 g; kidney = 0.6 ± 0.1 g; muscle = 0.4 × body mass. The average body mass was 32.4 ± 2.0 g (SD; n = 21).
[b] The peak value of radioactivity in urine was 27.7 ± 20.2% of initial dose at 60 min. The large standard deviation reflects the variation in urine production between individual animals.

TABLE II

PERCENTAGE INJECTED DOSE PER ORGAN VS TIME POSTINJECTION FOLLOWING SINGLE
INJECTION OF $[^{32}P]$-$[YEE(ahGalNAc)_3$-SMCC-AHT-
5'pTTTATAAGGGTCGATGTCCAT-$(\{^{35}S\}psA)_n]$ INTO MICE via TAIL VEIN[a,b]

Time postinjection (min)	Percentage initial dose per organ				
	Blood	Liver	Spleen	Kidney	Muscle
15	8.87 ± 2.98	46.2 ± 8.5	0.47 ± 0.13	5.96 ± 0.44	12.63 ± 6.50
30	6.62 ± 1.17	48.9 ± 14.0	0.36 ± 0.27	5.46 ± 2.72	11.12 ± 7.80
60	2.69	28.65	0.39	2.99	5.59
240	3.39 ± 0.53	20.1 ± 8.60	0.32 ± 0.02	1.35 ± 029	4.78 ± 2.05
1440	1.18 ± 0.64	6.48 ± 1.11	0.10 ± 0.01	0.55 ± 0.05	1.83 ± 0.26

[a] Where n is between one and four residues. Values are reported as the average of three trials ± one standard deviation. Approximately 0.5 μCi (30 pmol) of conjugate was injected into each animal. The mass of each organ was determined separately and was used to determine the percentage dose per organ from percentage dose per gram of tissue. Typical values for the mass of each organ or tissue are blood = 0.07 × body mass; liver = 1.55 ± 0.23 g; spleen = 0.10 ± 0.02 g; kidney = 0.5 ± 0.06 g; muscle = 0.4 × body mass. The average body mass was 31.68 ± 1.31 g (SD; n = 21).
[b] The peak value of radioactivity in urine was 14.2 ± 7.01% of initial dose at 60 min. The large standard deviation reflects the variation in urine production between individual animals.

blood > spleen. The peak value of radioactivity for the urine was 17% of the injected dose and was reached after 30 min (data not shown). The amount of radioactivity associated with the kidney and blood decreased over time, whereas that associated with the muscle increased twofold up to 24 hr.

As a control, mice were injected with conjugate **1c**, which lacked the three terminal GalNAc residues, and therefore should not be recognized by ASGP receptor. As anticipated, this deglyco form of the conjugate showed significantly less cellular uptake, which consequently resulted in a delocalization of the conjugate away from the liver. Radioactivity was distributed in the order muscle > blood > kidney > liver > spleen (data not shown). In contrast, the rank order of tissue distribution for conjugate **1** was liver ≫ muscle > kidney > blood > spleen. The effect of glycosidase treatment on conjugate **1** was to increase the fraction of radioactivity appearing in the urine due to the diminished uptake by the liver. The peak value of radioactivity in the urine was 40% at 30 min postinjection conjugate **1c** and 17% for conjugate **1** (data not shown). This highly selective targeting of the oligodoxynucleoside methyl phosphonate to the liver was effectively achieved because of the single covalent linkage between the oligomer and the ligand YEE(ahGalNAc)$_3$, which mimicks the soluble asialoglycoprotein receptor (ASGP) ligand.

It is important to note that the asialoglycoprotein receptor is not unique to hepatocytes, but it is also found on the surface of Kupffer cells.[31,32] However, it has been determined that Kupffer cells are unable to ingest soluble, molecular asialoglycoprotein.[31-36] Furthermore, these studies suggested that ligand uptake was triggered by insoluble particles with multiple binding sites and not by soluble glycoproteins. Indeed, the concentration of conjugate in the liver was 25-fold greater than that found in the blood and approximately 10-fold greater than in muscle based on whole tissue measurements (Table I). These results compared favorably, and were in some ways superior, to the outcome of similar experiments reported by Lu *et al.*,[37] where the delivery of a ^{32}P-labeled antisense oligonucleotide to rat liver was enhanced when compared to other tissues due to its complexation

[31] V. Kolg-Bachofen, J. Schlepper-Schaefer, and W. Vogell, *Cell* **29,** 859 (1982).
[32] J. Kuiper, H. F. Bakkeren, E. A. Biessen, and T. J. C. Van Berkel, *Biochem. J.* **299,** 285 (1994).
[33] G. Ashwell and A. G. Morell, *Adv. Enzymol.* **41,** 99 (1974).
[34] P. H. Schlesinger, T. W. Doebber, B. F. Mandell, R. White, C. Deschryver, J. S. Rodman, M. J. Miller, and P. Stahl, *Cell* **29,** 859 (1978).
[35] A. Hubbard and H. Stukenbrok, *J. Cell Biol.* **83,** 65 (1979).
[36] D. A. Wall, G. Wilson, and A. L. Hubbard, *Cell* **21,** 79 (1980).
[37] X. M. Lu, A. J. Fischman, S. L. Jyawook, K. Hendricks, R. G. Topkins, and M. L. Yarmush, *J. Nucleic Med.* **35,** 269 (1994).

with an asialoglycoprotein–poly (L-lysine) conjugate.[37] As noted by the authors, however, the preference of the complex for the liver was marginal because the spleen, lungs, and kidney accumulated the radiolabeled oligo(dN) as well (e.g., distribution for each tissue was ca. 6, 4, 2, and 2% of injected dose per gram, respectively, after 5 min postinjection[37]). It is of further interest to compare our results with those reported by Eichler *et al.*[38] where the biodistribution and rate of liver uptake were determined in rats for the hypolipidemic agent ansamycin, both alone and linked covalently to another triantennary ASGP ligand, *N*-[tris[(D-galactopyranosylosyl)methyl]methyl]-*N'*-(acetyl)glycinamide trisgalactosyl acetate. The authors reported that the liver uptake of the free drug and the conjugate was roughly equivalent, leading them to conclude that the triantennary ASGP ligand did not enhance the uptake of the drug by rat liver. This result is in contrast to our finding that mouse liver uptake was facilitated greatly by the covalent attachment of the ligand, YEE(ahGalNAc)$_3$.

Metabolism of Conjugates

To gain insight into the *in vitro* and *in vivo* metabolic fate of conjugate **1**, we examined extracts obtained from HepG2 cells grown in culture and from the liver and urine of mice by PAGE analysis. We noted that three classes of metabolites (classes I–III) were produced in HepG2 cells and in mouse liver, whereas only class I metabolites were isolated from mouse urine. Figure 5 shows the relative gel mobilities of the three classes of metabolites derived from liver homogenate extract samples of mice injected with ^{32}P-labeled conjugate **1**. At 15 min postinjection, a significant amount of intact conjugate **1** or conjugate **1c** remained. In an effort to identify these metabolites, their PAGE mobilities were compared to authentic materials that resulted from known enzymatic reactions of **1**. The first enzymatic digestion with *N*-acetylglucosaminidase (a glycosidase, known to remove the terminal GalNAc residues) produced a band that comigrated with metabolite **1c** (Fig. 1). The second metabolite of this class (**1d**) (Fig. 1), which comigrated with a compound produced by chymotrypsin digestion of **1**, was observed at each time point. It is reasonable to conclude, therefore, that these species migrating to the same region of the gel resulted from the degradation of the ligand and not from bond cleavage at other labile sites of **1**. For example, the hydrolysis of a single, aminohexyl side chain amide bond would yield **1e** with mass between **1c** and **1d** (Fig. 1).

Class II metabolites migrated considerably faster than those identified as class I. We have proposed that they arose due to unanticipated hydrolysis

[38] H.-G. Eichler, K. A. Menear, K. E. Dunnet, J. G. Hastewell, and P. W. Taylor, *Biochem. Pharm.* **44**, 2117 (1992).

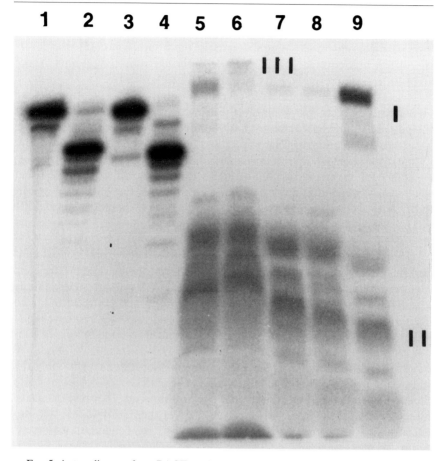

FIG. 5. Autoradiogram from PAGE analysis of the metabolites of **1** in mouse liver. Lane 1, **1**; lane 2, **1** treated with *N*-acetylglucosamidase; lane 3, **1** treated with chymotrypsin; lane 4, **1** treated with 0.1 *N* HCl; lanes 5 and 6, liver homogenate extracts at 2 hr postinjection; lanes 7 and 8, liver homogenate extracts at 1 hr postinjection; and lane 9, liver homogenate extracts at 15 min postinjection.

of the single phosphodiester linkage located at the 5′ terminus of the methyl phosphonate oligonucleoside. It was expected that this site would be stable toward cleavage by endonuclease activity[39] based on a model reaction conducted with snake venom phosphodiesterase in which no cleavage was observed. Cleavage at this site would release the terminal seven nucleotides

[39] B. S. Sproat, A. T. Lamond, B. Beijer, P. Neumer, and U. Ryder, *Nucleic Acids Res.* **17,** 3373 (1989).

of the methyl phosphonate oligonucleoside from the remainder of **1** and, most importantly, produce a relatively low molecular weight species bearing a single nucleotide containing the radiolabeled phosphate (**1f**) (Fig. 1). A fourth, yet unidentified, metabolite was observed at all time points except at 24 hr, migrating only slighty slower than **1d**.

The final class of metabolites (class III) was only observed in cultured HepG2 cell extracts. Interestingly, they migrated only a short distance into the gel, suggesting that they are high molecular weight species containing radioactive phosphorous. The release of radioactive phosphate from **1** and its subsequent incorporation into high molecular weight structures (nucleic acids or proteins) would account for this band. It is well documented that the endosomal compartment acidifies as it matures, reaching a pH as low as 5.5 before fusing with lysosomes.[40] Furthermore, the phosphoramidate linkage tying the methylphosphonate oligonucleoside ($\underline{T}_7$) to the ligand was prone to hydrolysis under acidic conditions to give **1g**.

The profile of metabolites observed in all extracts from HepG2 cells and liver homogenate included each class of metabolites. The early time points contained the majority of the radioactivity as class I species, mainly **1** and **1c**. At later time points, the distribution of metabolites shifted from class I to classes II and III, where at the last time point sampled, the majority of radioactive phosphorous resided with class III metabolites. This observation indicated that substantial hydrolysis of the P–N bond had occurred over the course of the experiment. Due to the fact that only the phosphorus at the P–N bond is labeled with ^{32}P, it was not possible to measure the metabolic fate of the oligonucleotide analog. Because extensive metabolism of the oligonucleotide would affect the ability to interact specifically with intracellular complementary nucleic acid sequences, adversely, future studies using oligonucleotide sequences labeled in other positions need to be performed. However, in any case, it is readily apparent that **1** was metabolized significantly once taken into HepG2 cells, suggesting that intracellular delivery of an antisense methylphosphonate oligonucleoside, or other agents, would be feasible by this method.

Antiviral Activity

With a demonstrated ability to enhance intracellular oligonucleotide concentrations, this versatile delivery system was then modified to contain antisense oligonucleotides directed against the surface antigen gene and the encapsidation site of the hepatitis B viral genome. Accordingly, these oligomers were linked covalently to the liver-specific ligand,

[40] A. L. Schwartz, G. J. Strous, J. W. Slot, and H. J. Geuze, *EMBO J.* **4**, 899 (1985).

YEE(ahGalNAc)$_3$, to form the corresponding conjugates. Antisense oligo-mers directed against these highly conserved key elements of HBV replica-tion have been shown to disrupt specific functions under conditions of preexisting chronic *in vitro* infection. In specific, it has been demonstrated that targeting the *sa* gene with ps oligomers resulted in up to an 85% inhibition of HBsAG accumulation in the media of HepG2 2.2.15 cells.[41,42] If the encapsidation site of the HBV pregenome is targeted, the results show a dramatic reduction of HBV virion DNA into the cell culture media.[21] The reported inhibition of HBV replicative elements by the oligomers described earlier was consistent, but varied in degree depending on the treatment regimen and concentration employed.

In order to assess the effects of these neoglycoconjugates on HBV gene expression, confluent monolayers of HepG2 2.2.15 cells were incubated for 96 hr in the presence of a single dose of either neoglycoconjugate or the corresponding unconjugated oligomer targeting either the surface antigen gene or the encapsidation signal. The specificity of binding was confirmed by treating cells with neoglycoconjugates containing random ps oligomers or the unconjugated random ps oligomers alone.

The results of these experiments proved that the ligand conjugation of oligomers increased their biological efficacy when targeted against specific elements of HBV replication. First, on treatment of HepG2 2.2.15 cells with conjugate **2**, surface antigen expression was reduced by 73% from 874.13 to 235.70 ng/10^6 cells at a concentration of 20 μM and by 43% to 505.14 ng/10^6 cells at 10 μM. In contrast, the corresponding unconjugated oligomer (**2b**), reduced HBsAG expression by 43% at 20 μM and by 30% at 10 μM. Neither the neoglycoconjugate nor the oligomer alone had any significant effect on HBsAG expression at concentrations below 10 μM (Table III).

Unlike the previous conjugate, the biological effect of the neoglycocon-jugate targeting the e site was more striking (Table III). In this case, the effectiveness of single concentrations was not only improved by the neogly-coconjugate, but the range of effective concentrations was also expanded. The administration of conjugate **3** to HepG2 2.2.15 cells reduced virion DNA expression from 2407.5 to 169.82 pg/10^6 cells at a concentration of 20 μM, a greater than 90% inhibition. The unconjugated form of this oligomer displayed similar results at 20 μM, reducing virion DNA expres-sion to 216.53 pg/10^6 cells, also a greater than 90% inhibition. However, as the concentration of the antisense oligomer decreased, the neoglycoconju-gate maintained a greater than 90% inhibition of virion DNA down to a

[41] G. Goodarzi, S. C. Gross, A. Tewari, and K. Watabe, *J. Gen. Virol.* **71,** 3021 (1990).
[42] W.-B. Offensperger, H. E. Blum, and W. Gerok, *Intervirology* **38,** 113 (1995).

TABLE III

EFFECT OF NEOGLYCOCONJUGATE TARGETING SURFACE ANTIGEN GENE ON ACCUMULATION OF HBsAG IN CULTURE MEDIA OF HepG2 2.2.15 CELLS[a]

Concentration (μM)	Surface antigen experiments		Encapsidation signal experiments	
	Conjugate 2 (ng HBsAG/10^6 cells)	Unconjugated 2b (ng HBsAG/10^6 cells)	Conjugate 3 (pgHBV virion DNA)	Unconjugated 3b (pgHBV virion DNA)
Untreated	874.13 ± 34	874.13 ± 34	2407.5 ± 195	2407.5 ± 195
20	235.7 ± 39.8	499.4 ± 55	169.82 ± 22	216.55 ± 17
10	505.14 ± 31.4	605.98 ± 35	150.99 ± 12.34	617.98 ± 65
5	607.31 ± 20.24	666.96 ± 63	159.26 ± 13.75	2039.7 ± 215.75
2	748 ± 86.81	776.91 ± 69.1	186.29 ± 20.01	2827.99 ± 256.98
1	747.48 ± 140	775.76 ± 64.66	312.82 ± 28.52	2628.29 ± 275.32

[a] SMCC-5'GTTCTCCATGTTCAG3' (conjugate 2), 5'GTTCTCCATGTTCAG3' (unconjugated 2b), and the effect of a neoglycoconjugate targeting the encapsidation site of HBV on accumulation of virion DNA of HBV in culture media of HepG2 2.2.15 cells: YEE(ahGal-NAc)₃-SMCC-5'AAAGCCACCCAAGGCA3' (conjugate 3) and 5'AAAGCCACCCAAG-GCA3' (unconjugated 3b). Cells were treated for 96 hr in all cases. Experiments using surface antigen or the encapsidation signal oligomeric forms were performed separately.

concentration of 1 μM. The effectiveness of the unconjugated oligomer diminished with decreasing concentration until untreated control levels were reached at 2 μM. Conjugate 3 treatments had no effect on HBsAG accumulation. This observation on conjugate 3 is a clear indication of the sequence specificity of these antisense agents.

In order to confirm the sequence specificity of the oligonucleotides described earlier, random ps oligomers, noncomplementary to any portion of the HBV genome, were synthesized. The conjugated and unconjugated forms of the random oligonucleotide (4 and 4b, respectively) exhibited no significant bioefficacy on either HBsAG or virion DNA expression at concentrations up to 30 μM (data not shown). The mechanism of this ligand-mediated enhancement of antiviral activity remains unknown. However, we have calculated previously that the delivery of potential antisense molecules by the defined neoglycoconjugate increased the intracellular concentration of the conjugate as much as 43 times in comparison to unconjugated oligomers.[8]

Toxicity

Toxicity of each treatment used in this study was determined by trypan blue exclusion. Measurements were made under culture conditions used for antiviral experiments. No significant toxicity at any concentration for any treatment was noted (data not shown).

Conclusions

Data presented here have demonstrated the beneficial effects of using a "simple" A-L-P construct consisting of a chemically defined, structurally homogeneous ligand–linker–oligonucleotide assembly.

In cell culture experiments, ligand conjugates of the A-L-P type enhanced tissue specificity of the receptor-mediated uptake, leading to a highly significant (20- to 40-fold) increase of intracellular oligonucleotide concentrations within hepatic cells (HepG2). In animal (mice) experiments, A-L-P conjugates led to a rapid (15 min) and large (40–80%) uptake to the animal liver when introduced into the blood system. Without the ligand, the oligomer is excreted rapidly into the urine. Once inside the hepatic cell or the liver, the A-L-P conjugate was metabolized because of the endosome–lysosome pathway, which is not expected to affect the integrity of the oligonucleotides constructed of enzyme-resistant backbones (such as nonionic methylphosphonates or, less effectively, phosphorothioates). In bioefficacy experiments, A-L-P conjugates indeed showed a sequence-dependent suppression of the integrated HBV viral expression in hepatoma cells (HepG2 2.2.15). Depending on the targeted gene, the suppression enhancement between conjugated and unconjugated oligomer forms can be at least 20-fold or more (encapsidation gene as measured by released HBV virion DNA). However, the true value for the bioefficacy with A-L-P conjugates cannot yet be fully demonstrated with an open, dynamic animal/human system, because without the covalently linked ligand, the oligomeric drug is excreted rapidly out of the system. This is in contrast to a static, closed system experiment, such as with cell culture demonstrations. All of the necessary elements are described and shown in this article, except the *in vivo* bioefficacy experiment. Finally, these results point to the possibility that the A-L-P assembly with a hepatic/asialoglycoprotein receptor-specific ligand can be valuable for the development of gene-specific, tissue-specific antiviral agents, particularly for HBV infection.

Acknowledgments

We gratefully acknowledge the following people for contributions to this work: Dr. Paul S. Miller, Dr. Reiko T. Lee, Dr. K. B. Lee, Dr. Manuel Oropeza, and Ms. Sarah Kipp. Additionally, we thank Dr. George Wu for the gift of HepG2 2.2.15 cells and Dr. Brent Korba for the gift of the HBV-specific probe. This work was supported by NIGMS grant award 1R01GM52984-01A1.

[18] Lipid-Based Formulations of Antisense Oligonucleotides for Systemic Delivery Applications

By SEAN C. SEMPLE, SANDRA K. KLIMUK, TROY O. HARASYM, and MICHAEL J. HOPE

Increasing the specificity of therapeutic drugs and improving their delivery to sites of disease are primary goals of today's pharmaceutical industry. One of the most exciting advances in recent years has been the development of antisense technologies, which are capable of modulating protein expression with exquisite specificity. Unfortunately, this class of drugs is particularly sensitive to nuclease degradation, is eliminated rapidly from the circulation after intravenous administration, and is severely limited in its ability to penetrate through cellular membranes unaided.[1-4] Attempts to address these problems through medicinal chemistry have produced several key advances. However, chemical alterations to improve one property of the molecule often affect other properties, potentially in a negative manner. For example, phosphorothioate oligonucleotides are more nuclease resistant than native phosphodiesters, but binding affinities to mRNA targets are lower and nonspecific interactions with plasma and cellular proteins are increased.[2,5,6]

The past decade has seen extensive use of liposomes and lipid-based delivery systems to improve the pharmacological properties of a variety of drugs. The principal benefits afforded therapeutic agents by liposomal encapsulation are enhanced plasma circulation lifetimes, increased delivery to sites of disease, and changes in tissue distribution, which can result in reduced toxic side effects. Liposomal preparations of doxorubicin (DOXIL) and daunorubicin (DaunoXome) have been approved for the treatment of HIV-associated Kaposi's sarcoma, whereas lipid-based formulations of amphotericin B (AmBisome, ABELCET, AMPHOTEC) are successful clinical products employed in the treatment of fungal infections. It is not surprising, therefore, that considerable interest has focused on developing lipid-based delivery systems to overcome the problems associated with the

[1] E. J. Wickstrom, *Biochem. Biophys, Methods* **13**, 97 (1986).
[2] S. Akhtar, R. Kole, and R. L. Juliano, *Life Sci.* **49**, 1793 (1991).
[3] S. Agrawal and R. Zhang, *Ciba Found. Symp.* **209**, 60 (1997).
[4] Y. Rojanasakul, *Adv. Drug Delivery Rev.* **18**, 115 (1996).
[5] M. Ghosh, K. Ghosh, and J. S. Cohen, *Nucleic Acids Symp. Ser.* **24**, 139 (1991).
[6] D. A. Brown, S. H. Kang, S. M. Gryaznov, L. DeDionisio, O. Heidenreich, S. Sullivan, X. Xu, and M. I. Nerenberg, *J. Biol. Chem.* **28**, 26801 (1994).

systemic administration of DNA- and RNA-based therapeutics. The intent of this article is to introduce the reader to the various lipid-based formulations applied to polynucleic acid drugs, with emphasis on the generation and characterization of delivery vehicles for intravenous applications. The majority of examples apply to antisense oligodeoxynucleotides (ASODN) as they represent the most widely available class of DNA-based drugs.

Lipid-Based Delivery Systems for Antisense Oligodeoxynucleotides

In general, the approaches used to formulate antisense oligonucleotides with lipids fall into three main categories. The first involves direct chemical conjugation of nucleic acids to lipid species such as cholesterol, alkyl chains of varying lengths, or phospholipid analogs. The rationale is to increase the lipid solubility of oligonucleotides through the addition of hydrophobic moieties in an attempt to enhance their penetration of cell membranes and improve intracellular delivery. Such modifications do not protect oligonucleotides from interactions with blood proteins or nucleases, nor are they expected to alter the plasma circulation lifetime significantly, which is strongly influenced by the presence of multiple negative charges in the oligonucleotide backbone.

The second lipid-based approach involves the formation of complexes between cationic lipids and oligonucleotides, often generically referred to as "lipoplexes." Mixing preformed cationic lipid vesicles (liposomes) with polynucleic acid at various charge ratios generates complexes, which are used extensively for the intracellular delivery of oligonucleotides and plasmids to cells in culture.[7,8] It is important to note that the nucleic acid in these systems is not encapsulated and, as such, would not typically be referred to in the context of "liposomal delivery." Cationic liposome–oligonucleotide complexes, while very effective *in vitro,* are not suitable for systemic delivery as they are eliminated rapidly from the blood, provide only partial nuclease protection, and are often toxic.[9–11] The third type of lipid-based delivery system, and the focus of this chapter, consists of antisense oligonucleotide encapsulated in the aqueous space enclosed within a lipid bilayer.

---

[7] J. Felgner, F. Bennett, and P. L. Felgner, *Methods* **5,** 67 (1993).
[8] M. J. Hope, B. Mui, S. Ansell, and Q. F. Ahkong, *Mol. Membr. Biol.* **15,** 1 (1998).
[9] C. Plank, K. Mechtler, F. C. Szoka, Jr., and E. Wagner, *Hum. Gene Ther.* **7,** 1437 (1996).
[10] M. C. Filion and N. C. Phillips, *Biochim. Biophys. Acta* **1329,** 345 (1997).
[11] D. C. Litzinger, *J. Liposome Res.* **7,** 51 (1997).

Methods for Encapsulating ASODN in Lipid-Based Delivery Systems

General Considerations

Encapsulation efficiency is an important characteristic of liposomal formulations. Strictly defined, this parameter indicates how much of the input drug, solute, or antisense is physically isolated inside a membrane during the formulation process. This parameter is often misused in describing systems composed of cationic lipid and DNA or RNA as the nucleic acid is also associated with the external surface of the particle. In the absence of specific methods to determine the size of this fraction, the term incorporation or association efficiency is probably more appropriate. Another physical characteristic of lipid-based delivery vehicles is the drug/lipid (D/L) ratio, usually reported on a weight/weight (w/w) basis for nucleic acids, but mol/mol is also common, especially for conventional drugs. In most instances, a high drug/lipid ratio is desirable in order to minimize the lipid doses administered with the drug. Retention of drug inside the carrier while in the circulation is another important requirement, as it is undesirable for drug to leak out of the vesicles before reaching the disease site. This is not a major concern for encapsulated anionic macromolecules such as antisense oligonucleotides and plasmids because of the large energy barrier preventing their diffusion through a lipid bilayer.[12] However, it does become a factor if the carrier membrane is unstable in plasma or a significant proportion of the drug is associated with the outside surface of the carrier. Finally, vesicle size has a significant impact on the pharmacokinetics of the delivery system. Large particles, with diameters >200 nm, are eliminated from the circulation much faster than small particles with diameters in the range of 80–100 nm.[13]

Numerous methods for encapsulating antisense oligonucleotides in lipid vesicles have been reported. The principal ones are discussed next and their relative merits are compared in Table I.

Passive Encapsulation by Dry Lipid Hydration

Passive encapsulation by dry lipid hydration is generally applied to the process whereby a solution of drug is used to hydrate a dry film or powder of lipids that spontaneously adopt a bilayer configuration in an aqueous environment. As liposomes form, a proportion of the solution is encapsulated, but no other physical or chemical mechanism drives the entrapment and the concentration of drug inside the liposome is equivalent to the

[12] S. Akhtar, S. Basu, E. Wickstrom, and R. L. Juliano, *Nucleic Acids Res.* **19**, 5551 (1991).
[13] J. H. Senior, *Crit. Rev. Ther. Drug Carrier Syst.* **3**, 123 (1987).

concentration outside. Encapsulation values are dependent on the concentration of lipid employed, with most lipid mixtures yielding 1–2 liter of entrapped solution per mole lipid. Interestingly, the size of the liposome does not influence the trapped volume as much as one might expect. This is because the bulk of the inner volume of large (multilamellar) liposomes is occupied by internal lipid lamellae. Consequently, the trapped volume per mole of lipid of a unilamellar vesicle with a diameter of 100 nm is similar to that of a multilamellar vesicle (MLV) with a diameter of several microns and net neutral charge. Making unilamellar vesicles suitable for drug delivery with diameters >200 nm has proven to be very difficult because of their inherent instability. The encapsulation volume of MLV can be enhanced by adding a surface charge or subjecting the liposome to cycles of freezing and thawing.[14] However, the preferred vesicle size for intravenous drug delivery is 100 nm and unilamellar vesicles of this size have a theoretical trapped volume of 2–3 liter/mole, depending on the molecular surface area occupied by the lipids employed. This volume is independent of the initial trapped volume of the parent MLV from which the unilamellar vesicles are made. In general, passive encapsulation is an inefficient process in which only 1–10% of the total drug is entrapped, resulting in low D/L ratios in the range of 0.01–0.10 (weight of drug/weight of lipid).

We have successfully employed passive encapsulation to entrap ASODN in phosphatidylcholine : cholesterol (PC : CH; 55 : 45, mol/mol) liposomes, employing the following procedure.[15] Individual lipids are dissolved in chloroform (50 mg/ml stocks), and the lipid components are aliquoted into a 16 × 100-mm glass culture tube to give 50–100 mg total lipid, in the desired molar ratio. The lipid mixture is vortexed and dried to a homogeneous film under a stream of nitrogen gas and low heat (40–50°). Residual solvent is removed by exposing the lipid film to high vacuum for 2–3 hr. Alternatively, the lipid film can be dissolved in benzene/methanol (70/30, v/v), frozen in liquid nitrogen, and lyophilized to a dry powder. A solution of antisense in phosphate-buffered saline (PBS) (10–250 mg in 1.0 ml) is added to the dry lipid and warmed to 50° for 1 hr, vortexing frequently. This process generates MLV and the sample is allowed to hydrate for a further 3–4 hr at room temperature or overnight at 4°. The sample is transferred to a cryovial and subjected to five freeze–thaw cycles by alternating between liquid nitrogen and a 50° water bath. MLV are reduced to a homogeneous population of unilamellar vesicles by passing the prepara-

[14] L. D. Mayer, M. B. Bally, M. J. Hope, and P. R. Cullis, *Chem. Phys. Lipids* **40**, 333 (1986).
[15] S. K. Klimuk, S. C. Semple, P. N. Nahirney, M. C. Mullen, C. F. Bennett, P. Scherrer, and M. J. Hope, submitted for publication.

TABLE I

LIPOSOME ENCAPSULATION PROCEDURES FOR ASODN

Formulation procedure	Lipid composition (molar ratio)	ODN		Encapsulation efficiency (%)	Final ODN/lipid ratio (w/w)	Particle size (nm)
		Length	Chemistry			
Passive[a]	DPPC:CH:SPDP-PE (65:34:1)	16	PO[b]	3	0.009[c]	~200
Passive[d]	EPC:CH:DMPG (49.4:32.4:18.2); EPC:DMPG (68:32)	4–14	PO, PS, MP	<2 (<30[c] if frozen and thawed)	<0.00001[c]	460 ± 200
Passive[e]	DPPC:CH:SPDP-PE (64:35:1)	16, 28	PO, PS	2–3	~0.005[c]	100–140[f]
Passive[g]	EPC:CH (60:40) ± 0.5 mol% folate-PEG-DSPE	15	PO, PS/PO	30–40	0.005[c]	100–140[f]
Passive[h]	PC:CH (80:20)	18	PS	<10	I.D.[i]	110 ± 40
Passive[j,k]	DOPE:CH:oleic acid (OA):palmitoyl-CD4 IgG (45:35:10:10)	20	PS	<10[f]	0.0025[c]	220 ± 55
Passive[l,m]	PC:CH:phosphatidylserine (61/26/13)	16	PO[b]	<10[f]	I.D.	Not indicated
Passive[n]	EPC:CH:folate-PEG-DSPE (54.5:36.4:9.1)	15	PO, PS/PO	15–20	I.D.	Not indicated
Passive[o]	DPPC:CH:DPPS OR DPPA (36.4:27.3:36.4)	30	PO[b]	24–32	<0.001[c]	~3200 (50–70 if extruded)
Passive[p]	EPC:CH (55:45)	20	PS	9–12	<0.10	100–140
Reverse phase[q]	DOPE:CH:OA (59:12:29)	15	PO[b]	10	0.0035[c]	170
Reverse phase[o]	DPPC:CH:DPPS or DPPA (4:3:4)	30	PO[b]	15–20	<0.007[c]	~325
Detergent dialysis[r]	DOPE:CH:CHEMS (60:20:20)	15	PO	11[c]	0.069[c]	150
Detergent dialysis[s]	DOPE:CH:oleic acid (40:40:20)	18	PO[b]	3–11.6	0.0002–0.0024[c]	Not indicated
Minimum volume entrapment[t]	PC:CH:CL (57:40:3)	15–20	PO, PS	50–60	0.06–0.07	Not indicated
Lipophilic[u]	EPC:CH:DMPG (60/20/20, w/w)	10	PS (unmodified); 2'-O-alkyl derivatives	5–7 (unmodified); 18–47[v] (modified)	0.000009[c]; (unmodified); <0.00006[c,v] (modified)	680
Lipophilic[w]	DOPC	16–19	MP, PO, PS	92[v] (MP); 72 (PO); 69 (PS)	<0.75[c,v]	2000–5000; reduced with sonication
Lipophilic[x]	DOPC	18	p-Ethoxy-	95[v]	0.75[c,v]	<900
Lipophilic[y]	DPPC:CH:SPDP-PE (64:35:1)	16 (tat)	Thiocholesteryl-PO	20–30[v]	<0.014[c,v]	100–140[D]
Lipophilic[z]	EPC	19	3'-thiocholesterol-2-O-methyl	8% (unmodified); 65%[v] (modified)	0.000026[c] (unmodified); 0.0002[c,v] (modified)	<100 nm[D]
Encapsulation with cationic lipids[aa]	DDAB:PC:CH (17.2:55.2:27.6)	15	PO, PS/PO	>90[v]	0.027[c]	1000–10000

Encapsulation with cationic lipids[bb]	DLPC:DOPE:TMAG (40:40:20) or DLPC:DOPE:phosphatidylserine (40:40:20)	17	PO	I.D.	I.D.	300–3000 (TMAG); ~200 (phosphatidylserine)
Encapsulation with cationic lipids[cc]	PC:DOPE:stearylamine (40:40:20)	18	PS	50.3	I.D.	200
Stabilized antisense-lipid particles (SALP)[dd]	PC:CH:DODAP:PEG-Cer-C14 or -C20 (20:45:25:10)	15–22	PS, PO	60–80 (PS); 50–60 (PO)	0.15–0.25	110 ± 30

[a] J. P. Leonetti, P. Machy, G. Degols, B. Lebleu, and L. Leserman, Proc. Natl. Acad. Sci. U.S.A. 87, 2448 (1990).

[b] Assumed.

[c] Calculated based on available data.

[d] S. Akhtar, S. Basu, E. Wickstrom, and R. L. Juliano, Nucleic Acids Res. 19, 5551 (1991).

[e] O. Zelphati, G. Zon, and L. Leserman, Antisense Res. Dev. 3, 323 (1993).

[f] Estimated based on preparation procedure.

[g] S. Wang, R. J. Lee, G. Cauchon, D. G. Gorenstein, and P. S. Low, Proc. Natl. Acad. Sci. U.S.A. 92, 3318 (1995).

[h] D. Wielbo, A. Simon, M. I. Phillips, and S. Toffolo, Hypertension 28, 147 (1996).

[i] I.D., insufficient data to calculate.

[j] A. Rahman, M. P. Selvam, R. Guirgus, and L. A. Liotta, J. Natl. Cancer Inst. 81, 1794 (1989).

[k] M. P. Selvam, S. M. Buck, R. A. Blay, R. E. Mayner, P. A. Mied, and J. S. Epstein, Antiviral Res. 33, 11 (1996).

[l] R. Morishita, G. H. Gibbons, K. E. Ellison, M. Nakajima, L. Zhang, Y. Kaneda, T. Ogihara, and V. J. Dzau, Proc. Natl. Acad. Sci. U.S.A. 90, 8474 (1993).

[m] M. Aoki, R. Morishita, J. Higaki, A. Moriguchi, I. Kida, S. Hayashi, H. Matsushita, Y. Kaneda, and T. Ogihara, Biochem. Biophys. Res. Commun. 231, 540 (1997).

[n] S. Li and L. Huang, J. Liposome Res. 8, 239 (1998).

[o] M. Fresta, R. Chillemi, S. Spampinato, S. Sciuto, and G. Puglisi, J. Pharm. Sci. 87, 616 (1998).

[p] S. K. Klimuk, S. C. Semple, P. N. Nahirney, M. C. Mullen, C. F. Bennett, P. Scherrer, and M. J. Hope, submitted for publication.

[q] C. Ropert, M. Lavignon, C. Dubernet, P. Couvreur, and C. Malvy, Biochem. Biophys. Res. Commun. 183, 879 (1992).

[r] P. G. Milhaud, J. P. Bongartz, and B. Lebleu, Drug Delivery 3, 67 (1996).

[s] D. D. Ma and A. Q. Wei, Leukocyte Res. 20, 925 (1996).

[t] A. R. Thierry and A. Dritschilo, Nucleic Acids Res. 20, 5691 (1992).

[u] J. A. Hughes, C. F. Bennett, P. D. Cook, C. J. Guinosso, C. K. Mirabelli, and R. L. Juliano, J. Pharm. Sci. 83, 597 (1994).

[v] Incorporation efficiency. Oligonucleotide is expected to be associated with both the inner and the outer monolayers of the lipid vesicles.

[w] A. M. Tari and G. Lopez-Berestein, J. Liposome Res. 7, 19 (1997).

[x] A. M. Tari, C. Stephens, M. Rosenblum, and G. Lopez-Berestein, J. Liposome Res. 8, 251 (1998).

[y] O. Zelphati, E. Wagner, and L. Leserman, Antiviral Res. 25, 13 (1994).

[z] B. Oberhauser and E. Wagner, Nucleic Acids Res. 20, 533 (1992).

[aa] P. C. Gokhale, V. Soldatenkov, F. H. Wang, A. Rahman, A. Dritschilo, and U. Kasid, Gene Ther. 4, 1289 (1997).

[bb] H. Arima, Y. Aramaki, and S. Tsuchiya, J. Pharm. Sci. 86, 438 (1997).

[cc] P. N. Soni, D. Brown, R. Saffie, K. Savage, D. Moore, G. Gregoriadis, and G. M. Dusheiko, Hepatology 28, 1402 (1998).

[dd] S. C. Semple, S. K. Klimuk, T. O. Harasym, N. Dos Santos, A. Sandhu, S. M. Ansell, P. Scherrer, and M. J. Hope, submitted for publication.

tion 10 times through two stacked polycarbonate filters (100-nm pore size), using an extrusion apparatus with a thermobarrel attachment maintained at 50°.[16] Depending on the viscosity of the sample, sequential size reduction through filters with 400-, 200-, and 100-nm pore sizes may be required. Nonencapsulated ODN is removed from the formulation by anion-exchange chromatography using DEAE-Sepharose CL-6B equilibrated in PBS.

It should be noted that the initial concentration of antisense oligonucleotide will affect the final antisense : lipid ratio but, in theory, will not influence the encapsulation efficiency. Increasing the lipid concentration enhances the encapsulation efficiency but also makes the solutions more viscous, which impedes the size reduction step. We have successfully prepared vesicles up to 250 mg total lipid and 200 mg of antisense (in 1.0 ml PBS); however, this is the upper limit for the technique.

Passive Encapsulation by Reverse-Phase Hydration

Reverse-phase hydration describes processes in which the lipid is hydrated directly from the organic solvent. The lipids are dissolved in a solvent that is partially miscible with aqueous solutions, typically ether or mixtures of chloroform and methanol. ASODN dissolved in a suitable aqueous buffer is added to the lipids to form an emulsion, thus providing a large interface between aqueous and organic phases where the lipid and nucleic acid come into contact. The emulsion can be sonicated gently to enhance lipid/ASODN mixing by increasing the surface area of the interface. The organic solvent is then removed by exposing the emulsion to either a stream of nitrogen or a partial vacuum using standard rotary evaporation equipment. As the solvent leaves, the lipid adopts bilayer structures in the aqueous solution and, in the process, encapsulates ASODN. Because liposomes prepared by reverse-phase evaporation tend to be smaller and contain fewer internal lamellae than when prepared by direct hydration, they exhibit higher trapped volumes per mole of lipid. In the absence of cationic lipids, encapsulation efficiencies as high as 20% have been reported.[17]

Encapsulation by Detergent Dialysis

This process is similar in concept to reverse-phase methods in that lipid/ASODN mixtures are cosolubilized in aqueous solutions containing detergent rather than an emulsion. Dry lipid mixtures are hydrated in buffers containing ASODN and sufficient detergent to ensure that the lipid

[16] M. J. Hope, M. B. Bally, G. Webb, and P. R. Cullis, *Biochim. Biophys. Acta* **812**, 55 (1985).
[17] M. Fresta, R. Chillemi, S. Spampinato, S. Sciuto, and G. Puglisi, *J. Pharm. Sci.* **87**, 616 (1998).

is solubilized in the form of mixed micelles. Octylglucoside is a detergent commonly employed because it has a high critical micellar concentration (CMC), making it easy to remove by dialysis. Moreover, it is nonionic and therefore does not interfere with the interaction between anionic oligonucleotides and charged lipids. In the absence of cationic lipids, low encapsulation efficiencies are typically observed (<10%).

Minimal Volume Entrapment

The terminology minimal volume entrapment (MVE) was used by Thierry and Dritschilo[18] to describe a passive hydration method that maximizes encapsulation efficiency. As the name implies, the approach is to hydrate a dry lipid film or powder with a small volume of concentrated antisense solution. Very little water is required for phospholipids to adopt a bilayer configuration so that it is possible to encapsulate as much as 70% of the hydrating solution. Once formed, the liposomes can be diluted without any loss of contents. However, size reduction by sonication or other methods must be carried out before dilution because the internal contents equilibrate with the external solution during this step. The formulation described by Thierry and Dritschilo consists of the neutral lipids PC, CH, and cardiolipin (CL), a negatively charged lipid. The presence of a negative surface charge increases the interlamellae spacing and therefore the encapsulated volume of the MLV. While good encapsulation efficiencies (50–70%) and reasonable drug : lipid ratios (0.06–0.07, w/w) are obtained, inclusion of the negatively charged lipid CL is not ideal for intravenous applications. Liposome compositions containing negatively charged lipids are eliminated rapidly from the circulation, activate complement, and may interfere with coagulation.[19–21]

Enhanced Lipophilicity

This method relies on the chemical modification of oligonucleotides to make them more lipophilic. Consequently, the modified antisense is poorly soluble in aqueous solution. The most common modifications are to the oligonucleotide backbone in which the nonbridging oxygen is substituted with a methyl group (methyl phosphonate ASODN) or ethoxy group (*p*-ethoxy-ASODN). The oligonucleotide is associated with lipid by cosolubilizing the two in an alcoholic solution (such as >90% *tert*-butanol) and then

[18] A. R. Thierry and A. Dritschilo, *Nucleic Acids Res.* **20**, 5691 (1992).
[19] A. Chonn, S. C. Semple, and P. R. Cullis, *J. Biol. Chem.* **267**, 18759 (1992).
[20] A. Chonn, P. R. Cullis, and D. V. Devine, *J. Immunol.* **146**, 4234 (1991).
[21] K. A. Mitropoulos, J. C. Martin, B. E. Reeves, and M. P. Esnouf, *Blood* **73**, 1525 (1989).

removing the solvent by lyophilization. When the dry powder is rehydrated, the hydrophobic ASODN associates with the lipid bilayer. This technique produces large liposomes (mean diameter ≤0.9 μm), with a reported entrapment efficiency of ~95%.[22–24] The size of these vesicle systems can be reduced by probe sonication or through the use of surfactants, although it has not been reported whether this occurs without a loss of material.

"Active" Encapsulation Employing Cationic Lipids

The inclusion of cationic lipids in liposome formulations has the obvious advantage of increasing electrostatic interactions between forming bilayers and polyanions, resulting in improved "active" encapsulation. While cationic lipid/ASODN vesicles can be formulated using any of the methods discussed earlier, the reference to encapsulation efficiency can become misleading. If the liposomal membrane contains a cationic lipid, polyanionic oligonucleotides may associate with both bilayer surfaces unless there is an asymmetric distribution of the lipid. Consequently, a substantial proportion of the formulated antisense may also adhere to the external surface of the liposome (or vesicle) and this fraction dissociates rapidly from the carrier in plasma. One example of a cationic lipid formulation of antisense that has been administered *in vivo* employs dimethyldioctadecylammonium bromide (DDAB). The total lipid mixture consists of PC/CH/DDAB in a molar ratio of 17:55:28 and >90% of the added oligonucleotide is associated with the lipid particles.[25] The size distribution of the resulting liposomes is heterogeneous and includes large structures 1–10 μm in diameter. The lipid particles tend to form small aggregates, most likely the result of electrostatic cross-linking by oligonucleotides. The particles were eliminated rapidly from the circulation but did show apparent activity *in vivo*. This was reflected as a 30–40% reduction in Raf-1 protein levels, relative to liposomal sense oligonucleotide, in radioresistant laryngeal squamous carcinoma cells grown as subcutaneous tumors in nude mice. Similar pharmacokinetics have been observed for antisense lipoplexes, including rapid plasma clearance, substantial accumulation in the liver and lungs, disassociation of lipid/ASODN components, and toxicities.[11,26]

[22] A. M. Tari, S. D. Tucker, A. Deisseroth, and G. Lopez-Berestein, *Blood* **84**, 601 (1994).

[23] A. M. Tari and G. Lopez-Berestein, *J. Liposome Res.* **7**, 19 (1997).

[24] A. M. Tari, C. Stephens, M. Rosenblum, and G. Lopez-Berestein, *J. Liposome Res.* **8**, 251 (1998).

[25] P. C. Gokhale, V. Soldatenkov, F. H. Wang, A. Rahman, A. Dritschilo, and U. Kasid, *Gene Ther.* **4**, 1289 (1997).

[26] D. C. Litzinger, J. M. Brown, I. Wala, S. A. Kaufman, G. Y. Van, C. L. Farrell, and D. Collins, *Biochim. Biophys. Acta* **1281**, 139 (1996).

Stabilized Antisense-Lipid Particles

We have developed stabilized antisense-lipid particles (SALP) as an intravenous delivery system, designed specifically for the treatment of systemic disease.[27] As mentioned earlier, cationic lipids are very effective in enhancing the association of oligonucleotides with lipid membranes but have several limitations when used *in vivo*. To take advantage of the enhanced ionic interactions between lipid and antisense, we employ an ionizable lipid that is cationic only at subphysiological pH. The other components include neutral structural lipids such as PC, CH and, in some instances, dioleoylphosphatidylethanolamine (DOPE). In addition, an exchangeable polyethylene glycol (PEG)-conjugated lipid is incorporated to provide a surface steric barrier that inhibits vesicle aggregation and fusion during formulation but leaves the delivery system rapidly after administration (described in more detail later).

A schematic representation of the formulation process is illustrated in Fig. 1. Lipid stocks are prepared in 100% ethanol and can be stored at $-70°$ for several weeks. Typical stock concentrations are 50 mg/ml for PC and 1,2-dioleoyl-3-dimethylammonium-propane (DODAP) and 20 mg/ml for PEG-ceramide and CH. CH is difficult to dissolve in 100% ethanol at >20 mg/ml. Brief (1–2 min) warming of the lipids at 65° may be required on removal from $-70°$ storage. To prepare a standard 1.0-ml formulation consisting of DSPC, CH, DODAP, and PEG-ceramide-C14 (20:45:25:10, mol/mol), 10 mg total lipid is added to a glass culture tube to a final ethanol volume of 0.4 ml. The lipids are mixed and warmed to 65° for 1–2 min. In a separate tube, phosphorothioate ASODN (0.5–2.5 mg, as assessed by OD_{260}) is dissolved in 0.6 ml of 300 mM citrate buffer (pH 4.0) and also warmed to 65° for 1–2 minutes. The ASODN solution is prepared immediately prior to use, therefore minimizing the potential for depurination. During formulation, it is essential that the pH is maintained below the pK of the ionizable aminolipid. This provides the necessary positive charge to facilitate an electrostatic interaction between lipid and oligonucleotide. The antisense solution is mixed vigorously while the lipids are added in a dropwise manner using a Pasteur pipette. For a 1.0-ml sample it should take >30 sec to complete the addition. If the lipid is added too quickly, some aggregation/precipitation may be observed. The resulting mixture is extruded under nitrogen pressure at 65° through three stacked 100-nm polycarbonate filters using an extrusion apparatus with a thermobarrel attachment.[16] Nitrogen pressures need not exceed 500 psi and are more

[27] S. C. Semple, S. K. Klimuk, T. O. Harasym, N. Dos Santos, A. Sandhu, S. M. Ansell, P. Scherrer, and M. J. Hope, submitted for publication.

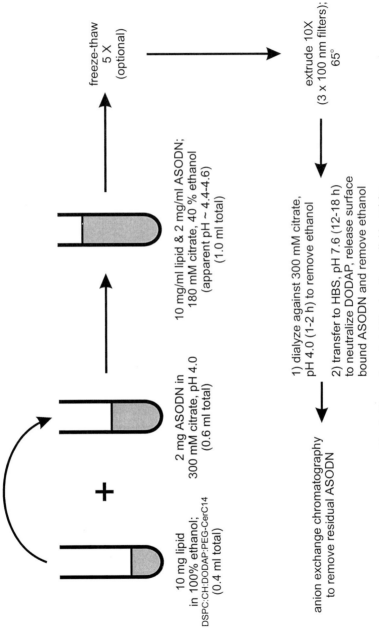

FIG. 1. Generation of stabilized antisense-lipid particles.

commonly less than 150 psi. Following extrusion, the sample is transferred immediately into prerinsed dialysis tubing (molecular weight cutoff, 12,000–14,000). Allow approximately 75–100% extra length for expansion. Dialyze in >1000 volumes of 300 mM citrate, pH 4.0, for 1 hr to remove residual ethanol. Transfer the sample to >1000 volumes HEPES-buffered saline (pH 7.6) and continue dialysis for at least 12–18 hr to equilibrate the pH and remove residual ethanol. To remove nonencapsulated ASODN, the sample is passed through a DEAE-Sepharose CL-6B anion-exchange column.

The SALP formulation procedure can be scaled up in equal volumetric ratios as required. It normally generates a population of vesicles with mean diameters that range from 100 to 130 nm. Vesicle size is usually confirmed by quasi-elastic light scattering, although methods such as freeze-fracture and cryoelectron microscopy have also been used. The encapsulation efficiency is typically determined as the oligonucleotide/lipid ratio compared before and after ion-exchange chromatography. For most oligonucleotides, encapsulation efficiencies of 65–80% and final drug/lipid ratios of 0.15 to 0.2 (w/w) are readily achieved.

Characterization of SALP and Lipid-Based Delivery Systems

Lipid-based delivery systems can enhance the pharmacological and therapeutic properties of nucleic acid drugs in four general ways. (1) Encapsulation in a lipid bilayer protects the drug from enzymes, which may either degrade or inactivate the drug in the biological milieu.[28,29] (2) Appropriately designed lipid carriers enhance the circulation lifetime of polynucleic acids from minutes to several hours, ensuring the delivery of an intact sequence to a variety of tissue sites. (3) Delivery systems on the order of 100 nm in diameter, which exhibit long circulation times, extravasate through leaky vasculature common to sites of inflammation, infection, and developing tumors, increasing the delivery of intact drug to diseased areas.[30] (4) Encapsulation prevents polyanionic drugs from binding plasma proteins, therefore reducing hemodynamic toxicities. The following section describes some of the methods used to measure these characteristics. The examples given are for SALP but the same techniques can be applied to the characterization

[28] A. R. Thierry and G. B. Takle, in "Delivery Strategies for Antisense Oligonucleotide Therapeutics" (S. Akhtar, ed.), p. 199. CRC Press, Boca Raton, FL, 1995.
[29] J. J. Wheeler, L. Palmer, M. Ossanlou, I. MacLachlan, R. W. Graham, Y. P. Zhang, M. J. Hope, P. Scherrer, and P. R. Cullis, Gene Ther., in press.
[30] S. K. Hobbs, W. L. Monsky, F. Yuan, W. G. Roberts, L. Griffith, V. P. Torchilin, and R. K. Jain, Proc. Natl. Acad. Sci. U.S.A. 95, 4607 (1998).

of any lipid-based delivery system for antisense oligonucleotides or plasmid DNA.

Serum and Nuclease Stability Assays

To demonstrate that oligonucleotides encapsulated in SALP are protected from nuclease degradation, we incubated SALP in both serum and serum-free media containing purified nuclease. Serum assays are important because protein and lipoprotein components interact with liposomes *in vivo* and can disrupt membrane structure, exposing the contents of the carrier.[13] SALP were prepared employing 1.0 mg of 5'-[^{32}P]phosphodiester or phosphorothioate antisense oligonucleotide (specific activity, 9×10^4 cpm/μl) in 0.5 ml of HBS and incubated with 2.5 ml of normal mouse serum at 37°. These conditions were chosen to mimic the actual plasma concentration of encapsulated oligonucleotide immediately following the intravenous administration of a 40-mg/kg dose into a 27-g mouse (200-μl injection volume). At various times, 10-μl aliquots were removed and combined with 90 μl of gel-loading buffer (94% formamide, 20 mM EDTA, 0.05% bromphenol blue). Samples were frozen in liquid nitrogen and stored at -70° until analysis. The integrity of the oligonucleotide was assessed on a 20% denaturing polyacrylamide sequencing gel containing 7 M urea. The samples were run at 600 V ($\sim$60 W) until the bromphenol blue-tracking dye migrated two-thirds down the gel (1.5–2 hr). The gel was covered in plastic wrap, placed on a phosphorimager screen overnight, and analyzed using a Molecular Dynamics Storm 840 imager. As an alternative to using ^{32}P-labeled ASODN, gels can be incubated for 15–30 min in SYBR Green I (Molecular Probes) nucleic acid gel stain (1 : 5000 dilution in TBE buffer, according to the manufacturer's instructions), followed by UV illumination to visualize the bands. However, the intensity of bands decreases with lower molecular weight degradation products. Intact oligonucleotide is observed for up to 48 hr when encapsulated, whereas no parent oligonucleotide remains in the nonencapsulated (free) sample after 8 hr (Fig. 2). Similar results are obtained for samples isolated from the plasma of animals after the intravenous administration of free or encapsulated antisense.

Another assay we employ frequently to evaluate the degree of oligonucleotide protection by lipid-based formulations involves incubating particles with fungal S1 nuclease. This enzyme effectively digests both phosphodiester and phosphorothioate ASODN. Free or encapsulated oligonucleotide is incubated at 55° with excess S1 nuclease (100U nuclease/μg ODN), according to the manufacturer's instructions (GIBCO-BRL, Gaithersburg, MD). Some samples are incubated with a detergent such as Triton X-100 (1% final concentration, v/v) to solubilize the lipid membrane and allow

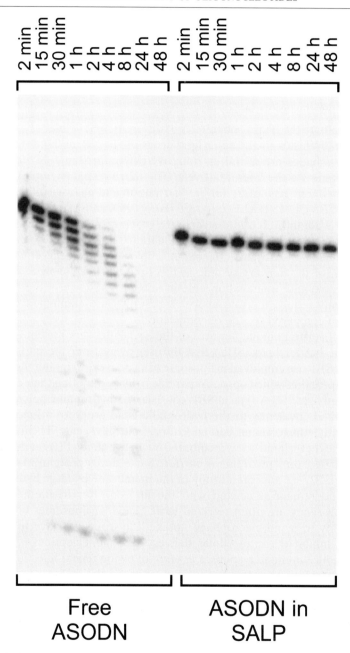

Free
ASODN

ASODN in
SALP

Fig. 2. Serum stability of phosphodiester ASODN encapsulated in SALP. The stability of 5′-[³²P]phosphodiester ASODN was evaluated in normal mouse serum as described in the text.

access to the nuclease. This demonstrates that the ASODN is only protected as long as the encapsulating membrane remains intact. At various times, aliquots are removed and processed as described earlier. Typically, free antisense is degraded within 5 min under these conditions (data not shown).

Plasma Pharmacokinetics

To evaluate the pharmacokinetics of different lipid delivery systems, we typically administer single, bolus lipid doses of 50–100 mg/kg. Intravenous injection volumes are 10 ml/kg body weight (approximately 200 μl per mouse) made via the lateral tail vein. The murine strains employed most commonly in our laboratory for biodistribution and pharmacokinetic studies are ICR, CD-1, BALB/c, or C57BL/6. An equivalent dose of free ASODN is administered into a separate group of mice for comparison to the formulated material. To measure clearance kinetics we use either ^{3}H- or ^{35}S-labeled ASODN and [^{3}H]- or [^{14}C]cholesterylhexadecyl ether (CHE) as a label for the delivery system, as this lipid has been shown to be non-exchangeable and nonmetabolizable in mice.[31] As shown in Fig. 3, SALP dramatically reduce the rate at which ASODN are eliminated from the circulation compared to free oligonucleotide.

One of several unique features of the SALP delivery system is the use of a steric barrier lipid (SBL) that diffuses away from the particle after injection. SBL are employed in liposome drug delivery to inhibit the binding of serum proteins, which opsonize particles, thus promoting their clearance by macrophages.[32,33] A typical SBL consists of a hydrophilic polymer (e.g., PEG) linked covalently to the hydrophilic head group of a lipid such as phosphatidylethanolamine, ceramide, or simple alkyl chain. The hydrophobic moieties of the lipid act as membrane anchors holding PEG at the surface of the carrier, thus providing a barrier that reduces protein interactions. Typically SBL are anchored firmly in the membrane by large hydrophobic groups and remain associated with the liposome for its lifetime in the circulation. They are often referred to as "stealth" liposomes because of their ability to avoid macrophage uptake. However, we have found that SBL can inhibit the intracellular delivery of oligonucleotides *in vitro* as well as induce immune responses *in vivo* following repeat injection-dosing schedules.[34] Moreover, despite the increase in circulation half-life observed

[31] Y. Stein, G. Halperin, and O. Stein, *FEBS Lett.* **111**, 104 (1980).

[32] S. C. Semple, A. Chonn, and P. R. Cullis, *Adv. Drug Delivery Rev.* **32**, 3 (1998).

[33] A. J. Bradley, D. V. Devine, S. M. Ansell, J. Janzen, and D. E. Brooks, *Arch. Biochem. Biophys.* **357**, 185 (1998).

[34] S. C. Semple, T. O. Harasym, K. Clow, S. M. Ansell, S. K. Klimuk, and M. J. Hope, submitted for publication.

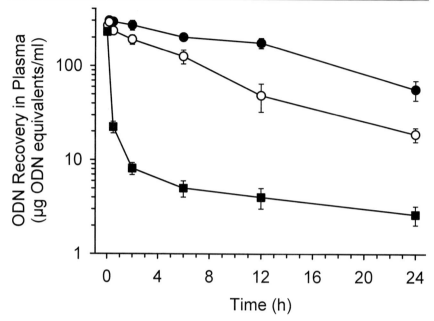

FIG. 3. Plasma clearance of SALP. The plasma clearance of vascular endothelial growth factor (VEGF) ASODN (■), SALP (PEG-Cer-C14, ○), and SALP (PEG-Cer-C20, ●) was evaluated at various times in ICR mice. ASODN was administered intravenously as a bolus dose of 15 mg/kg (lipid dose ~80 mg/kg). Each data point represents the mean and standard deviation of four animals.

for PEG-coated particles, this does not appear to enhance disease site targeting.[35] Therefore, SBL are not required for therapeutic activity but are necessary for the SALP formulation process because they prevent particle aggregation and fusion.

Consequently, we have developed SBL that associate with lipid/oligonucleotide particles during their formation and prevent undesirable particle–particle interactions but diffuse away from the carrier on injection. This is accomplished by reducing the size of the hydrophobic anchor.[36] We employ PEG-ceramides as the preferred SBL. By altering the length of the acyl chain linked to the sphingosine backbone, the rate at which PEG diffuses away from the outer surface of particles *in vivo* can be adjusted from minutes to many hours. The half-lives for PEG-ceramide-C14 and PEG-ceramide-C20 in SALP are ~5 min and 24 hr, respectively.[34] Therefore,

[35] S. K. Klimuk, S. C. Semple, P. Scherrer, and M. J. Hope, *Biochim. Biophys. Acta,* in press.
[36] M. S. Webb, D. Saxon, F. M. Wong, H. J. Lim, Z. Wang, M. B. Bally, L. S. Choi, P. R. Cullis, and L. D. Mayer, *Biochim. Biophys. Acta* **1372,** 272 (1998).

SALP containing PEG-ceramide-C14 experience more rapid clearance from the circulation than SALP containing the C20 chain, but it is still about 100-fold greater than free ASODN (Fig. 3).

Delivery to Sites of Inflammation and Tumors

We have characterized liposome delivery and extravasation to a site of inflammation using a murine model of contact hypersensitivity.[35] Mice previously sensitized to the chemical irritant 2,4-dinitrofluorobenzene (DNFB) undergo a reproducible and measurable inflammatory response when rechallenged with DNFB. This model has proven to be very convenient for assessing the disease site delivery of free and lipid-based formulations of antisense oligonucleotides. Induction of inflammation is achieved as follows:

Day 1. Shave the abdomen of ICR or BALB/c mice closely using an electric razor. Apply 25 μl of 0.5% DNFB solution (in acetone : olive oil, 4 : 1, v/v) and allow to penetrate for 1–2 min before replacing animals in cage.

Day 2. Make a second 25-μl application of a fresh 0.5% DNFB solution.

Day 6. Anesthetize the animals (halothane). Record baseline ear thickness measurements using an engineer's micrometer and apply 10 μl of 0.1% DNFB to the left ear (leaving the right ear as a control). Drug is then administered via the lateral tail vein (up to 10 ml/kg injection volume) once the mice have recovered from anesthesia. Radiolabeled or fluorescent lipids and oligonucleotides can be used to quantitate delivery to the inflamed ear.

A comparison between inflamed and noninflamed ears is used to measure the enhanced delivery to the disease site (Fig. 4A). Radiolabeled ASODN is administered as described earlier. At preferred times after initiation of inflammation (6–24 hr), mice are anesthetized with ketamine/xylazine (164 : 4, mg/kg). To remove residual blood from the inflamed tissue, mice are perfused with 10 ml of saline using a 10-ml syringe and 21-gauge needle inserted in the left ventricle of the heart. Mice are sacrificed by cervical dislocation and the ears removed with iris scissors by dissecting around the pinnae. Whole ears are placed in 7-ml glass scintillation vials. To prepare the ears for scintillation counting, 0.5–1.0 ml of Solvable tissue solubilizer is added and the ears are allowed to digest for 24–48 hr at 60–70° (small amounts of cartilage will likely remain). The digested sample is decolorized with 200 μl of 30% hydrogen peroxide for 2–3 hr at room temperature until a clear solution is obtained. Then 5.0 ml of Pico-Fluor40 scintillation cocktail (developed for tissue analyses) is added and the samples are analyzed by standard liquid scintillation counting.

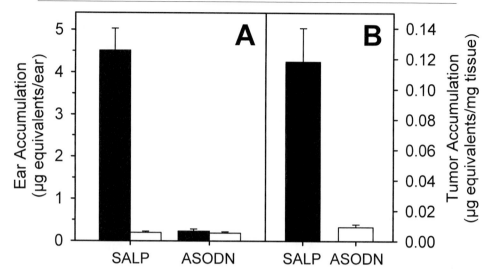

FIG. 4. Delivery of SALP to sites of disease: inflammation and tumors. The accumulation of phosphorothioate ASODN was evaluated in a murine DNFB-induced contact hypersensitivity model [(A) inflamed ear, filled bars; noninflamed ear, open bars] and in a Lewis lung carcinoma model (B) as described in the text. ASODN or SALP were administered intravenously as a single bolus dose (20 mg/kg ASODN). Each data point represents the mean and standard deviation of five animals.

Similar studies can be performed with tumor models. For example, the murine Lewis lung carcinoma has proven to be useful for evaluating liposome delivery as it exhibits rapid growth, grows well as a subcutaneous tumor, and is well vascularized. In a typical study, 5×10^5 cells are injected in the flank of C57BL/6 mice and allowed to grow to ~500 mg. A labeled sample is administered at 50- to 100-mg/kg lipid doses and delivery is evaluated at various times. Because liposomes and SALP can circulate for extended periods, it is important to exsanguinate the animals and remove as much blood as possible by perfusion. Alternatively, a correction for residual blood in the tissue can be made by measuring blood volumes using ^{51}Cr-lableled red cells, labeled liposomes, or other techniques.[37] The tumor is excised with a scalpel and collected into FastPrep tubes. Distilled water or saline is added (500 μl) to the tube and the sample is homogenized for three cycles of 8 sec at setting 5 using a FastPrep FP120 apparatus (BIO 101, Inc.). A 200-μl aliquot of the sample is added to scintillation vials and digested and processed as described for the ear samples.

[37] M. B. Bally, L. D. Mayer, M. J. Hope, and R. Nayar, in "Liposome Technology" (G. Gregoriadis, ed.), p. 27. CRC Press, Boca Raton, FL, 1993.

If the ASODN is ^{32}P- or ^{35}S-labeled, aliquots of the tissue homogenate can be extracted using standard phenol/chloroform/isoamyl alcohol extraction methods.[38] The sequence integrity of free and encapsulated ASODN can then be assessed using denaturing PAGE methods as described previously under "Serum and Nuclease Stability Assays." Injected doses of 5–10 μCi per animal enhances detection significantly, particularly in tissues where <1% of the injected dose is expected to accumulate; however, less radioactivity can still provide useful data, although the rate of detection may be considerably longer.

Enhanced Efficacy in Animal Models of Disease

The contact hypersensitization model described in the previous section is convenient for evaluating the anti-inflammatory activity of ASODN. We have studied the activity of ASODN directed against murine intercellular adhesion molecule-1 (ICAM-1). The procedure for establishing inflammation is as described earlier, except that 10 μCi/animal of [^{3}H]methylthymidine (diluted in sterile 0.9% saline) is administered intraperitoneally (500 μl) on day 5 postsensitization to prelabel bone marrow cells and circulating leukocytes. Efficacy can be monitored by evaluating the ratio of infiltrating leukocytes in the inflamed versus control ear or by inhibition of ear swelling (edema) as determined by ear thickness measurements or ear weights. Significant increases in anti-inflammatory activity were observed when the antisense was encapsulated in either conventional 100-nm PC/CH liposomes prepared by passive encapsulation or SALP.[15,27]

Toxicity

Many of the delivery systems described in previous sections, with the exception of neutral liposomes and SALP, exhibit toxicities when administered intravenously. Cationic liposome–oligonucleotide complexes can manifest nonspecific pharmacological activity as a result of complement activation, induction of cytokines, and damage to liver and other tissues. The majority of liposomes and lipid-based delivery systems are ultimately removed by fixed macrophages in the liver and spleen, whereas free ASODN distributes primarily to liver and kidney. Therefore, it is useful to evaluate clinical chemistry parameters that correlate with liver and kidney damage. After intravenous administration, we typically evaluate the plasma levels of leakage enzymes (these leak out of cells as a result of membrane damage) such as alanine aminotransferase, aspartate aminotransferase, and lactate dehydrogenase, as well as production enzymes (these provide an

[38] J. Temsamani, M. Kubert, and S. Agrawal, *Anal. Biochem.* **215**, 54 (1993).

indication of the protein synthesis and metabolic capability of cells), such as alkaline phosphatase, to monitor liver function. Kidney function can be evaluated by monitoring plasma creatinine and urea nitrogen levels. For these analyses, we have used an automated clinical chemistry analyzer (Vitros DT60II and DTSCII module; Johnson & Johnson) or similar instruments. The advantage of this method is that the instrument uses "dry chemistries" and has very low sample requirements (10 μl/test). This is particularly useful for studies involving mice. However, we have also used commercial kits (Sigma Diagnostics) with very similar results, although sample volume requirements are greater.

Because complement activation is a concern for polyanionic drugs such as antisense, we also employ a complement consumption assay as part of the lipid formulation screening process. This assay is a relatively straightforward modification of the standard CH_{50} assay used to determine complement hemolytic titer clinically. The assay is described adequately elsewhere, but basically involves an *in vitro* incubation of test sample with fresh human or rat serum, followed by an incubation with antibody (hemolysin)-sensitized sheep red blood cells (EA cells).[9,20,33] Residual complement activity is determined based on the lysis of red blood cells and the spectrophotometric determination of hemoglobin release. It is critical that appropriate controls be included in the assay, particularly at lipid concentrations >2–3 mM. These would include incubations of the diluted test sample directly with red blood cells. Direct lysis of red cells, in the absence of complement, can be a problem when employing cationic lipids. A background absorbance blank is also required (i.e., in the absence of red cells) because the lipid will cause light scattering and influence absorbance measurements at higher concentrations. The assay can also be used to evaluate formulations that have been administered directly into animals. Blood can be recovered at 30 min postinjection, allowed to clot, and the serum collected. Dilutions of serum are incubated with EA cells for 30 min at 37°, centrifuged, and the supernatant is assayed for hemoglobin release. The assay works well with rat serum but not with mouse serum.

[19] *In Vitro* Transport and Delivery of Antisense Oligonucleotides

By J. Hughes, Anna Astriab, Hoon Yoo, Suresh Alahari, Earvin Liang, Dmitri Sergueev, Barbara Ramsey Shaw, and R. L. Juliano

Introduction

Antisense oligonucleotides must enter cells and interact with pre-mRNA and mRNA in the nucleus and cytoplasm in order to exert their pharmacological and biological effects. Usually, pharmacological activity results from the binding of the antisense molecule to a target RNA, followed by degradation of the RNA by the enzyme RNase H,[1,2] although there are also examples of RNase H-independent antisense mechanisms.[3–6] In general, it seems that most types of oligonucleotides are taken up by cells through endocytosis and initially accumulate in an endosomal–lysososomal compartment.[7] A number of potential cell surface receptors for oligonucleotides have been described, including some as yet undefined proteins,[8] the MAC-1 integrin,[9] scavenger receptors,[10] and a protein that may act as a transporter for oligonucleotides.[11] Once in the cell, some of the accumulated oligonucleotide is gradually released from endosomes and enters the cytoplasm through a poorly understood mechanism. From the cytoplasm the antisense oligonucleotide can enter the nucleus rap-

[1] C. F. Bennett, *Biochem. Pharmacol.* **55,** 9 (1998).

[2] S. T. Crooke, *Adv. Pharmacol.* **40,** 1 (1997).

[3] B. F. Baker, S. S. Lot, T. P. Condon, S. Cheng-Flournoy, E. A. Lesnik, H. M. Sasmor, and C. F. Bennett, *J. Biol. Chem.* **272,** 11994 (1997).

[4] N. Mignet and S. M. Gryaznov, *Nucleic Acids Res.* **26,** 431 (1998).

[5] J. M. Kean, S. A. Kipp, P. S. Miller, M. Kulka, and L. Aurelian, *Biochemistry* **34,** 14617 (1995).

[6] H. Sierakowska, M. J. Sambade, S. Agrawal, and R. Kole, *Proc. Natl. Acad. Sci. U.S.A.* **93,** 12840 (1996).

[7] S. Akhtar and R. L. Juliano, *Trends Cell Biol.* **2,** 139 (1992).

[8] G. Q. Yao, S. Corrias, and Y. C. Cheng, *Biochem. Pharmacol.* **51,** 431 (1996).

[9] L. Benimetskaya, J. D. Loike, Z. Khaled, G. Loike, S. C. Silverstein, L. Cao, J. el Khoury, T. Q. Cai, and C. A. Stein, *Nature Med.* **3,** 414 (1997).

[10] E. A. Biessen, H. Vietsch, J. Kuiper, M. K. Bijsterbosch, and T. J. Berkel, *Mol. Pharmacol.* **53,** 262 (1998).

[11] B. Hanss, E. Leal-Pinto, L. A. Bruggeman, T. D. Copeland, and P. E. Klotman, *Proc. Natl. Acad. Sci. U.S.A.* **95,** 1921 (1998).

idly,[12,13] permitting interactions with RNA species there. Thus, in most cases, antisense molecules initially accumulate in an endosomal–lysosomal compartment that represents a pharmacological dead end. Therefore, agents or techniques that promote the initial cellular interaction between the oligonucleotide and the cell membrane, promote the transfer of oligonucleotides from endosomes to the cytosol, or allow the oligonucleotides to bypass the endosomal compartment should enhance the pharmacological effectiveness of antisense molecules.

In most cell culture studies, free antisense oligonucleotides are initially ineffective and only become active in the presence of an appropriate facilitator or delivery agent.[1,14] It may be possible to attain antisense effects without a delivery agent by utilizing high concentrations of stable oligonucleotides,[15] but this is the exception rather than the rule. A number of approaches have been used to enhance the cytoplasmic and nuclear delivery of antisense oligonucleotides. These will be discussed individually in the following sections. However, there is clearly a need for improved technologies for enhancing the delivery and pharmacological effectiveness of oligonucleotides. For example, many of the agents currently used in laboratory studies are unsuited to *in vivo* application: they are too toxic, they form very large complexes that would be cleared from the circulation, or they are not able to function in the presence of plasma proteins. It has been quite difficult to identify improved delivery agents for oligonucleotides; one problem has been that many of the assays for antisense activity are arduous. A major issue then is the development and use of appropriate screens to evaluate agents or strategies designed to enhance antisense delivery and therapeutic effects.

Methodology

Rapid Functional Assay for Antisense Delivery

In order to devise new approaches for enhancing the delivery and pharmacological effectiveness of antisense oligonucleotides, it is important to have a rapid, precise, and reproducible assay for evaluating antisense actions. In the past this has been problematic, as assays of antisense effects were based on the reduction of a signal from control levels. Because the

[12] J. P. Leonetti, N. Mechti, G. Degols, C. Gagnor, and B. LeBleu, *Proc. Natl. Acad. Sci. U.S.A.* **88,** 2702 (1991).

[13] T. L. Fisher, T. Terhorst, X. Cao, and R. W. Wagner, *Nucleic Acids Res.* **21,** 3857 (1993).

[14] J. A. Hughes, A. V. Avrutska, A. Aronson, and R. L. Juliano, *Pharmacol. Res.* **13,** 404 (1996).

[15] P. J. Gonzalez-Cabrera, P. L. Iversen, M. F. Liu, M. A. Scofield, and W. B. Jeffries, *Mol. Pharmacol.* **53,** 1034 (1998).

antisense effect was often only partial, the assay would involve measuring the difference between two large numbers, which is always an inaccurate process. Further, because a negative effect was being measured, it was often difficult to discriminate true antisense effects from simple cytotoxicity. Kole and colleagues[6,16] have devised a new assay for antisense activity that provides a positive readout. This assay is described more fully in another article of this volume. In brief, it involves using an antisense compound to correct improper splicing of an intron inserted into the gene for luciferase. Therefore, if the antisense reaches its target, binds the RNA, and corrects splicing, active enzyme is made and the signal can be quantitiated on a luminometer. In our experience, this is an excellent assay to evaluate the impact of various delivery agents. It is simple, rapid, quantitative, and not subject to concerns about cytotoxicity, as an increase in enzyme levels can only take place in a viable cell. A description of our current use of this assay follows.

HeLa cells transfected with a reporter gene construct containing a β-globin intron placed in the luciferase gene are used[16]; these cells are termed HeLa line Luc/705. These are plated in a six-well tray at a density of 3×10^5 cells per well in 3 ml of 10% fetal bovine serum (FBS)/DMEM (Dulbecco's modified Eagle's medium) and antibiotics (100 μg/ml of strep-tomycin). Cells are maintained for 24 hr at 37° in a humidified incubator (5% CO_2/95% air). Typically a 100-μl aliquot of oligonucleotide (2'-O-Me-phosphorothioate 5'-CCTCTTACCTCAGTTACA-3') at a given concen-tration in Opti-MEM is mixed with 100 μl of Opti-MEM containing various concentrations of delivery agent; alternatively, the oligonucleotide might be linked chemically to a delivery moiety. It is important to use an RNase H-independent analog of this type to allow correct processing of the mes-sage. After being mixed briefly, the preparation is left undisturbed at room temperature for about 15 min, followed by dilution to 1 ml with Opti-MEM before being layered on the cells. The cells are incubated for 6 hr, and subsequently the medium is replaced with 10% FBS/DMEM. After 18 hr, the cells are rinsed with phosphate-buffered saline (PBS) and lysed in 100 μl of lysis buffer (200 mM Tris–HCl, pH 7.8, 2 mM EDTA, 0.05% Triton X-100) on ice for 15 min. Following centrifugation (13,000 rpm, 2 min), 5 μl of supernatant cell extract is mixed with 100 μl of luciferase assay buffer [3 mM ATP, 15 mM MgSO$_4$, 30 mM Tricine, 10 mM dithiothreitol (DTT), pH 7.8] and 100 μl of luciferase substrate (1 mM D-luciferin, pH 6.8). The light emission is quantified for 10 sec using a Monolight 2010 luminometer (Analytical Luminescence Laboratory). Luciferase activity is expressed as relative light units (RLU) per well. Light emission is normalized to the

[16] S. H. Kang, M. J. Cho, and R. Kole, *Biochemistry* **37**, 6235 (1998).

protein concentration of each sample, determined according to the bicin-chonic acid assay (Pierce Chemical Co.) for protein concentration. Prior to actually testing the effect of the delivery agent using this assay, it is important to evaluate the toxicity of the agent (e.g., by trypan blue staining or by effects on cell growth rates); typically, concentrations of the delivery agent that cause less than 10% cytotoxicity should be chosen.

Use of Fluorescence Techniques to Monitor Oligonucleotide Uptake

In testing new delivery agents it is important to evaluate the total cellular accumulation of oligonucleotides, as well as their subcellular distribution. We have found fluorescence techniques to be most suited to this purpose. Although there has been concern about whether monitoring a fluorophore label will provide an accurate evaluation of the behavior of the oligonucleo-tide itself, we have generally found judicious application of fluorescent techniques to be extremely useful.[17,18]

Flow Cytometry. To measure total cell uptake, flow cytometry is invalu-able because it allows precise quantitation as well as the ability to examine individual members of the entire cell population. As an example of typical use of flow cytometery in antisense studies, we provide conditions used in the analysis of oligonucleotide uptake promoted by a peptide delivery agent described later. After exposure of cells to fluorophore-labeled oligonucleo-tides (FITC, green fluorescence), signals emitted from the samples are analyzed using a Becton-Dickinson FACSort (San Jose, CA). Green fluo-rescence is monitored with a 530/30-nm bandpass filter, and photomultiplier tube pulses are amplified logarithmically. Ten thousand cells are counted at a flow rate between 100 and 200 cells per second. Live cells are gated using their morphological properties (forward scatter and side scatter, set on logarithmic mode), and signals from dead cells are rejected. The mean fluorescence intensity of the populations of oligonucleotide treated cells is calculated using histograms and is expressed in arbitrary units using a LYSYS II software program (Becton-Dickinson). Dual fluorescence (red/green) could also be used, for example, to compare total fluorescent oligonu-cleotide uptake (green channel) to oligonucleotide effect on the expression of a target protein detected with a fluor-tagged antibody (red channel); see Vaughn *et al.*[19] for a good example of this approach.

[17] S. K. Alahari, N. M. Dean, M. H. Fisher, R. DeLong, M. Manoharan, K. L. Tivel, and R. L. Juliano, *Mol. Pharmacol.* **50,** 808 (1996).
[18] Y. Shoji, S. Akhtar, A. Periasamy, B. Herman, and R. L. Juliano, *Nucleic Acids Res.* **19,** 5543 (1991).
[19] J. P. Vaughn, J. Stekler, S. Demirdji, J. K. Mills, M. H. Caruthers, J. D. Iglehart, and J. R. Marks, *Nucleic Acids Res.* **24,** 4558 (1996).

Fluorescence Microscopy. To evaluate subcellular distribution, we feel that digital fluorescence microscopy is far more reliable than other approaches (such as cell fractionation). Although confocal microscopy is preferred, as it can use optical sectioning to discriminate externally bound from internalized fluorophore, in practice, standard fluorescence microscopy can also provide very useful information.

As a typical example of the use of fluorescence microscopy in antisense research, we describe the conditions used to study the subcellular distribution of some peptide–oligonucleotide conjugates that are described further later. In brief, the cells are plated in growth medium on alcohol-washed glass coverslips, allowed several hours to attach and spread, and then incubated with fluorophore-tagged oligonucleotides. Typically, oligonucleotide concentrations in the 100 nM to 1 μM range would be used, with incubation at 37° for several hours in serum-free culture medium. The cells are then washed several times with buffer and may be fixed if desired. Coverslips are then mounted on standard glass slides and are viewed on a Zeiss Axioscop vertical microscope using a 40× oil immersion objective. Images are captured using a Zeiss CCD video camera system, and image analysis is done with the MetaMorph imaging system (Universal Imaging Corp., West Chester, PA). It should be noted that fixation (2% formaldehyde in PBS for 30 min) can be used to retain fluorescent oligonucleotide in cells for both microscopy and flow cytometry.

Delivery Systems for Oligonucleotides

Liposomal Systems. Oligonucleotides can be encapsulated in the liposome interior, bound onto the liposome surface, and, in the case of cationic liposomes, form an electrostatic complex. Advantages of the ODN–liposome interactions are twofold: they can protect the oligonucleotide from degradation (e.g., exo- and endonucleases) and they can increase the amount of oligonucleotide that is internalized by cells. These aspects are true for either normal encapsulation or complexation with cationic lipids. The major problem with liposomes composed of natural lipids (e.g., neutral and anionic lipids) is that they are associated with low encapsulation efficiency. The low encapsulation is due to the negative charge of the oligonucleotide and its relatively high molecular weight. Oligonucleotides with hydrophobic attachments (e.g., cholesterol) have greater interactions with the lipophilic portions of liposomes and generally have greater encapsulation efficiency.[20]

For standard liposomes (neutral and anionic), two approaches may be

[20] O. Zelphati, E. Wagner, and L. Leserman, *Antiviral Res.* **25**, 13 (1994).

attempted to improve oligonucleotide encapsulation. First, a high lipid-to-oligonucleotide ratio can be used in the formation of liposomes. The major problem with this approach is that often a high percentage of the liposomes is large and multilamellar, which significantly reduces the fraction of internal aqueous space, limiting the amount of oligonucleotide per liposome. Another problem encountered during entrapment is that when oligonucleotides are added to preformed lipid films, they are not well distributed with respect to the internal/external space of the liposome. This imbalance results in a lower percentage of the oligonucleotide in the interior aqueous phase as compared to the bulk solution. Rupturing and reforming the liposomes can rectify this imbalance. Commonly employed methods include freeze thawing, freeze-drying, and rehydration; during these processes the bilayers open and close and the asymmetry of the oligonucleotide distribution is reduced. Overall, these methods can increase the ratio of amount of oligonucleotides entrapped within anionic liposomes. By working with 150 to 100 mM lipids, typically 10 to 40% encapsulation efficiency can be achieved with oligonucleotide concentrations of several milligrams per milliliter following these methods. Other methods, such as reverse-phase evaporation,[21] can entrap as much as 70% of hydrophilic molecules such as oligonucleotides.[22]

ANIONIC LIPOSOMES. A simple protocol is presented for the formation of anionic liposomes containing oligonucleotides. We have described only one formulation; multiple lipids/solvents and other modifications can be used to produce liposomes for the desired purpose. We usually obtain phospholipids from Avanti Polar Lipids (Alabaster, AL). The purity of the lipids should be checked by thin-layer chromatography before use.[23] In this formulation the following lipids were used: egg L-α-lecithin [phosphatidylcholine (PC)], a neutral lipid, dimyristoyl phosphatidylglycerol (DMPG), a negatively charged lipid, and cholesterol (CH); the molar ratio of lipids is PC/DMPG/CH, 6:2:2.[24] Addition of a negatively charged lipid decreases the rate at which the liposomes aggregate. Dissolve the lipids in chloroform (other organic solvents, such as ethanol/methanol, can be used to reduce the likelihood of toxicity). Evaporate the organic solvent carefully using a dry nitrogen stream or a rotary evaporator. Resuspend the dry lipid film using an aqueous buffer containing the oligonucleotides (typically concentration 0.5–5 mM). Suspend the lipid mixture in the aqueous buffer and

[21] F. Szoka, Jr., and D. Papahadjopoulos, *Proc. Natl. Acad. Sci. U.S.A.* **9,** 4194 (1978).

[22] B. Oberhauser and E. Wagner, *Nucleic Acids Res.* **20,** 533 (1992).

[23] R. R. C. New, Liposomes: A Practical Approach. IRL Press, Oxford, 1990.

[24] J. A. Hughes, C. F. Bennett, P. D. Cook, C. J. Guinosso, C. K. Mirabelli, and R. L. Juliano, *J. Pharm. Sci.* **83,** 597 (1994).

allow the mixture to hydrate above the transition temperature of the lipid for 30–60 min (with mild shaking). This procedure yields large, multilamellar vesicles (MLV). After formation of the liposomes, the entire formulation can be freeze-thawed (from dry ice/acetone, to 37° water) several times to increase oligonucleotide entrapment efficiency. If a particular size of liposome is required, size reduction can be accomplished via extrusion through polycarbonate membranes (Poretics, Livermore, CA) of desired size. Liposomes can be formulated easily in size ranges from 800 to 100 nm using this method. Liposomes may have to be extruded several times to obtain the desired size. If very small liposomes (100 nm) are required, serial passage through membranes should be used, starting with the largest pore size and working to smaller ones. When sizing is complete, nonencapsulated oligonucleotides can be removed by gel chromatography (Sephadex G-25 or G-50) or by centrifugation. The centrifugation speed will depend on the final size of liposome and its composition.

CATIONIC LIPOSOMES. Another possibility for the delivery of oligonucleotides is electrostatic binding of these negatively charged molecules to cationic liposomes (also termed "cytofectins"). This, the "gold standard" and most widely used delivery approach for *in vitro* oligonucleotide delivery, involves complexation of the oligonucleotides with commercially available cationic lipids such as Lipofectin, as described originally by Bennett *et al.*[25] Liposomes containing cationic lipids have been used since the late 1980s for the delivery of high molecular weight nucleic acids.[26] In this approach, the cationic lipid (e.g., DOTMA) binds the negatively charged DNA and causes a condensation between the plasmid DNA and the liposomes. However, because oligonucleotides are much smaller than plasmid DNA and cannot condense the cationic vesicles into small particles, they adsorb onto liposomes and eventually lead to their aggregation and fusion. This process can result in very large liposomes and/or lamellar particulates that precipitate. It is thus fairly important to use the cationic lipid/oligonucleotide complex within 30 min after initial mixing. The ionic strength of the incubation solution will also influence the interaction rate and final complex size. A variety of commercially obtainable cytofectins are available (Table I). In most cases, the preformed cationic lipid is mixed with the oligonucleotide and allowed to interact, and then the complex is either diluted or applied directly onto tissue culture cells. Depending on the cytofectin used, either a low serum or a serum-containing tissue culture

[25] C. F. Bennett, M. Y. Chiang, H. Chan, J. E. Shoemaker, and C. K. Mirabelli, *Mol. Pharmacol.* **41,** 1023 (1992).
[26] P. L. Felgner, T. R. Gadek, M. Holm, R. Roman, H. W. Chan, M. Wenz, J. P. Northrop, G. M. Ringold, and M. Danielsen, *Proc. Natl. Acad. Sci. U.S.A.* **84,** 7413 (1987).

TABLE I
CATIONIC LIPIDS

Cationic lipid	Commercial source
LipofectACE	Life Technologies
Lipofection	Life Technologies
LipifectAMINE	Life Technologies
CellFECTIN	Life Technologies
DMRIE-C	Life Technologies
DDAB	Sigma
DC-Chol	Sigma
DOTAP	Boehringer Manneheim, Avanti Polar Lipids
MRX-230 and MRX-220	Avanti Polar Lipids
Transfectam	Promega
TransFast	Promega
Tfx-10, Tfx-20 and Tfx-50	Promega
ProFection-CaPO$_4$	Promega
Profection-DEAE-Dextran	Promega

medium is used. Unfortunately, at this time the best way to select an appropriate cationic lipid is through trial and error. Each cell line can demonstrate individual responses in terms of both effectiveness and toxicity for the various cationic lipids. Researchers can also produce their own cationic lipid formulations from several readily purchased cationic lipids (this is less expensive than buying premade cytofectin preparations). Unlike formulations that use anionic or neutral liposomes that encapsulate the oligonucleotides, cationic lipids are preformed before being mixed with anionic oligonucleotides. The methods of formation of cationic liposomes are quite simple and are amenable to many laboratories.

In addition to the cationic lipid, a second helper lipid is often added to the formulation. In most cases the helper lipid is dioleoylphosphatidyletha-nolamine (DOPE). This lipid is thought to assist in the release of oligonucle-otides from the endosome compartments of the cell. The exact nature of the release mechanism is under investigation, but most likely involves the tendency of DOPE to form inverse hexagonal lipid structures.[27] Other helper lipids, such as cholesterol, have also been reported to be effective.[28] In most cases the ratio of cationic lipid to helper lipid is between 1:1 to 3:1. It should be stressed that all cationic lipids do not require a helper lipid for activity.

---

[27] O. Zelphati and F. C. Szoka, Jr., *Pharm. Res.* **13,** 1367 (1996).
[28] P. C. Gokhale, V. Soldatenkov, F. H. Wang, A. Rahman, A. Dritschilo, and U. Kasid, *Gene Ther.* **4,** 1289 (1997).

Typically, the two lipids are dissolved in an organic solvent (e.g., chloroform or methanol). One of the simplest combinations would be a DOTAP/DOPE 1:1 molar ratio. The solvent is removed either under vacuum or by blowing an inert gas over the solution followed by rehydration with sterile water, tissue culture media, or dextrose 5%. A typical concentration would be between 0.5 and 2 mg of cationic lipid per milliliter. The heterogeneous population of liposomes is then subject to particle size reduction. The two most common techniques used are sonication and membrane extrusion. Whichever method is used, liposomes or particles close to 100 nm are desired. Unlike the anionic liposome formulation, sonication can be used in particle size reduction. Sonication does not harm the liposomes but has been shown to degrade oligonucleotides in bulk solution. This type of formulation with cationic lipids is better described as a cationic particle than as a liposome, as normal bilayers are often not obtained. After production, the cationic particles are stable for several weeks at 4°. Cationic lipid particles are mixed with the aqueous oligonucleotide solution at a particular weight ratio (ranges from 1–10 to 1); lipid:oligonucleotide complexes are then incubated for a period of time (between 15 and 30 min).

Novel Cationic Amphiphiles. We have been interested in a novel class of cationic molecules called umbrella amphiphiles as potential oligonucleotide delivery agents.[28a] These molecules have two or more cholate surfactant moieties coupled to a polycationic tail that provides a binding site for oligonucleotides. The synthesis of these molecules has been described elsewhere.[29] We have used umbrella compounds for oligonucleotide delivery purposes as follows. A 100-μl aliquot of an oligonucleotide in Opti-MEM (Life Sciences, Grand Island, NY) is mixed with 100 μl of Opti-MEM-containing umbrella compound. The umbrella molecules are maintained as a stock (1 mg/ml) in ethanol prior to dilution into Opti-MEM. After being mixed briefly, the preparation is left at room temperature for about 15 min, followed by dilution to 1 ml with Opti-MEM or serum-containing Opti-MEM. This material is then used to treat cells: typically the time of exposure would be 6–24 hr; the final oligonucleotide concentration is typically 10 nM to 1 μM; the final amphiphile concentration is in the micromolar range. One interesting characteristic of the use of umbrella amphiphiles as delivery agents is their ability to function in the presence of serum proteins. As shown in Fig. 1, an oligonucleotide/umbrella complex elicits quite a strong antisense response, even in the presence of 50% serum. Another interesting aspect of umbrella compounds is that they do not require the

[28a] DeLong *et al.,* submitted for publication.
[29] V. Janout, M. Lanier, G. Deng, and S. L. Regen, *Bioconjug. Chem.* **8,** 891 (1997).

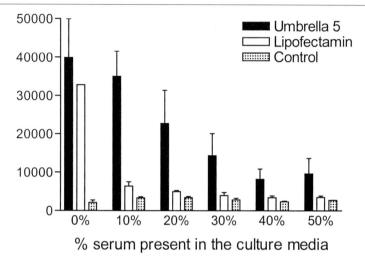

Fig. 1. Activation of luciferase activity in HeLa Luc/705 cells by oligonucleotides in free form or complexed with Lipofectamine or with an umbrella amphiphile evaluated as a function of serum concentration. Data illustrate the effect of various concentrations of serum on the effectiveness of an oligonucleotide in activating luciferase expession. For comparison sake, the same oligonucleotide was complexed with Lipofectamine.

use of a "helper" lipid such as DOPE, in contrast to most commercial cationic lipid preparations.

Biodegradable pH-Sensitive Surfactants. Oligonucleotides and their delivery systems accumulate in endosomes during their intracellular transport. Therefore, a factor limiting the pharmacological effectiveness of oligonucleotide is their inefficient transport from endosomes to their sites of action in the cytoplasm or nucleus. We have taken advantage of the acidification process that occurs during endocytosis as a trigger to increase nucleic acid release. Investigators have used several methods to enhance nucleic acid release, including viral peptides[30] and α helical-forming peptides.[14] We have attempted to use a synthetic approach to facilitate nucleic acid release from endosomes. Consequently, we have developed accessory compounds that enhance endosome to cytoplasmic transfer; this may be vital to antisense therapy. Biodegradable pH-sensitive surfactants (BPS) were developed to address potential pitfalls that cationic liposome systems might have and to further enhance the cellular delivery of oligonucleotides. Dodecyl-2-(1'-imidazolyl)propionate (DIP), a member of the BPS family, was demonstrated previously to increase the biological effect of oligonucleotides[14] and

[30] J.-P. Bongartz, A.-M. Aubertin, P. G. Milhaud, and B. Lebleu, *Nucleic Acids Res.* **22,** 4681 (1994).

of plasmid DNA.[31] Due to the properties of BPS compounds, they can be incorporated easily into anionic, neutral, or cationic liposomes. Because cationic liposomes also demonstrate the ability to facilitate the endosome transfer of nucleic acids, there may be a greater rational for use of BPS in anionic and neutral liposome formulations. We have used fluorescent methodologies to demonstrate the effectiveness of BPS agents on oligonucleotide intracellular trafficking.

Two different lipid formulations were utilized: L-α-lecithin liposomes and DIP plus L-α-lecithin liposomes (molar ratio 0.3). The lipid rehydration method described earlier was used to form neutral liposomes containing fluorescein isothiocyanate (FITC)-labeled oligonucleotides in the aqueous phase (0.5 mM). To increase encapsulation efficiency, five freeze–thaw cycles were employed after a 30-min period of shaking the tubes by hand (see earlier discussion). The liposomes were then passed three times through 600-nm polycarbonate membranes (Poretics, Livermore, CA) using a high-pressure extruder (Lipex Biomembrane Inc., Vancouver, Canada). The size of the liposomes, as measured by a volume–weight Gaussian distribution, was determined to be 630 ± 165 nm (standard deviation) by a dynamic light-scattering method using a NICOMP 380 ZLS Zeta potential/particle sizer (Santa Barbara, CA).

The liposomal preparations of oligonucleotides were added to CV-1 (monkey kidney fibroblast) cells grown on coverslips. After 4 hr of incubation in serum-free medium, 10% fetal bovine serum was added. At the end of each sampling time (4, 8, and 24 hr), the wells were washed with PBS and cells were fixed in 2% formaldehyde for 30 min. The coverslips were then transferred to the slides with Gel/Mount (Biomeda). Cellular uptake and distribution of the FITC-oligonucleotides were then evaluated. Cells incubated with FITC-labeled ONs were imaged using a Bio-Rad MRC-600 scanning confocal system equipped with a krypton/argon laser. Images were collected on an Olympus IMT-2 inverted microscope using the 488/568-nm line at which the excitation light was attenuated with a 1% neutral density filter to minimize photobleaching and photodamage. Figure 2 demonstrates the effect of the incorporation of DIP into a liposome formulation containing fluorescently labeled oligonucleotides. The photomicrograph indicates that a substantial redistribution of the oligonucleotides to cytoplasmic and nuclear compartments occurs within the cells treated with DIP containing liposomes.

Dendrimers. There has been considerable interest in dendrimers as delivery agents for antisense compounds. Cationic dendrimers, which are repetitively branched polymers usually terminating in amino groups, have

[31] E. Liang and J. A. Hughes, *Biochem. Biophys. Acta* **1369,** 39 (1998).

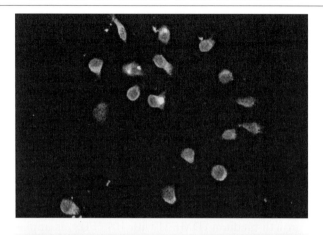

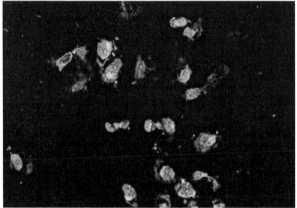

FIG. 2. The effect of pH-dependent surfactant DIP on liposome-mediated delivery of oligonucleotides to cells. (Top) Cells treated with control liposomes containing FITC oligonucleotide. (Bottom) Cells treated with DIP–liposomes and FITC oligonucleotide.

been used as delivery agents for both genes and oligonucleotides.[32–36] One interesting aspect of dendrimers is that the delivery complex can be of relatively modest molecular size,[34] as opposed to the situation with cationic liposomes. Dendrimers can also be modified covalently with peptides or

[32] J. Haensler and F. C. Szoka, Jr., *Bioconjugate Chem.* **4,** 372 (1993).

[33] M. X. Tang, C. T. Redemann, and F. C. Szoka, Jr., *Bioconjug. Chem.* **7,** 703 (1996).

[34] R. DeLong, K. Stephenson, T. Loftus, S. K. Alahari, M. H. Fisher, and R. L. Juliano, *J. Pharm. Sci.* **86,** 762 (1997).

[35] S. K. Alahari, R. K. DeLong, M. Fisher, N. M. Dean, P. Viliet, and R. J. Juliano, *J. Pharm. Exper. Ther.* **286,** 419 (1998).

[36] S. W. Poxon, P. M. Mitchell, E. Liang, and J. A. Hughes, *Drug Delivery* **3,** 255 (1996).

carbohydrates,[37,38] thus providing possible opportunities to target the complex to different cell types. We have explored the use of several types of dendrimers as delivery agents, evaluating the results both by use of the luciferase assay described earlier and by fluorescence studies. The oligonucleotide/dendrimer complexes were formed by adding oligonucleotide in 100 μl of Opti-MEM to an equal volume of dendrimer (Dendritech, Midland, MI) (generations four to seven) in Opti-MEM and incubating for 15 min at room temperature. Dendrimers are usually stored as aqueous solutions at concentrations of 20 mg/ml. Complexes (200 μl) containing oligonucleotide and varying proportions of a given generation dendrimer are diluted to 1 ml with serum-free medium (the final concentration of dendrimer is typically 2–4 μM). Cells are incubated at 37° for 6 hr with the dendrimer/oligonucleotide complexes and then washed with Opti-MEM and 2 ml of growth medium (DMEM with 10% FBS) is applied. After an additional incubation period of 18 hr, cells are lysed and analyzed for expression of luciferase as described earlier. Relative concentrations of cationic dendrimer and oligonucleotide are treated as weight ratios (cationic polymer μg/oligonucleotide μg) or, if the primary amine content is known, as charge ratios (cationic polymer primary amines/anionic oligonucleotide phosphates). Typically charge ratios (cationic to anionic) of about 30/1 have provided effective delivery complexes in serum-free medium. Dendrimers are also capable of delivering antisense compounds in the presence of serum. A commercial "fractured" dendrimer preparation (Superfect, Qiagen) is available for gene and oligonucleotide delivery.

Transport-Enhancing Peptides. Most oligonucleotide delivery strategies involve the formation of rather large molecular complexes between the oligonucleotide and the delivery moiety. An alternative strategy for oligonucleotide delivery is to couple the oligonucleotide to a relatively small molecule that can enhance its entry into cells. Although peptides and proteins do not penetrate across cell membranes, a number of unusual polypeptides have been identified that cannot only enter cells readily but can even carry other covalently attached molecules along. There are several possible types of these "transport-enhancing peptides." One type comprises pH-sensitive fusogenic peptides based on sequences derived from viral proteins such as the influenza virus hemagglutinin.[39,40] Another type comprises designed pH-sensitive amphipathic helix peptides, with the prototype being

[37] J. C. Spetzler and J. P. Tam, *Pept. Res.* **9**, 290 (1996).
[38] J. P. Thompson and C. L. Schengrund, *Glycoconj. J.* **14**, 837 (1997).
[39] H. C. Ha, N. S. Sirisoma, P. Kuppusamy, J. L. Zweier, P. M. Woster, and R. A. Casero, *Proc. Natl. Acad. Sci. U.S.A.* **95**, 11140 (1998).
[40] C. Pichon, I. Freulon, P. Midoux, R. Mayer, M. Monsigny, and A. C. Roche, *Antisense Nucleic Acid Drug Dev.* **7**, 335 (1997).

the "GALA" peptide.[41,42] Both of these types of peptide can cause destabilization and fusion of bilayer membranes in a pH-sensitive manner and have been used with some success in gene transfection studies[43] and for oligonucleotide delivery.[30,40] Several other peptides have been identified that are obviously not amphipathic helices, but that also seem able to penetrate membranes and carry other substances into the cytoplasm of cells. This includes a 35 amino acid sequence (TAT) from the human immunodeficiency virus (HIV) Tat protein,[44,45] a 16 amino acid sequence (ANT) from the *Drosophila* antennapedia protein,[46,47] and a short sequence from the signal peptide segment of a FGF.[48,49] Although the precise mechanisms involved are still unclear, these peptides have shown a remarkable ability to enhance the cytoplasmic delivery of substances that do not ordinarily enter cells. Thus, transport-enhancing peptides may be very promising tools for improving the delivery of oligonucleotides both in cell culture and in the *in vivo* setting.

The Tat protein from HIV is a transactivator that modulates transcription through its interaction with the TAR element in the HIV LTR. A peptide from Tat (49–57) that contains the basic sequence RKKRRQRRR has been shown to possess membrane binding and uptake activity. The mechanism by which Tat traverses a membrane and the precise intracellular location of this event remain unclear. However, Tat binds efficiently to cells around 10^7 sites per cell and is then internalized by an adsorptive endocytosis process.[50] The ability of HIV 1 Tat to traverse a membrane was used to transport macromolecules such as β-galactosidase, horseradish peroxidase, RNase A, and domain III *Pseudomonas* exotoxin A.[44,45]

We decided to use peptide from Tat (49–57) that contains the basic sequence as a delivery agent for antisense oligonucleotides. As a first trial,

[41] R. A. Parente, L. Nadasdi, N. K. Subbarao, and F. C. Szoka, Jr., *Biochemistry* **29**, 8713 (1990).
[42] T. B. Wyman, F. Nicol, O. Zelphati, P. V. Scaria, C. Plank, and J. Szoka, *Biochemistry* **36**, 3008 (1997).
[43] C. Plank, B. Oberhauser, K. Mechtler, C. Koch, and E. Wagner, *J. Biol. Chem.* **269**, 12918 (1994).
[44] S. Fawell, J. Seery, Y. Daikh, C. Moore, L. L. Chen, B. Pepinsky, and J. Barsoum, *Proc. Natl. Acad. Sci. U.S.A.* **91**, 664 (1994).
[45] L. L. Chen, A. D. Frankel, J. L. Harder, S. Fawell, J. Barsoum, and B. Pepinsky, *Anal. Biochem.* **227**, 168 (1995).
[46] D. Derossi, A. H. Joliot, G. Chassaing, and A. Prochiantz, *J. Biol. Chem.* **269**, 10444 (1994).
[47] D. Derossi, G. Chassaing, and A. Prochiantz, *Trends Cell Biol.* **8**, 84 (1998).
[48] Y.-Z. Lin, S. Y. Yao, R. A. Veach, T. R. Torgerson, and J. Hawiger, *J. Biol. Chem.* **270**, 14255 (1995).
[49] S. Dokka, D. Toledo-Velasquez, X. Shi, L. Wang, and Y. Rojanasakul, *Pharm. Res.* **14**, 1759 (1997).
[50] D. A. Mann and A. D. Frankel, *EMBO J.* **10**, 1733 (1991).

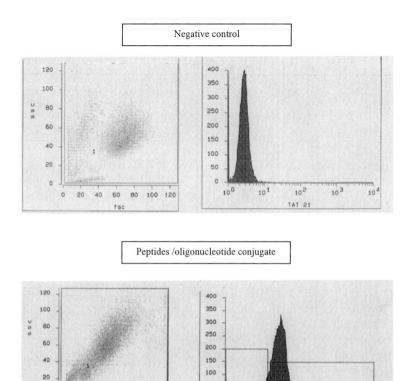

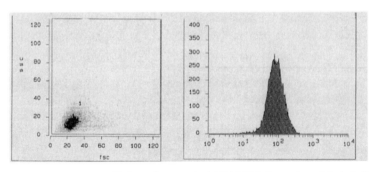

FIG. 3. Flow cytometry of HeLa cells treated with TAMRA fluorophore-labeled oligonucle-otides. Data illustrate the effect of Lipofectin or a TAT peptide–oligonucleotide conjugate on the cell uptake of oligonucleotide. In each case the left-hand panel shows the light scatter characteristics of the cell population, whereas the right-hand panel indicates the total cell uptake measured with the TAMRA fluor that is linked covalently to the oligonucleotide. The negative control represents cells treated with a TAMRA oligonucleotide but no delivery agent.

a 2′-O-Me 21-mer oligonucleotide targeted to the 705 splice site[16] was conjugated with RKKRRQRRRPPQC from the HIV Tat protein and labeled with a TAMRA fluorescence marker. This peptide/oligo conjugate was used in HeLa cells to evaluate the cellular uptake by flow cytometry and intracellular distribution using fluorescence microscopy. It is also possible to examine the biological activity of the conjugate using the Luciferase-based splice correction assay described earlier.

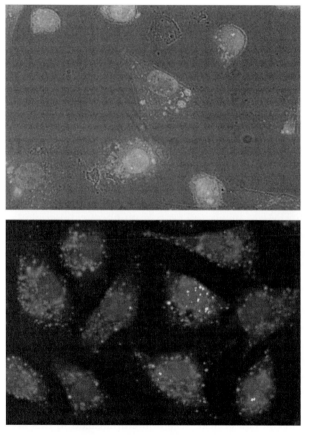

FIG. 4. Fluorescence micrographs of HeLa cells treated with TAMRA oligonucleotide plus Lipofectamine (top) or a peptide–oligonucleotide–TAMRA conjugate (bottom). Data illustrate the effect of conjugation with a TAT peptide on the subcellular distribution of oligonucleotide in the HeLa Luc/705 cell line. For comparison sake, the same oligonucleotide was complexed with Lipofectamine. As shown, both the cationic lipid and the delivery peptide moiety result in substantial nuclear accumulation. A free oligonucleotide would show far less total uptake and little nuclear localization (not shown).

Preparation of the peptide/oligonucleotide conjugate included synthesis and high-performance liquid chromatography (HPLC) purification of 5'-RS-S, 3'-NH$_2$-modified oligonucleotides. The next procedures were activation of the RS-S-5' group of the oligonucleotides with a 2-pyridylthio moiety, coupling the resulting PyS-S-5'-oligo-3'-NH$_2$ with a N-hydroxysuccinimidyl ester of the red fluor TAMRA, and subsequent isolation of the product from the excess of unreacted dye. Finally, the conjugation of a cysteine-terminated peptide with an oligonucleotide via an S-S linkage was done, followed by HPLC purification.

Further experimental protocols were similar to those described earlier or previously.[35] Briefly, HeLa cells line Luc/705 were grown in 162-mm flasks to 95% confluency and then seeded into 100-mm dishes at 5×10^6/dish in 10% FCS/DMEM and incubated for 24 hr. The cells were washed twice with PBS. Peptide/oligonucleotide conjugates, oligonucleotide complexed with Lipofectin (20 μg/ml) as a positive control, and oligonucleotides with no delivery agent as negative control were mixed in Opti-MEM and incubated with cells at 37° overnight. The final concentration for all oligonucleotides was 1 μM. The cells were removed with trypsin/EDTA and split for fluorescence microscopy or flow cytometry analysis (Fig. 3). For flow cytometry, half of the cells were resuspended in 500 μl PBS and measured for the accumulation of TAMRA marker using a Becton-Dickinson flow cytometry with Cicero software. Half of the cells were resuspended in 1 ml 10% FCS/DMEM incubated for 6 hr on fibronectin-coated cover slides. Microscopic analysis (Fig. 4) was performed using Zeiss fluorescence microscope (oil objective 40×) with a color CD camera connected to a PC with image analysis software, as described earlier.

Summary

A variety of techniques are currently available to enhance the cellular uptake and pharmacological effectiveness of antisense oligonucleotides in the *in vitro* setting. The choice of technique will depend on the context of investigation, the likelihood of cytotoxicity due to the delivery agents, and the ease and convenience of the approach. The considerations for the delivery of antisense molecules in the *in vivo* setting[51] are likely to be quite different from the cell culture situation emphasized in this article.

[51] R. L. Juliano, S. Alahari, H. Yoo, R. Kole, and M. Cho, *Pharm. Res.,* in press.

[20] *In Vitro* and *in Vivo* Delivery of Antisense Oligodeoxynucleotides Using Lipofection: Application of Antisense Technique to Growth Suppression of Experimental Glioma

By Akira Matsuno, Tadashi Nagashima, Haruko Katayama, and Akira Tamura

Introduction

Antisense strategy for the prevention of the function of the target gene was first described by Belikova *et al.* in 1967.[1] Since then, antisense strategy using synthetic oligodeoxynucleotides has been applied to the suppression of specific gene expression, related to tumorigenesis, growth, or development of tumors.[2–5] Antisense strategy using synthetic oligodeoxynucleotides can also suppress the biological activity of the corresponding gene, serving as one of the effective therapeutic choices.[6,7] These works range from a pioneer in cell-free extracts and cultured cells[2,8] to more recent *in vivo* experiments.[9,10] Because synthetic oligodeoxynucleotides have more advantages over synthetic oligoribonucleotides, the antisense technique using synthetic oligodeoxynucleotides has become a widely used method for the modulation of various gene expressions and its biological activity. This article describes an antisense technique using an *in vitro* and *in vivo* delivery system of lipofection and demonstrates the growth suppression of rat C6 glioma cells utilizing the antisense synthetic oligodeoxynucleotide for mi-

[1] A. M. Belikova, V. F. Zarytova, and N. I. Grineva, *Tetrahedron Lett,* **37,** 3557 (1967).

[2] P. C. Zamecnik and M. L. Stephenson, *Proc. Natl. Acad. Sci. U.S.A.* **75,** 280 (1978).

[3] J. Goodchild, *in* "Oligodeoxynucleotides: Antisense Inhibitors of Gene Expression" (J. S. Cohen, ed.), Vol. 12, p. 53. Macmillan, London, 1989.

[4] J. S. Cohen, *in* "Gene Regulation: Biology of Antisense RNA and DNA" (R. P. Erickson and J. G. Izant, eds.), Vol. 1, p. 247. Raven, New York, 1991.

[5] L. Neckers, L. Whitesell, A. Rosolen, and D. A. Geselowitz, *CRC Crit. Rev. Oncogen.* **3,** 175 (1992).

[6] M. Rothenberg, G. Johnson, C. Laughlin, I. Green, J. Cradock, N. Sarver, and J. S. Cohen, *J. Natl. Cancer Inst.* **81,** 1539 (1989).

[7] C. A. Stein and Y. C. Cheng, *Science* **261,** 1004 (1993).

[8] P. S. Miller, L. T. Braiterman, and P. O. Ts'o, *Biochemistry* **16,** 1988 (1977).

[9] W. B. Offensperger, S. Offensperger, E. Walter, K. Teubner, G. Igloi, H. E. Blum, and W. Gerok, *EMBO J.* **12,** 1257 (1993).

[10] T. Skorski, D. Perrotti, M. Nieborowska-Skorska, S. Gryaznov, and B. Calabretta, *Proc. Natl. Acad. Sci. U.S.A.* **94,** 3966 (1997).

crotubule-associated protein (MAP) 1A mRNA with emphasis on the utility of this system.

Antisense Strategy

Aim of Antisense Strategy: Loss of Gene Function. Two different methods have been applied to the analysis of various gene function, i.e., gain of function and loss of function. Antisense strategy, together with a gene knockout model and dominant-negative method, is oriented for loss of gene function. Antisense synthetic oligodeoxynucleotide-mediated gene inhibition can be used to identify the physiological and pathophysiological role of the target gene. It has various advantages, such as technical simplicity and possible application for medical treatment. Antisense synthetic oligodeoxynucleotide-mediated gene inhibition has only a transient efficiency; however, this property serves for the analysis of genes related to development and differentiation through studies on the sequel evoked by a transient switch-off of genes.

Mechanism of Loss of Gene Function by Antisense Strategy. Antisense synthetic oligodeoxynucleotides complementary to a particular DNA are considered to bind to the gene, including this sequence through Hoogsteen base pairing, creating a triple-helical structure that cannot be transcribed. Similarly, antisense synthetic oligodeoxynucleotides complementary to a targeted mRNA sequence prevent its translation into the final protein product. RNA:DNA hybrids, i.e., duplexes of mRNA and antisense synthetic oligodeoxynucleotide, are known to be digested by the activity of intracellular RNase H, which is a ubiquitous enzyme that recognizes RNA:DNA hybrids and cleaves RNA moiety. The use of unmodified oligodeoxynucleotides by the pioneers led to the discovery that this activity of RNase H plays an essential role in translation inhibition.[11,12] The other mechanisms of antisense synthetic oligodeoxynucleotides are also reported. Antisense synthetic oligodeoxynucleotides can bind to the opened double-helix structure and inhibit the transcription of the targeted gene. Antisense synthetic oligodeoxynucleotides, including the correspondent sequence of the intron, are supposed to bind the immature RNA and interfere with the splicing of RNA. They can suppress the intracellular translocation of mRNA from the nucleus to the cytoplasm.

Rational Sequence Design of Antisense Synthetic Oligodeoxynucleotides. Three important characteristics, i.e., selectivity, stability, and solubility,

[11] C. Cazenave, C. A. Stein, N. Loreau, N. T. Thuong, L. M. Neckers, C. Subasinghe, C. Helene, J. S. Cohen, and J. J. Toulme, *Nucleic Acids Res.* **19,** 4255 (1989).
[12] J. Minshull and T. Hunt, *Nucleic Acids Res.* **14,** 6433 (1986).

should be required for antisense synthetic oligodeoxynucleotides. The rational sequence of antisense synthetic oligodeoxynucleotides depends on the target sequence of the gene. For suppression of the gene through creating a triple-helical structure, antisense synthetic oligodeoxynucleotides should be designed complementary to the A + G-rich sequence that is supposed to be important for the function of the gene. For the inhibition of translation, antisense synthetic oligodeoxynucleotides corresponding to the region of translational initiation are commonly utilized.

To increase the chances of targeting a unique sequence, one may make the antisense oligodeoxynucleotides long enough. However, this long sequence leads to decreased specificity. The longer antisense sequence may generate a lot of partially complementary sequences, and this long duplex may yield mismatched complexes that can be a substrate for RNase H. Therefore, the ideal length of antisense synthetic oligodeoxynucleotides is considered to be 15- to 30-mer because this length of sequence has a benefit in the efficiency of cellular uptake. A computer-assisted homology search is essential for the design of antisense synthetic oligodeoxynucleotides. The effectiveness and selectivity of antisense synthetic oligodeoxynucleotides are dependent critically on the recognition of the most suitable target sequence, which can be determined by a commercially available service (Biognostik GmbH, Göttingen, Germany).

Purification of Antisense Synthetic Oligodeoxynucleotides. Antisense synthetic oligodeoxynucleotides are purified after being synthesized using a DNA synthesizer. A simple high-performance liquid chromatography (HPLC) purification is insufficient. They should be purified through at least two HPLC steps, electrophoresis, and ultrafiltration.

Stability and Modification of Antisense Synthetic Oligodeoxynucleotides. In the early days of antisense technology, ordinary oligodeoxynucleotides were used, which are degenerated rapidly through exonuclease within cells and culture medium. This short-life property of ordinary oligodeoxynucleotides prompted the development of modified, nuclease-resistant oligodeoxynucleotides. Phosphodiester oligodeoxynucleotides, similar to native DNA and RNA, are degenerated easily by serum and intracellular nucleases with a half-life of around 20 min.[13] Thereafter, synthetic oligodeoxynucleotides resistant to nucleases have been created by successful manipulation of the backbone. One of them is phosphorothioate oligodeoxynucleotide, created by isosteric substitution of sulfur for oxygen on the phosphorus residue that retains the negative charge of phosphodiester oligodeoxynucleotide. In serum and tissue culture experiments, this molecule has a half-life of

[13] H. Sands, L. J. Gorey-Feret, A. J. Cocuzza, F. W. Hobbs, and D. Chidester, *Mol. Pharmacol.* **45**, 932 (1994).

over 12 hr.[13,14] This analog is used mostly for antisense application[7] because of its good nuclease resistance and the ability to elicit RNase H activity.

Lipofection as Delivery System of Antisense Synthetic Oligodeoxynucleotides

Lipofection is one of the recently developed methods for the transfection of oligodeoxynucleotides. This method has several advantages, such as technical simplicity, relatively high transfection efficiency, and lower cytotoxicity. Lipofection has a 10–40% efficiency of transfection, which is higher than that of calcium phosphate coprecipitation.[15] From the viewpoint of efficiency, microinjection and electroporation that have higher rate of 50–100% and 40–80%, respectively, may be advantageous. Nevertheless, we usually utilize the lipofection method because of its technical simplicity. Cationic lipid vesicles encircle and bind to anionic oligonucleotides. This lipid–oligonucleotide complex binds to the anionic cell membrane and is considered to be taken up through phagocytosis or a membranous fusion mechanism and to be finally transported to the nucleus. Various reagents are available commercially.

Control Experiment of Antisense Strategy

To interpret precisely the biological effect induced by antisense synthetic oligodeoxynucleotides, several negative control studies are required. It is essential to confirm the negative effect in the experiments using both sense and scramble sequences of oligodeoxynucleotides. The sense sequence should not be used as the only control, and the scramble sequence in which the same composition is arranged in a different order should be utilized as one of the negative control studies. The other alternatives of negative control studies are experiments using mismatched and mixed sequences of oligodeoxynucleotides. One or several mismatches introduced in the antisense sequence can weaken the biological effect. The introduction of the four bases at each position can also be expected as a good negative control.

Growth Suppression of Experimental Glioma Cells Using Antisense Synthetic Oligodeoxynucleotides

Antisense Strategy Targeted to MAP 1A mRNA. We have applied the antisense synthetic oligodeoxynucleotides technique using the delivery sys-

[14] J. M. Campbell, T. A. Bacon, and E. Wickstrom, *J. Biochem. Biophys. Methods* **20**, 259 (1990).
[15] M. Strauss, *in* "Antisense Technology, a Practical Approach" (C. Lichtenstein and W. Nellen, eds.), p. 229. IRL Press, New York, 1997.

tem of lipofection to the experiment of growth suppression of rat glioma cells. The target gene to which we have applied the antisense strategy is MAP 1A.

Microtubules, which are composed of tubulin and MAPs, are the ubiquitous cytoskeletal structural components that are involved in intracellular transport. MAPs are known to mediate the binding of membranous organelles, actin, and intermediate filaments to microtubules. They constitute spindle fibers that are present in the mitotic stage of cells and are considered to be important for various cell cycle processes.[16] MAP 1A, one of the well-known high molecular weight MAPs found in microtubules isolated from brain tissue, is more generally distributed than MAP 2 and is found both in dendrites and axons of neurons and in glial cells.[17,18] MAP 1A is also known to be present in various cultured glioma cells, such as rat C6 glioma cells.[19,20,21] MAP 1A is believed to be involved in stabilizing microtubules[16] and, therefore, it is considered that the inhibition of MAP 1A may result in the instability of microtubules and may influence various kinetic behaviors of cells. The suppression of MAP 1A mRNA expression using an antisense oligodeoxynucleotide can be considered to inhibit the proliferation of rat C6 glioma cells.

The sequence of the antisense phosphorothioate oligodeoxynucleotide for MAP 1A mRNA is GAA CTC AGC CAC ACC ATC CAT, corresponding to the sequence of the translational initiation region of MAP 1A mRNA.[22] The phosphorothioate oligodeoxynucleotides of antisense, sense, and scramble sequence were synthesized with a DNA synthesizer (Applied Biosystems Model 392; Foster City, CA).

Preparation of C6 Glioma Cells. The rat C6 glioma cell is a well-established cell line and has been well characterized. Two thousand C6 glioma cells are seeded into 35 × 10-mm tissue culture dishes (Becton Dickinson and Company, Lincoln Park, NJ) and chamber slides (Nunc, Inc., Naperville, IL) containing Eagle's minimum essential medium supplemented with 10% (v/v) fetal bovine serum (FBS), 1× nonessential amino acids, 300 μg/ml L-glutamate, 10% (w/v) pyruvate, 0.14% (w/v) sodium bicarbonate, and

[16] G. S. Bloom, *in* "Tubulin and Associated Proteins: Guidebook to the Cytoskeletal and Motor Proteins" (T. Kreis and R. Vale, eds.), p. 108. Oxford Univ. Press, New York, 1993.
[17] G. S. Bloom, T. A. Schoenfeld, and R. B. Vallee, *J. Cell Biol.* **98,** 320 (1984).
[18] G. Huber and A. Matus, *J. Neurosci.* **4,** 151 (1984).
[19] C. Koszka, F. E. Leichtfried, and G. Wiche, *Eur. Cell. Biol.* **38,** 149 (1985).
[20] G. Wiche, and M. A. Baker, *Exp. Cell. Res.* **138,** 15 (1982).
[21] G. Wiche, E. Briones, C. Koszka, U. Artlieb, and R. Krepler, *EMBO J.* **3,** 991 (1984).
[22] A. Langkopf, J. A. Hammerback, R. Muller, R. B. Vallee, and C. C. Garner, *J. Biol. Chem.* **15,** 16561 (1992).

streptomycin (0.5 mg/ml)–ampicillin (500 units/ml). The cells are incubated at 37° in a humidified atmosphere of 5% (v/v) CO_2 in air.

In Vitro Treatment of C6 Glioma Cells with Phosphorothioate Oligodeoxynucleotide

After incubation in the just-mentioned culture medium for 3 days at 37° in a humidified atmosphere of 5% CO_2 in air, C6 glioma cells are cultured in serum- and antibiotics-free medium containing 1 μM antisense phosphorothioate oligodeoxynucleotide for MAP 1A mRNA and 5 $\mu g/ml$ Lipofectin (GIBCO-BRL, Gaithersburg, MD). The control experiments are C6 glioma cells incubated with phosphorothioate oligodeoxynucleotides of the sense and scramble sequence and those without oligodeoxynucleotide treatment. For the colony-forming assay, C6 glioma cells in the seven 35×10-mm tissue culture dishes of each group are incubated for 12 and 34 hr, respectively. C6 glioma cells incubated for 34 hr in the chamber slide of each group are used for the light microscopic immunohistochemical staining of MAP 1A. Cells treated with fluorescein isothiocyanate (FITC)-labeled phosphorothioate oligodeoxynucleotides are used for the evaluation of the efficiency of their uptake by C6 glioma cells.

Evaluation of Efficiency of Uptake of Phosphorothioate Oligodeoxynucleotides by C6 Glioma Cells. C6 glioma cells are incubated with FITC-labeled phosphorothioate oligodeoxynucleotides for 34 hr in the Nunc chamber slide. They are inspected under light microscopy, and the immunofluorescence signals and their cellular distribution are observed.

The immunofluorescence signals are observed to be concentrated in nuclei of the C6 glioma cells that are incubated with FITC-labeled antisense, sense, and scramble phosphorothioate oligodeoxynucleotides (Fig. 1a). Only background signals are observed in C6 glioma cells without treatment of FITC-labeled phosphorothioate oligodeoxynucleotides (Fig. 1b).

Suppression of Expression of MAP1A: Light Microscopic Immunohistochemical Staining of MAP 1A. C6 glioma cells incubated in Nunc chamber slides are used for immunohistochemical staining of MAP 1A. The endogenous peroxidase activity is blocked with 0.3% H_2O_2 in methanol for 30 min. After pretreatment with normal sheep sera, the slides are immunostained with commercially available antibodies, anti-MAP-1A antibody [mouse, monoclonal antibody, 1:100 diluted in bovine serum albumin (BSA)–0.01 M phosphate-buffered saline, pH 7.4 (PBS), Sigma Immunochemicals, St. Louis, MO] at room temperature for 1 hr. The horseradish peroxidase (HRP)-linked antibody against mouse immunoglobulin (IgG) (sheep, 1:50 diluted in BSA-PBS, Amersham International plc., Buckinghamshire, UK) is applied as the secondary antibody for 30 min. The immu-

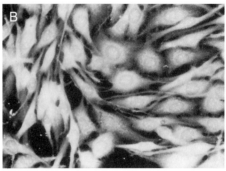

FIG. 1. Immunofluorescence signals were observed to be concentrated in nuclei of C6 glioma cells that were incubated with FITC-labeled antisense (a), sense, and scramble phosphorothioate oligodeoxynucleotides. Only background signals were observed in C6 glioma cells without the treatment of FITC-labeled phosphorothioate oligodeoxynucleotides (b).

nohistochemical localization of MAP 1A is developed with freshly prepared 3,3′-diaminobenzidine tetrahydrochloride (DAB) dissolved in 0.05 M Tris–HCl, pH 7.6, and 0.017% H_2O_2 for 7 min. Nuclear staining is carried out with methyl green. Normal rat brain tissue is used for the positive control study of immunohistochemical staining of MAP 1A; for the negative control study, normal murine serum is substituted for the primary antibody.

In cells without oligodeoxynucleotide treatment and those treated with oligodeoxynucleotides of sense and scramble sequences, MAP 1A is stained immunohistochemically in the cytoplasm (Figs. 2a–2c). The immunoreactivity of MAP 1A and the number of MAP 1A immunopositive cells are decreased markedly in antisense oligodeoxynucleotide-treated cells compared to those in sense, scramble oligodeoxynucleotide-treated cells and nontreated cells (Fig. 2d). The negative control study with substitution of normal murine serum for the primary antibody yields no immunopositivity (Fig. 2e). This immunohistochemical study confirmed the suppression of MAP 1A expression in C6 gliomas induced by the transfection of antisense synthetic oligodeoxynucleotides for MAP 1A mRNA.

Colony-Forming Assay. C6 glioma cells are fixed with 100% methanol for 20 min and stained with 33% Giemsa's staining solution. The number of colonies containing more than 50 cells from each group is counted, and the mean and standard deviation (SD) of the number of colonies of each group are calculated and examined by Student's t test.

The numbers of colonies of C6 glioma cells treated for 12 hr with oligodeoxynucleotides of antisense, sense, and scramble sequences are 41.43 (SD: 8.75), 76.71 (SD: 23.44), and 101.43 (SD: 27.05), in average, respectively (Fig. 3a). There is a statistically significant difference between the antisense

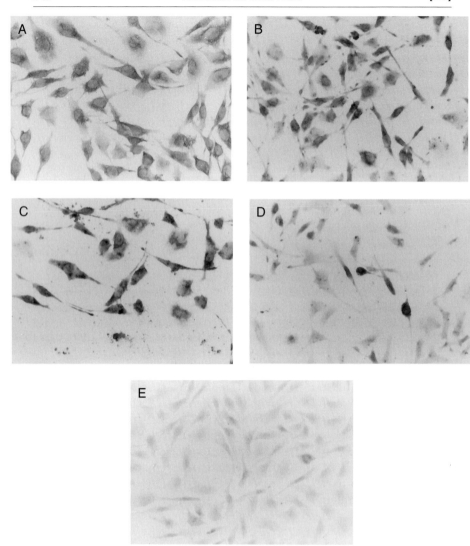

FIG. 2. In cells without oligodeoxynucleotide treatment (a) and those treated with oligo-deoxynucleotides of sense (b) and scramble (c) sequences, MAP 1A was stained immunohisto-chemically in the cytoplasm. The immunoreactivity of MAP 1A and the number of MAP 1A-immunopositive cells were decreased markedly in antisense oligodeoxynucleotide-treated cells (d) compared with those in sense and scramble oligodeoxynucleotide-treated cells and in nontreated cells. The negative control study with the substitution of normal murine serum for the primary antibody yielded no immunopositivity (e).

oligodeoxynucleotide-treated group and sense or scramble oligodeoxy-nucleotide-treated groups ($p < 0.01, p < 0.001$).

The numbers of colonies of C6 glioma cells treated for 34 hr with oligodeoxynucleotides of antisense, sense, and scramble sequences are 1.29 (SD: 0.95), 14.43 (SD: 7.48), and 11.71 (SD: 7.72), in average, respectively. There is a statistically significant difference between the antisense oligo-deoxynucleotide-treated group and sense or scramble oligodeoxynucleo-tide-treated groups ($p < 0.001, p < 0.01$) (Fig. 3b).

Flow Cytometrical Analysis on Cell Cycle of C6 Glioma Cells. C6 glioma cells are washed with 0.01 M PBS (Ca^{2+}- and Mg^{2+}-free), treated with 0.02% trypsin/0.1% ethylenediaminetetraacetic acid (EDTA), and centrifuged at 1500 rpm for 5 min at 4°. The pellets are washed twice with 0.01 M PBS (Ca- and Mg-free). The cells are adjusted in number (2×10^6/ml) and fixed with 70% (v/v) ethanol at 4°. They are stained with 50 μg/ml propidium iodide solution containing 100 U/ml RNase A, and the percentages of cells in G_1, S, and G_2 + M phases are analyzed with a flow cytometer (Fax Caliver, Becton Dickinson and Company, Lincoln Park, NJ). The flow cytometrical analysis is examined for every three samples from each group. The mean and SD of cells in G_1, S, and G_2 + M phases of each group are calculated and examined by Student's t test.

The percentages in G_1, S, and G_2 + M phases of cells treated with antisense oligodeoxynucleotide for MAP 1A are 65.17 (SD 2.45), 18.93 (SD 0.81), and 15.9 (SD 2.52), respectively (Fig. 4). The corresponding percentages of cells treated with sense, scramble oligodeoxynucleotides, and nontreated cells are 59.37 (SD 1.33), 22.37 (SD 1.10), and 18.3 (SD 2.36); 58.0 (SD 0.72), 21.83 (SD 2.14), and 20.1 (SD 1.85); 60.63 (SD 1.61), 20.03 (SD 1.40), and 19.3 (SD 3.01), respectively (Fig. 4). There is a statisti-cally significant difference in the percentages of G_1 phase cells between antisense oligodeoxynucleotide-treated cells and cells treated with oligo-deoxynucleotides of sense or scramble sequence or nontreated cells ($p < 0.001, p < 0.01, p < 0.05$) (Fig. 4). These data suggest G_1 arrest of C6 glioma cells induced by the antisense oligodeoxynucleotide for MAP 1A mRNA.

In Vivo Transfection of Phosphorothioate Oligodeoxynucleotide to C6 Glioma Cells Using Lipofection Delivery System and Miniosmotic Pump

Implantation of C6 Glioma Cells within Rat Brain and Delivery of Phos-phorothioate Oligodeoxynucleotide Using a Miniosmotic Pump. Under an-esthesia with ethyl ether and pentobarbiturate, C6 glioma cells with a concentration of 10^6 in 10 μl culture medium without FBS are implanted stereotactically within rat brain tissue, with the depth of 5 mm through a

a

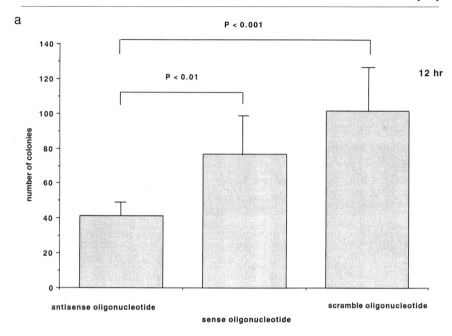

b

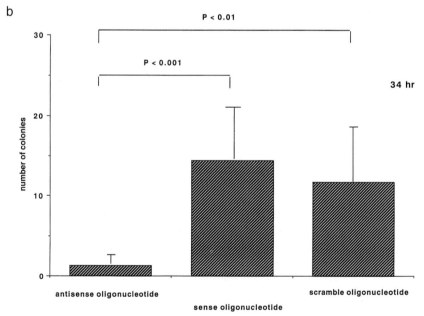

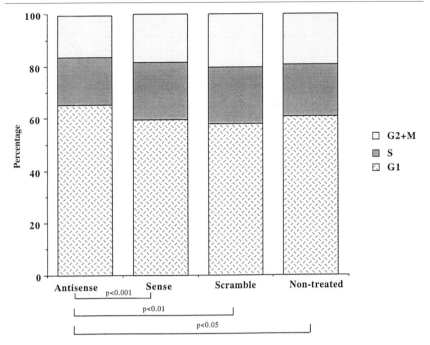

FIG. 4. Percentages of G_1, S, G_2 + M phases of cells treated with antisense oligodeoxy-nucleotide for MAP 1A were 65.17 (SD 2.45), 18.93 (SD 0.81), and 15.9 (SD 2.52), respectively. Corresponding percentages of cells treated with sense, scramble oligodeoxynu-cleotides, and nontreated cells were 59.37 (SD 1.33), 22.37 (SD 1.10), and 18.3 (SD 2.36); 58.0 (SD 0.72), 21.83 (SD 2.14), and 20.1 (SD 1.85); and 60.63 (SD 1.61), 20.03 (SD 1.40), and 19.3 (SD 3.01), respectively. There was a statistically significant difference in percentages in the G_1 phase of cells between antisense oligodeoxynucleotide-treated cells and cells treated with oligodeoxynucleotide of sense or scramble sequence or nontreated cells ($p < 0.001$, $p < 0.01$, $p < 0.05$).

FIG. 3. (a) Numbers of colonies of C6 glioma cells treated for 12 hr with oligodeoxynucleo-tides of antisense, sense, and scramble sequences were 41.43 (SD: 8.75), 76.71 (SD: 23.44), and 101.43 (SD: 27.05), in average, respectively. There was a statistically significant difference between the antisense oligodeoxynucleotide-treated group and sense or scramble oligodeoxy-nucleotide-treated groups ($p < 0.01$, $p < 0.001$). (b) Numbers of colonies of C6 glioma cells treated for 34 hr with oligodeoxynucleotides of antisense, sense, and scramble sequences were 1.29 (SD: 0.95), 14.43 (SD: 7.48), and 11.71 (SD: 7.72), in average, respectively. There was a statistically significant difference between the antisense oligodeoxynucleotide-treated group and sense or scramble oligodeoxynucleotide-treated groups ($p < 0.001$, $p < 0.01$).

burr hole opened in the right parietal bone. After 7 days, through the same burr hole, an angled cannula connected to a miniosmotic pump (Alzet Model 1007D, Alza Corp., Palo Alto, CA) containing 1 mM antisense, sense, or scramble phosphorothioate oligodeoxynucleotide in 100 μl sterile saline is inserted into the established tumor of C6 glioma cells in each rat brain tissue. The effectiveness of antisense phosphorothioate oligodeoxynucleotide for MAP 1A mRNA can be determined by calculating and comparing the survival time of each rat.

Discussion

Oligodeoxynucleotides should meet several requirements if they are applicable for clinical use. First of all, they are to be stable *in vivo*. For this purpose, synthetic oligodeoxynucleotides resistant to nucleases have been created by successful manipulation of the backbone. One of them is the phosphorothioate oligodeoxynucleotide, created by the isosteric substitution of sulfur for oxygen on the phosphorus residue that retains the negative charge of the phosphodiester oligodeoxynucleotide. The second requirement is the ability to enter the cell and interact with the intracellular target. For the purpose of enhancing this ability, fat-soluble delivery molecules such as lipofectin can be used to improve the cytoplasmic and nuclear delivery of charged oligodeoxynucleotides.[23] Third, antisense binding must interfere with the transcription or translation of the targeted DNA or mRNA sufficiently in order to obtain a satisfactory influence on the intracellular level of the encoded protein. Targets such as initiation sequences, termination sequences, and splice regions are generally considered to produce the most effective inhibition. In addition, the chronic delivery of oligodeoxynucleotides will be required for obtaining the continuous effect because their effect is transient in general. Clinical application may be problematic due to this property; however, an *in vivo* delivery system using a miniosmotic pump would be helpful for the continuous suppression of the target genes. Finally, the antisense oligodeoxynucleotide should interact solely with the desired target. Charged oligodeoxynucleotides can interact with a number of cellular proteins. The mechanism of these nonspecific interaction is not fully understood. For this reason, negative control experiments using sense and scramble sequence oligodeoxynucleotides are essential for the precise evaluation of the effect of antisense oligodeoxynucleotide.

The present study has demonstrated, by a colony-forming assay, that the antisense oligodeoxynucleotide for MAP 1A has suppressed the *in vitro*

---

[23] T. Gura, *Science* **270**, 575 (1995).

proliferation of rat C6 glioma cells significantly. The effect of the antisense oligodeoxynucleotide on the expression of MAP 1A has been demonstrated with immunohistochemical staining. This report will be the first one describing the efficiency of the antisense oligodeoxynucleotide for MAP 1A on the growth suppression of rat C6 glioma cells. Previously, there have been some reports concerning the therapy of malignant tumors using drugs affecting the microtubules. For example, taxol, which promotes the assembly of microtubules, is known to inhibit mitosis and has been suggested as a promising anticancer drug.[24,25] Antimicrotubule therapy mainly using paclitaxel (taxol) has been applied to ovarian, breast, and non-small cell lung cancers.[26] Its efficiency on the proliferative activity of malignant tumors has been shown to be enhanced when combined with radiotherapy. Taxol has significant potency *in vitro* against malignant brain tumors and that the activity occurs at concentrations of taxol that have been shown previously to be effective for several tumors against which the drug is currently being evaluated clinically.[27–29] However, its accessibility to the central nervous system is controversial because it cannot penetrate the blood–brain barrier.[30] Another report suggests the acceleration of tumor invasion by taxol.[31] Because the antitumor therapy of gliomas using taxol has not been promising so far, our new therapeutic modality using the antisense oligodeoxynucleotide for MAP 1A will be a choice in the treatment of gliomas.

The effect of the antisense oligodeoxynucleotide for MAP 1A on the cell cycle of rat C6 glioma cells has also been shown by flow cytometrical analysis. This study has shown that the G_1 arrest of C6 glioma cells is induced by the antisense oligodeoxynucleotide for MAP 1A. MAP 4, which is one of the high molecular weight MAPs, plays a role in the polymerization and stability of microtubules in interphase and mitotic cells.[9] Ookata *et al.*[32] have shown that the p34cdc2/cyclin B complex associates with microtu-

[24] K. L. Crossin and D. H. Carney, *Cell* **27**, 341 (1981).

[25] P. B. Schiff, J. Fant, and S. B. Horwits, *Nature* **277**, 665 (1979).

[26] E. K. Rowinsky, W. P. McGuire, and R. C. Donehower, *in* "Principles and Practice of Gynecologic Oncology Updates" (W. J. Huskins, C. A. Perez, and R. C. Young, eds.), Vol. 1, p. 1. Lippincott, Philadelphia, 1993.

[27] K. A. Walter, M. A. Cahan, A. Gur, B. Tyler, J. Hilton, O. M. Colvin, P. C. Burger, A. Domb, and H. Brem, *Cancer Res.* **54**, 2207 (1994).

[28] M. A. Cahan, K. A. Walter, O. M. Colvin, and H. Brem, *Cancer Chemother. Pharmacol.* **33**, 441 (1994).

[29] C. Hough, F. Fukamauchi, and D. M. Chuang, *J. Neurochem.* **62**, 421 (1994).

[30] J. J. Heimans, J. B. Vermorken, J. G. Wolbers, C. M. Eeltink, O. W. Meijer, M. J. Taphoorn, and J. H. Beijnen, *Ann. Oncol.* **5**, 951 (1994).

[31] D. L. Silbergeld, M. R. Chicoine, and C. L. Madsen, *Anticancer Drugs* **6**, 270 (1995).

[32] K. Ookata, S. Hisanaga, J. C. Bulinski, H. Murofushi, H. Aizawa, T. J. Itoh, H. Hotani, E. Okumura, K. Tachiban, and T. Kishimoto, *J. Cell. Biol.* **128**, 849 (1995).

bules in the mitotic spindle and premeitotic aster in starfish oocytes and that MAP 4 might be responsible for this interaction. In glioma cells, the expression of MAP 4 is still unknown. The present study has shown that the suppression of MAP 1A results in the G_1 arrest of C6 glioma cells and strongly suggests an important role of MAP 1A in the cell cycle. The suppression of MAP 1A using an antisense oligodeoxynucleotide can be established as a novel antitumor therapy for gliomas.

Acknowledgments

This study was partly supported by the Grant-in-Aid for Scientific Research (No. 09671450) from the Ministry of Education, Science, and Culture of Japan and by the Grant for Scientific Research from the Japan Brain Foundation.

[21] Preparation and Application of Liposome-Incorporated Oligodeoxynucleotides

By Ana M. Tari

Antisense oligodeoxynucleotides have been used to study the functional roles of various proteins. However, nuclease susceptibility, low cellular uptake, and poor intracellular delivery have been the major limitations of the application of antisense oligonucleotides.[1-4] To overcome some of these limitations, antisense oligonucleotides made of various nuclease-resistant phosphodiester analogs, such as phosphorothioates, methyl phosphonates, and P-ethoxys, have been synthesized. These analogs have substitutions at the nonbridging oxygen atom of the phosphate bond: phosphorothioates with a sulfur atom, methyl phosphonates with a methyl group, and P-ethoxys with an ethyl group added to the oxygen atom (Fig. 1). Phosphorothioates, like phosphodiesters, are hydrophilic, whereas methyl phosphonates and P-ethoxys are hydrophobic.

Typically, antisense oligonucleotides, ranging between 15 and 24 bases in length, were used. With this length, the antisense sequences are long and unique enough to bind to the target mRNA selectively. RNase H may be activated to digest the target mRNA when it is bound with an antisense

[1] R. Wagner, *Nature* **372**, 333 (1994).
[2] C. Stein and R. Narayanan, *Curr. Opin. Oncol.* **6**, 587 (1994).
[3] A. Thierry and A. Dritschilo, *Nucleic Acids Res.* **20**, 5691 (1992).
[4] J. Tonkinson and C. Stein, *Nucleic Acids Res.* **22**, 4268 (1994).

DNA. This may potentiate the effectiveness of antisense oligonucleotides.[1,2] However, RNase H is only activated by antisense DNA made of phosphorothioates but not methyl phosphonates or P-ethoxys.

Carriers for Antisense Oligonucleotides

   Although the synthesis of nuclease-resistant oligonucleotide analogs has improved the application of antisense oligonucleotides, effective clinical application of these agents may require carriers that can further their intracellular delivery. To improve the intracellular delivery of antisense oligonucleotides, we have been using liposomes as carriers for these agents. Liposomes are lipid vesicles that are biodegradable, nontoxic, and easy to prepare. They are highly versatile and can be used to entrap both hydrophobic and hydrophilic compounds, although at a higher efficiency with the former than with the latter. Liposomes target naturally to organs, such as liver and spleen, and to a lower extent to bone marrow. Proteins or antibodies, derivatized with fatty acid chains, can be incorporated into liposomal membranes so that they can bind to cell surface receptors or antigens and may give these carriers targeting features.

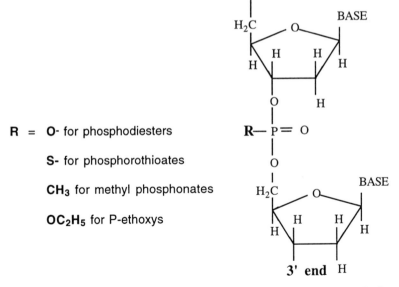

FIG. 1. Structures of phosphodiester and its nuclease-resistant analogs. Phosphodiesters and phosphorothioates are hydrophilic, whereas methyl phosphonates and P-ethoxys are hydrophobic.

Preparation of Liposome-Incorporated Oligonucleotides

Labeling of Oligonucleotides with ^{32}P Radioisotope

Hydrophilic and hydrophobic oligonucleotides, made of the same sequence, were incorporated into liposomes.[5-7] The oligonucleotide sequence was 5' GGGCTTTTGAACTCTGCT 3'. To measure the incorporation efficiencies of oligonucleotides in liposomes, oligonucleotides were labeled at 37° with [γ-^{32}P]ATP (Amersham Life Sciences, Cleveland, OH) at the 5' end by T4 polynucleotide kinase (Boehringer Mannheim Corp., Indianapolis, IN) as follows[7]:

Oligonucleotides (~0.2 μmol)	140 μl
[γ-^{32}P]ATP (1 mCi)	100 μl
T4 kinase	10 μl
10× kinase buffer	35 μl
Distilled H$_2$O	65 μl
Total volume	350 μl

The volume of dimethyl sulfotide (DMSO) present should be at least 25% (v/v) for the labeling reactions of methyl phosphonates and P-ethoxys. This ensures that these hydrophobic oligonucleotides will remain soluble during the labeling reactions. The labeling reaction was then incubated at 37° for 2 hr for phosphodiesters, about 2–4 hr for phosphorothioates, and 16–24 hr for methyl phosphonates and P-ethoxys. After the labeling reactions were finished, oligonucleotides were filtered twice with a Microcon-3 filter (Amicon, Beverly, MA) at 7000 rpm for approximately 2 hr to separate the ^{32}P-labeled oligonucleotides from free [^{32}P]ATP. The purity of the separation was ascertained using 20% polyacrylamide gel electrophoresis. Both phosphodiesters and phosphorothioates can enter polyacrylamide gels. However, free [^{32}P]ATP will migrate much faster than ^{32}P-labeled phosphodiesters or phosphorothioates because the former are lower in molecular weight. Even though methyl phosphonates and P-ethoxy oligonucleotides are neutral and cannot enter polyacrylamide gels efficiently, free [γ-^{32}P]ATP can. As a result, ^{32}P-labeled methyl phosphonates and P-ethoxy oligonucleotides will be retarded in polyacrylamide gels whereas free ^{32}P will migrate much faster. That way, one can still observe how much ^{32}P is associated with the labeled oligonucleotides. Typically, after purification, there is <1% of free ^{32}P associated with oligonucleotides.

[5] A. M. Tari, S. Tucker, A. Deisseroth, and G. Lopez-Berestein, *Blood* **84,** 601 (1994).
[6] A. M. Tari, M. Khodadadian, D. Ellerson, A. Deisseroth, and G. Lopez-Berestein, *Leuk. Lymphoma* **21,** 93 (1996).
[7] A. M. Tari and G. Lopez-Berestein, *J. Liposomes Res.* **7,** 19 (1997).

Incorporation of Radiolabeled Oligonucleotides into Liposomes

Dioleoylphosphatidylcholine (DOPC) lipids were purchased from Avanti Polar Lipids (Alabaster, AL). Radiolabeled oligonucleotides were incorporated into DOPC liposomes as follows[7]:

^{32}P-labeled oligonucleotides (0.2 μmol)	140 μl
DOPC (4 μmol)	157 μl
tert-Butanol	3 ml

DOPC is dissolved in *tert*-butanol at 20 mg/ml. Oligonucleotides are dissolved in water or DMSO at ~8 mg/ml. Oligonucleotides and DOPC are aliquoted and mixed well before adding excess *tert*-butanol. Because DMSO is present, *tert*-butanol should be added and for at least 95% (v/v) so that the mixture can be well frozen in an acetone/dry ice bath before being lyophilized overnight. The lyophilized preparation is hydrated with 0.9% normal saline at a final oligonucleotide concentration of 0.1 m*M*.

Separation of Free Unincorporated Oligonucleotides from Liposomal-Incorporated Oligonucleotides

Dialysis is used to separate unincorporated hydrophilic oligonucleotides from those incorporated in liposomes, whereas centrifugation over a Ficoll solution is used to separate unincorporated hydrophobic oligonucleotides from those incorporated in liposomes.[7]

Liposomal mixtures containing free and incorporated hydrophilic (phosphodiesters and phosphorothioates) oligonucleotides are dialyzed (molecular weight cutoff 12,000–14,000) against a 1000-fold excess of HEPES-buffered saline at room temperature overnight. Aliquots of liposomal oligonucleotides are taken before and after dialysis for liquid scintillation counting to assess the incorporation of phosphodiesters or phosphorothioates oligonucleotides in liposomes. Typically, phosphodiesters and phosphorothioates are incorporated into liposomes at 30–70%.[7]

The separation of unincorporated hydrophobic (methyl phosphonates and P-ethoxys) oligonucleotides from those incorporated in liposomes is done using a 10% (w/v) Ficoll solution. One milliliter of the liposomal mixture, which contains liposomal oligonucleotides and unincorporated oligonucleotides, is layered over 2–3 ml of the Ficoll solution and centrifuged for 10 min at 2000 rpm. Liposomal oligonucleotides are recovered on top of the Ficoll solution, whereas unincorporated oligonucleotides are collected in the pellet. Aliquots of liposomal oligonucleotides are taken before and after Ficoll separation and are sent to liquid scintillation counting to assess the incorporation of methyl phosphonates or P-ethoxy oligonucleotides in liposomes. Methyl phosphonates and P-ethoxys are incorporated into liposomes at 90–95% efficiency.[7] Because hydrophobic oligonucleotides can be incorporated at such a high efficiency, liposome-incorporated hydrophobic oligonucleotides are used routinely without going through the separation procedure every time. Thus, we have decided to mainly use liposome-incorporated hydrophobic oligonucleotides for our studies.

Cellular Localization of Oligonucleotides

We then studied the cellular localization of free and liposome-incorporated oligonucleotide.[8] A fluorescent P-ethoxy oligonucleotide with the sequence 5' TCGTATCTATGCGCTC 3' was synthesized by conjugating rhodamine to the 5' end of the oligonucleotide (Oligos Etc., Willsonville, OR). After the labeling procedure, fluorescent P-ethoxy oligonucleotides are separated from the free rhodamine molecule, and the purity of fluorescent P-ethoxys is ascertained to be ≥90% by HPLC. Rhodamine-conjugated P-ethoxys are incorporated into liposomes as described earlier, except that radiolabeled oligonucleotides are replaced with rhodamine-conjugated oligonucleotides.

Acute lymphocytic leukemic (ALL-1) cells are plated at 2.5×10^5 cells/ 0.5 ml of RPMI medium in a 24-well plate. They are incubated with 4 μM of rhodamine-labeled free or liposomal P-ethoxys for 0 to 48 hr. They are washed with Dulbecco's phosphate-buffered saline twice before being viewed under a laser-powered confocal fluorescent microscope.

Very weak cytoplasmic fluorescence can be seen in ALL-1 cells even after being incubated with free P-ethoxys for 24 hr.[9] The uptake of free P-ethoxys is not extended even after 48 hr of incubation (Fig. 2A). In contrast, very intense fluorescence can be seen within ALL-1 cells when incubated with liposomal P-ethoxys (Fig. 2B). In fact, when fluorescent liposomal oligonucleotides are used, fluorescence can be detected within the cellular cytoplasm as early as 2.5 hr.[8] However, the fluorescence appears to level off between 10 and 24 hr.[8]

Fluorescent liposomal P-ethoxys are found mainly in the cytoplasm, but not the nucleus. Fluorescence appears to be distributed throughout the whole cytoplasm in a diffuse, but not punctate, manner. This diffuse manner indicates that liposomal P-ethoxys are not confined to vesicular structures, such as the endosomes or the lysosomes. Similar data are obtained with fluorescent liposomal methyl phosphonates.[8] These data demonstrate that liposomes can increase the cytoplasmic delivery of hydrophobic oligonucle-

---

[8] A. M. Tari, M. Andreeff, H.-D. Kleine, and G. Lopez-Berestein, *J. Mol. Med.* **74**, 623 (1996).

[9] A. M. Tari, N. Neamati, M. Andreeff, and G. Lopez-Berestein, *in* "Targeting of Drugs" (G. Gregoriadis and B. McCormack, eds.), p. 163. Plenum Press, New York, 1996.

A

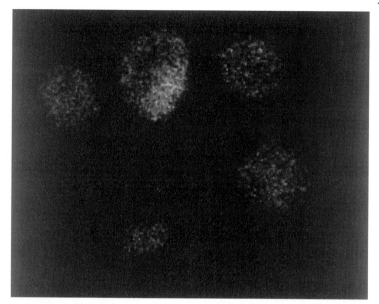

B

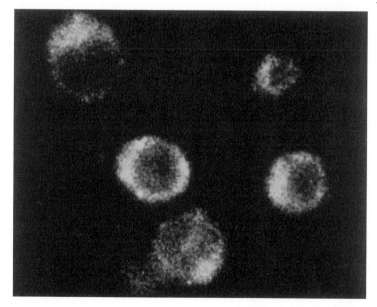

Fɪɢ. 2. Fluorescent microscopy of liposomal P-ethoxy oligonucleotides. ALL-1 cells were incubated with 4 μM of free (A) or liposome-incorporated (B) fluorescent P-ethoxy oligonucleotides for 48 hr, washed, and viewed under a laser-powered confocal fluorescent microscope.

otides. To test whether liposomal antisense oligonucleotides could indeed inhibit proteins expression selectively, antisense sequences were then incorporated into liposomes and delivered to various cell lines.

Specific Inhibition of Bcl-2 Protein Expression by Liposomal Bcl-2 Antisense Oligonucleotides

Antisense and control Bcl-2 oligonucleotides, made of P-ethoxy oligonucleotides (Oligos Etc.), are incorporated into liposomes. An oligonucleotide specific for the translation initiation site of human BCL-2 mRNA—5′ CAGCGTGCGCCATCCTTC 3′—is used as the Bcl-2 antisense oligonucleotide, whereas a scrambled version of Bcl-2 antisense oligo—5′ ACGGT-CCGCCACTCCTTCCC 3′—is used as the Bcl-2 control oligonucleotide.[10] These oligonucleotides incorporated into liposomes as described earlier, except that all oligonucleotides used are not labeled with any radioisotope or fluorescent dye. Liposomal Bcl-2 antisense oligonucleotides (L-Bcl-2 AS) and liposomal Bcl-2 control oligonucleotides are delivered to a human transformed follicular lymphoma cell line, CJ.

One hundred thousand CJ cells/well, seeded in a 6-well plate in 3 ml of medium, are incubated with 2, 3, and 4 μM of liposomal oligonucleotides for 3 days. Both treated and untreated cells are lysed in 100 μl of lysis buffer (1% Triton, 150 mM NaCl, and 25 mM Tris, pH 7.4) at 0° for 30 min. After centrifugation at 12,000g for 10 min, the supernatants are recovered and normalized for total protein content. Five and 25 μg of cell lysates are used for Bcl-2 and Actin protein analysis, respectively. The lysates are mixed with sample buffer containing 1% SDS and 1% 2-mercaptoethanol and boiled for 5 min. SDS–PAGE is run on 10% polyacrylamide gels, transferred electrophoretically to nitrocellulose membranes, and blocked in 10% nonfat dry milk. Membrane filters are incubated with a 6C8 hamster antihuman Bcl-2 monoclonal antibody or with a mouse antiactin monoclonal antibody (Amersham Life Sciences, Cleveland, OH) at room temperature for 2 hr. After washing and incubating with a peroxidase-labeled antihamster (Kirkegaard & Perry Laboratories Inc., Gaithersburg, MD) or are antimouse (Amersham) secondary antibody, blots are developed using an enhanced chemiluminescence system (Amersham). Densitometric scans are performed on Western blots using a Gilford Response Gel Scanner (CIBA Corning, Medfield, MA). An area integration of absorbance peaks at 500 nm is used to determine Bcl-2 protein inhibition by estimating the protein ratios of Bcl-2/actin.

[10] M. Tormo, A. M. Tari, T. McDonnell, F. Cabanillas, J. Garcia-Conde, and G. Lopez Berestein, *Leuk. Lymphoma* **30,** 367 (1998).

When CJ cells are incubated with 2, 3, and 4 μM of L-Bcl-2 AS, the levels of Bcl-2 protein decrease in a dose-dependent manner (Fig. 3). However, when the same cell lysates are used, the levels of the control actin protein are not decreased. A densitometric scan of the ratios of Bcl-2/actin protein is performed; the ratios of Bcl-2/actin protein are inhibited by 28, 57, and 64%, respectively. Neither Bcl-2 nor actin protein expression is inhibited in cells treated with the same doses of liposomal Bcl-2 control oligonucleotides (Fig. 3). These data indicate that Bcl-2 antisense oligonucleotides, when incorporated into liposomes, could specifically inhibit Bcl-2 protein expression.[10]

Specific Inhibition of Crkl and Grb2 Protein Expression by Liposomal Crkl and Liposomal Grb2 Antisense Oligonucleotides

We then tested other liposomal antisense oligonucleotides. The Crkl antisense oligonucleotide—5'GTCGAACCTGGCGGAGGA 3'—and the Grb2 antisenses oligonucleotide—5'ATATTTGGCGATGGCTTC 3' (Oligos Etc.) were incorporated into liposomes. The control oligonucleotide—5'GAAGGGCTTCTGCGTC 3' (Oligos Etc.)—was incorporated into liposomes as well.[11] BV173 and K562 cells are seeded in a six-well plate in 2 ml of medium and treated with 4–8 μM of liposomal oligonucleotides. Untreated cells are also maintained in culture. After incubating with liposomal Crkl antisense (L-Crkl AS) oligonucleotides or liposomal Grb2 antisense (L-Grb2 AS) oligonucleotide, for 2–3 days, cell lysates are prepared as described earlier. Five to 10 μg of protein lysates is used for Western blots. Membrane filters are incubated with anti-Grb2 monoclonal antibody (Transduction Laboratories, Lexington, KY) or with rabbit anti-Crkl polyclonal antibody (Santa Cruz Laboratory, Santa Cruz, CA). After washing and incubation with a peroxidase-labeled antimouse or antirabbit secondary antibody (Amersham Life Sciences, Cleveland, OH), blots are developed by an enhanced chemiluminescence system (Amersham). When the production of Grb2 protein is targeted to be inhibited, Crkl protein is used as the control protein. Conversely, when Crkl protein is to be inhibited, Grb2 protein is used as control.

Cell lysates are prepared after incubating K562 cells with 4–8 μM of L-Crkl AS or liposomal control oligonucleotides for 2 days (Fig. 4A). Crkl protein inhibition is seen only with L-Crkl AS oligonucleotides but not with liposomal control oligonucleotides. At 8 μM L-Crkl AS oligonucleotide concentration, approximately 75% of Crkl protein inhibition is observed.[11]

[11] A. M. Tari, R. Arlinghaus, and G. Lopez-Berestein, *Biochem. Biophys. Res. Commun.* **235,** 383 (1997).

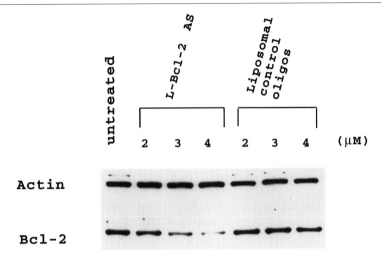

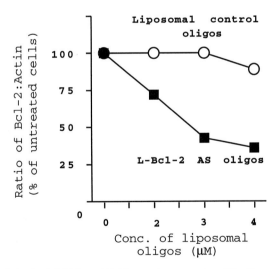

FIG. 3. Selective inhibition of Bcl-2 protein expression. CJ cells were incubated with 2–4 μM of L-Bcl-2 AS oligonucleotides and liposomal control oligonucleotides. After 3 days of culture, protein-containing lysates were prepared and subjected to SDS–PAGE and transferred to nitrocellulose membranes. Blots were incubated with antibodies specific for Bcl-2 (target) or actin (control) proteins.

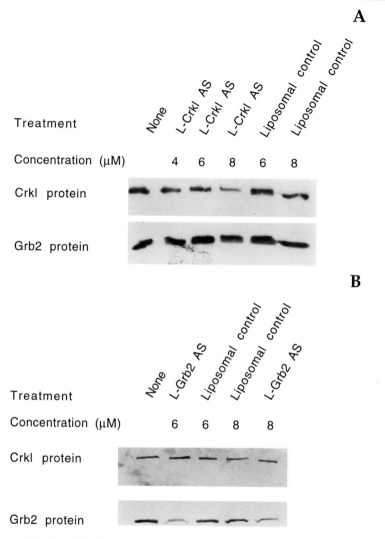

FIG. 4. Selective inhibition of the production of Crkl and Grb2 proteins. K562 (A) and BV173 (B) cells were incubated with 4–8 μM of L-Crkl AS oligonucleotides (A) and L-Grb2 AS oligonucleotides (B), respectively. Cells were incubated with equal concentrations of liposomal control oligonucleotides. After 2 (K562 cells) to 3 (BV173) days of culture, protein-containing lysates were prepared and subjected to SDS–PAGE and transferred to nitrocellulose membranes. Blots were cut into sections and incubated with antibodies specific for Crkl or Grb2 protein.

Neither L-Crkl AS nor liposomal control oligonucleotides inhibited the level of the control Grb2 protein (Fig. 4A).

L-Grb2 AS oligonucleotides could also selectively inhibit Grb2 protein expression (Fig. 4B). The expression of Grb2 protein in BV173 cells is decreased by 50% when L-Grb2 AS, but not liposomal control, oligonucleotides are used.[11] Neither L-Grb2 AS nor liposomal control oligonucleotides inhibit the level of the control Crkl protein (Fig. 4B). These data indicate that, similar to L-Bcl-2 AS oligonucleotides, L-Crkl AS and L-Grb2 AS oligonucleotides could also specifically inhibit the expression of their target proteins, i.e., Crkl and Grb2 proteins, respectively.[11]

Selective Growth Inhibition of Philadelphia Chromosome-Positive Leukemic Cells by L-Grb2 AS Oligonucleotides

Because liposomal antisense oligonucleotides could selectively inhibit protein expression, they could then be used to investigate the cellular functions of proteins. Grb2 AS oligonucleotide—5′ ATATTTGGC-GATGGCTTC 3′—and control oligonucleotide—5′ TCGCCACTC-GATCCTGCCCG 3′—are incorporated into liposomes and delivered to Philadelphia chromosome-positive (Ph+) leukemic cells—ALL-1, BV173, and K562 cells—and Philadelphia chromosome-negative leukemic cells—HL60 cells. K562 cells are seeded at 5×10^3 cells/well whereas all other cells are seeded at 10×10^3 cells/well in a 96-well plate in 0.1 ml of medium. Cells are incubated with a 0–12 μM final concentration of L-Grb2 AS oligonucleotides or liposomal control oligonucleotides for 4–5 days.[11] The viability of the leukemic cells is measured by the alamarBlue dye (Alamar Sciences, Sacramento, CA) incorporation method. Fifty microliters of cells/well is aliquoted and added to 130 μl of fresh medium. Twenty microliters of alamarBlue dye is added to each well. After incubation at 37° (K562 and HL60 cells for 6–8 hr and ALL-1 and BV173 cells for 12–14 hr), the plates are read directly on a microplate reader (Molecular Devices Corp., Menlo Park, CA) at 570 and 595 nm. The difference in absorbance between 570 and 595 nm is taken as the overall absorbance value of the cells. The viabilities of cells treated with liposomal oligonucleotides are compared with those of untreated controls. Growth inhibitory effects are defined as a decrease in viability due to reduced incorporation of the alamarBlue dye by the leukemic cells.

Ph+ leukemic cells are incubated with increasing concentrations of L-Grb2 AS oligonucleotides (Fig. 5). A decrease in viabilities is found with all three Ph+ leukemic cell lines. The IC$_{50}$ of L-Grb2 AS oligonucleotides in ALL-1, BV173, and K562 cells is 4, 8, and 8 $\mu M,$ respectively. However,

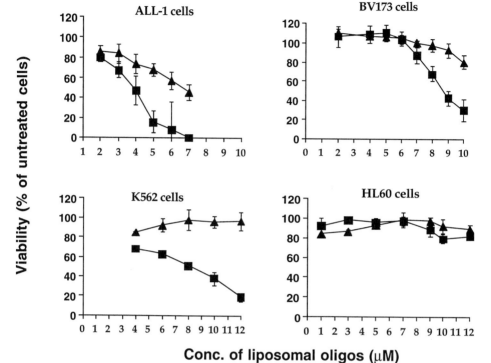

FIG. 5. Inhibition of the proliferation of Ph⁺ leukemic cells by L-Grb2 AS oligonucleotides. Ph⁺ leukemias, such as ALL-1, BV173, and K562 cells, and Philadelphia chromosome-negative HL60 cells were incubated with 0–12 μM of L-Grb2 AS oligonucleotides (■) and liposomal control oligonucleotides (▲) for 4 (ALL-1 cells) to 5 (all other cells) days. The alamarBlue dye incorporation method was used to measure the viability of leukemic cells. Viabilities of cells treated with liposomal oligonucleotides were compared with those of untreated controls.

when Ph⁺ cells are incubated with liposomal control oligonucleotides under identical conditions, the viability of Ph⁺ cells is affected only slightly, except for ALL-1 cells, which show 40% inhibition at the highest concentration of liposomal control oligonucleotides used (Fig. 5). These results demonstrate that L-Grb2 AS oligonucleotides can inhibit the proliferation of Ph⁺ cells by 70–100% at optimal concentrations.[11] However, L-Grb2 AS and liposomal control oligonucleotides did not inhibit the proliferation of the negative control HL60 cells (Fig. 5). Therefore, by using L-Grb2 AS oligonucleotides, we could demonstrate that Grb2 protein is important for the proliferation of Ph⁺ leukemic cells.

In Vivo Behavior of Liposomal P-Ethoxy Oligonucleotides

Safety Studies

ICR mice weighing 25–30 g are obtained from Harlan Sprague–Dawley (Indianapolis, IN). DOPC and P-ethoxy oligos are from Avanti Polar Lipids (Alabaster, AL) and Oligos Etc., respectively. The oligonucleotide sequence utilized is 5′GGGCTTTTGAACTCTGCT 3′ which had previously been shown by us to selectively inhibit Bcr-Abl protein production.[5,9] Bcr-Abl protein is only expressed in human Ph⁺ leukemias and is not expressed in mice. Therefore, this oligonucleotide sequence should not induce any toxicity unless P-ethoxy oligonucleotides themselves are toxic to mice.

Groups of 8–10 mice each are injected via the tail vein with multiple doses (1 dose/day for 5 consecutive days) of liposomal P-ethoxy oligonucleotides.[12] Mice are injected intravenously with liposomal P-ethoxy oligonucleotides at individual doses of 3.8–15.0 mg P-ethoxy oligonucleotides/kg of body weight, which is equivalent to a total dose of 19–75 mg P-ethoxy oligonucleotides/kg of body weight. Blood samples are obtained 2 weeks after the first day of injection and analyzed for hematologic parameters and for biochemical tests of hepatic and renal functions. At these doses, liposomal P-ethoxy oligonucleotides do not result in any abnormalities on hematologic parameters nor on renal and hepatic function tests (Table I).

Before using radiolabeled liposomal P-ethoxy oligonucleotides for pharmacokinetics and organ distribution studies, we had studied the serum stability of P-ethoxy oligonucleotides and the serum stability of the association of the radiolabels with the oligonucleotides. Liposomal P-ethoxy oligonucleotides are labeled with ^{32}P as described earlier. After incubating radiolabeled liposomal P-ethoxy oligonucleotides in mouse and rat serum, aliquots of radiolabeled liposomal oligonucleotides are extracted and run on a 20% polyacrylamide gel. We found that radiolabels and liposomal P-ethoxy oligonucleotides remain intact for up to 96 hr.

Extraction of ^{32}P-Labeled Liposomal Oligonucleotides
from Whole Blood and Tissues

Because the color of whole blood could interfere with liquid scintillation counting, whole blood and tissue samples are treated and decolorized with hyamine hydroxide (ICN Pharmaceuticals, Costa Meca, CA) prior to liquid scintillation counting.[12] The treatment procedure is done according to manufacturer's protocol. Whole blood (0.1 ml) or tissue (50–100 mg) samples

---

[12] A. M. Tari, C. Stephens, M. Rosenblum, and G. Lopez-Berestein, *J. Liposome Res.* **8**, 251 (1998).

TABLE I
LIPOSOMAL P-ETHOXY OLIGONUCLEOTIDES NOT TOXIC TO ICR MICE[a]

	Hematological parameters[b]					
Oligonucleotide dose (mg/kg)	WBC ($\times 10^3/\mu$l)	RBC ($\times 10^6/\mu$l)	Hemoglobin (g/dl)	Hematocrit (%)	MCV (fl)	Platelets ($\times 10^3/\mu$l)
0	5.0 ± 1.9	8.3 ± 0.4	14.1 ± 0.8	37.3 ± 2.2	44.3 ± 1.3	862 ± 240
3.8	4.3 ± 2.1	7.7 ± 0.6	13.5 ± 1.0	36.0 ± 2.5	45.8 ± 1.5	828 ± 247
7.5	5.0 ± 1.7	8.1 ± 0.5	14.2 ± 0.7	37.1 ± 2.1	45.9 ± 1.0	991 ± 314
15.0	4.9 ± 1.8	8.0 ± 0.5	13.8 ± 0.7	36.5 ± 2.1	45.8 ± 0.9	1274 ± 239

	Renal and hepatic functions[b]			
Oligonucleotide dose (mg/kg)	Creatinine (mg/dl)	BUN (mg/dl)	SGOT (IU/liter)	Alkaline phosphatase (IU/liter)
0	0.44 ± 0.05	30.8 ± 1.9	212.7 ± 121.2	67.0 ± 17.0
3.8	0.45 ± 0.05	28.2 ± 3.7	228.3 ± 124.9	72.5 ± 41.2
7.5	0.40 ± 0.05	27.1 ± 2.4	192.0 ± 182.6	54.9 ± 14.2
15.0	0.42 ± 0.09	27.6 ± 3.2	190.9 ± 178.2	79.7 ± 30.6

[a] Five male and five female ICR mice were used.
[b] All parameters and functions were expressed as mean values ± SD.

are digested with 0.5–1.0 ml of hyamine hydroxide:ethanol (1:2, v/v) at 60° overnight. Then 0.5 ml of 30% H_2O_2 and 15 ml of liquid scintillation cocktail are added to the digested samples. The amount of ^{32}P associated with the samples is counted by liquid scintillation. ^{32}P counting efficiency is approximately 70%.

Pharmacokinetics

Male Lewis rats weighing approximately 400 g obtained from Harlan Sprague–Dawley are anesthetized with sodium thiopental (35 mg/kg of body weight intraperitoneally) during the entire period of the experiment.[12] The right femoral artery and left femoral vein are surgically exposed and cannulated. Fifteen minutes after this procedure, three rats each are injected intravenously with liposomal P-ethoxy oligonucleotides at a dose of 7.5 mg of P-ethoxy oligonucleotides per kg of body weight. Blood samples of approximately 0.2 ml each are withdrawn at 2, 5, 10, 15, 20, 30, 60, 120, 180, and 240 min after injection. After each withdrawal, the catheter is flushed with sterile saline containing 10% sodium heparin. An aliquot of the injected dose is maintained as a control sample. Whole blood samples are extracted and treated as described earlier before assaying for ^{32}P radioactivity by liquid scintillation. Pharmacokinetic parameters are determined by nonlinear regression analysis (Rstrip, Micro Math, Inc.). Data are best fit to a two-compartment model:

$$C_t = Ae^{-\alpha t} + Be^{-\beta t}$$

where C_t equals concentration at time t, A and B are the y axis intercepts, α and β are the constants for distribution and elimination, respectively, and t is time. $t_{1/2}\alpha$ and $t_{1/2}\beta$ are calculated from $\ln 2/\alpha$ and $\ln 2/\beta$, respectively. C_0 is calculated from the equation at time zero; therefore, C_0 equals $A + B$. V_D is calculated as the ratio of initial dose to C_0.

The clearance of liposomal P-ethoxy oligonucleotides from plasma closely fit a two-compartment mathematical model (correlation $r^2 > 0.98$). The initial distribution phase occurs over the first 10 min after injection $(t_{1/2}\alpha = 6.7 \pm 0.2 \text{ min})$; the terminal phase half-life $(t_{1/2}\beta)$ is 436.1 $\pm$ 23.9 min (Table II). The immediate apparent volume of distribution (V_D) approximates slightly higher than the total blood volume (20.7 $\pm$ 0.7 ml) for rats of this size.[12]

Organ Distribution of Liposomal Oligonucleotides

Twenty ICR mice, weighing an average of 30 g, are from Harlan Sprague–Dawley. Mice are divided into five groups of four mice each. One group of mice is untreated whereas the other four groups are injected intravenously with liposome-incorporated radiolabeled oligonucleotides at a dose of 15 mg of P-ethoxy oligonucleotides per kg of body weight.[12] These four groups of four mice are sacrificed at 4, 24, 48, or 72 hr. Organs are dissected, and 50- to 100-mg samples from each mouse are weighed and processed as described earlier before radioactivity is counted in a scintillation counter.

TABLE II
SINGLE-DOSE PHARMACOKINETICS OF LIPOSOMAL P-ETHOXY OLIGONUCLEOTIDES

Animal	$t_{1/2}\alpha$ (min)	$t_{1/2}\beta$ (min)	$C_0{}^a$ (mg/ml)	$V_D{}^b$ (ml)	$C \times t^c$ (mg $\times$ min/ml)
1	6.9	461.1	0.13	20.0	20.0
2	6.6	413.4	0.12	21.4	21.4
3	6.7	433.8	0.13	20.8	23.1
Mean $\pm$ SD	6.7 $\pm$ 0.2	436.1 $\pm$ 23.9	0.13 $\pm$ 0.01	20.7 $\pm$ 0.7	21.5 $\pm$ 1.6

[a] Concentration at time zero.
[b] Volume of distribution.
[c] Area under the concentration curve = $A/\alpha + B/\beta$.

Liposomal P-ethoxy oligonucleotides could be detected in various organs, such as liver, spleen, heart, lungs, and kidneys, and in the bone marrow at a much lower extent.[12] However, at all time points measured, the highest tissue concentrations of liposomal P-ethoxy oligonucleotides are detected in spleen and liver, which are the major organs of leukemia and lymphoma manifestation (Table III). After 4 hr of injection, oligonucleotide equivalent concentrations in liver and spleen range between 30 and 34 μg/g of tissues, whereas oligonucleotide equivalent concentrations in heart, lungs, and kidneys range between 11 and 17 μg/g of tissues. Oligonucleotide concentrations from various organs decrease over time. By 72 hr, oligonucleotide concentrations in liver and spleen had decreased to 12–14 μg/g of tissues, as compared to 7–10 μg of oligonucleotides of heart, lungs, and kidney tissues. Oligonucleotide clearance from most organs is at a $t_{1/2}$ of approximately 72 hr, except in the liver where the $t_{1/2}$ is approximately 48 hr, A much lower concentration of P-ethoxy oligonucleotides (~0.2 μg) can be detected in the bone marrow where it remains relatively constant during the first 48 hr and then decreases slightly at 72 hr.

In summary, hydrophobic oligonucleotides, such as P-ethoxy oligonucleotides, can be incorporated into liposomes at 90% or greater efficiency in a one-step procedure that does not involve any further separation or purification. Because liposomes can deliver P-ethoxy oligonucleotides intracellularly at high concentrations, liposomal P-ethoxy oligonucleotides can be used reliably as antisense oligonucleotides to investigate protein function. Because liposomal P-ethoxy oligonucleotides are safe to use and can reach liver, spleen, and bone marrow at high concentrations, L-Bcl-2 AS, L-Crkl AS, and L-Grb2 AS may be used as novel therapeutic modalities to treat leukemias and lymphomas.

TABLE III

ORGAN DISTRIBUTION OF LIPOSOMAL P-ETHOXY OLIGONUCLEOTIDES IN MICE

Time (h)	Organ (mean μg of oligonucleotide equivalents[a]/g of organ)						
	Spleen	Liver	Lungs	Kidneys	Heart	BM[b]	Blood[c]
4	33.7 ± 3.5	29.6 ± 4.4	17.0 ± 2.2	16.7 ± 2.4	10.8 ± 1.7	0.18 ± 0.06	0.34 ± 0.01
24	29.4 ± 1.6	22.6 ± 0.8	13.5 ± 1.2	15.9 ± 1.2	10.7 ± 0.8	0.23 ± 0.06	0.30 ± 0.03
48	17.8 ± 3.2	14.4 ± 2.5	10.6 ± 2.0	12.7 ± 2.5	9.2 ± 2.1	0.26 ± 0.07	0.24 ± 0.03
72	14.1 ± 0.9	11.6 ± 0.5	8.5 ± 1.0	9.7 ± 0.6	7.1 ± 0.6	0.17 ± 0.01	0.16 ± 0.02

[a] Liposomal P-ethoxy oligonucleotides were expressed as mean ± SD.
[b] Two femurs.
[c] 0.2 ml of blood volume.

Acknowledgments

This work was supported in part by the National Leukemia Research Foundation and the University of Texas MD Anderson Cancer Center Breast Cancer Research Program.

[22] Cell-Specific Optimization of Phosphorothioate Antisense Oligodeoxynucleotide Delivery by Cationic Lipids

By Shirley A. Williams and Jeffrey S. Buzby

The ability to inhibit gene expression is an important tool for studying the normal function of specific gene products within a cell. Antisense (AS) nucleic acids have been utilized successfully as one means of downregulating intracellular mRNA levels,[1-3] either through endonucleolytic cleavage by RNase H[4] or by blocking initiation or elongation of translation.[5] Antisense nucleic acid is most commonly delivered in the form of short (<30 nucleotide) oligodeoxynucleotides (ODNs).[6] These techniques have been used to show that cellular protooncogenes, especially transcription factors, are sensitive to AS inhibition and that they function in regulating proliferation and differentiation.[7,8]

Translocation of the c-*myc* oncogene on chromosome 8 with the immunoglobulin loci on chromosome 2, 14, or 22, resulting in c-*myc* overexpression, is characteristic of Burkitt's lymphoma.[9,10] In the most frequent translocation [t(8;14)] of the c-*myc* gene (8q24) with the Ig heavy chain gene (IgH; 14q32), two forms of the disease arise: the sporadic form in which the breakpoint is located within the gene, commonly near the first intron of the c-*myc* gene, and the endemic form in which the breakpoint is located outside of the gene, commonly in the 5'-untranslated region of the c-*myc*

[1] C. A. Stein and Y. C. Cheng, *Science* **261,** 1004 (1993).
[2] S. T. Crooke and C. F. Bennett, *Annu. Rev. Pharmacol. Toxicol.* **36,** 107 (1996).
[3] R. W. Wagner and W. M. Flanagan, *Mol. Med. Today* **3,** 31 (1997).
[4] S. T. Crooke, *Antisense Nucleic Acid Drug Dev.* **8,** 133 (1998).
[5] S. T. Crooke, *Antisense Nucleic Acid Drug Dev.* **8,** 115 (1998).
[6] G. Sczakiel, *J. Hematother.* **3,** 305 (1994).
[7] F. E. Cotter, *Hematol. Oncol.* **15,** 3 (1997).
[8] A. M. Gewirtz, *Crit. Rev. Oncog.* **8,** 93 (1997).
[9] C. M. Croce and P. C. Nowell, *Blood* **65,** 1 (1985).
[10] I. Magrath, *Adv. Cancer Res.* **55,** 133 (1990).

gene.[11,12] Translocation of the c-*myc* gene results in its overexpression due to disruption of its transcriptional regulation.[13,14]

Antisense ODNs have been shown to block c-*myc* gene expression in a number of different cell lines.[15] In the promyelocytic cell line, HL-60, AS to the c-*myc* gene was shown to reduce proliferation, inhibit tumor formation, and induce cellular differentiation.[16–19] A number of ODN modifications have been demonstrated to increase AS ODN stability,[20–23] thus increasing the efficiency of downregulation. By using phosphorothioated AS ODNs, we were able to reduce survival of HL-60 to the same extent as unmodified AS ODN but at a 10-fold lower concentration.[24] This has a practical benefit, as high concentrations of AS ODNs can generate nonspecific inhibition of cell proliferation due to general toxicity and not as a direct result of specific hybridization and consequent antisense inhibition.[1]

The combination of AS ODN modification and facilitation of delivery, including the use of viral- or lipid-based vectors, has been exploited to enhance the efficacy and applicability of AS ODNs.[25,26] The use of synthetic

[11] P.-G. Pelicci, D. M. Knowles II, I. Magrath, and R. Dalla-Favera, *Proc. Natl. Acad. Sci. U.S.A.* **83,** 2984 (1986).

[12] B. Shiramizu, F. Barriga, J. Neequaye, A. Jafri, R. Dalla-Favera, A. Neri, M. Guttierez, P. Levine, and I. Magrath, *Blood* **77,** 1516 (1991).

[13] R. Dalla-Favera, S. Martinotti, R. C. Gallo, J. Erickson, and C. M. Croce, *Science* **219,** 963 (1983).

[14] R. Taub, C. Moulding, J. Battey, W. Murphy, T. Vasicek, G. M. Lenoir, and P. Leder, *Cell* **36,** 339 (1984).

[15] B. Calabretta and T. Skorski, *Anticancer Drug Des.* **12,** 373 (1997).

[16] A. Harel-Bellan, D. K. Ferris, M. Vinocour, J. T. Holt, and W. L. Farrar, *J. Immunol.* **140,** 2431 (1988).

[17] J. Holt, R. Redner, and A. Nienhuis, *Mol. Cell. Biol.* **8,** 963 (1988).

[18] E. L. Wickstrom, T. Bacon, A. Gonzalez, D. Freeman, G. Lyman, and E. Wickstrom, *Proc. Natl. Acad. Sci. U.S.A.* **85,** 1028 (1988).

[19] M. E. McManaway, L. M. Neckers, S. L. Loke, A. A. Al-Nassar, R. L. Redner, B. T. Shiramizu, W. L. Goldschmidts, B. E. Huber, K. Bhatia, and I. T. Magrath, *Lancet* **335,** 808 (1990).

[20] C. A. Stein and J. S. Cohen, *Cancer Res.* **48,** 2659 (1988).

[21] P. Miller, *in* "Oligodeoxynucleotides: Antisense Inhibitors of Gene Expression" (J. Cohen, ed.), p. 79. Macmillan, London, 1989.

[22] L. Neckers, L. Whitesell, A. Rosolen, and D. Geselowitz, *Crit. Rev. Oncog.* **3,** 175 (1992).

[23] N. M. Dean and R. H. Griffey, *Antisense Nucleic Acid Drug Dev.* **7,** 229 (1997).

[24] S. A. Williams, E. R. Gillan, E. Knoppel, J. S. Buzby, Y. Suen, and M. S. Cairo, *Ann. Oncol.* **8,** S25 (1997).

[25] A. H. Conrad, M. A. Behlke, T. Jaffredo, and G. W. Conrad, *Antisense Nucleic Acid Drug Dev.* **8,** 427 (1998).

[26] M. J. Hope, B. Mui, S. Ansell, and Q. F. Ahkong, *Mol. Membr. Biol.* **15,** 1 (1998).

cationic lipids as a vector was first described by Felgner *et al.*[27] We have shown a dramatic reduction in AS ODN delivery time from 96 to 5 hr through the use of cationic lipid delivery in Burkitt's lymphoma cells.[28] However, we also found that cationic lipid reagents differed in their toxicity and ODN delivery efficiency among individual Burkitt's lymphoma cell lines and translocations. Therefore, careful selection and optimization of cationic lipid reagent, including concentration as well as specific lot number, are imperative.

The focus of this article is to detail the methodology required to efficiently utilize cationic lipids for delivering phosphorothioate AS ODNs into several different lymphoma cell lines and, in particular, on the joint optimization of conditions for two cell lines acting as controls for each other during the same AS experiments.

Cell Culture

Two sporadic type (CA46 and ST486) and two endemic type (Daudi and EB3) Burkitt's lymphoma cell lines (American Type Tissue Collection) are cultured in RPMI 1640 medium (Sigma, St. Louis, MO) supplemented with 15% heat-inactivated fetal bovine serum (Gemini Bioproducts, Calabasas, CA), 1% penicillin/streptomycin (GIBCO-BRL, Gaithersburg, MD), and 1% GlutaMAX (GIBCO-BRL) at 37°, 95% air/5% CO_2. Cells are subcultured twice weekly to a final cell density of 4×10^5 viable cells/ml, as determined by counting of viable cells with a hemacytometer using trypan blue staining (Sigma). Experiments are performed within 20 passages from frozen culture stock and begin 18–20 hr after subculturing. The time after subculture is critical for the consistency of results with cationic lipid-mediated ODN delivery (data not shown).

Oligodeoxynucleotides

Phosphorothioated ODNs (Oligos etc, Wilsonville, OR) specific for translocations in the 5'-untranslated region of the c-*myc* gene characteristic of sporadic Burkitt's lymphoma tumors include AS ODN directed to the 5'-cap site[29] (5'-AS), 5'-GCAGCACAGCGGGGGT-3'; the complementary strand of 5'-AS or sense (5'-S), 5'-ACCCCCGCGCTGTGCTGC-3' to control for the structure of the 5'-AS; and the 5'-AS sequence with reversed

[27] P. L. Felgner, T. R. Gadek, M. Holm, R. Roman, H. W. Chan, M. Wenz, J. P. Northrop, G. M. Ringold, and M. Danielson, *Proc. Natl. Acad. Sci. U.S.A.* **84,** 7413 (1987).
[28] S. A. Williams, L. Chang, J. S. Buzby, Y. Suen, and M. S. Cairo, *Leukemia* **10,** 1980 (1996).
[29] T. A. Bacon and E. Wickstrom, *Oncogene Res.* **6,** 13 (1991).

polarity (5'-revAS), 5'-TGGGGGCTCGACACGACG-3' to control for the sequence and base composition of 5'-AS. Similarly, phosphorothioated ODNs specific for translocations in the first intron of the c-*myc* gene characteristic of endemic Burkitt's lymphoma tumors include AS ODN directed to the unspliced first intron[19] (1i-AS), 5'-GGCTGCTGGAGCGGGGCA-CAC-3'; the complementary strand of 1i-AS or sense (1i-S), 5'-GTGTGCCCCGCTCCAGCAGCC-3'; and the 1i-AS sequence with reversed polarity (1i-revAS), 5'-CACACGGGGCGAGGTCGTCGG-3'. ODNs were purified by desalting by the manufacturer (Oligo etc).

General Methodology for Antisense Oligodeoxynucleotide Delivery by Cationic Lipids

Lipofectamine or Lipofectin reagent (GIBCO-BRL) and ODNs are each diluted separately into 0.1 ml of RPMI 1640 medium (Sigma) without serum or antibiotics. The diluted ODNs and cationic lipid are combined, mixed gently, and incubated for 30 min at room temperature to allow the formation of cationic lipid/ODN complexes. Viable Burkitt's lymphoma cells are counted using trypan blue staining, pelleted in a swinging bucket centrifuge (200g for 10 minutes), and resuspended in serum-free, antibiotic-free RPMI 1640 medium to an estimated concentration of 3.125×10^6 cells/ml, assuming complete sedimentation. The cationic lipid/ODN complex solution (0.2 ml) is added to 0.8 ml of cells. Thus, 2.5×10^6 cells/ml are incubated with various concentrations of ODNs in serum-free medium containing varying amounts of cationic lipid for 5 hr at 37° with 5% CO_2.

All incubations with Lipofectamine or Lipofectin are performed in 14-ml, 17×100-mm, snap-cap, polystyrene tubes (Falcon, Becton-Dickinson) to decrease cationic lipid adherence to the vessel wall.[30]

Cationic Lipid Optimization

The first step involves selecting a cationic lipid reagent and optimizing its concentration for each individual cell line. Lipofectamine is selected over Lipofectin to deliver ODN to these particular cells for reasons that will be presented in the following section. The maximum amount of reagent used can be limited by its toxicity to each cell line, which varies considerably, depending on both the reagent and its concentration. Furthermore, specific lots of Lipofectamine are found to perform differently, requiring reoptimization for each lot (data not shown). Therefore, it is best to obtain a sufficient supply of a single lot for an entire study.

[30] P. L. Felgner and M. Holm, *Focus* **11,** 21 (1989).

The toxicity of Lipofectamine is tested on each cell line for a range of concentrations. Two, 5, 10, or 20 μl of Lipofectamine (corresponding to final concentrations of 3.4, 8.5, 17, or 34 μM, respectively) is diluted into 100 μl of serum-, antibiotic-free RPMI 1640 medium in polystyrene tubes. Note that maintaining these final concentrations does not necessarily assure reproducibility from lot to lot. Because no ODN is added during cationic lipid optimization, an additional 100 μl of the just-described medium is added to the tubes. Aliquots of 0.8 ml cells in RPMI 1640 medium without serum or antibiotics are distributed into sterile polystyrene tubes, and the cationic lipid solution (200 μl) is added to a final cell concentration of 2.5×10^6 cells/ml. The cells are mixed with the cationic lipid solution by gentle rotation and then incubated at 37° for 5 hr. Cultures are analyzed for total viable cells compared to untreated controls using trypan blue staining.

The initial goal was to determine the maximum tolerable cationic lipid levels, defined as ≥80% of untreated cell survival for each cell line. Although both sporadic (CA46 and ST486) and endemic (EB3 and Daudi) cell lines exhibit similar levels of tolerance to 3.4 μM of Lipofectamine, different patterns emerge at higher concentrations (Fig. 1). EB3 tolerates levels up to 17 μM, whereas Daudi can only tolerate 8.5 μM (Fig. 1). ST486 also tolerates up to 8.5 μM, but CA46 can only tolerate the lowest concentration (3.4 μM) tested (Fig. 1). Note the nonlinear dose–response curve, which

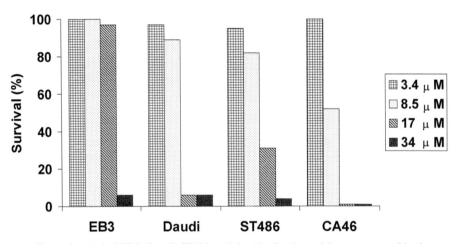

FIG. 1. Survival of EB3, Daudi, ST486, and CA46 cells after a 5-hr exposure to Lipofectamine. Cells were treated with 3.4, 8.5, 17, or 34 μM Lipofectamine, as described, and counted immediately after the 5-hr treatment. Survival was defined as the percentage of viable cells compared to untreated controls. Data shown are representative of at least two independent experiments.

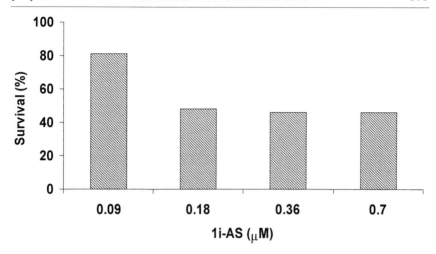

FIG. 2. Survival of ST486 cells after a 5-hr exposure to 8.5 μM Lipofectamine and 1i-AS. Cells were treated with 0.09, 0.18, 0.36, or 0.7 μM 1i-AS + Lipofectamine, as described, and counted immediately after the 5-hr treatment. Survival was defined as the percentage of viable cells compared to untreated controls (no Lipofectamine or 1i-AS). Data shown are representative of two independent experiments.

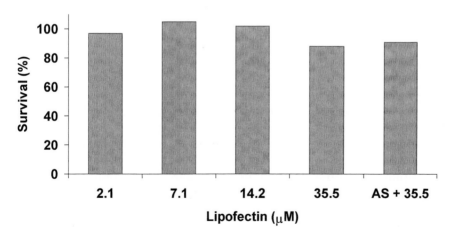

FIG. 3. Survival of ST486 cells after a 5-hr exposure to Lipofectin. Cells were treated with 2.1, 7.1, 14.2, or 35.5 μM Lipofectin alone or with 0.36 μM 1i-AS + 35.5 μM Lipofectin in the last lane, as described, and counted immediately after the 5-hr treatment. Survival was defined as the percentage of viable cells compared to untreated controls. Data shown are representative of two independent experiments.

A

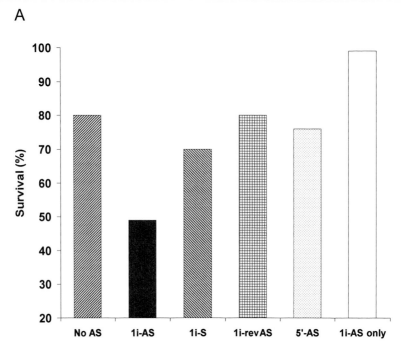

FIG. 4. Survival of (A) ST486 and (B) Daudi cells after a 5-hr exposure to ODNs and Lipofectamine. All cells were treated with 8.5 μM Lipofectamine except for 1i-AS only or 5'-AS only. All cells were also treated with 0.36 μM ODN in addition to Lipofectamine except for the No AS lane, which was a control for the toxicity of 8.5 μM Lipofectamine alone. Cells were counted immediately after the 5-hr treatment. Survival was defined as the percentage of viable cells compared to untreated controls (no Lipofectamine or ODN). Data shown are representative of two independent experiments.

is particularly evident for the endemic cell lines where a two-fold increase in Lipofectamine concentration reduces the level of survival from >90 to <10%. Thus, even a slight change in Lipofectamine concentration can cause a significant decrease in tolerance, necessitating careful handling of the cationic lipid to avoid even small pipetting errors that could affect an entire experiment. ST486 and Daudi are selected as representative of sporadic and endemic Burkitt's lymphoma cells, respectively, for testing c-*myc* AS inhibition, as both exhibit similar patterns of tolerance up to 8.5 μM Lipofectamine.

Facilitation of AS ODN Delivery by Lipofectamine

Using 8.5 μM Lipofectamine, four different concentrations of c-*myc* AS ODN (1i-AS) were tested for their ability to specifically decrease c-*myc*-

B

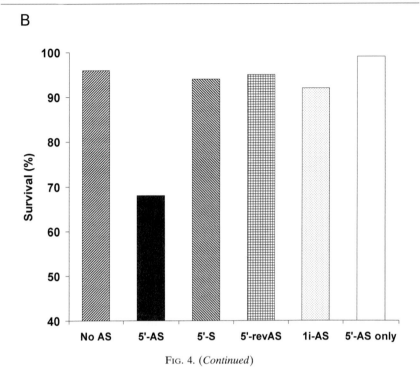

FIG. 4. (*Continued*)

dependent survival of ST486. While 0.09 μM 1i-AS had little effect on survival, the three concentrations ≥ 0.18 μM all reduced survival by 50% (Fig. 2). A concentration of 0.36 μM 1i-AS was chosen for subsequent experiments in order to use the least amount of AS ODN without approaching the 0.09 μM concentration, which exhibited less inhibition.

Since it appears that cationic lipid coats the ODNs,[31] the amount of cationic lipid used may require further optimization for a particular ODN concentration. Therefore, lower amounts of Lipofectamine were tested to determine if it was possible to decrease the potential for toxicity by decreasing the cationic lipid to be combined with the previously optimized 1i-AS concentration. However, decreasing the dose of Lipofectamine from 8.5 to 6.8 μM rendered the treatment ineffective (data not shown), confirming the necessity for the accurate measurement of cationic lipid dose as well.

In order to determine if another cationic lipid formulation could facilitate AS ODN delivery with less toxicity than Lipofectamine, the sensitivity of ST486 to increasing amounts of Lipofectin was examined. No concentra-

[31] P. L. Felgner and G. M. Ringold, *Nature* **337,** 387 (1989).

tion tested (2.8, 7.1, 14.2, or 35.5 μM, corresponding to 2, 5, 10, or 25 μl) showed any toxicity to the ST486 cell line (Fig. 3). However, when 35.5 μM Lipofectin was tested in combination with 0.36 μM 1i-AS, no decrease in survival was observed (Fig. 3), as was also the case for AS ODN doses of 0.09, 0.18, 0.7, 1.4, or 2.8 μM.

Use of AS ODN Controls

In order to test the specificity of the AS response in two Burkitt's lymphoma cell lines with different translocations, the cationic lipid and ODN concentrations established for ST486 were tested on Daudi. The Daudi AS ODN (5'-AS) was directed at the transcriptional start site,[29] whereas the ST486 AS ODN (1i-AS) was directed at the first intron.[19] Thus, the two cell lines acted as controls for specificity of the response to their respective specific AS ODNs. Additionally, both AS ODNs contain a sequence of four contiguous guanosine residues (G-quartet), which has been shown to promote nonspecific binding.[32] Therefore, reverse AS ODNs, which maintained base composition, position, and the G-quartet structure, were included among the controls to eliminate the possibility of nonspecific G-quartet effects, one of the potential artifacts encountered when using AS technology.

ST486 showed a decrease in survival with both Lipofectamine and 1i-AS, whereas the three control ODNs showed no appreciable inhibitory effects compared to untreated cells (Fig. 4A). Note that 0.36 μM 1i-AS without Lipofectamine had no effect on ST486 survival after 5 hr, as expected from previous studies.[28] Daudi showed a decrease in survival with Lipofectamine and 5'-AS, whereas none of the other treatments decreased survival by more than 10% (Fig. 4B). Once again, 0.36 μM 5'-AS without Lipofectamine had no effect on Daudi survival after 5 hr, as expected.[28]

Data show that for at least some individual cell types, it may be necessary to optimize all aspects of ODN delivery, including cationic lipid concentration, AS ODN concentration, and cationic lipid formulation, in addition to identifying an effective AS ODN. Thus, the use of cationic lipids adds another variable to the AS ODN delivery process that must be controlled by optimization. However, the dramatic decreases in uptake time and efficiency[28] make this approach very worthwhile for many experimental applications.

[32] P. Yaswen, M. R. Stampfer, K. Ghosh, and J. S. Cohen, Antisense Res. Dev. 3, 67 (1993).

Acknowledgments

The authors thank Drs. Mitchell Cairo and Diane Nugent for support, laboratory facilities, and helpful discussions.

Section III

RNA Studies

[23] Intracellular Expression and Function of Antisense Catalytic RNAs

By DANIELA CASTANOTTO, MICHAELA SCHERR, and JOHN J. ROSSI

Ribozymes represent a relatively new approach for specific inactivation of gene expression. To date, ribozymes have been reported to be effective in cell culture against a variety of RNA targets, including viral, endogenous, and transfected sequences. Ribozymes have been utilized to functionally destroy mRNAs as well as for the repair of mutant RNAs.[1,2] These RNA catalysts have progressed to the stage where both hammerhead and hairpin ribozymes are currently being tested in human clinical trials. The effective use of ribozymes *in vivo* depends on effective target site selection, knowledge of the rules governing intracellular localization, effective delivery to target cells, and maintaining a high and long-term intracellular expression of the engineered gene.

Over the last decade ribozymes have been widely employed in a variety of systems. Their use as tools to control gene expression is both enticing and simple in concept. Indeed, ribozymes can be designed to specifically base pair with and cleave any given RNA target, with the result of producing an irreversible inactivation of the targeted RNA. Because these catalytic RNAs can be designed to have exquisite specificity and are inherently devoid of coding capacities, there is little likelihood for undesired cellular toxicities. For the popular hammerhead and hairpin ribozymes, which are usually less than 60 nucleotides in length, multiple ribozymes can be expressed from the same transcriptional units.[3–5] Ribozymes can also be expressed from a variety of promoters.[6] The delivery of ribozyme genes can be accomplished via several routes, including viral vectors, transfection, and lipofection. Thus, the combination of specificity, the possibilities of endogenous synthesis or exogenous delivery, and the capability to be

[1] R. E. Christoffersen and J. J. Marr, *J. Med Chem.* **38**, 2023 (1995).

[2] J. J. Rossi, *Trends Genet.* **14**, 295 (1998).

[3] J. Ohkawa, N. Yuyama, Y. Takebe, S. Nishikawa, and K. Taira, *Proc. Natl. Acad. Sci. U.S.A.* **90**, 11302 (1993).

[4] J. Ohkawa, N. Yuyama, Y. Takebe, S. Nisikawa, M. Homann, G. Sczakiel, and K. Taira, *Nucleic Acids Sym. Ser.* **29**, 121 (1993).

[5] C. J. Chen, A. C. Banerjea, G. G. Harmison, K. Haglund, and M. Schubert, *Nucleic Acids Res.* **20**, 4581 (1992).

[6] E. Bertrand, D. Castanotto, C. Zhou, C. Carbonnelle, N. S. Lee, P. Good, S. Chatterjee, T. Grange, R. Pictet, D. Kohn, D. Engelke, and J. J. Rossi, *RNA* **3**, 75 (1997).

adapted for the destruction or modification of any RNA species make ribozymes extremely attractive genetic tools and potential therapeutic agents.

The major limitations to effective intracellular ribozyme use are colocalization of ribozymes with target RNAs[7] and target site accessibility.[8,9] We will address these specific problems in this work. In addition, strategies and protocols for expression and delivery of ribozyme genes are addressed.

Catalytic Motifs

Six catalytic motifs have been utilized as *trans*-acting, catalytic, antisense agents. These are the group I intron, RNase P, the self-cleaving domains of plant viroids (composed of the hammerhead and hairpin ribozyme, illustrated in Fig. 1), the self-cleaving domain of hepatitis delta virus,[10–12] and the *Neurospora* VS RNA.[13,14] Group I introns are defined by complex, but phylogenetically conserved, sequence homologies and secondary structures, but many, if not all, require protein factors to splice efficiently *in vivo*.[15] The best characterized group I intron, and perhaps the most thoroughly studied ribozyme, is the intervening sequence of *Tetrahymena thermophila*. This was the first ribozyme for which the *cis*-cleaving reaction was converted to a *trans* reaction.[15,16] A structure at 5 Å resolution of the *Tetrahymena* intron has been solved by X-ray crystallography and has revealed a striking similarity between its active site and the active sites found in protein enzymes.[17] Although this RNA enzyme has been engineered to cleave a variety of substrates, including single-stranded DNA,[18] its application in biological systems has been limited by its complexity and a lack of specificity in substrate pairing. These problems have not yet been solved, but recent application of the group I intron in *trans* splicing of two different messenger

[7] B. A. Sullenger and T. R. Cech, *Science* **262,** 1566 (1993).
[8] A. Lieber and M. Strauss, *Mol. Cell Biol.* **15,** 540 (1995).
[9] A. Lieber, M. Rohde, and M. Strauss, *Methods Mol. Biol.* **74,** 45 (1997).
[10] J. J. Rossi, *Curr. Biol.* **4,** 469 (1994).
[11] J. J. Rossi, *Curr. Opin. Biotechnol.* **3,** 3 (1992).
[12] D. Castanotto, J. J. Rossi, and J. O. Deshler, *Crit. Rev. Eukaryot. Gene Expr.* **2,** 331 (1992).
[13] B. J. Saville and R. A. Collins, *Proc. Natl. Acad. Sci. U.S.A.* **88,** 8826 (1991).
[14] T. Rastogi and R. A. Collins, *J. Mol. Biol.* **277,** 215 (1998).
[15] T. R. Cech, *Annu. Rev. Biochem.* **59,** 543 (1990).
[16] D. Herschlag and T. R. Cech, *Biochemistry* **29,** 10159 (1990).
[17] B. L. Golden, A. R. Gooding, E. R. Podell, and T. R. Cech, *Science* **282,** 259 (1998).
[18] D. Herschlag and T. R. Cech, *Nature* **344,** 405 (1990).

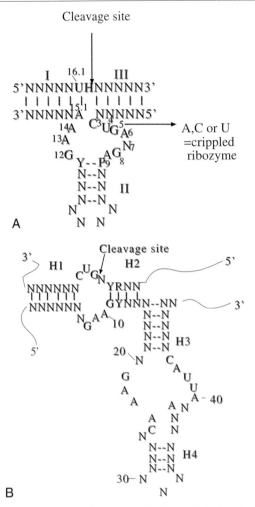

FIG. 1. Generalized depictions of hammerhead (A) and hairpin (B) ribozymes. N, any base; Y, pyrimidine; P, purine.

RNAs has renewed interest in the biological applications of this ribozyme.[19–21]

RNase P is a ubiquitous enzyme whose primary cellular function is processing the 5' leader sequence from pre-tRNA transcripts.[22] The enzyme

[19] J. T. Jones, S. W. Lee, and B. A. Sullenger, *Nature Med.* **2,** 643 (1996).
[20] J. T. Jones, S. W. Lee, and B. A. Sullenger, *Methods Mol. Biol.* **74,** 341 (1997).
[21] N. Lan, R. P. Howrey, S. W. Lee, C. A. Smith, and B. A. Sullenger, *Science* **280,** 1593 (1998).
[22] S. Altman, *Biotechnology* **13,** 327 (1995).

itself does not need to be modified to carry out reactions in *trans,* as this is the only ribozyme known to bind and cleave free substrate molecules in nature. In contrast to other ribozymes, the substrate needs to be modified to create a structure that mimics the pre-tRNA transcripts normally cleaved by this enzyme. RNAs, which are used for this purpose, have been designated as external guide sequences.[23,24] RNase P can be used for the site-specific cleavage of intracellular RNAs that form substrates with external guide sequences,[25–27] provided that the external guide sequence is added in *trans* and is able to bind to the target RNA.

The hammerhead, the hairpin, and the HDV are smaller and simpler ribozymes than the group I and RNAse P enzymes. For intracellular applications, hammerhead and hairpin ribozymes have thus far been proven to be the most popular. The structure of the hammerhead ribozyme has been well understood for quite some time.[28,29] This ribozyme is the simplest and has been the most popular for intracelluar applications. The catalytic core of the hairpin ribozyme is slightly more complex than the hammerhead,[30–33] but lends itself well to *trans* cleavage reactions. With the exclusion of few minimal requirements and the obligatory G at the 3′ end of the cleavage site (Fig. 1B), there is total flexibility in the substrate–guide sequence combination. The hairpin also catalyzes a more efficient ligation reaction than the hammerhead and thus may be more amenable to *in vitro* evolution and selection schemes.[34,35]

A high-resolution HDV crystal structure has been solved,[36] providing a better understanding of the catalytic mechanism of this ribozyme. The HDV ribozyme has evolved within a mammalian intracellular environment, making it an attractive candidate for use in human and mammalian systems. Better understanding of its function and substrate specificities should foster the use of this ribozyme in mammalian systems.

[23] Y. Yuan and S. Altman, *EMBO J.* **14,** 159 (1995).
[24] A. C. Forster and S. Altman, *Science* **249,** 783 (1990).
[25] F. Liu and S. Altman, *Genes Dev.* **9,** 471 (1995).
[26] Y. Li and S. Altman, *Nucleic Acids Res.* **24,** 835 (1996).
[27] Y. Yuan, E. S. Hwang, and S. Altman, *Proc. Natl. Acad. Sci. U.S.A.* **89,** 8006 (1992).
[28] W. G. Scott and A. Klug, *Trends Biochem. Sci.* **21,** 220 (1996).
[29] H. W. Pley, K. M. Flaherty, and D. B. McKay, *Nature* **372,** 111 (1994).
[30] A. Hampel, *Prog. Nucleic Acid Res. Mol. Biol.* **58,** 1 (1998).
[31] T. Tuschl, J. B. Thomson, and F. Eckstein, *Curr. Opin. Struct. Biol.* **5,** 296 (1995).
[32] N. G. Walter, E. Albinson, and J. M. Burke, *Nucleic Acids Symp. Ser.* **36,** 175 (1997).
[33] F. Walter, A. I. H. Murchie, and D. M. J. Lilley, *Biochemistry* **37,** 17629 (1980).
[34] A. Berzal-Herranz and J. M. Burke, *Methods Mol. Biol.* **74,** 349 (1997).
[35] S. Joseph and J. M. Burke, *J. Biol. Chem.* **268,** 24515 (1993).
[36] A. R. Ferre-D'Amare, K. Zhou, and J. A. Doudna, *Nature* **395,** 567 (1998).

Choice or Ribozyme Cleavage Targets and Intracellular Expression
of Ribozymes

Determining Accessible Target Sites

In order for ribozymes to be effective in downregulating gene expression, the targeted region(s) of the RNA needs to be accessible to the binding and subsequent cleavage of the ribozyme. RNA within the cell is complexed with heterogeneous ribonuclear proteins (hnRNPs) and small nuclear ribonuclear proteins (snRNPs) within the nucleus and with hnRNPs and the translational apparatus in the cytoplasm. Portions of RNA transcripts may also be sequestered in secondary and tertiary structures, thus making them inaccessible to ribozyme interaction. Predicting sites that are accessible to ribozyme binding and cleavage is problematic. The use of computer RNA-folding programs such as MFOLD can be useful in determining regions of strong secondary structure, but these programs cannot predict the most accessible sites in the target RNA. For most ribozyme experiments, the rate-limiting step is the choice of a good cleavage site. The use of random ribozyme libraries in cells and cell extracts is the most direct approach to target identification,[8] but this is a laborious procedure, which requires cloning and sequence analysis of reverse transcription-polymerase chain reaction (RT-PCR) products. Mapping of the target structure can be performed *in vitro* using enzyme or chemical reagents.[37] Another *in vitro* approach utilizes the binding of DNA oligonucleotides to *in vitro* prepared RNA followed by incubation with the enzyme RNase H, which cleaves the RNA only within DNA–RNA hybrids. Based on quantitation of the cleaved fragments, sites most accessible to oligonucleotide hybridization can be identified.[38]

Ideally, the approach for identifying a ribozyme cleavage site should take advantage of the RNA in its native, protein-associated state. We have devised a strategy for identifying accessible sites in a message utilizing native mRNAs in cellular extracts.[39] This is a straightforward approach, which is applicable to any target RNA. The underlying principle is that RNAs associated with proteins in their native state provide the best substrates for identifying regions, which can base pair with oligonucleotides such as antisense DNAs and ribozymes. The use of DNA oligonucleotide targeted to potential ribozyme cleavage sites can be used as a facile screening procedure prior to making ribozymes to the targets. DNA oligonucleotides are inexpensive to make, and a large number of potential sites can

---

[37] T. B. Campbell, C. K. McDonald, and M. Hagen, *Nucleic Acids Res.* **25**, 4985 (1997).
[38] K. R. Birikh, Y. A. Berlin, H. Soreq, and F. Eckstein, *RNA* **3**, 429 (1997).
[39] M. Scherr and J. J. Rossi, *Nucleic Acids Res.* **26**, 5079 (1998).

be screened rapidly by this procedure. The most accessible target sites are expected to generate the strongest cleavage products. Thus far, we have observed a good correlation between results obtained using the DNA oligonucleotides in extracts with the site accessibility for ribozymes in extracts and *in vivo*. The protocol we recommend for the identification of potential ribozyme cleavage sites is described next. The most important component is the use of cell extracts prepared from cells, which express the RNA of interest. The extract preparation we utilize was originally developed for Pol III transcription.[40]

Preparation of Cellular Extracts

1. Pellet approximately 8×10^7 cells and wash twice in phosphate-buffered saline.
2. Resuspend pellets in two times the volume of the cell pellet in hypotonic swelling buffer (7 mM Tris–HCl, pH 7.5; 7 mM KCl; 1 mM MgCl$_2$; 1 mM 2-mercaptoethanol) and, after a 10-min incubation on ice, transfer to a Dounce homogenizer (VWR, San Diego, CA).
3. Apply 20 strokes with a tight pestle B and add 1/10th of the final volume-neutralizing buffer (21 mM Tris–HCl, pH 7.5; 116 mM KCl; 3.6 mM MgCl$_2$; 6 mM 2-mercaptoethanol).
4. Centrifuge the homogenate at 20,000g for 10 min at 4°.
5. Transfer the supernatants, which are rich in RNA-binding proteins and RNAse H activity and contain endogenous mRNAs, to an Eppendorf tube on ice and use then immediately or store at −70° in hypotonic buffer containing 45% glycerol.

RNase H Digestion of DNA/RNA Hybrids with Endogenous RNase H in Cell Extracts

The design of oligonucleotides for testing in these extracts can be based on the localization of potential NUH (N = A,C,G, or U; H = A,C, or U) ribozyme cleavage sites or random oligonucleotide libraries can be used. In the former case, the oligonucleotide length should match the desired total ribozyme flanking sequence length, usually 14 to 20 bases. If a random library is utilized, we have found that only 9 bases of diester DNA are enough to stimulate endogenous RNase H activity in a variety of cell extracts. Depending on whether defined sequence or random sequence oligonucleotides are used, the time of incubation in the extracts will vary. For defined oligonucleotides, under the following conditions, only a couple of minutes are required. For random oligonucleotides, the concentration

[40] P. A. Weil, J. Segall, B. Harris, S. Y. Ng, and R. G. Roeder, *J. Biol. Chem.* **254**, 6163 (1979).

of the pool and the length of time of incubation will have to be determined empirically.

This protocol can be used to directly map accessible sites on native target transcripts in Pol III extracts. Specific oligonucleotides are added to the extract containing the native targeted RNA, and products generated by endogeneous RNAse H cleavage are subsequently analyzed and quantified.

1. Carry out RNAse H-mediated cleavage experiments in a total volume of 30 μl, containing 20 μl polymerase III extract, 1 mM dithiothreitol (DTT), 20–40 units RNase inhibitor (Promega, Madison, WI), and 50 nM each of the defined sequence antisense oligonucleotides (ODNs). If a random library is used, the concentration and time of incubation need to be determined empirically based on the complexity of the pool.
2. Incubate ODNs and RNA for 2–10 min at 37°. To stop the reaction, digest the mixture with DNase I for 45 min at 37° (this destroys the antisense ODNs and any residual chromosomal DNA in the extracts).
3. Phenol extract and ethanol precipitate the RNA.
4. Isolate the remaining RNA and use it as a substrate for reverse transcription and subsequent DNA PCR. For random oligonucleotide libraries, the sites of cleavage need to be identified by RT-LMPCR. The description of this procedure is beyond the scope of this article. The protocols described in Ref. 41 and 42 describe this procedure in great detail.

The reverse primer used to amplify the target mRNA should be complementary to a region outside of the defined nucleotide targets. This primer can be 5′-end-labeled with rhodamine or fluorescein. As an internal standard, an endogenous, nontargeted mRNA such as β-actin should be coamplified. The reverse primer for the internal standard can be 5′ labeled with a different fluorescent dye than the target sequence primer.

5. MMV reverse transcription is carried out according to the protocol specified by the supplier (Life Technology, Grand Island, NY) using 50 ng random hexamer primer and 10 units of reverse transcriptase.
6. Amplify an aliquot (18 μl) of the RT reaction using specific primers for the selected target and the remaining aliquot (2 μl) using, for example, β-actin primers.
7. Electrophorese aliquots (0.5 μl) from the PCR reactions in a denaturing polyacrylamide gel, and quantitate the fluorescein- and

[41] J. Komura and A. D. Riggs, *Nucleic Acids Res.* **26,** 1807 (1998).
[42] E. Bertrand, M. Fromont-Racine, R. Pictet, and T. Grange, *Proc. Natl. Acad. Sci. U.S.A.* **90,** 3496 (1993).

rhodamine-tagged products via a Genescan analysis on an Applied Biosciences prism 377 DNA sequencer using Genescan analysis software version 2.1 (ABI, Weiterstadt, Germany). Other systems for analyzing fluorescently tagged, amplified products can also be utilized.

Confirmation of ODN Efficacy by RNase H Digestion of DNA/ Endogenous RNA Hybrids with Endogenous RNAse H in Vivo: Optional but Useful

For this protocol, 5'-fluorescein and 3' terminal phosphorthioate groups should be added to the oligonucleotides to allow detection and to increase intracellular stability. Transfect each of the antisense ODNs as well as a sense control ODN into cultured cells using a cationic lipid.

For NIH 3T3 cells the following protocol has been utilized. Conditions for lipid-mediated transfection will vary with the cell line being transfected.

1. Cells are grown to a density of 10^6 cells/ml in 100-mm dishes in Dulbecco's modified Eagle's medium (DMEM) containing 10% fetal calf serum (FCS) for 24 hr.
2. Transfect different fluorescein-labeled antisense phosphorothioate ODNs (0.1 μM) using the cationic liposome Lipofectamine (2 $\mu g/ \mu l$, Life Technology) in OPTI-MEM (Life Technology).
3. Remove the liposome/DNA complexes after 16 hr of incubation, and add DMEM media containing 10% FCS to the cells.
4. After 8 hr, transfect the cells once again with the same lipid–ODN concentration as described earlier and incubate for another 16 hr.
5. Wash the transfected cells with phosphate-buffered saline prior to RNA extraction.
6. Isolate total RNA and carry out RNase protection assays to monitor the effect of the ODNs. In parallel, carry out an RNase H protection assay on an endogenous transcript (such as GAPDH or β-actin) as an internal standard. Alternatively, the RT-PCR approach described earlier can be utilized. The cellular uptake of the fluorescently labeled ODNs should be monitored microscopically to ensure that adequate numbers of cells have taken up the oligonucleotides.

Testing Ribozymes on Native mRNAs in Cell Extracts

Sites in the target RNA for which the ODNs provided the greatest levels of target RNA reduction should be tested with hammerhead ribozymes in cell extracts. The simplest procedure for accomplishing this task is to synthesize oligonucleotides corresponding to the ribozyme sequence and clone these in a plasmid vector with T7 and T3 promoters. The ribozymes are transcribed *in vitro* and tested in the extracts as follows.

1. Ribozyme-mediated cleavage experiments with the native target RNA are carried out in a total volume of 30 μl, containing 20 μl polymerase III extract, 40 units RNase inhibitor (Promega), and 50–200 nM each of the various ribozymes. The ribozymes and RNA are incubated for 20 min at 37° before adding 50–200 nM of oligonucleotide complementary to the ribozyme sequence to allow the endogenous RNase H to destroy the ribozyme prior to further RNA analyses.
2. The mixture is digested with 25 units DNase I (Boehringer Mannheim, Germany) for 45 min at 37° followed by phenol extraction and ethanol precipitation in the presence of 1 μg glycogen carrier (Boehringer Mannheim).
3. Reverse transcription is performed according the manufacturer's protocol (Life Technology) using 50 ng of random hexamer primer and 50 units of Moloney murine virus reverse transcriptase. The DNA PCR analysis and gene scan analysis are performed as described earlier and in Scherr and Rossi.[39]

Expression Cassettes

Once verification of the ribozyme activity to the chosen target sites in cell extracts is obtained, the next step is cloning of the ribozyme gene in the appropriate expression cassette. Selecting an appropriate cassette for intracellular gene expression is an important step for obtaining intracellular efficacy. The use of strong promoters to express catalytic RNAs may aid in generating higher intracellular ribozyme levels, but may not necessarily translate into effective intracellular ribozyme function. One of the main obstacles that must be overcome is the ability to maintain long-term expression in stably transduced or transfected cells. Gene expression from randomly integrated DNA sequences is largely affected by epigenetic silencing mechanisms, and maintaining levels of expression suitable for ribozyme function can become problematic.

Choice of Promoters

The expression of catalytic RNAs has been directed successfully by Pol II and Pol III promoters. Inducible, repressible, or tissue-specific promoters can be used to confer temporal, cell-specific expression. Each of these systems has some unique feature that distinguishes it from the others. For Pol II promoters, the addition of a 5' cap and 3' poly(A) tail may be important for stability and intracellular cytoplasmic localization. However, a disadvantage of Pol II expression cassettes is that the sequence elements required to achieve efficient transcription often extend for hundreds of nucleotides. Appended 5′ UTR and 3′ UTR sequences could affect proper folding of the ribozyme. A possible compromise can be achieved with mammalian snRNA Pol II promoters such as from the U1 snRNA gene.[6] This is a relatively strong promoter that is expressed ubiquitously in a variety of cell types of both human and murine origin. Even though the U1 promoter normally drives transcription of a small nuclear RNA, it can be used for transcribing capped ribozyme transcripts either by replacing the U1 sequence with the ribozyme sequence or by inserting the ribozyme within the stable U1 coding region.[6,43] The intracellular distribution of U1-promoted ribozymes embedded in the U1 sequence is both cytoplasmic and nuclear. If the appropriate signals in U1 snRNA are included, the ribozyme transcript will shuttle to the cytoplasm and back into the nucleus. A potential disadvantage is that transcripts devoid of U1 snRNA flanking 5′ and 3′ secondary structures tend to be unstable.[6]

Expressing ribozymes from Pol III cassettes can lead to high intracellular levels of the ribozymes. Depending on the type of Pol III expression unit chosen, transcripts can be localized to either the nucleus or the cytoplasm. As examples, ribozymes expressed from the human U6 snRNA promoter can be engineered to have either nuclear or cytoplasmic localization.[6,44]

Figure 2 summarizes these different constructs. The U6 promoter is a somewhat unusual Pol III promoter as all of the regulatory elements are contained within sequences 5′ of the transcribed U6 sequence. This allows engineering of a variety of ribozyme containing transcripts to be expressed from this strong promoter. The only constraint is that Pol III terminates within a stretch of four or five transcribed U's, therefore, such stretches must be avoided in the ribozyme design. The U6 transcript is normally capped with a γ methyl phosphate, provided that the first 27 bases of the U6 sequence are included in the transcript.[45] By eliminating only the last eight nucleotides, uncapped transcripts can be made,[44] which can sometimes prove to be useful. Another interesting feature of the U6 promoter is that some of the transcriptional regulatory elements are interchangeable with those of the U1 promoter. Addition of the U6 TATA element and flanking sequences to the U1 promoter convert it to a Pol III-driven unit. Conversely, removal of the TATA element from the U6 promoter converts this from

[43] A. Michienzi, S. Prislei, and I. Bozzoni, *Proc. Natl. Acad. Sci. U.S.A.* **93,** 7219 (1996).
[44] P. D. Good, A. J. Krikos, S. X. Li, E. Bertrand, N. S. Lee, L. Giver, A. Ellington, J. A. Zaia, J. J. Rossi, and D. R. Engelke, *Gene Ther.* **4,** 45 (1997).
[45] R. Singh, S. Gupta, and R. Reddy, *Mol. Cell Biol.* **10,** 939 (1990).

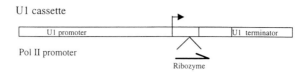

U1 cassette

Pol II promoter

Ribozyme

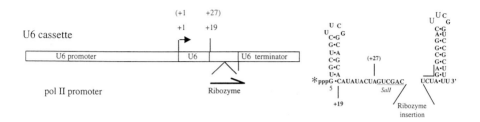

U6 cassette

pol II promoter   Ribozyme

Ribozyme
insertion

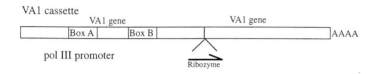

VA1 cassette

pol III promoter   Ribozyme

Nonprocessing tRNA Val cassette

Modified tRNA VAl   Ribozyme

pol III promoter

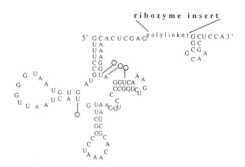

Fig. 2. Ribozyme expression cassettes. snRNA U1 and U6 cassettes are described in Bertrand *et al.*,[6] the VA1 gene cassette is described in Cagnon *et al.*,[48] and the nonprocessing tRNA Val cassette is described by Hampel.[30] Box A and box B represent the two essential Pol III promoter elements. The lillipop structure in the tRNA Val figure points to modifications in the primary sequence of tRNA Val, which prevent 5' and 3' end processing of the transcripts.

a Pol III to a Pol II expression system.[46] Therefore, the polymerase specificities of these promoters can be interchanged and some their unique features may be combined.

Most other Pol III promoters harbor transcriptional regulatory elements within the transcribed region. This is true for tRNAs and the adenoviral VA1 RNA, both of which have been utilized for ribozyme expression. The adenoviral VA1 promoter has been engineered to allow the incorporation of ribozymes or antisense RNAs within a highly structured stem–loop.[47,48] Substitution of a ribozyme for the loop and part of the stem eliminates the cellular kinase inhibitory domain, thus making this a useful vector for ribozyme expression. An important feature of the VA system utilized in this way is that the transcripts are predominantly cytoplasmic.[47,48]

Transfer RNAs also maintain the Pol III regulatory elements within the mature coding sequence of the tRNA. For use in ribozyme expression, the ribozyme is most often inserted either within the anticodon loop or in the aminoacyl acceptor stem.[6,49–51] A very useful Pol III expression system is a modified version of mammalian tRNA Val.[30] This Pol III cassette allows cloning of the ribozyme downstream of the structural portion of the tRNA. The tRNA itself is not end processed as there are several base changes that perturb its tertiary structure while still allowing expression. Figure 2 depicts the essential features of each of these promoters.

Insulators

A high level of intracellular gene expression has been attained with several expression cassettes; however, continuous maintenance of gene expression has been one of the major obstacles to successful gene therapy. New strategies need to be employed to protect the promoter from epigenetic silencing. An emerging idea is to protect the promoter with insulator sequences. Insulators are protein-binding DNA sequences that function unidirectionally in many cellular environments. These sequences lack significant stimulatory or inhibitory effects of their own and are usually placed between promoter and enhancer sequences. Their function is to isolate transcriptional units from nearby regulatory elements and to modulate enhancer–

[46] I. W. Mattaj, N. A. Dathan, H. D. Parry, P. Carbon, and A. Krol, *Cell* **55**, 435 (1988).
[47] S. Prislei, S. B. Buonomo, A. Michienzi, and I. Bozzoni, *RNA* **3**, 677 (1997).
[48] L. Cagnon, M. Cucchiarini, J. C. Lefebvre, and A. Doglio, *J. Acquir. Immune Defic. Syndr. Hum. Retrovirol.* **9**, 349 (1995).
[49] M. Cotton and M. L. Birnstiel, *EMBO J.* **8**, 3861 (1989).
[50] N. Yuyama, J. Ohkawa, Y. Inokuchi, M. Shirai, A. Sato, S. Nishikawa, and K. Taira, *Biochem. Biophys. Res. Commun.* **186**, 1271 (1992).
[51] J. D. Thompson, D. F. Ayers, T. A. Malmstrom, T. L. McKenzie, L. Ganousis, B. M. Chowrira, L. Couture, and D. T. Stinchomb, *Nucleic Acids Res.* **23**, 2259 (1995).

promoter interactions. The earliest and best-characterized insulators are from *Drosophila*. In 1991, Kellum and Schedl[52] found that DNase hypersensitive sites located at the ends of the heat shock locus can insulate reporter genes from the effects of neighboring regulatory sequences. Shortly thereafter it was found that the *Drosophila* gypsy retrotransposon harbors insulating activity from position effects as well.[53,54] More insulator sequences have now been characterized and some have been identified in vertebrates.[55,56]

Other elements that may be prove useful for long-term expression are the scaffold attachment regions (SAR), which are thought to define boundaries of chromatin domains. It has been shown that SAR elements can enhance the expression of heterologous genes in stable transfection experiments and in transgenic mice,[57,58] although in some cases they do not confer position-independent expression.[59] A human interferon-β gene-derived SAR was demonstrated to enhance gene expression from moloney murine leukemia virus (MoMuLV) vector in primary T cells.[60] Another potentially useful sequence for gene therapy is the Thy-1 fragment. This element has been shown to prevent DNA methylation in immature B cells, mouse embryos, and embryonic stem cells.[61,62] The use of insulators in situations requiring a long-term expression of ribozymes should be explored.

Auxiliary Sequences

For ribozyme expression, the transcriptional context within which the nontranslated ribozyme is expressed is very important for efficacy. Adding extra sequences to the ribozyme transcript could alter the proper folding and binding of the ribozyme to its target. However, having the ribozyme as part of a transcript, which includes a translatable message, may be important for transcript stability and/or ribozyme efficacy.[6] Thus, it may be useful to compare the intracellular stability and effectiveness of ribozyme-containing RNAs when expressed separately or within the context of a

[52] R. Kellum and P. Schedl, *Cell* **64**, 941 (1991).
[53] R. R. Roseman, V. Pirrotta, and P. K. Geyer, *EMBO J.* **12**, 435 (1993).
[54] P. K. Geyer and V. G. Corces, *Genes Dev.* **6**, 1865 (1992).
[55] X. P. Zhong and M. S. Krangel, *Proc. Natl. Acad. Sci. U.S.A.* **94**, 5219 (1997).
[56] J. H. Chung, M. Whiteley, and G. Felsenfeld, *Cell* **74**, 505 (1993).
[57] D. Klehr, K. Maass, and J. Bode, *Biochemistry* **30**, 1264 (1991).
[58] M. Kalos and R. E. Fournier, *Mol. Cell Biol.* **15**, 198 (1995).
[59] L. Poljak, C. Seum, T. Mattioni, and U. K. Laemmli, *Nucleic Acids Res.* **22**, 4386 (1994).
[60] M. Agarwal, T. W. Austin, F. Morel, J. Chen, E. Bohnlein, and I. Plavec, *J. Virol.* **72**, 3720 (1998).
[61] Y. Shimizu, Y. Oka, H. Ogawa, T. Kishimoto, and H. Sugiyama, *Immunol. Invest.* **21**, 183 (1992).
[62] M. Szyf, G. Tanigawa, and P. L. McCarthy, Jr., *Mol. Cell Biol.* **10**, 4396 (1990).

protein-coding sequence.[6] Other elements can be included in the design of an expression cassette to provide stability to the intracellular transcripts. For instance, stem–loop structures may be added at both 5′ and 3′ ends of the ribozyme.

Successful ribozyme-mediated inhibition is also dependent on the colocalization of the ribozyme transcripts with the targeted RNAs (for a review, see Ref. 63). Intracellular localization and subcompartmentalization can be manipulated by adding specific signals to the RNA transcripts. These localization signals, which are often found in the 3′ UTR of certain messenger RNAs,[64] can be attached to the 3′ end of the ribozyme transcript to direct its intracellular trafficking.

In summary, in designing a cassette for the endogenous expression of ribozymes, the following should be considered.

1. The choice of promoter-expression cassette: Pol II versus Pol III. This choice should be based on desired levels of expression, desired intracellular localization, and whether the ribozyme is to be part of a chimeric transcript.
2. The choice of the promoter should take into account the delivery vehicle. The type of vector to be used for ribozyme expression, transfected plasmid versus viral vector will have an impact on the choice of promoter. In retroviral vectors, Pol III or the U1 Pol II promoters can be cloned as double copy inserts in the U3 region of the long terminal repeats (LTR) (Fig. 3). In AAV or adenoviral vectors, consideration must be given to the amount of DNA that can be included and still allow packaging of the vectors (for reviews of viral vectors, see Ref. 65).
3. The choice of sequences appended to the ribozyme transcript will dictate the trafficking and target colocalization of the ribozyme. Inclusions of snRNA coding regions[6,43] or tethering the ribozymes to viral sequences can be used to affect specific intracellular compartmentalization and colocalization with the target RNA.[7,66]
4. Insulator sequences may be inserted within the vector that will be used for ribozyme gene transduction or transfection. Such sequences may be important when long-term expression under nonselective conditions is required.

[63] G. M. Arndt and G. H. Rank, *Genome* **40,** 785 (1997).
[64] Y. Oleynikov and R. H. Singer, *Trends Cell Biol.* **8,** 381 (1998).
[65] R. A. Morgan and W. F. Anderson, *Annu. Rev. Biochem.* **62,** 191 (1993).
[66] B. K. Pal, L. Scherer, L. Zelby, E. Bertrand, and J. J. Rossi, *J. Virol.* **72,** 8349 (1998).

Ribozyme expression cassette inserted between LTRs

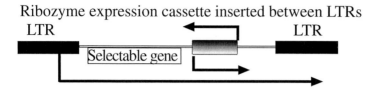

Double copy inserts of ribozyme cassette in either orientation

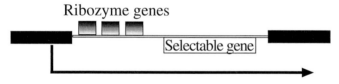

Multiple ribozyme genes transcribed from LTR promoter

Ribozyme genes

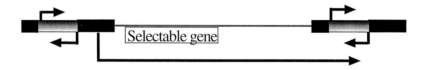

Fig. 3. Schematic of strategies for ribozyme expression from a retroviral vector. Arrows indicate the directions of transcription. Small expression units such as those depicted in Fig. 2 can be inserted as double copy inserts in either orientation within the U3 region of the long terminal repeats (LTR). Cassettes inserted between the LTRs can act as independent transcription units as well as being part of LTR-promoted transcripts.

Delivery of Ribozyme Genes

There are two basic approaches for the delivery of ribozyme genes into cells: (1) viral vector-mediated transduction of the ribozyme genes into the desired cells or (2) transfection of the DNA encoding the ribozyme gene(s).

Viral Vectors

Several classes of viral vectors are being exploited for the delivery of genes *in vivo,* including DNA (adenoviruses, herpes virus, adeno-associated virus) and RNA retroviruses. It is beyond the scope of this article to review the advantages and disadvantages of each system. For an overview, the reader is referred to the review article by Morgan and Anderson.[65] General concerns are associated with the use of viral vectors such as residual infectivity, toxicity, and rescue of infectivity by recombination. Each vector system has its own set of advantages and disadvantages that ultimately dictate its use in a specific application. To date, the most extensively utilized viral vectors have been murine-derived amphotropic retroviruses. These viral vectors can infect a wide variety of cell types; they integrate into the host chromosome and, thus, are potentially capable of long-term expression. The integration process requires cell replication, restricting the use of these vectors to actively dividing cells. Nevertheless, retroviral vectors are currently the preferred method for *in vivo* gene transfer because of the efficiency of transduction and integration in the host genome. The retroviral genome includes only few genes (*gag, pol,* and *env*), the promoter and enhancer sequences contained in the LTR, splicing acceptor and donor sites, and the packaging signal. Thus, the design of a retroviral vector is relatively simple. The expression cassette containing the selected ribozyme gene(s) can be cloned into the U3 region of the 3′ LTR to generate a double copy insert or in between the two LTR sequences (Fig. 3).[6,67,68] Although potentially capable of long-term expression, the LTR viral promoters are subject to progressive silencing after chromosomal integration. Specific modifications have been made to a Moloney murine-derived retroviral vector in an attempt to alleviate this problem.[69,70]

In these vectors the LTR enhancer region of MoMuLV was substituted with the corresponding LTR region from the myeloproliferative sarcoma virus (MPSV). The negative control region (NCR), which binds the YY1 transcription factor, was deleted and the primer-binding site of MoMuLV was mutated or substituted with the primer-binding site of the dl587 rev strain. These modifications resulted in enhanced expression of an inserted gene in murine embryonic carcinoma cells (EC), embryonic stem cells, and fibroblast cells.[69,71] Substitution of the primer-binding site and the enhancer

[67] P. A. Hantzopoulos, B. A. Sullenger, G. Ungers, and E. Gilboa, *Proc. Natl. Acad. Sci. U.S.A.* **86**, 3519 (1989).

[68] B. A. Sullenger, T. C. Lee, C. A. Smith, G. E. Ungersand, and E. Gilboa, *Mol. Cell Biol.* **10**, 6512 (1990).

[69] P. M. Challita, D. Skelton, A. el-Khoueiry, X. J. Yu, K. Weinberg, and D. B. Kohn, *J. Virol.* **69**, 748 (1995).

[70] W. J. Krall, D. C. Skelton, X. J. Yu, I., Riviere, P. Lehn, R. C. Mulligan, and D. B. Kohn, *Gene Ther.* **3**, 37 (1996).

[71] P. B. Robbins, X. J. Yu, D. M. Skelton, K. A. Pepper, R. M. Wasserman, L. Zhu, and D. B. Kohn, *J. Virol.* **71**, 9466 (1997).

region are also important for expression from internal promoters, as there is evidence that these elements can reduce the expression of heterologous promoters placed between and within the LTR regions.[72,73]

Following the construction of the retroviral viral vector, it must be packaged into virions, which can be transduced into the appropriate target cells. Packaging can be carried out either by transient transfection[74] or by establishing permanent packaging cell lines.[75] Although transient transfection is very time efficient and allows the use of human cell lines as the producer of viral particles, variations in transfection efficiencies may provide an undesirable variable when comparing several constructs. The use of packaging cell lines and cocultivation of the packaging cells and the transductants is more time-consuming, but provides more reproducible transduction efficiencies.

Nonviral Delivery

For the delivery of ribozyme genes into tissue culture cells, standard calcium phosphate or electoporation transfection systems are generally employed. These techniques are well described in a variety of laboratory manuals. Alternatively, the use of cationic lipids for ribozyme gene delivery or even for delivery of presynthesized ribozymes can be very effective.[76–78] The following protocol is used for encapsulating DNA or RNA molecules in cationic lipids for cell delivery. Many different commercially available reagents are suitable for nucleic acid delivery into a variety of cell types. The protocol we describe is based on the use of Lipofectin (Life Technology), which has been used successfully for plasmid DNA, performed antisense ODN, and ribozyme delivery in both human and murine cells. It is recommended that the use of a plasmid containing the green fluorescent protein (GFP) or other readily assayable marker be utilized as a test of delivery efficiency.

1. Plate exponentially growing cells in tissue culture dishes at 5×10^5 cells/well and grow overnight in a CO_2 incubator at 37° to 80% confluency.

[72] C. M. Gorman, P. W. Rigby, and D. P. Lane, *Cell* **42,** 519 (1985).

[73] D. D. Bowtell, S. Cory, G. R. Johnson, and T. J. Gonda, *J. Virol.* **62,** 2464 (1988).

[74] F. Grignani, T. Kinsella, A. Mencarelli, M. Valtieri, D. Riganelli, L. Lanfrancone, C. Peschle, G. P. Nolan, and P. G. Pelicci, *Cancer Res.* **58,** 14 (1998).

[75] C. M. Lynch and A. D. Miller, *J. Virol.* **65,** 3887 (1991).

[76] R. W. Malone, P. L. Felgner, and I. M. Verma, *Proc. Natl. Acad. Sci. U.S.A.* **86,** 6077 (1989).

[77] N. Duzgunes and P. L. Felgner, *Methods Enzymol.* **221,** 303 (1993).

[78] P. L. Felgner, Y. J. Tsai, L. Sukhu, C. J. Wheeler, M. Manthorpe, J. Marshall, and S. H. Cheng, *Ann. N.Y. Acad. Sci.* **772,** 126 (1995).

2. Dilute the presynthesized DNA or RNA molecules and the Lipofectin reagent with OPTI-MEM I (Life Technology) medium. The amount of nucleic acids and lipid suspension needs to be optimized for each cell type. This should be done using a reporter system such as GFP or β-galactosidase.

3. Mix the diluted reagent from the previous step, vortex gently, and incubate for 5–10 min at room temperature (prepare this complex in a polystyrene tube because it can stick to polypropylene). If RNA is used, you may perform this step on ice to avoid chemical and enzymatic degradation.

4. Wash the cells three times with serum-free medium.

5. Add the lipid complex and incubate the cells at 37° in a CO_2 incubator (5–10% CO_2) for 3–6 hr. In general, transfection efficiency increases with time, although toxic conditions may develop after 8 hr.

6. Add 3 ml medium with 20% of serum (the serum is dependent on the cell type; FCS serum may be used).

7. Incubate the cells for 24–48 hr at 37° in a CO_2 incubator.

8. Harvest cells and assay for gene activity.

Assaying Intracellular Ribozyme Function

A mutant, noncleaving form of the ribozyme should always be used as a control for antisense activity and, when possible, the reduction in the RNA level should be confirmed by a reduction in the level of the encoded protein. Noncleaving, crippled ribozymes are created by mutating one or more nucleotides in the catalytic domains of the hammerhead or hairpin ribozymes.[79,80] Standard techniques such as Northern gel, RNase protection, primer extension, and RT-PCR can be performed to analyze *in vivo* ribozyme function. The RNA analysis is not by itself conclusive because it is possible for cleavage to take place during the RNA extraction. The use of stable, ribozyme-expressing clones allows more accurate determinations than those obtained from transiently transfected cells. In both cases, a nontargeted RNA (and protein if possible) should be used as internal standards.

RNase Protection

1. To establish a ^{32}P-labeled RNA probe to detect ribozyme expression, runoff transcription of an antisense sequence of the target is carried

[79] D. E. Ruffner, G. D. Stormo, and O. C. Uhlenbeck, *Biochemistry* **29**, 10695 (1990).
[80] A. Berzal-Herranz, S. Joseph, B. M. Chowrira, S. E. Butcher, and J. M. Burke, *EMBO J.* **12**, 2567 (1993).

out using standard *in vitro* transcription with T7 or T3 RNA polymerase (New England Biolabs, Beverly, MA).

2. To generate an internal control and a loading control, a linearized plasmid such as pTRI-GAPDH provided in the Direct Protect kit (Ambion, Austin, TX) should also be labeled with one or more ^{32}P-labeled nucleotides.

3. Ten to 15 μg of total RNA isolated from treated cells is mixed with 4–7×10^5 cpm ribozyme or target probe and a predetermined dilution of the internal standard probe to perform the RNase protection assay according to the instructions provided with the Direct Protect kit (Ambion). Protected fragments are separated on an appropriate denaturing gel system and the protected fragments should be quantitated on a PhosphorImager or by other means (direct counting of gel slices).

Primer Extension Analysis

1. The primers should be positioned downstream (to the 3' side) of the sequence cleaved by the ribozyme. As an internal standard and loading control, a ubiquitously expressed RNA such as the human U6 snRNA should be utilized. For the human snRNA, we use 0.1 pmol of ^{32}P-labeled U6 primer with the following sequence d(TATGGAACGCTTCACGAATTTG).

2. One picomole of ^{32}P-labeled ribozyme primer and 15–20 μg total RNA are heated for 3 min at 80° in reaction buffer containing 0.5 M Tris–HCl, pH 8.3, 0.6 M NaCl, and 0.1 M DTT. After cooling to room temperature, magnesium acetate is added to a final concentration of 6 mM and combined with dNTPs (2 mM each NTP) and avian mycloblastosis virus (AMV) reverse transcriptase (1 unit/μl)(Life Technology) at 48° for 45 min. Reactions are quenched with formamide-loading buffer and loaded onto a denaturing polyacrylamide gel. Gels are analyzed and quantitated by phosphorimage analysis or direct counting of gel slices.

Concluding Remarks

Much has been learned about the structure–function relationships of catalytic RNAs, and ribozymes have proven to be useful for a variety of tasks, including inactivation of gene expression, the repair of mutant transcripts, and probing of various aspects of RNA metabolism. Despite many successful examples, the *in vivo* use of ribozymes is nonetheless problematic, largely due to uncertainties about target site accessibility and intracellular colocalization. It is strongly recommended that the investigator test more than one ribozyme target site and two or more expression systems to achieve maximal efficacy. The choice of cleavage site is the most important first step, and the use of new strategies to identify targets in their native intracellular state will aid greatly in this process. Ribozymes can potentially be employed as efficient surrogate genetic tools and hold promise as therapeutic agents. By being aware of the potential problems for ribozyme utilization, the design and implementation of ever more effective ribozyme strategies should be possible.

Acknowledgments

This work was supported by National Institutes of Health (NIH) Grants AI 138592, AI 29329, and AI 42552. We thank W. F. Fitzgerald for technical assistance.

[24] Selective Degradation of Targeted mRNAs Using Partially Modified Oligonucleotides

By JOHN M. DAGLE and DANIEL L. WEEKS

Introduction

In 1978, Stephenson and Zamecnik described the first successful use of synthetic oligonucleotides to alter gene expression.[1,2] Since that time the field of antisense technology has grown dramatically. Naive expectations that antisense oligonucleotides would immediately become useful in the treatment of human diseases resulted in early unmerited pessimism. It must be kept in mind that a great deal has been learned over the last two decades regarding the uptake,[3,4] degradation,[5–11] specificity,[12] toxic-

[1] M. Stephenson and P. Zamecnik, *Proc. Natl. Acad. Sci. U.S.A.* **75,** 285 (1978).
[2] P. Zamecnik and M. Stephenson, *Proc. Natl. Acad. Sci. U.S.A.* **75,** 280 (1978).
[3] J. T. Holt, R. L. Redner, and A. W. Nienhuis, *Mol. Cell. Biol.* **8,** 963 (1988).
[4] S. L. Loke *et al., Proc. Natl. Acad. Sci. U.S.A.* **86,** 3474 (1989).
[5] C. Cazenave, M. Chevrier, N. T. Thuong, and C. Helene, *Nucleic Acids Res.* **15,** 10507 (1987).
[6] J. M. Dagle, J. Walder, and D. L. Weeks, *Nucleic Acids Res.* **18,** 4751 (1990).
[7] J. M. Dagle, D. L. Weeks, and J. Walder, *Antisense Res. Dev.* **1,** 11 (1991).
[8] J. M. Dagle, M. E. Andracki, R. J. DeVine, and J. A. Walder, *Nucleic Acids Res.* **19,** 1805 (1991).
[9] M. R. Rebagliati and D. A. Melton, *Cell* **48,** 599 (1987).
[10] T. Woolf, C. Jennings, M. Rebagliati, and D. Melton, *Nucleic Acids Res.* **18,** 1763 (1990).
[11] E. Wickstrom, *J. Biochem. Biophys. Methods* **13,** 97 (1986).
[12] T. M. Woolf, D. A. Melton, and C. Jennings, *Proc. Natl. Acad. Sci. U.S.A.* **89** (1992).

ity,[6,7,10,13] and efficacy[6,7,10] of antisense oligonucleotides. The information derived from asking these complex mechanistic questions has driven the development of several new classes of modified oligonucleotides. Modifications involving the phosphate bond, base, and sugar moiety of oligonucleotides have created numerous compounds with varying physical and biological properties. The approval by the FDA of Vitravene (Isis Pharmaceuticals), a modified oligonucleotide, for the treatment of cytomegalovirus retinitis in patients suffering from acquired immunodeficiency syndrome (AIDS) heralds the beginning of antisense therapeutics. As hurdles related to drug delivery are overcome, antisense oligonucleotides will possess an increasingly important role in medicine. This article discusses the design, synthesis, and utilization of chimeric, phosphoramidate-modified antisense oligonucleotides for *in vivo* studies in the amphibian *Xenopus laevis*.

Xenopus oocytes and embryos provide a unique opportunity to examine the *in vivo* efficacy of oligonucleotide-based approaches in different cellular environments. Stage VI oocytes, which are removed surgically from the adult female frog, are nondividing cells in G_2 meiotic arrest.[14] They have little active transcription of endogenous genes encoding mRNA, and although most of the mRNA is not being actively translated, it is apparently stabilized through association with proteins found in the oocyte. Thus, levels of endogenous messages can be monitored for turnover without having to consider new mRNA synthesis. Although the stage VI oocyte is not actively synthesizing endogenous mRNA, it will support the transcription of genes introduced via the injection of plasmid into the nucleus. Thus, specificity of oligonucleotide action can be monitored using messages transcribed by plasmids injected into the oocyte.

Xenopus embryos are obtained by the fertilization of eggs laid by the adult female. These embryos do not transcribe endogenous genes until the 4000 cell stage, approximately 7 hr after fertilization.[15,16] Importantly, all developmental decisions up to that point are made using maternal molecules present in the embryo prior to fertilization. In experiments designed to examine the loss of a particular mRNA, therefore, oligonucleotide-mediated mRNA degradation cannot be compensated by transcription during early development. In addition, the normal, rapid developmental progression through cleavage stages and gastrulation are very sensitive to a variety

[13] R. Smith *et al.*, *Development* **110**, 769 (1990).
[14] J. N. Dumont, *J. Morphol.* **136**, 155 (1972).
[15] J. Newport and M. Kirschner, *Cell* **30**, 675 (1982).
[16] J. Newport and M. Kirschner, *Cell* **30**, 687 (1982).

of toxic molecules. These processes serve as rigorous tests of potential nonspecific (sequence-independent) oligonucleotide effects.

Principles of Oligonucleotide Design

The key to designing oligonucleotides for use *in vivo* is to attenuate or prevent their degradation by active intracellular nucleases without affecting antisense efficacy. Because almost any modification of the phosphate bond greatly increases the stability of that individual linkage, early attempts to produce nuclease-resistant compounds involved chemical modification of all of the internucleoside linkages in an oligonucleotide. A wide variety of modifications were developed and tested. Fully modified oligonucleotides, possessing no native phosphodiester linkages, demonstrated a greatly increased resistance to nucleolytic degradation.[17-21] Interestingly, these very complex compounds, which bore little similarity to anything normally seen by a cell, had little or no biological activity. This apparent paradox was understood in the context of the mechanism through which most antisense oligonucleotides act. Following hybridization of an oligonucleotide to a complementary target message, the fate of the resulting DNA : RNA heteroduplex depends on the nature of the component oligonucleotide. Oligonucleotides possessing anionic internucleoside linkages, either unmodified phosphate diesters or sulfur-modified phosphorothioates, form heteroduplexes that are substrates for the ubiquitous enzyme ribonuclease H (RNase H).[22-24] RNase H cleaves the RNA component of the heteroduplex, rendering the message permanently untranslatable. In fact, the two RNA cleavage fragments produced from this reaction are generally unstable in the cell and are degraded rapidly. The oligonucleotide, however, is unaffected by this reaction and may then dissociate from the cleaved RNA and bind to another target message. Therefore, the potential exists for a few oligonucleotide molecules to completely eliminate an entire pool of target messages.

A second possible mode of action occurs when the DNA : RNA heteroduplex is not a substrate for intracellular RNase H. This occurs when the

[17] P. S. Miller, K. N. Fang, N. S. Kondo, and P. O. P. Ts'o, *J. Am. Chem. Soc.* **93,** 6657 (1971).

[18] P. S. Miller, K. B. McParkland, K. Jayaraman, and P. O. P. Ts'o, *Biochemistry* **20,** 1874 (1981).

[19] R. L. Letsinger, S. A. Bach, and J. S. Eadie, *Nucleic Acids Res.* **14,** 3487 (1986).

[20] D. E. Jensen and D. J. Reed, *Biochemistry* **17,** 5098 (1978).

[21] D. M. Tidd and H. M. Warenius, *Br. J. Cancer* **60,** 343 (1989).

[22] C. A. Stein, C. Subasinghe, K. Shinozuka, and J. S. Cohen, *Nucleic Acids Res.* **16,** 3209 (1988).

[23] R. Y. Walder and J. A. Walder, *Proc. Natl. Acad. Sci. U.S.A.* **85,** 5011 (1988).

[24] J. Shuttleworth and A. Colman, *EMBO J.* **7,** 427 (1988).

heteroduplex contains an oligonucleotide with no anionic region, i.e., a fully modified oligonucleotide. Translation of the target message to a polypeptide is thought to be inhibited by physically preventing passage of the ribosome through the region of the heteroduplex. Obviously, no intrinsic enzymatic activity would be required for this pathway of inhibition. Studies using the methyl phosphonate-modified oligonucleotides developed by Miller and Ts'o[17,25] have shown that steric inhibition of translation can occur.[26,27] The inhibition reported, however, was seen only at relatively high concentrations of oligonucleotide. In a direct comparison of anionic and nonionic oligonucleotides, Walder and Walder[23] demonstrated clearly that oligonucleotide-directed cleavage of a message by RNase H is by far the most effective means of antisense inhibition. In general, it appears that oligonucleotides that do not act through an RNase H-dependent mechanism are ineffective antisense agents. An exception to this general rule may be the morpholino-modified oligonucleotides described by Summerton and Weller[28] that are reported to effectively inhibit translation in a specific manner.

We view an antisense oligonucleotide as a cofactor for RNase H that is necessary for the recognition and cleavage of a complementary message by the enzyme. Our oligonucleotides are designed to be nuclease resistant while retaining the ability to direct the RNase H-mediated cleavage of targeted messages in *Xenopus* oocytes and embryos. The methoxyethyl phosphoramidate modification, which we have characterized, and the general structure of two chimeric oligonucleotides, which will be discussed further, are shown in Fig. 1. Each oligonucleotide consists of a central core of unmodified DNA that is flanked by regions possessing phosphoramidate-modified internucleoside linkages. These regions were designed to serve distinct proposes. The central DNA core serves as a recognition and active site for RNase H when bound to a target message. The uncharged flanking regions, which participate in the hybridization of the oligonucleotide to the target message, are completely nuclease resistant. They serve to block the active exonucleases present in *Xenopus* oocytes and embryos from degrading the oligonucleotide. Although one terminal modification should adequately prevent exonuclease activity, maximizing the extent of modification is desirable to limit sites available for degradation by latent intracellular endonucleases. As discussed earlier, however, modification of all phosphate linkages limits antisense activity severely.

[25] P. S. Miller, J. C. Barrett, and P. O. P. Ts'o, *Biochemistry* **13**, 4887 (1974).
[26] L. J. Maher and B. J. Dolnick, *Nucleic Acids Res.* **16**, 3341 (1988).
[27] K. R. Blake *et al.*, *Biochemistry* **24**, 6139 (1985).
[28] J. Summerton and D. Weller, *Antisense Nucleic Acid Drug Dev.* **7**, 187 (1997).

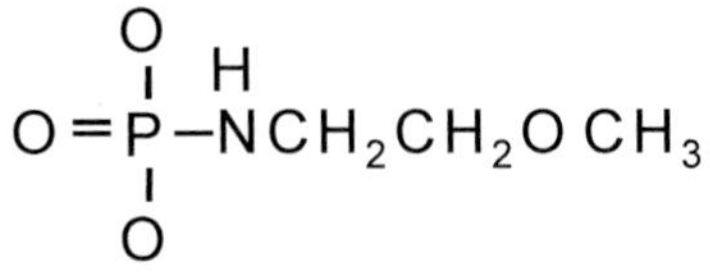

$$O = \overset{\overset{O}{|}}{\underset{\underset{O}{|}}{P}} - \overset{H}{N} CH_2 CH_2 O CH_3$$

2-Methoxyethyl-
Phosphoramidate *

C*T*G*A C A A C A T G A*C*T*G*C   An2 NON

T*G*G*T A T A T C C A G*T*G*A*T   Con NON

FIG. 1. Phosphoramidate modification and oligonucleotide sequences. The methoxyethyl phosphoramidate modification is shown along with two oligonucleotides possessing a chimeric structure. Phosphoramidate linkages present in the flanking regions are represented by an asterisk. Each oligonucleotide contains an internal unmodified diester region. This structure is designated NON.

The end result of effective oligonucleotide design is to develop a compound that is recognized by one nuclease, RNase H, but not by other nucleases, namely those involved in the degradation of single-stranded DNA. Figure 2 demonstrates the increased stability of an internally labeled oligonucleotide possessing several 3'- and 5'-terminal methoxyethyl phosphoramidate modifications. An internally labeled unmodified oligonucleotide (labeled Unmod in Fig. 2) shows some degradation at 0 min, the 20 to 40 sec it takes to inject the compound and immediately place the embryo on dry ice. The unmodified oligonucleotide is undetectable 10 min after injection. The terminally modified oligonucleotide, which migrates more slowly as a result of removing the negative charge from the phosphate linkages, is much more stable. This compound shows a slow, steady disappearance, but is still present 30 min after injection. The first lane (uninj in Fig. 2) represents a mixture of modified and unmodified oligonucleotides prior to injection. The oligonucleotides (both modified and unmodified) used in these degradation studies are labeled internally by an enzymatic process of 5'-end-labeling and subsequent ligation, a strategy that limits the degree of possible modification. This was done instead of simply 5'-end-labeling a full-length oligonucleotide because the 5'-hydroxyl group of an oligonucleotide is not a substrate for T4 polynucleotide kinase if the 5'-terminal internucleoside linkage contains a phosphoramidate bond. In addition, using this strategy, the disappearance of signal cannot be attributed to the action of an intracellular phosphatase removing an accessible

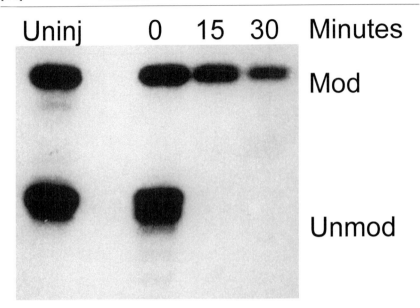

FIG. 2. Oligonucleotide degradation *in vivo*. A mixture of unmodified (Unmod) and chimeric phosphoramidate-modified oligonucleotides (Mod) was injected into single cell *Xenopus* embryos. Immediately after injection (0) and at 15 and 30 min postinjection the embryos were lysed in phenol and the oligonucleotides were extracted. Equal amounts of radioactivity were analyzed by electrophoresis on a 20% polyacrylamide–7 *M* urea gel and visualized by autoradiography. The uninjected mix is shown in the lane marked Uninj.

phosphate. The oligonucleotides we actually use for *in vivo* antisense studies always contain a greater number of modified linkages and would, therefore, be expected to be even more stable than suggested by the data presented here.[6]

To determine the effect of the extent of modification on *in vivo* antisense activity, a series of oligonucleotides targeting the maternal *Xenopus* mRNA An2 were synthesized possessing different lengths of unmodified central diesters.[6,29] All of the oligonucleotides were 16-mers possessing the same sequence as An2 NON, seen in Fig. 1.[7] These compounds were injected individually into single cell *Xenopus* embryos. Figure 3 is a ribonuclease protection assay measuring levels of An2 mRNA, demonstrating that chimeric oligonucleotides can indeed be modified beyond the point of biological utility. Oligonucleotides with at least six consecutive unmodified diesters were effective at eliminating the target message An2. There is a distinct change in An2 message levels, however, when the number of unmodified

[29] D. L. Weeks and D. A. Melton, *Proc. Natl. Acad. Sci. U.S.A.* **84**, 2798 (1987).

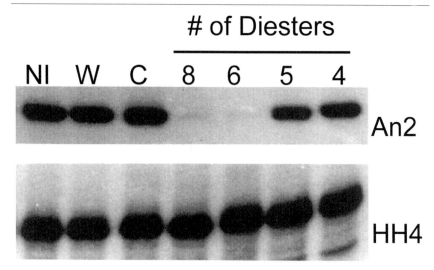

FIG. 3. Substrate specificity of *Xenopus* RNase H. Single cell embryos were injected with 5 ng of a chimeric oligonucleotide targeting the message An2, each possessing the indicated number of unmodified phosphodiesters. Total RNA was isolated 4 hr after injection, and levels of An2 and histone H4 (HH4) were quantified by RNase protection. Protected bands were analyzed by electrophoresis on a 6% polyacrylamide–7 M urea gel and visualized by autoradiography. NI, no injection; W, water injected; C, control oligonucleotide.

phosphate diesters is decreased from six to five. The level of An2 mRNA following the injection of an oligonucleotide possessing only four consecutive unmodified diesters was similar to the level seen in embryos that were uninjected (NI), water injected (W), or injected with a control oligonucleotide (C). Importantly, none of the oligonucleotide conditions had any effect on an unrelated control mRNA, histone H4 (HH4). In *Xenopus* embryos, at least six consecutive unmodified internucleoside linkages are required if an injected antisense oligonucleotide is to remain a suitable substrate for RNase H.

In addition to increasing the intracellular lifetime of the oligonucleotide, terminal modifications may actually increase the specificity of the antisense oligonucleotide. An unmodified 16-mer can bind to a target message and direct RNase H to cleave at a number of sites over the entire heteroduplex. If the unmodified oligonucleotide binds to a related, nontarget message, differing from the target by only a few nucleotides, the resulting hetero-duplex could also be cleaved at several sites. By limiting the active site of the heteroduplex to the absolute minimum required for RNase H activity, we make mismatches between the antisense oligonucleotide and the target message in the central unmodified region incompatible with mRNA cleav-

TABLE I
SEQUENCE OF ANTISENSE OLIGONUCLEOTIDES

Compound	Sequence[a]
Cyclin A1 mRNA	5' A A U A U G A A G A A A U C U A C 3'
Cyclin B1 and B2 mRNA	5' A A U A U G A A G A G A U G U A C 3'
Oligonucleotide	3' T*T*A*T*A C T T C T C T A*C*A*T*G

[a] *Key to symbols:* ● represents mismatches between cyclin A and B mRNAs; | represents base pairing between oligonucleotide and mRNA; and * represents a phosphoramidate internucleoside linkage.

age. The specificity of the chimeric oligonucleotides was tested by targeting the elimination of three related maternal messages: cyclin B1, B2, and A1.[30] As seen in Table I, over the span of the antisense oligonucleotide utilized, cyclin B1 and B2 have identical sequences. The sequence of cyclin A1 contains two mismatches with respect to the antisense oligonucleotide. One of the mismatches occurs in the modified terminal region, whereas the other mismatch is present in the central unmodified region of the heteroduplex. Following the injection of the chimeric antisense oligonucleotide (seen in Table I), total RNA was isolated and evaluated by Northern analysis. Levels of cyclin B1, B2, and A1 mRNA, which were quantified by sequential probing of the membrane, are shown in Fig. 4. Although there is significant similarity among the three messages of interest, only cyclin B1 and B2 were targeted for elimination by the antisense oligonucleotide. The level of cyclin A1 mRNA, which contains a single mismatch (A–C) in the limited cleavage site of the heteroduplex, was not altered significantly. By minimizing the size of the heteroduplex active site, we increase the relative importance of the base pairing in that region. Injection of a similarly modified control oligonucleotide had no effect on any of the messages evaluated.

In any antisense experiment, especially those employing chemically modified oligonucleotides, the inclusion of control oligonucleotides in the experimental design is vital. Interpretation of antisense data is impossible if the relative contribution of specific and nonspecific effects cannot be determined. The assumption that all effects are due to elimination of the target message is naïve and scientifically imprecise. Although there is a general consensus that control oligonucleotides must be used, there is little agreement as to the appropriate sequence of the control. In general, four types of control oligonucleotides are currently in use. The sense control

[30] D. L. Weeks, J. A. Walder, and J. N. Dagle, *Development* **111**, 1173 (1991).

2 hours 5 hours

N W C B N W C B

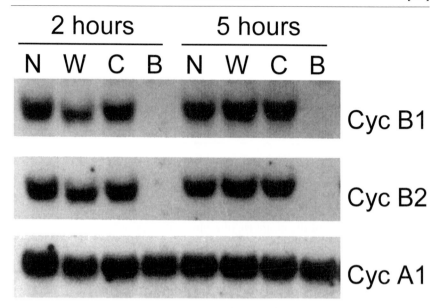

Cyc B1

Cyc B2

Cyc A1

FIG. 4. Specificity of chimeric antisense oligonucleotides *in vivo*. Single cell *Xenopus* embryos were injected with 5 ng of the chimeric cyclin antisense oligonucleotide shown in Table I or with a control oligonucleotide. At 2 and 5 hr after injection, total RNA was isolated from five embryos, separated by size on a formaldehyde/agarose gel, and transferred to a nylon membrane. The membrane was probed sequentially with end-labeled oligonucleotides specific for cyclin B1 (Cyc B1), cyclin B2 (Cyc B2), and then with a labeled cyclin A cDNA (Cyc A1). Signal was detected by autoradiography, and the membrane was stripped of radioactivity between each successive hybridization. N, no injection; W, water injected; C, control oligonucleotide; B, cyclin B antisense oligonucleotide.

has the same sequence as the target message. The scrambled control has the exact base composition of the antisense oligonucleotide, but in a different order. The random control has a completely random sequence with an equal fraction of each base. Finally, a control oligonucleotide can be an antisense compound directed against a specific message that is not expressed in the cells under investigation. The control oligonucleotide should have the same type and extent of modification as the antisense compound. A final control for those using modified oligonucleotides is an unmodified oligonucleotide of the same sequence. This allows a direct comparison of the antisense properties of the two oligonucleotides.

Oligonucleotide Synthesis

Oligonucleotides are synthesized (3' to 5') on an Applied Biosystems PCR-Mate DNA synthesizer using the hydrogen phosphonate chemistry developed by Froehler *et al.*[31,32] All reagents for DNA synthesis are purchased from Glen Research (Sterling, VA). Other organic reagents are purchased from Aldrich (Milwaukee, WI). When using H-phosphonate chemistry, trivalent phosphorous oxidation to the pentavalent state is performed after the oligonucleotide (or desired oligonucleotide segment of the same modification class) is synthesized. This is a major difference from standard phosphoramidite chemistry in which oxidation is performed every cycle, after monomer coupling. Chimeric oligonucleotides are synthesized in blocks, where each member of the block is either modified identically or unmodified. The desired number of 3' residues is coupled using the H-phosphonate synthesis program recommended by the manufacturer. The controlled pore glass solid support containing the 3' segment of oligonucleotide is removed from the synthesizer and dried under vacuum.

Oxidative amidation of hydrogen phosphonate diesters is performed manually in 5-ml polypropylene/polyethylene syringes containing no rubber parts (Aldrich). The oligonucleotide is treated with 3.3 ml of a 10% solution of 2-methoxyethylamine (stored over molecular sieves) in anhydrous carbon tetrachloride (stored over calcium hydride) for 40–60 min. The solution is moved back and forth, between the syringes, every few minutes. The solid support is then rinsed with 1% triethylamine in acetonitrile, dried under vacuum, and placed back on the DNA synthesizer. The second block of nucleotides, which will become the unmodified region, is added next. Oxidation of this segment is performed on the synthesizer according to the manufacturer's protocol. Briefly, the solid support is treated with 5% (w/v) iodine in tetrahydrofuran:pyridine:water (15:2:2) for 4 min and then with the same solution diluted 1:1 with 8% triethylamine in tetrahydrofuran:water (43:3) for 3 min. The oxidation of hydrogen phosphonate diesters requires more stringent conditions than the oxidation of linkages produced using phosphoramidite chemistry. We therefore make our iodine solution as needed so that it is never more than 2 days old. The final segment of the oligonucleotide is synthesized on the DNA synthesizer and oxidatively amidated manually, as described earlier. The 5'-dimethoxytrityl (DMT) moiety is not removed immediately after synthesis (as is often done using phosphoramidite chemistry) as it serves as a chromatographic handle during high-performance liquid chromatography (HPLC) purification. The oligonucleotide is removed from the controlled pore glass support and the protecting groups present on the exocyclic amines of A, G, and C residues are removed during an 18-hr incubation in 1 ml of ammonium hydroxide at room temperature. We have alternatively used an 8-hr incubation at 60°

[31] B. C. Froehler, *Tetrahedron Lett.* **27**, 5575 (1986).
[32] B. C. Froehler, P. G. Ng, and M. D. Matteucci, *Nucleic Acids Res.* **14**, 5399 (1986).

with success. After the addition of 5–10 μl of triethylamine to prevent any inadvertent detritylation during subsequent manipulations, the crude oligonucleotide/ammonia solution is transferred to plastic microtubes and dried in a vacuum centrifuge.

A major advantage of H-phosphonate chemistry over the standard phosphoramidite approach is the great flexibility it allows. The H-phosphonate diester produced following monomer condensation can be oxidized to a native phosphate diester, phosphorothioate, phosphoramidate, or phosphotriester linkage. The ability to generate oligonucleotides containing several types of modifications, either alone or mixed, is simply not possible with other synthetic strategies.

The major disadvantage of H-phosphonate chemistry is the relatively poor cycle yields compared with other methods of DNA synthesis. Although the coupling efficiency of standard phosphoramidite chemistry is greater than 99%, our coupling efficiency using H-phosphonate chemistry is at best 95–96%. This results both in a lower final yield of oligonucleotide and in a concomitant increase in the amount of undesirable truncation products. For example, when synthesizing an 18-mer with a coupling efficiency of 96% per cycle, the maximum theoretical yield is 50% of the starting material. Additional losses also accrue during purification of the full-length oligonucleotide away from the truncated failure sequences.

Oligonucleotide Purification

Rigorous purification of oligonucleotides is an important process to reduce or eliminate nonspecific (sequence independent) cellular toxicity. Many of the solvents and chemicals used in the synthesis of oligonucleotides can seriously interfere with normal cellular processes. Removal of biologically inactive truncation sequences is also necessary as there appears to be a limit to the total amount of oligonucleotide that can be injected without seeing adverse effects on embryonic development. Injection of more than 15–20 ng of unmodified or chimeric oligonucleotide into *Xenopus* embryos, for example, consistently results in an abnormal early phenotype and lethality prior to gastrulation.

Our oligonucleotides are initially purified using reversed-phase HPLC using a UV detector. The crude oligonucleotide solution, which was dried previously under vacuum, is resuspended in 300 μl of HPLC buffer A (5% acetonitrile, 10 mM ethylenediamine acetate, pH 8.0). This solution is loaded onto a PRP-1 column (Hamilton) and eluted with a 5–50% gradient of acetonitrile (see Table II for a HPLC gradient program). The absolute retention time of the modified oligonucleotides depends on the extent and type of modification. In general, truncated failure sequences elute in a

TABLE II
HPLC GRADIENT PROFILE[a]

Time (min)	Flow rate (ml/min)	% buffer B	Duration (min)
0	1	0	
5	1	50	30
35	1	100	5
45	1	0	2
55	0	0	1

[a] Buffer A: 5% acetonitrile, 10 mM ethylenediamine acetate, pH 8.0; buffer B: 50% acetonitrile, 10 mM ethylenediamine acetate, pH 8.0.

lower concentration of acetonitrile than the full-length oligonucleotide that still possesses the hydrophobic 5'-DMT group. The full-length oligonucleotide is collected and dried under vacuum. The oligonucleotide residue is resuspended in 300 μl of water and dried again under vacuum to remove any remaining ethylenediamine from the HPLC buffer. The residue is resuspended in 400 μl of 80% acetic acid in water to remove the acid-labile DMT group. Following a 30-min incubation, the acetic acid is removed by centrifugation under a strong vacuum. Note that the prolonged exposure of oligonucleotides to acidic solutions results in depurination of A and G residues. The oligonucleotide residue is resuspended in HPLC buffer A and is subjected again to reversed-phase HPLC using the gradient described earlier. A full-length oligonucleotide without the DMT group is collected and dried under vacuum. The oligonucleotide residue is resuspended in 500 μl of 10 mM KCl. This is done to exchange any anions that are bound to the oligonucleotide with chloride and to exchange any bound cations with the potassium ion. The solution is applied to a NAP-5 column (Pharmacia, Piscataway, NJ), prewashed with at least 15 ml of water, and eluted in 1 ml of sterile, distilled water. The oligonucleotide solution is dried under vacuum, resuspended in 500 μl of water, and applied again to a NAP-5 column. The oligonucleotide, now devoid of low molecular weight contaminants, is dried under vacuum, resuspended in 100 μl of water, and quantified by UV absorbance at 260 nm. For injection into oocytes or embryos, stock oligonucleotide solutions are prepared at a concentration of either 0.5 or 0.25 μg/μl. Oligonucleotide identity and purity are confirmed using MALDI-TOF mass spectrometry.

This purification strategy consistently provides high-purity oligonucleotide with a low incidence of nonspecific toxicity. Our total yields, which approach 100 nmol, are rather low compared to standard oligonucleotide chemistry, which requires minimal purification. The final amount of oligonucleotide produced is more than adequate for direct microinjection into hundreds of *Xenopus* oocytes and embryos. These yields, however, greatly limit the use of modified oligonucleotides for many cell culture applications in which the compounds are by necessity diluted into relatively large volumes of culture media.

Injection into Oocytes and Embryos

Stage VI oocytes are obtained from mature female *Xenopus* frogs. Following surgical removal, the desired ovarian tissue is rinsed in several washes of buffer OR-2 [82.5 mM NaCl, 2.5 mM KCl, 1 mM Na$_2$PO$_4$, 5 mM HEPES (pH 7.8), 1 mM CaCl$_2$, 1 mM MgCl$_2$] and teased apart manually. The tissue is then incubated in OR-2 with 2% collagenase (Sigma) for 2–3 hr to separate groups of oocytes from connective tissue. Using fine forceps, the translucent single layer of follicle cells is stripped off individual oocytes. The oocytes are stored in OR-2 at 18° until injected (less than 24 hr).

Xenopus eggs are isolated and fertilized as described previously.[33] Briefly, mature females are injected with 100 U of gonadatropin present in pregnant mare serum (Sigma, St. Louis, MO) 2–6 days prior to the harvesting of eggs. They are then injected with 1000 U of human chorionic gonadotropin (Sigma) 8–10 hr prior to harvest. Eggs are stripped gently from the peritoneal cavity into a high salt buffer (100 mM NaCl, 2 mM KCl, 0.2 mM MgSO$_4$, 0.4 mM CaCl$_2$, 5 mM Tris, pH 7.5) to prevent premature hydration of the surrounding jelly coats. The high salt solution is removed and the eggs are fertilized with *Xenopus* testes macerated in 0.1× MMR (1× MMR = 100 mM NaCl, 2 mM KCl, 2 mM CaCl$_2$, 1 mM MgCl$_2$, 5 mM HEPES, pH 7.4). After 30 sec of gentle agitation, water is added to the eggs so that the jelly coats will hydrate, thus preventing polyspermy. Eggs that have been fertilized will predictably orient animal (pigmented) hemisphere up, so any eggs that have not oriented this way are discarded. Prior to injection, the embryos are treated briefly with a solution of 2% cysteine (NaOH added to pH 8) to remove the jelly coats. This allows easier penetration of the injection needle into the embryo. The eggs are rinsed several times with charcoal-filtered water and placed in 1× MMR containing 4% Ficoll prior to injection.

Oligonucleotides are injected into the cytoplasm of oocytes and embryos as described previously.[6] We use a Singer MK-1 micromanipulator and a Geneve Inject+matic pressure injector. Injections are done using glass needles made from pulled capillary tubes. During injection, oocytes and embryos are supported against the side of a glass slide resting on top of,

[33] M. R. Rebagliati, D. L. Weeks, R. P. Harvey, and D. A. Melton, *Cell* **42,** 769 (1985).

and positioned slightly offset of, a second glass slide. The injection volumes are typically 10–20 nl. Nonspecific, sequence-independent oligonucleotide toxicity is related directly to both the amount of chimeric oligonucleotide injected and the purity of the compound. We routinely see embryos injected with up to 10–12 ng of oligonucleotide development normally into swimming tadpoles. Injection of more than 15–20 ng, however, consistently results in slow rates of cleavage and embryonic death prior to gastrulation.

Following injection, oocytes are placed into OR-2 solution and incubated at 18° until they are harvested for mRNA analysis. Embryos are placed in 1× MMR containing 4% Ficoll until 4–5 hr after injection. This solution minimizes the oozing of cytoplasmic material and yolk platelets from the injection site. The embryos must be moved into a solution of lower osmolarity, such as 0.1× MMR or water, well before gastrulation for normal development to proceed.

RNA Analysis

Embryos and oocytes are harvested into 1.5-ml microtubes at the desired time points and stored at −80° until they can be processed. RNA is isolated using a modification of the procedure described by Chomczynski and Sacchi.[34] Between 5 and 10 embryos or oocytes are thawed by pulsed sonication in 500 μl of RNA isolation buffer [4 M guanidinium isothiocyanate, 0.5% sarcosyl, 100 mM 2-mercaptoethanol, 25 mM sodium citrate, pH 7.0 in diethyl pyrocarbonate (DEPC)-treated water]. To this is added 50 μl of 2 M sodium acetate, pH 4.0. After the addition of 500 μl of water-saturated (nonbuffered) phenol and 100 μl of chloroform : isoamyl alcohol (49 : 1), the solution is agitated vigorously for several seconds. The samples are incubated on ice for 10 min and then centrifuged at maximum speed in a refrigerated microcentrifuge for 15 min. The aqueous phase is transferred to a new tube and mixed with 700 μl of 2-propanol. The samples are stored at −20° for several hours, and the RNA is pelleted by centrifugation at maximum speed in a refrigerated microcentrifuge for 12 min. The RNA pellet is resuspended in 50 μl of RNA isolation buffer. After the addition of 150 μl of ethanol, the RNA is pelleted as described earlier. The RNA pellet is washed with 180 μl of 70% ethanol, resuspended in DEPC water, and stored at −80° until analysis.

The levels of target and control messages are determined using either Northern analysis or ribonuclease protection analysis. These standard methods have been described extensively elsewhere.[9,33]

[34] P. Chomczynski and N. Sacchi, *Anal. Biochem.* **162,** 156 (1987).

Enhanced Efficacy of Chimeric Oligonucleotide

Using the localized *Xenopus* message An2 as a target, we have demonstrated that chimeric antisense oligonucleotides are effective in directing RNase H-mediated mRNA elimination. In order to directly compare the antisense activity of a chimeric oligonucleotide to that of an unmodified oligonucleotide of the same sequence, we again chose An2 as a model target. Single cell embryos were injected with either 10 ng of the unmodified oligonucleotide or with decreasing amounts of the modified compound, from 10 to 0.02 ng. Total RNA was isolated and message levels were examined by RNase protection. Dose–response data, seen in Fig. 5, show the expected decrease in residual mRNA with increasing amounts of modified oligonucleotide. In contrast, the antisense activity of the unmodified oligonucleotide is strikingly ineffective. We estimate that 10 ng of unmodified oligonucleotide is roughly equivalent to 0.2 ng of the chimeric oligonucleotide in the ability to decrease An2 mRNA levels. Neither compound has an effect on the level of the control message histone H4, again demonstrating the specificity of the reaction. In our experience, it has not been

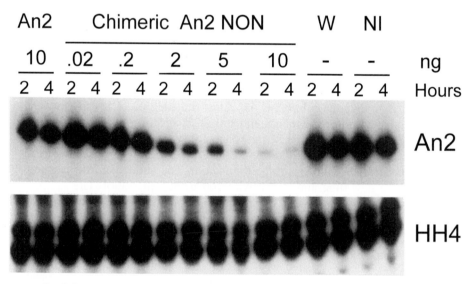

FIG. 5. *In vivo* comparison of modified and unmodified oligonucleotides. Single cell *Xenopus* embryos were injected with either 10 ng of unmodified An2 oligonucleotide or the indicated amount of the chimeric An2 oligonucleotide and harvested at 2 and 4 hr. Total RNA was isolated, and levels of An2 and histone H4 (HH4) were quantified by a ribonuclease protection assay. Protected bands were analyzed by electrophoresis on a 6% polyacrylamide–7 *M* urea gel and visualized by autoradiography. An2, Unmodified An2 oligonucleotide; NI, not injected; W, water injected.

possible to inject a single cell embryo with enough unmodified oligonucleotide to lower the levels of any target mRNA significantly without affecting development adversely. The rapid nucleolytic degradation of the unmodified oligonucleotide is the most likely explanation for the lack of antisense activity. This has previously been a major hurdle in the use of oligonucleotides to create a null phenotype in developing *Xenopus* embryos.[10,24]

The use of oligonucleotides in *Xenopus* oocytes has been more straightforward. Because oocytes neither replicate their DNA nor execute a complex developmental plan, they are much more tolerant of injecting relatively large amounts of oligonucleotide. Injection of as much as 50 ng of oligonucleotide per oocyte does not result in any overt signs of toxicity. These large amounts are sufficient to direct mRNA elimination, even when using unmodified oligonucleotides, despite the active nucleases present in oocytes. Therefore, although the chimeric phosphoramidate-modified oligonucleotides possess superior antisense activity compared to the unmodified compounds in oocytes, they offer little practical advantage for most studies in the *Xenopus* oocyte. Modified oligonucleotides definitely have a role in cases where it is critical to reduce and maintain the levels of the target message as low as possible for an extended period of time.

Conclusions

At a theoretical level, the utilization of an oligonucleotide complementary to a target message to prevent translation sounds extraordinarily simple. There are, however, significant hurdles to the use of antisense oligonucleotides *in vivo*. Issues of nuclease stability and RNase H activity, which were presented here, must be considered in any discussion involving antisense oligonucleotides. A third important issue, cellular uptake of oligonucleotides, has not been discussed. We have circumvented this complex and poorly understood area by directly microinjecting chimeric oligonucleotides into single cells.

The ability to microinject oligonucleotides into *Xenopus* oocytes and embryos has made this an ideal system to ask important mechanistic questions involving oligonucleotide metabolism and mode of action. It is not possible, for example, to determine the rate of intracellular oligonucleotide degradation in cell or tissue culture because the rate of internalization is generally unknown. Additionally, it is not possible to compare the efficacy of two oligonucleotides adequately if they have different rates of internalization.

When performing antisense experiments, it is vital to show specificity of action. It is currently not acceptable, for example, to use cell doubling time as the only method to assay the effect of an antisense oligonucleotide targeting a growth factor. Two important principles must be demonstrated. (1) The antisense oligonucleotide must not affect the level of at least one control mRNA. This demonstrates that the oligonucleotide is not affecting all message levels by more global mechanisms involving overall transcription or mRNA degradation. (2) A similarly modified control oligonucleotide must not affect the target message. This demonstrates that the level of target message is not particularly sensitive to the intracellular presence of any modified oligonucleotide. We have shown that chimeric phosphoramidate-modified antisense oligonucleotides fulfill these important criteria when used in *X. laevis* oocytes and embryos.

[25] Reversible Depletion of Specific RNAs by Antisense Oligodeoxynucleotide-Targeted Degradation in Frog Oocytes

By GEORGE L. ELICEIRI

Introduction

Reversible, specific depletion of an RNA species is essential in studying the function of that molecule. RNA depletion in frog oocytes is particularly important for RNA species unknown in yeast. Experiments in whole cells are crucial for processes that cannot be reproduced yet in cell-free systems, such as rRNA precursor cleavage processing steps that are dependent on the addition of an exogenous small nucleolar RNA.[1]

Antisense oligodeoxynucleotides target RNA degradation by RNase H. Their capacity to target specific RNA degradation has been shown for various RNA species.[2-14] The instability of unmodified oligodeoxynucleo-

[1] C. A. Enright, E. S. Maxwell, G. L. Eliceiri, and B. Sollner-Webb, *RNA* **2,** 1094 (1996). Erratum: *RNA* **2,** 1318 (1996).

[2] P. Dash, I. Lotan, M. Knapp, E. R. Kandel, and P. Goelet, *Proc. Natl. Acad. Sci. USA* **84,** 7896 (1987).

[3] C. Jessus, C. Cazenave, R. Ozon, and C. Hélène, *Nucleic Acids Res.* **16,** 2225 (1988).

[4] N. Sagata, M. Oskarsson, T. Copeland, J. Brumbaugh, and G. F. Vande Woude, *Nature* **335,** 519 (1988).

[5] S.-Q. Pan and C. Prives, *Science* **241,** 1328 (1988).

[6] S. Saxena and E. J. Ackerman, *J. Biol. Chem.* **265,** 3263 (1990).

[7] R. Savino and S. A. Gerbi, *EMBO J.* **9,** 2299 (1990).

[8] B. A. Peculis and J. A. Steitz, *Cell* **73,** 1233 (1993).

[9] B. A. Peculis and J. A. Steitz, *Genes Dev.* **8,** 2241 (1994).

tides *in vivo*[15] makes them ideal for reversible depletion experiments because the oligonucleotide is already degraded by the time the exogenous RNA is introduced. There is no ribozyme that, when transfected, efficiently degrades another RNA molecule and whose cellular level can be manipulated easily.

In contrast to liposomes, microinjection provides efficient delivery of nucleic acids into cells. Microinjection is particularly important when an RNA species needs to be delivered to the nucleus, and this RNA is not transported efficiently to the nucleus when introduced into the cytoplasm. RNA processing that requires a different RNA molecule can be studied in cotransfected somatic cells when it is known where both molecules can be mutated so that processing of the mutated substrate RNA is dependent on the mutated form of the *trans*-acting RNA molecule.[16] Specific endogenous RNA depletion, followed by injection of that RNA, is essential at the beginning of a study, when this information is unavailable.

It is possible to do biochemical analysis from a single *Xenopus laevis* oocyte, as it is much larger than a somatic cell. Late stages of *X. laevis* oocytes, such as V and VI,[17] have low levels of endogenous transcription by RNA polymerase II. Then, for most RNA species, after their degradation there is no significant replenishment by endogenous biosynthesis within the relatively short time periods of microinjection experiments. Because oligodeoxynucleotides have half-lives of a few minutes in frog oocytes,[15] shortly after injection their cellular levels are too low to affect a newly injected RNA. Microinjection into frog oocytes is now the only *in vivo* metazoan experimental system to do reversible depletion of an RNA species. Antisense oligonucleotides may target the cleavage of a second RNA species in addition to the intended RNA.[18] It is essential to show reversion of the observed phenotype by specifically raising the intracellular level of the intended RNA species. This can be done easily by injection in frog oocytes. There have been excellent articles on the isolation and injection

---

[10] K. T. Tycowski, M.-D. Shu, and J. A. Steitz, *Science* **266**, 1558 (1994).
[11] R. K. Mishra and G. L. Eliceiri, *Proc. Natl. Acad. Sci. USA* **94**, 4972 (1997).
[12] K. T. Tycowski, C. M. Smith, and J. A. Steitz, *Proc. Natl. Acad. Sci. USA* **93**, 14480 (1996).
[13] D. A. Dunbar and S. J. Baserga, *RNA* **4**, 195 (1998).
[14] M. Bellini and J. G. Gall, *Mol. Biol. Cell* **9**, 2987 (1998).
[15] C. Prives and D. Foukal, *in* "Methods in Cell Biology" (B. K. Kay and H. B. Peng, eds.), Vol. 36, p. 185. Academic Press, New York, 1991.
[16] Y. Zhuang and A. M. Weiner, *Cell* **46**, 827 (1986).
[17] J. N. Dumont, *J. Morphol.* **136**, 153 (1972).
[18] D. A. Dunbar, V. C. Ware, and S. J. Baserga, *RNA* **2**, 324 (1996).

of frog oocytes[19-21] and on the use of antisense oligodeoxynucleotides in frog oocytes.[15]

Equipment

Microinjector

We use an automatic microinjector whose plunger, through an oil column, pushes the sample from the injection needle (Nanoject, Drummond Scientific Co., Broomall, PA). The volumes delivered usually are 10–50 nl. A micromanipulator is used to move the glass needle.

Stereomicroscope

A microscope with adjustable magnification and a fiber optic light source are needed to see the oocyte injections and the isolation of nuclei. We use a Bausch and Lomb StereoZoom 7 microscope.

Needle Puller

We pull glass capillaries so that the needle has a taper of ~7–8 mm,[22] using a Pul-1 needle puller (World Precision Instruments, Sarasota, FL) and glass capillaries from Drummond. After pulling, the needles are cut with a razor blade at an angle of ~20° to produce a tip inner diameter of ~10 μm (~13 μm outer diameter).[22] The tip diameter is measured with a compound microscope that has 400× magnification and a calibrated reticle in one ocular. We use needles whose tips were not ground with a needle beveler.

Incubator

Oocytes are kept at ~18°. An inexpensive incubator can be made with a bench-top refrigerator, a heating strip, a thermostat, and a small fan.

[19] A. Colman, in "Transcription and Translation. A Practical Approach" (B. D. Hames and S. J. Higgins, eds.), p. 49. Oxford-IRL Press, Washington, DC, 1984.
[20] A. Colman, in "Transcription and Translation. A Practical Approach" (B. D. Hames and S. J. Higgins, eds.), p. 271. Oxford-IRL Press, Washington, DC, 1984.
[21] M. P. Terns and D. S. Goldfarb, in "Methods in Cell Biology" (M. Berrios, ed.), Vol. 53, p. 559. Academic Press, New York, 1998.
[22] D. L. Stephens, T. J. Miller, L. Silver, D. Zipser, and J. E. Mertz, Anal. Biochem. **114,** 299 (1981).

Materials

Oligonucleotides

Unmodified oligodeoxynucleotides are used. They are complementary to some segments of the RNA species that need to be depleted. The selection of the RNA segment to be targeted can be based on (a) the accessibility of that RNA segment in a cell extract when incubated with an antisense oligonucleotide and RNase H; (b) the apparent presence of that RNA segment in a single-stranded region of the molecule, based on sequence phylogeny or chemical or enzymatic probing experiments; or (c) the evolutionary conservation of the sequence of that RNA segment. Antisense oligonucleotides approximately 10–23 nucleotides long have been used successfully.[2–15] If insufficient RNA degradation is observed, shorter or longer versions of the oligonucleotide may need to be tested. Desalted oligonucleotides can be adequate, but better results may be obtained with some purification, such as reversed-phase column chromatography followed by gel filtration.

Modified Barth Solution (MBS)

This solution consists of 88 mM NaCl, 1 mM KCl, 2.4 mM NaHCO$_3$, 10 mM N-2-hydroxymethylpiperazine-N'-2-ethanesulfonic acid (HEPES) (pH 7.5), 0.82 mM MgSO$_4$, 0.33 mM Ca(NO$_3$)$_2$, and 0.41 mM CaCl$_2$. A 10× stock solution is autoclaved and stored at 4° or −20°. After dilution for use, concentrated stocks of penicillin and streptomycin are added to a final concentration of 10 μg/ml. Oocytes are maintained in MBS.

Frogs

We purchase female $X.$ $laevis$ frogs that are at least 2 years old and never injected with hormones from Nasco (Fort Atkinson, WI) or Xenopus I (Ann Arbor, MI). They are fed frog brittle (Nasco) three times a week and are kept in water at ~18° with 12-hr light–dark cycles.

Methods

In Vitro Synthesis of RNA

Enzymatic synthesis is commonly used to make the RNA species that is going to be injected into oocytes that were depleted of that RNA. The DNA template is the desired sequence, cloned or amplified by the polymerase chain reaction, preceded by a bacteriophage RNA polymerase promoter, such as T7, T3, or SP6; it is transcribed with the corresponding RNA polymerase.[23] Unlabeled RNA is typically used for "rescue" experiments. Uncapped RNA is made if the RNA species of interest is naturally uncapped and is relatively stable after injection into frog oocytes. If not, a m^{7}G cap can be added to the 5′ of the RNA by doing the transcription in the presence of an excess of the commercially available cap analog m^{7}GpppG over GTP (e.g., at a final concentration of 40 μM m^{7}GpppG, 10 μM GTP).

Removal of Ovarian Tissue

Frogs are anesthetized by immersion in 0.4% 3-aminobenzoic acid ethyl ester. A ~1-cm incision is made on the ventral lateral area of the frog. The *X. laevis* ovary has about 24 lobes. Depending on the number of oocytes needed within the following few days, one or more lobes are removed. The muscle and skin layers can be sutured together, with two to three separate stitches. For example, we use a 3-0 braided silk suture attached to a taper BB needle (Ethicon, Somerville, NJ).

Isolation of Oocytes

Microinjection experiments are done with the largest oocytes, stages V and VI, which are ~1–1.2 mm in diameter.[17] Separated individual oocytes are used for microinjection. We isolate oocytes manually, holding a clump of ovarian tissue with a pair of blunt forceps and plucking each individual oocyte, held by the membrane that attaches it to the rest, with a pair of fine-tipped forceps. Undamaged oocytes are transferred to fresh MBS with a wide-bore pipette and stored ~18°. The solution in which the oocytes are stored should be replaced with fresh MBS at least twice a day, discarding all damaged oocytes each time. Healthy oocytes have homogeneous pigmentation on their animal (dark) hemisphere. Oocytes with uneven pigmentation on their animal hemisphere are discarded.

Injection of Antisense Oligonucleotides

Antisense oligonucleotides are usually injected into the cytoplasm as it is easier and more effective in targeting the degradation of several nuclear RNAs.[8–12,15] Negative control oligonucleotides (other sequences) are also injected. Volumes of up to 50 nl are injected into the vegetal (light) hemisphere. Between 2 and 14 pmol of oligonucleotide per oocyte has been used successfully.[2–15] Squares of plastic mesh with an opening of ~0.8–1 mm (e.g., Spectrum, Houston, TX) are attached to the bottom of 3.5-cm

[23] P. A. Krieg and D. A. Melton, *Methods Enzymol.* **155**, 397 (1987).

petri dishes with drops of household cement or chloroform. The wells of the mesh hold the oocytes in place during injection. After injection, oocytes are usually incubated for at least 2–4 hr, to make sure that most of the oligonucleotide has been degraded before injection of RNA. The targeted RNA is cleaved in that time, but a more complete degradation may require overnight incubation.[15] Coinjection of two oligonucleotides, complementary to different segments of the target RNA, may be necessary for a more complete depletion.[13] Degradation of the targeted RNA, relative to the fate of negative control RNA species, during postinjection oocyte incubation can be monitored by RNA gel blot analysis or primer extension analysis.

Analysis of Affected Function

The function expected to be affected sometimes can be monitored on an endogenous macromolecule, such as in the processing of a wild-type, full-length rRNA precursor. Then, a suitable precursor, such as $[\alpha\text{-}^{32}P]GTP$, can be injected into the cytoplasm of oocytes that were incubated for an appropriate number of hours after oligonucleotide injection. Injection of a gene or RNA may be necessary for functions that need to be monitored on partial-length or otherwise mutated macromolecules.

Injection of RNA

Copies of the RNA that was depleted, as well as negative control RNAs (other sequences), are injected. To "rescue" function, a nuclear RNA species is injected into the nucleus if it is not transported efficiently to the nucleus after cytoplasmic injection. Between 60 and 110 fmol of an RNA species per oocyte has been used successfully.[9–11] For nuclear injections, volumes of up to 20 nl per oocyte are injected into the animal hemisphere. The needle is inserted at the center of the dark hemisphere so that the needle tip reaches a depth of about one-quarter the diameter of the oocyte. To verify later that the sample was delivered into the nucleus, each sample to be injected is mixed with Blue dextran (2×10^6 molecular weight; Sigma, St. Louis, MO) to a final concentration of 20 mg/ml.[24] After RNA injection, oocytes are incubated for the minimum number of hours needed to see maximal recovery of the affected function. If the *in vitro*-synthesized RNA requires nucleotide modifications to be functional, oocytes should be incubated for at least 6 hr after RNA injection.[25] In some cases, harvesting whole oocytes is sufficient. Oocytes are fractionated into nucleus and cytoplasm (a) if verification of nuclear delivery is needed, after nuclear injection of Blue

[24] A. Jarmolowski, W. C. Boelens, E. Izaurralde, and I. W. Mattaj, *J. Cell Biol.* **124,** 627 (1994).
[25] Y.-T. Yu, M.-D. Shu, and J. A. Steitz, *EMBO J.* **17,** 5783 (1998).

dextran-containing samples (then, isolated nuclei will appear slightly blue), and/or (b) if subsequent analysis is easier in isolated nuclei. We isolate nuclei from oocytes under mineral oil by making a hole in the animal hemisphere with a 27-gauge hypodermic needle.[26,27]

Concluding Remarks

Antisense olgonucleotide-targeted degradation of specific RNAs in *X. laevis* oocytes is a unique experimental system. Its value has been proven for various RNA molecules. For RNA species whose yeast homolog is unknown, it is often the only available system to test the functional effect of the reversible depletion of a specific RNA molecule. There are many metazoan non-mRNA species whose function is unknown and probably many more yet to be discovered; the yeast functional homolog of some of them may be difficult to recognize or may not exist. In these cases, RNA depletion experiments in frog oocytes may be very valuable. This experimental system may be particularly useful in testing the function of multiple RNA mutants or RNA fragments, including competition experiments.

Acknowledgments

I thank Brenda A. Peculis and Elsebet Lund for their help. Work in this laboratory was supported by a grant from the National Institute of General Medical Sciences.

[26] E. Lund and P. L. Paine, *Methods Enzymol.* **181**, 36 (1990).
[27] P. L. Paine, M. E. Johnson, T. T. Lau, L. J. M. Tluczek, and D. S. Miller, *BioTechniques* **13**, 238 (1992).

[26] Inactivation of Gene Expression Using Ribonuclease P and External Guide Sequences

By CECILIA GUERRIER-TAKADA and SIDNEY ALTMAN

Introduction

Experiments performed in the late 1970s[1] indicated that antisense oligonucleotides could be used to affect gene expression. Their activity is often associated with their ability to induce RNase H-mediated cleavage of the

[1] P. C. Zamecnik and M. L. Stephenson, *Proc. Natl. Acad. Sci. U.S.A.* **75**, 280 (1978).

target RNA via formation of a DNA/RNA hybrid. The discovery that RNAs[2,3] could act as enzymes (a function previously ascribed only to proteins) by catalyzing self-cleavage or cleavage of another RNA molecule has extended antisense-targeted technology. Most methods that employ self-cleaving ribozymes, such as those that utilize the hammerhead or hairpin ribozymes, are based on the fact that the structural domain of these enzymes can be divided into two separate entities (or oligonucleotides): one is the catalytic moiety and the other, containing the cleavage site, is referred to as the substrate. The sequence of these two oligonucleotides allows them to form hydrogen (and/or non-hydrogen)-bonded interactions, resulting in the formation of the structural domain capable of carrying out the cleavage reaction. In general, a particular sequence in the target RNA is designated as the substrate, and the ribozyme (antisense RNA) is custom designed to form an active complex with the substrate through conventional Watson–Crick hydrogen bonding. The sequence encoding the enzyme is cloned between the appropriate promoter and transcription termination signal and is delivered to the cells (for a review, see Bertrand and Rossi[4] and Couture and Stinchcomb[5]).

Another variation of the antisense concept, which will be discussed in this article, is based on the substrate recognition properties of the ubiquitous enzyme ribonuclease P (RNase P). To understand the development of such a technology, it is important to be familiar with some of the properties of RNase P (for a review, see Gopalan et al.[6] and Frank and Pace[7]).

Properties of RNase P: Substrate Recognition

The main physiological role of RNase P consists of one-step cleavage of precursor tRNAs (ptRNAs) to generate 5′ termini of mature tRNA molecules. This cleavage does not require the presence of a specific sequence at the cleavage site, but does depend on a higher order structure in ptRNA substrates. Much of our knowledge about RNase P is derived from studies of the enzyme from *Escherichia coli* and *Bacillus subtilis*. In *E. coli,* RNase P is composed of an RNA subunit (M1 RNA, 377 nucleotides) and a small, basic protein (C5 protein, 119 amino acid residues). *In*

[2] K. Kruger, P. J. Grabowski, A. J. Zaug, J. Sands, D. E. Gottschling, and T. R. Cech, *Cell* **31,** 147 (1982).
[3] C. Guerrier-Takada, K. Gardiner, T. Marsh, N. Pace, and S. Altman, *Cell* **35,** 849 (1983).
[4] E. Bertrand and J. Rossi, *Nucleic Acids Mol. Biol.* **10,** 301 (1996).
[5] L. A. Couture and D. T. Stinchcomb, *Trends Genet.* **12,** 510 (1996).
[6] V. Gopalan, S. J. Talbot, and S. Altman, *in* "RNA-Protein Interactions" (K. Nagai and I. Mattaj, eds), p. 103. Oxford Univ. Press, Oxford, 1994.
[7] D. N. Frank and N. R. Pace, *Annu. Rev. Biochem.* **67,** 153 (1998).

vivo they are both essential for the activity of RNAse P[8]; however, *in vitro*, M1 RNA can function as a true enzyme, accurately and efficiently catalyzing the hydrolysis of ptRNAs in the absence of the protein moiety.[3] In contrast, human RNase P is far more complex: it has a higher protein-to-RNA ratio than the bacterial enzyme (at least six protein subunits copurify with the enzyme activity in highly purified RNase P from HeLa cells),[9] and its RNA moiety has not yet been found to be catalytically active in the absence of its protein subunits.

Experiments to identify which domains of a ptRNA could be deleted without abolishing cleavage by RNase P from *E. coli* have been carried out. The resulting model substrate lacked not only the D stem and loop, but also the AC stem and loop domains: it consisted of a stem–loop structure resembling the T stem stacked on the acceptor stem of a tRNA (Fig. 1).[10] Furthermore, Forster and Altman[11] went on to show that the loop could be removed from the minimal structure. The two oligonucleotides pairing with each other remained a good substrate for the enzyme. The oligonucleotide that was cleaved (5' proximal) was the substrate, whereas the other (3' proximal), named external guide sequence (or EGS), must hybridize with the substrate to form a complex that is recognized by RNase P (Fig. 1).

Human RNase P, which recognizes a narrower range of substrates, was unable to cleave these complexes. In order to overcome this problem, Altman suggested the design of an EGS that looks like most of a ptRNA molecule when in a complex with an RNA target. The strategy was successful: a small RNA fragment, pAva, containing the 5' leader sequence and the first 14 nucleotides from the tRNA^{Tyr} from *E. coli* was efficiently cleaved by human RNase P when another RNA (EGS) corresponding to the remaining 3' proximal sequence of tRNA^{Tyr} was hybridized to it (Fig. 1). These new EGSs were named 3/4 EGSs because they contain about three-fourths of a tRNA sequence, lacking only the 5' proximal sequence of the acceptor stem and D stem.[12] Further studies revealed that even an mRNA, rather than a fragment of a ptRNA, could become a substrate for human RNase P in the presence of a specific 3/4 EGS.[12] These observations suggested that, in principle, any RNA could be targeted by a custom-designed EGS for specific cleavage by RNase P. This article describes the methodology used to achieve inactivation of gene expression using RNase P and external guide sequences.

[8] R. Kole, M. F. Baer, B. C. Stark, and S. Altman, *Cell* **19**, 881 (1980).
[9] N. Jarrous, P. S. Eder, C. Guerrier-Takada, C. Hoog, and S. Altman, *RNA* **4**, 407 (1998).
[10] W. H. McClain, C. Guerrier-Takada, and S. Altman, *Science* **238**, 527 (1987).
[11] A. C. Forster and S. Altman, *Science* **249**, 783 (1990).
[12] Y. Yuan, E. S. Hwang, and S. Altman, *Proc. Natl. Acad. Sci. U.S.A.* **89**, 8006 (1992).

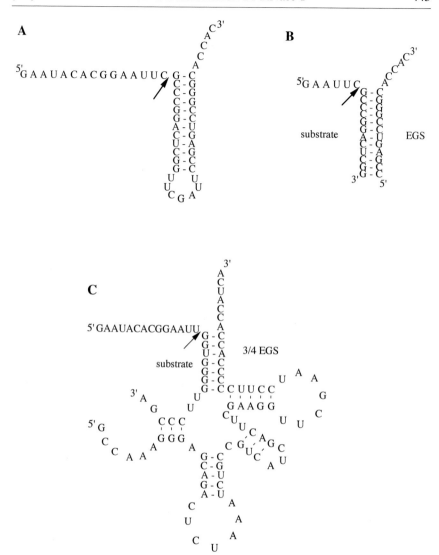

FIG. 1. Model substrates for *E. coli* (A and B) and human RNase P (C). (A) pAT1, (B) substrate (target RNA)–EGS complex (derived from pAT1), and (C) substrate (pAva)–3/4 EGS complex (derived from ptRNA^Tyr). Arrows indicate the sites of cleavage by RNase P.

Outline of Method

Steps involved in RNA inactivation experiments with RNase P are (1) choice of the appropriate target site, (2) design of EGSs and testing of the activity of the system *in vitro,* (3) cloning of the EGSs into expression vectors and their introduction into cells, and (4) performance of assays to determine the degree of inhibition of the targeted gene expression and correlation of such inhibition with RNase P activity.

Choice of Target Site

It is important to identify accessible single-stranded regions in the target mRNA with which EGSs can base pair. In buffers containing Mg^{2+}, RNase T1 cleaves preferentially after single-stranded guanosine residues; this enzyme is used, therefore, to map the region around the translation initiation site of mRNAs because this region must be accessible to ribosomes. 5'-end-labeled RNAs (prepared by transcription *in vitro*) are resuspended in 20 μl of buffer A [50 mM Tris–HCl (pH 7.5), 100 mM NH$_4$Cl, 10 mM MgCl$_2$] and placed on ice. One microliter of RNase T1 (50 units/ml) is added and the reaction is left to proceed for 30 sec. The reaction is stopped by adding 2 μl phenol, and the RNA is precipitated with 2 volumes of ethanol after the addition of sodium acetate to a final concentration of 0.3 M and glycogen (20 μg/ml) as carrier. After centrifugation of the mixture, the RNA pellet is collected, washed twice with 75% ethanol, and dried. The RNA is then resuspended in loading dye buffer [9 M urea, 0.05% xylene cyanol (XC), 0.05% bromphenol blue (BPB)] and electrophoresed on a sequencing gel in a lane adjacent to a partial alkali digest ladder of the same RNA. Accessible regions (as judged by the presence of cleavage) are mapped. Guanosine residues are chosen as the base 3' to the site for cleavage by RNase P (the majority of tRNAs have a guanosine at position +1). Other desirable features of the targeted region are a pyrimidine at position −1, a single-stranded region in the target RNA upstream from the cleavage site (avoid G residues at positions −2 and −3 because they can base pair with the 3' CCA sequence of the EGS), and a uridine residue at position +8 (only for 3/4 EGSs, see later).

The partial RNase T1 digest is carried out on fragments of mRNA transcripts and may not reflect the folding of the mRNA species found in the cells. We have found, however, that most of the sites accessible *in vitro* are also available for base pairing with their specific EGSs *in vivo*.[12–14] In

[13] D. Plehn-Dujowich and S. Altman, *Proc. Natl. Acad. Sci. U.S.A.* **95**, 7327 (1998).
[14] Y. Li, C. Guerrier-Takada, and S. Altman, *Proc. Natl. Acad. Sci. U.S.A.* **89**, 3185 (1992).

some studies, mapping *in vivo* with dimethyl sulfate (DMS) is used to determine accessible regions in the mRNA.[15,16]

Design of EGSs and Cleavage Assays

Design of Stem EGSs

Based on the results of several experiments with RNase P from *E. coli*, we designed stem EGSs (see Fig. 2) to form 13 to 16 bp with the target RNA.[14,17–19] All stem EGSs were designed with a 3′-terminal ACCA sequence because 3′-terminal nucleotides of small model substrates (pAT1, see Fig. 1) are important for cleavage by RNase P from *E. coli* (deletion of the 3′ ACCA sequence of pAT1 abolished cleavage by the enzyme). To construct genes for EGSs under the control of bacteriophage T7 RNA polymerase, DNA oligonucleotides containing sequences for the T7 promoter, for the various EGSs used, and for a hammerhead core were ligated to a *Bam*HI–*Hin*dIII fragment containing the T7 terminator sequence (this DNA fragment was obtained from pET3040[20]), and the insert was cloned into pUC 19, which had been digested with *Eco*RI and *Hin*dIII (Fig. 2). Plasmid DNAs that contained the new inserts were obtained from transformants and subjected to sequence analysis to ascertain that the expected sequence was present. These plasmid DNAs were digested with *Bst*NI and used as template for transcription *in vitro* using T7 RNA polymerase to obtain EGSs to be used in assays *in vitro*.

Design of 3/4 EGSs

For experiments with human RNase P, 3/4 EGSs targeting the *cat* gene are obtained by polymerase chain reaction (PCR), using a plasmid encoding the gene for ptRNA[Tyr] (pTyr) as the template, and with oligonucleotides EC-1A 5′GCCAAACTGAGCAGACTC3′ and EC-1B 5′GCCGAAGCT-TTAAATGGTGAGGCATGAAGGATTCGAACA3′ as forward and reverse primers, respectively (Fig. 3). The PCR product is digested with *Hin*dIII, purified on a nondenaturing polyacrylamide gel, ligated to DNA oligonucleotides containing the sequences for the T7 promoter (see earlier

[15] S. C. Climie and J. D. Friesen, *J. Biol. Chem.* **263**, 15166 (1988).
[16] A. J. Zaug and T. R. Cech, *RNA* **1**, 363 (1995).
[17] C. Guerrier-Takada, Y. Li, and S. Altman, *Proc. Natl. Acad. Sci. U.S.A.* **92**, 11115 (1995).
[18] C. Guerrier-Takada, R. Salavati, and S. Altman, *Proc. Natl. Acad. Sci. U.S.A.* **94**, 8468 (1997).
[19] Y. Li and S. Altman, *Nucleic Acids Res.* **24**, 835 (1996).
[20] A. H. Rosenberg, B. N. Lade, D.-S. Chui, S.-W. Lin, J. J. Dunn, and F. W. Studier, *Gene* **56**, 125 (1987).

A

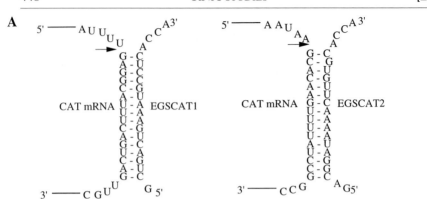

B

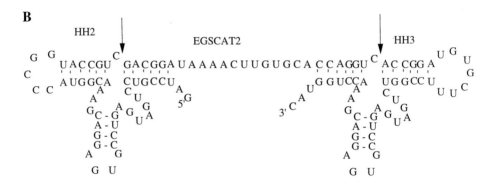

C

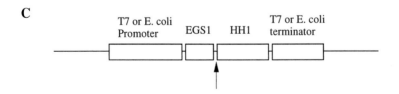

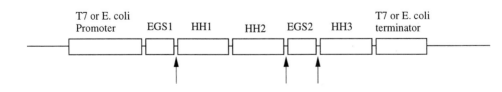

discussion), and inserted into pUC 19 (digested with *Eco*RI and *Hin*dIII). All plasmid DNAs are sequenced to confirm the presence of the correct insert. These plasmids are digested with *Dra*I, and 3/4 EGS RNAs are synthesized *in vitro* by transcription with T7 RNA polymerase. Many studies, including *in vitro* selection experiments, have been carried out to find EGSs that are more efficient in directing human RNase P to cleave a specific target RNA.[21–24] Among those, 3/4 EGSs, from which the anticodon stem and loop domain were deleted (ΔACEGS), have been used successfully in gene targeting experiments (Fig. 3). 3/4 EGSs that can be used as control when expressed in cells in tissue culture have either the UUCGAAU sequence in the T loop replaced with GGAUCCG or a point mutation (C to G) in this loop (Fig. 3). These EGSs are inactive in assays carried out *in vitro* and *in vivo* (see later).

Assay of RNase P Activity on EGS–Substrate (mRNA) Complex

The various mRNA substrates obtained by transcription *in vitro* using T7 RNA polymerase are labeled internally using $[\alpha\text{-}^{32}P]GTP$ or 5′-end-labeled using $[\gamma\text{-}^{32}P]ATP$ and T4 polynucleotide kinase. RNA substrate and its specific stem EGS (also transcribed by T7 RNA polymerase, see earlier) are incubated at 37° in the presence of M1 RNA in buffer B [50 mM Tris–HCl (pH 7.5), 100 mM MgCl$_2$, 100 mM NH$_4$Cl, 4% (w/v) polyethylene glycol (PEG)] or in the presence of RNase P holoenzyme (M1 RNA +C5 protein) in buffer C [20 mM HEPES–KOH (pH 8.0), 400 mM ammonium acetate, 10 mM magnesium acetate, 5% glycerol]. For experiments using partially purified human RNase P from Hela cells, mRNA substrates and their specific 3/4 EGSs are incubated at 37° in buffer A. For most of the EGS:mRNA complexes, an annealing step [4 min at 80° in 100 mM Tris–HCl (pH 8.3), followed by quick cooling on ice] prior

[21] Y. Yuan and S. Altman, *Science* **263,** 1269 (1994).
[22] D. Kawa, J. Wang, Y. Yuan, and F. Liu, *RNA* **4,** 1397 (1998).
[23] M. Werner, E. Rosa, J. L. Nordstrom, A. R. Goldberg, and S. T. George, *RNA* **4,** 847 (1998).
[24] M. Y. Ma, B. Jacob-Samuel, J. C. Dignam, U. Pace, A. R. Goldberg, and S. T. George, *Antisense Nucleic Acid Drug Dev.* **8,** 415 (1998).

FIG. 2. (A) Structure of external guide sequences (EGSCAT1 and EGSCAT2) directed against *cat* mRNA. Only a segment of the mRNA to which the EGS hybridizes is shown. Arrows indicate the sites of cleavage by RNase P. (B) Structure of the cloning cassette composed of two hammerhead sequences (HH2 and HH3) and an EGS (EGSCAT2). (C) Schematic representation of cloning cassettes of synthetic genes for either one EGS or two EGSs in tandem. Arrows in (B) and (C) indicate the sites of cleavage by the hammerhead ribozyme.

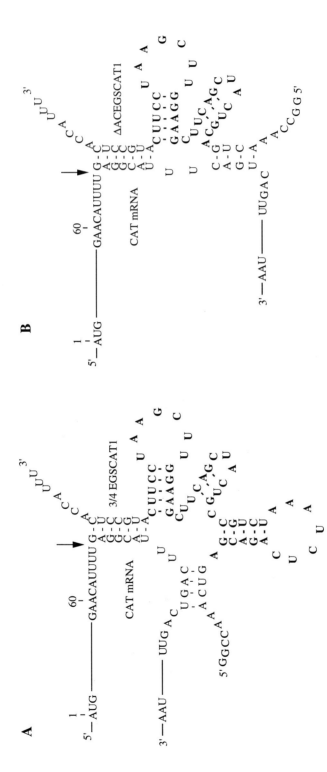

Fig. 3. Proposed secondary structures of complexes of *Cat* mRNA and various EGSs. (A) 3/4 EGSCAT1. (B) ΔACEGSCAT1. Boldface letters indicate the sequence derived from the pTyr DNA template. Arrows indicate the site of cleavage by RNase P.

to the addition of enzyme did not improve the efficiency of the cleavage reaction. For each EGS:mRNA pair, the optimum ratio of EGS to target RNA, as well as the concentration of RNase P, should be determined. After incubation, the cleavage reaction is stopped by the addition of an equal volume of loading dye buffer and the samples are analyzed on denaturing gels containing 7 M urea. Results indicate that both *E. coli* and human RNase P cleave the mRNA substrate at the expected site.

Cloning of EGSs into Expression Vectors

Stem EGSs

When used in experiments carried out in *E. coli* cells, these EGSs were cloned downstream from an *E. coli* as well as from T7 promoter. EGSs and HH sequences (similar to the ones described earlier) were cloned between *Bam*HI and *Hin*dIII sites of pKB283 (Lawrence and Altman, unpublished), which is a derivative of pUC 19 with a 283-bp insert that contains the promoter region for the gene encoding M1 RNA (*rnpB*). The T7 terminator sequence was replaced by an M1 terminator sequence obtained from the plasmid pM1P[25] (Fig. 2). The hammerhead sequence was inserted to ensure that the RNA transcribed *in vivo* would undergo self-cleavage and release EGSs that contained only four nucleotides beyond the 3' CCA sequence. (The efficiency of cleavage of substrates by RNase P is reduced significantly by a long 3'-flanking sequence.) When two different EGSs targeted to the same mRNA were used to improve the level of inhibition, a "cloning cassette" containing three HH (HH1, HH2, and HH3) sequences was employed (Fig. 2). Each element of these cassettes was obtained by annealing two 5'-phosphorylated single-stranded DNA oligonucleotides with staggered ends that facilitated the cloning of these elements in the appropriate sequence and orientation.

Cloning of 3/4 EGSs under Control of Mouse U6 snRNA Promoter

To test these EGSs *in vivo* (tissue culture cells), the synthetic gene for the guide sequence was obtained by PCR using EC-1A (see earlier) and EC-1C 5' GCGCGGTACCAAAAATGGTGAGGCATGAAGG 3' as primers. The DNA fragment obtained was digested with *Kpn*I and inserted into the plasmid pmU6(-315/1) between *Pst*I (blunted) and *Kpn*I sites. This plasmid contains the promoter for the gene for the mouse U6 small nuclear RNA and a signal for transcription termination by RNA polymerase III.[26]

---

[25] U. Lundberg and S. Altman, *RNA* **1,** 327 (1995).
[26] G. Das, D. Henning, D. Wright, and R. Reddy, *EMBO J.* **7,** 503 (1988).

The advantages of choosing this promoter are that U6 snRNA has a very strong pol III promoter, the regulatory elements for U6 snRNA are located solely in the upstream sequence of the gene (the length of EGSs can be controlled precisely and extra flanking sequences that may impair the targeting activity of the EGS can be avoided), and RNA transcripts remain in the nucleus where RNase P is located. If the promoter to be used will generate EGSs with long 5′ and 3′ flanking sequences, it may be useful to clone these EGSs between HH sequences using a "cloning cassette" similar to the one described earlier. These plasmids expressing synthetic genes for EGSs were introduced into eukaryotic cells by transient transfection or they were first cloned into a retroviral vector that was then used to generate stable cell lines expressing EGSs.

Expression of EGSs in *Escherichia coli*

The expression of stem EGSs in *E. coli* could cause inactivation of gene expression through an antisense effect or RNase III cleavage of the target mRNA. In order to correlate the inhibition of gene expression with RNase P activity, we used an *E. coli* strain that was temperature sensitive for RNase P but expressed β-galactosidase and alkaline phosphatase activities (our first test targets) at wild-type levels. T7A49 was constructed by phage P1 transduction, with NHY322[27] as the donor and BL21(DE3)[28] as the recipient, with selection not only for tetracycline resistance and temperature sensitivity (donor's markers), but also for β-galactosidase and T7 RNA polymerase activities (recipient's markers).

Plasmids harboring synthetic genes for EGSs (pNT7DS1HH and pNT7APHH) under the control of the T7 promoter are introduced into T7A49. Control EGSs (pNT7CIHH and pNT7NHH) are directed against phage λ CI and N mRNA, respectively. These cells are incubated overnight at 30° in LBC (LB medium supplemented with carbenicillin at 100 μg/ml). Overnight cultures are diluted to OD_{600} of 0.05 in fresh LBC medium and are incubated at 30° for about 3 hr. Cells are harvested and resuspended in phosphate-free P medium (for induction of alkaline phosphatase activity) supplemented with carbenicillin and isopropyl-β-D-thiogalactoside (IPTG) (2 mM final, for induction of T7 RNA polymerase and β-galactosidase activity). One-half of each culture is incubated at 30° and the other half at 43° for another hour. Cultures are placed on ice to stop cell growth and three aliquots are taken from each culture: one for assays of β-galactosidase,

[27] L. A. Kirsebom, M. F. Baer, and S. Altman, *J. Mol. Biol.* **204,** 879 (1988).
[28] F. W. Studier and B. A. Moffatt, *J. Mol. Biol.* **189,** 113 (1986).

alkaline, and acid phosphatase, one for assays of RNase P activity, and one for extraction of total RNAs.

The results obtained indicate clearly that inhibition of alkaline phosphatase or β-galactosidase activity *in vivo* using EGS technology is dependent on RNase P activity. Indeed, the extent of inhibition at nonpermissive temperature (RNase P activity was not detected in crude extracts prepared from these cells) is much lower (0–11%) than that observed at permissive temperature (50–60%). Northern blot analysis indicated that cleavage by the hammerhead ribozyme was efficient *in vivo*, thereby generating stable EGS RNAs of the appropriate size; furthermore, the steady-state levels of EGSs at nonpermissive temperature were five- to eightfold higher than the level at permissive temperature, thus indicating that the inhibition of gene expression due to an antisense effect, or due to RNase III cleavage, is negligible. The low level of inhibition observed at 43° could be explained by some residual RNase P activity present when the cells were transferred from 30° to 43°.[17] We have not been able to achieve complete inhibition of gene expression in *E. coli.* This may be due to several factors, such as a limit of the steady-state copy number of EGSs in our constructs and the inaccessibility of a fraction of the mRNA population due to the tight coupling of transcription and translation in *E. coli.* We explored the system further to determine whether the method can be used efficiently to alter levels of resistance to antibiotics (chloramphenicol) in *E. coli* with wild-type RNase P activity.

Two different stem EGSs (EGSCAT1 and EGSCAT2) were designed to bind to *cat* mRNA as described earlier and introduced into *E. coli* strain BL21(DE3) harboring the plasmid pACYC184 [BL21(DE3)/pA-CYC184]. A third plasmid had both EGSs cloned in tandem as described earlier (EGSCAT1+2) (Fig. 2). The control plasmid was pKB283, with no EGS. Our strategy was to increase the EGS:target ratio by cloning each under the control of different promoters and by using more than one EGS against the same target. Our results indicated that, under the same growth conditions, phenotypic conversion was more efficient in cells that harbor EGS genes under T7 control than in cells with EGSs under the control of an *E. coli* promoter, probably because the steady-state copy number of EGS RNA is higher in the former cells. With either type of promoter, the inhibition of Cat activity was more efficient when two EGSs were present in the cells. To increase the EGS-to-target ratio in a way different from the one mentioned earlier, plasmids with EGSs under *E. coli* control were introduced into a strain (7027)[29] with the drug resistance marker (Cm^R) in single copy on the host chromosome. In these

[29] K. F. Wertman, A. R. Wyman, and D. Botstein, *Gene* **49,** 253 (1986).

experiments, a drastic inhibition of cell growth (>90%) was observed even at 5 μg/ml of Cm in the medium.[18] Additionally, cells expressing EGSs that were grown in medium with Cm (70 μg/ml) had low viability even on plates that contained no Cm (Fig. 4). Overnight cultures of *E. coli* 7027 (transformed with plasmids that harbored different EGSs) in LBCarb/Cm (LB medium supplemented with 50 μg/ml carbenicillin and 5 μg/ml chloramphenicol) were diluted to A_{600} of 0.05 in LBCarb/Cm and the cultures were incubated at 37°. Cell growth was followed by measuring A_{600}. Aliquots of all cultures were taken after 2 hr and 45 min of incubation and diluted to $A_{600} = 5 \times 10^{-6}$. One hundred microliters of each dilution was plated on LB, LBAmp (100 μg/ml), and LBCm (5 μg/ml) plates, and the average number of colonies was calculated for each culture. Compared to cells not expressing EGSs (100% survival rate), cells expressing EGSCAT1 had a 35% survival rate and those expressing EGSCAT2 had a 70% survival rate. Cells expressing both EGSs had only a 6% survival rate. The low viability was not due to a

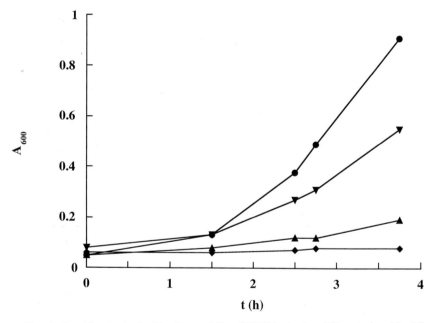

FIG. 4. Growth rates in liquid culture of *E. coli* 7027 harboring different plasmids. All cultures had Carb (50 μg/ml) and Cm (5 μg/ml). The plasmids used were pKB283 (no EGS control) (●), pKBEGSCAT1 (▲), pKBEGSCAT2 (▼), and pKBEGSCAT1 + 2 (♦).

nonspecific effect of EGS expression because cells that expressed EGSs, but that harbored no *Cat* genes, were perfectly viable in the absence of Cm.

Expression of EGSs in Cells in Tissue Culture

Gene regulation and expression in mammalian cell have been advanced by the use of techniques that allow the introduction of nucleic acids into eukaryotic cells. An ideal method to transfer genes (transfection or infection) should have the following properties: (a) reproducibility, (b) a high yield of transfer, (c) low toxicity, and (d) suitability for applications *in vitro* and *in vivo*. We are not going to describe any specific method in details, but only to report some results obtained with cells in tissue culture. In these experiments, the plasmids were introduced into the cells either by transient transfection or were cloned into retroviral vectors that were used to make stable cell lines. *Cat* mRNA was chosen as the first target for human RNase P because the gene coding for *Cat* can be assayed easily and the gene product can be expressed readily in cells in tissue culture.[12] Studies *in vitro* had shown that an EGS (here referred to as 3/4 EGSCAT1 because it targets the same site as EGSCAT1 described earlier), which can base pair with nucleotides 67–73 and 76–79 of *Cat* mRNA (Fig. 3), can direct human RNase P to cleave that mRNA at position 67 as expected. To examine whether 3/4 EGSCAT1 can function *in vivo,* we monitored *Cat* activity in cells in tissue culture that expressed both *Cat* mRNA and 3/4 EGSCAT1 RNA. Plasmid pCAT (Promega, Madison, WI), the plasmid that contained the synthetic gene for 3/4 EGSCAT1 under the control of the mouse U6 promoter (pmU6-3/4EGSCAT1), and the plasmid pXGH5 [which contains the gene for human growth hormone (hGH)] were used simultaneously to transfect human lung cancer cells (A549, American Type Culture Collection, Rockville, MD). In control cells, pmU6-3/4EGSCAT1 was replaced with pmU6(-315/1). The level of hGH secreted into the medium remained the same in cells with or without pmU6-3/4EGSCAT1, indicating that the EGSs were specifically targeting the *Cat* gene. Northern blot analysis revealed that 3/4EGSCAT1 RNA is expressed in the cells. The maximum inhibition of the CAT activity was about 60% at 48 hr posttransfection, when 1 μg pmU6-3/4EGSCAT1 was used: after that time the kinetics were no longer reproducible because the transfected cells started to lose plasmids. This level of inhibition of *Cat* expression was similar to the one obtained in *E. coli.* However, it is possible to improve the level of inhibition if EGSs lacking the anticodon domain are used.

EGSs were designed to target the mRNA encoding the gene for thymidine kinase (TK) of herpes simplex virus 1. The expression of TK protein was reduced by 50% with 3/4 EGS and by 75% with ΔACEGS.[22] In experiments carried out to inhibit the production of influenza (flu) viruses in cells in tissue culture, EGSs were first cloned into a retroviral vector that was then used to generate stable cell lines: a new strategy used in these studies was to express two EGSs simultaneously targeting mRNAs of the *PB2* (polymerase subunit 2) and *NP* (nucleocapsid) genes. Both protein and particle production were inhibited between 90 and 100% in cells that expressed the two EGSs.[13] In these latter two studies, the use of an inactive 3/4 EGS with mutations in the T loop ruled out the possibility of inhibition of gene expression by an antisense effect.

Concluding Remarks

The RNase P approach described in this article offers several advantages that are not present in other systems: RNase P is present and abundant in every living cell, it does not require a specific sequence in the target, it destroys the target RNA irreversibly, and it uses short oligonucleotides or long RNA transcripts as EGSs. M1 RNA, the RNA component from RNase P from *E. coli*, when linked covalently to a stem EGS, is able to cleave a targeted RNA even in buffers containing a low concentration of MG^{2+}.[17,19,30] We have not included this approach in this article because only a few experiments have been carried out so far. RNase P from *E. coli* (in combination with stem EGSs) was as efficient as eukaryotic RNase P (in combination with 3/4 EGSs) in cleaving the targeted RNA. However, further research is needed to improve the nuclease resistance of EGSs *in vivo*, to facilitate EGSs to find and interact efficiently with their target RNAs (forming a tRNA-like substrate), and to colocalize these complexes with RNase P for efficient cleavage.

[30] F. Liu and S. Altman, *Genes Dev.* **9**, 471 (1995).

[27] Disruption of mRNA–RNP Formation and Sorting to Dendritic Synapses by Antisense Oligodeoxynucleotides

By EMMA R. JAKOI and W. L. SEVERT

Introduction

Cellular differentiation requires a heterogeneous distribution of proteins to establish and maintain regions of specialized function. Two mechanisms have been identified that provide for spatial localization of proteins: the first involves directed targeting of the polypeptide and the second utilizes sorting of the encoding mRNA for localized protein synthesis. Although the localization of mRNA is well established as an important mechanism for generating polarity during embryogenesis,[1] it has only recently emerged as an important means for establishing regional differentiation in somatic cells, such as neurons,[2] fibroblasts,[3–5] and skeletal myoblasts,[6] where the local regulation of function may be critical. This process involves the formation of ribonucleoprotein particles (RNP), which govern sorting and anchoring of the mRNA. In general, two components are involved: a *cis*-acting mRNA sequence(s) that encodes spatial information and *trans*-acting proteins that confer stability, govern translation, and mediate tethering to cytoskeletal elements and/or movement. Identities of the *cis*-acting and *trans*-acting components that mediate mRNA localization in somatic cells remain largely unknown.

Given the implied important role that RNPs play in establishing cellular differentiation and in plasticity, methods used for the analysis of RNP composition and localization are of increasing interest. This article describes the use of antisense oligodeoxynucleotides (anti-ODN) in functional assays of specific RNP components: (1) the assay of *cis*-acting sequences and *trans*-acting proteins in RNP formation in cell extracts by gel mobility shift and (2) the assay of *cis*-acting sequences in mRNA sorting by fluorescence *in situ* hybridization (FISH) and confocal microscopy. Using these assays, we have shown that the *cis*-acting sequence encoding the testis–brain RNA-binding protein motif (Y element) in ligatin mRNA is involved in RNP formation and sorting in hippocampal neurons.[7] To achieve this status, we selectively inactivate mRNA using anti-ODN.

Anti-ODN is a short synthetic DNA molecule capable of entering cells and inhibiting expression of a given gene. It is usually designed to anneal with high specificity to its complement sequence on mRNAs to suppress synthesis of the cognate proteins. Inhibition of translation typically results from either ribonuclease H degradation of the RNA:DNA hybrid[8,9] or

[1] D. R. Micklem, *Dev. Biol.* **172**, 377 (1995).
[2] O. Steward, *Neuron* **18**, 9 (1997).
[3] C. L. Sundell and R. H. Singer, *Science* **253**, 1275 (1991).
[4] E. H. Kislauski, Z. F. Singer, and K. L. Taneja, *J. Cell Biol.* **123**, 165 (1993).
[5] E. H. Kislauski, X. C. Zhu, and R. H. Singer, *J. Cell Biol.* **127**, 441 (1993).
[6] M. A. Hill and P. Gunning, *J. Cell Biol.* **122**, 825 (1993).
[7] W. L. Severt, Ph.D. thesis (1998).
[8] G. D. Hoke, K. Drapper, S. M. Freier, C. Gonzalez, V. B. Driver, M. C. Zounes, and D. J. Ecker, *Nucleic Acid Res.* **19**, 5743.
[9] R. Y. Walder and J. A. Walder, *Proc. Natl. Acad. Sci. U.S.A.* **86**, 5011 (1988).

inhibition of the translation complex.[10] The anti-ODN can also be directed to a *cis*-acting sequence encoded within the untranslated 3' region (3'UTR) of a localized mRNA. Because specific regions of the 3'UTR govern stability, translation, and/or targeting of the mRNA, perturbation of this region can alter the location and/or expression levels of a given protein. Thus anti-ODN can be used to generate a cell deficient in a specific protein without affecting the synthesis of other cellular proteins, providing an alternate approach to transgenic knockouts. Moreover, genes required for viability can be studied because the anti-ODN effect is usually 50–70% complete.[10,11]

The biological effectiveness of the anti-ODN depends on its design. To retain specificity and to optimize cellular uptake, ODN of 11–25 nucleotides in length are used.[12–14] The choice of nucleotide sequence to be targeted remains in part empirical because its accessibility for hybridization is unpredictable due to the folding of the mRNA. To enhance the efficacy of an ODN, sequences that form duplex structures, such as stable homodimers or stem–loops, at 50° and lower are avoided. In addition, sequences that include repeats of four guanidine (G) residues[15,16] and with GC:AT ratios >60–70%[17] are not used due to increased nonspecific binding to DNA and RNA. To disrupt RNP formation and mRNA localization in hippocampal neurons, we targeted the known binding motif (Y element) of the TB-RBP with an anti-ODN 13 nucleotides in length.

Stability of the ODN directly affects its delivery and hence activity. Synthetic ODN has a phosphodiester backbone that renders it unstable in the blood and in culture media containing serum with a half-life of minutes to a few hours.[17–19] With phosphorothioate or methyl phosphonate derivatives, the half-life of the ODN increases to several days; however, nonspe-

[10] C. Wahlestedt, E. Golanov, S. Yamamoto, F. Yee, H. Ericson, H. Yoo, C. E. Inturrisi, and D. J. Reis, *Nature* **363**, 260 (1993).
[11] K. M. Standifer, C.-C. Chien, C. Wahlestedt, G. P. Brown, and G. W. Pasternak, *Neuron* **12**, 805 (1994).
[12] S. T. Crooke, *Annu. Rev. Pharmacol. Toxicol.* **32**, 329 (1992).
[13] A. Colman, *J. Cell Sci.* **97**, 399 (1990).
[14] S. L. Loke, C. A. Stein, X. H. Zhang, K. Mori, M. Nakanishi, C. Subasinghe, J. S. Cohen, and L. M. Neckers, *Proc. Natl. Acad. Sci. U.S.A.* **86**, 3474.
[15] Y. Saijo, B. Uchiyama, T. Abe, K. Satoh, and T. Nukiwa, *Jpn J. Cancer Res.* **88**, 26 (1997).
[16] Y. Castier, E. Chemla, J. Nierat, D. Heudes, M. A. Vaseur, C. Rojnoch, B. Bruneval. A. Carpentier, and J. N. Fabiani, *J. Cardiovasc. Surg.* **39**, 1 (1998).
[17] C. Wahlestedt, *Trends Pharmacol.* **15**, 42 (1994).
[18] S. T. Crooke, *FASEB J.* **7**, 533 (1993).
[19] E. Wickstrom, *J. Biochem. Biophys. Methods* **13**, 97 (1986).

cific effects also increase.[12,20] Phosphorothioate–ODN binds nonspecifically to protein with one- to threefold higher affinity than the underivatized phosphodiester ODN.[21,22] Additionally, derivatization alters the charge of the ODN, altering its cellular uptake and effective working concentration. To reduce nonspecific effects and yet increase stability, the terminal three bases of the ODN (3' and 5' ends) can be derivatized and the derivatized ODN administered at low concentrations (<1 μM) every other day or two. Because the effectiveness of the ODN is dependent on the sequence targeted, a concentration-response curve should be generated for each analog. In our experience, RNP formation can be disrupted in primary cultures of hippocampal neurons grown in defined media by phosphodiester-ODN (Operon, Inc.) administered once daily at low concentrations (1 μM).

Sequence-independent effects have been reported with this experimental approach, necessitating the use of several controls. Most commonly used is sense-ODN that contains the same sequence as the targeted mRNA sequence. However, in some instances, sense-ODN may alter transcription[22] and hence alter function. As alternatives, reverse-ODN (i.e., 3' to 5' sequence of the anti-ODN transcribed as 5' to 3' sequence) or scrambled anti-ODN (i.e., > -4 nucleotides of the anti-ODN mismatched) is used as a control.[18] All sequences are checked against the GenBank database to ensure that an encoding, identifiable sequence has not been selected. As an additional control, we examine whether the biological effect is reversed on removal of the anti-ODN. In our experience, reversal occurs in 4–5 days.

Gel Shift Assay for RNP Formation

The gel mobility shift assay is a widely used technique for measuring the interactions of nucleic acid and proteins in a sequence-specific manner. This assay allows detection of RNA-binding proteins in crude extracts and permits their identification using specific antibodies to further retard the mobility of the RNA–protein complex (super shift). The sequence specificity of the RNA–protein interaction can be determined using competition

[20] F. Morvan, H. Porumb, G. Degols, I. Lefebvre, A. Pompon, B. S. Sproat, B. Rayner, C. Malvy, B. Lebleu, and J. L. Imbach, *J. Med. Chem.* **36**, 280 (1993).
[21] J. R. Perez, Y. Li, C. A. Stein, S. Majunder, A. van Oorschot, and R. Narayanan, *Proc. Natl. Acad. Sci. U.S.A.* **91**, 5959 (1994).
[22] B. J. Chiasson, J. N. Armstrong, M. L. Hooper, P. R. Murphy, and H. A. Robertson, *Cell Mol. Neurobiol.* **14**, 507 (1994).
[23] K. Yokoyama and F. Imamoto, *Proc. Natl. Acad. Sci. U.S.A.* **84**, 7363 (1987).

assays. For competition, formation of the RNA–protein complex (RNP) is disrupted by preincubating either the lysate with unlabeled sense-ODN to form a DNA–protein complex or the radiolabeled RNA probe with anti-ODN to form a radiolabeled RNA : DNA hybrid.

Cell Culture Reagents

Glial Feed

Dulbecco's minimal medium (MEM, GIBCO) containing
Glutamine	2 mM
Horse serum	10%
Conditioned medium from 2-week-old glial beds	20%

Neuronal Feed

Dulbecco's minimal medium (MEM) containing HEPES	15 mM
Glutamine	2 mM
Glucose	0.06%
Insulin	5.0 μg/ml
Transferrin	100 μg/ml
Putrescine	100 μM
Sodium selenite	30 nM
Progesterone	20 nM
Triiodothyronine	20 ng/ml
Sodium pyruvate	1 mM
Ovalbumin	0.01%

Gel Shift Reagents

Cell lysis buffer: 25 mM Tris-HCl, pH 7.9, 0.5 mM EDTA, and 0.1 mM phenylmethylsulfonyl fluoride (PMSF)

Second Shift buffer: 20 mM HEPES, pH 7.9, 50 mM KCL, 0.2 mM EDTA, 0.5 mM PMSF, 0.5 mM dithiothreitol (DTT), and 10% glycerol

10× mix: 40 mM ATP, 70.0 mM MgCl$_2$, 1.0 mM EDTA, and 6.0 mM DTT

Polyvinyl alcohol (PVA) (cold soluble): 10% (w/v)

Neuronal Cell Cultures

Rat hippocampi from neonate day 2 are digested with trypsin (0.025%, 20 min at 37°) in saline containing 1.0% penicillin/streptomycin. The tissue is triturated three times and plated (10^6 cells per 100-mm plate) onto a confluent 2-week-old glial feeder bed (0.5 × 10^6 cells per 100-mm plate)

in neuronal feed. After 1 day, the medium is replaced with neuronal feed. Cells are maintained for 15–18 days before use at 37°.

Preparation of Whole Cell Extracts

Cells are washed once with ice-cold phosphate-buffered saline (PBS). Ice-cold cell lysis buffer (200 μl) is added and the cells are scraped into a 1.5-ml centrifuge tube. Cell lysates are prepared by repetitive (3×) freeze–thaw lysis. The lysate is centrifuged at 15,000g (4°) for 15 min. The supernatant is aliquoted, frozen in dry ice, and stored at $-70°$.

Labeling of RNA Probe

Radiolabeled and unlabeled RNA transcripts are synthesized *in vitro* with SP6 (or T7) polymerase using the Riboprobe *In Vitro* Transcription System (Promega Madison, WI). For each experiment, a radiolabeled (hot) probe and an unlabeled (cold) specific competitor are synthesized. The reaction mix is prepared at room temperature according to the manufacturer's recommendations. To test for RNP formation with ligatin mRNA, we routinely use RNA transcripts derived from ligatin UTR500 cDNA as probes. The plasmid is linearized and the UTR500 consisting of 500-nt and 59-nt of polylinker is used to synthesize transcripts. For control transcripts of unrelated sequence, the *Eco*RI-linearized Riboprobe pGEM-11Z $(-)$ polylinker (58-nt) is used. For specific competition with the TB-RBP Y element, sense RNA transcripts are synthesized from a 67-nt protamine 3'UTR template (pGEMc) consisting of 42-nt of 3'UTR and 25-nt of poly-linker[24] using SP6 polymerase.

SP6 transcription reaction (Promega):

	Hot (μl)	Cold (μl)
5× SP6 transcription buffer	4	8
100 mM DTT	2	4
RNase inhibitor (RNasin)	2	4
10 mM GTP	1	2
10 mM UTP	1	2
10 mM CTP	1	2
10 mM ATP	—	2
0.3 mM ATP	1	—
[α-^{32}P]ATP (10 mCi/μl)	5	—
Linearized plasmid (2.0 μg/μl)	2	4
SP6 polymerase (16 U/μl)	1	2
H$_2$O (RNase free)	—	10

[24] Y. K. Kwon and N. B. Hecht, *Proc. Natl. Acad. Sci. U.S.A.* **88**, 3584 (1991).

The reaction is incubated at 37° for 60 min. RNase-free DNase (1 μg DNase/μl template) is added and incubation is continued for 15 min at 37°. Subsequently, 70 μl of sterile H_2O and 10 μl of 3.0 M sodium acetate (RNase free) are added and the reaction mix is extracted with an equal volume of phenol/chloroform. Unincorporated free trinucleotides are removed by centrifugation on a Select-D G25 spin column (TE midi, 5'-3' Prime, Inc.). RNA is recovered by ethanol precipitation. The labeled precipitate is dissolved in 100 μl H_2O. An aliquot (2 μl) is removed and counted. The number of counts in the total volume is calculated. For each reaction mix, 100,000 to 250,000 cpm are used.

RNA Binding Assay

Premix per sample in ice
0.5 μg tRNA	1.0 μl
10% PVA	2.5 μl
10× mix	2.5 μl
RNasin	1.0 μl
(aliquot 7 μl of premix per reaction)	
Cell extract	6.0 μl (12 μg of protein)
RNA (or ODN) competitor	5.0 μl
Radiolabeled probe	2.0 μl
Make up to total volume with second buffer	25 μl

Prior to use, the radiolabeled probe (1–2 ng), in the absence or presence of competitor ODN (or RNA), is heated at 65° for 10 min and then slow cooled at room temperature. The reaction mix is prepared in ice. The probe is added in ice. The gel mobility shift assay is performed according to Gillis and Malter.[25] The binding reaction mixes are incubated at 30° for 10 min in a total volume of 25 μl. Heparin (40 μg) is added. The reaction mixture is vortexed, incubated in ice (10 min), and then digested with ribonuclease T1 (1U) at 37° (30 min). The sample (25 μl) is applied to a nondenaturing gel and the formed complexes are resolved by electrophoresis.

In lysates of hippocampal neurons, a single shift band is resolved using transcripts of UTR500. The formation of this RNP is sequence specific as shown by competition assays. Two assays are performed: in the first, the radiolabeled probe is preincubated with the cold competitor at 5 to 50-fold excess. We use both RNA and ODN as specific competitors. In our experience, RNA encoding the TB-RBP Y element inhibits the formation of RNP at 10-fold excess. The TB-RBP Y element anti-ODN (specific competitor ODN, 5'-AGCCCAGAGCTTG-3') prehybridized to the

[25] P. Gillis and J. S. Malter, *J. Biochem.* **266**, 3172 (1991).

radiolabeled RNA probe at 5-fold excess also disrupts RNP formation. As controls, we use the TB-RBP Y element sense ODN (5'-CAAGCTCTGGGCT-3') and the nonspecific ODN sequence (5'-TCTGTATAGACATGATGGCTG-3') at 5-fold excess. In the second assay, the lysate is preincubated with the competitor RNA or ODN to inactivate the binding activity. Competitor (or noncompetitor) RNA is added at 10- and 50-fold excess. No shift band is formed in the presence of competitor RNA. The addition of noncompetitor RNAs (tRNA or nonspecific sequence RNA) at 50-fold excess does not affect RNP formation. In this competition assay, anti-ODN to the TB-RBP Y element prevents the formation of the shift band but the TB-RBP Y element sense-ODN does not.

Polyacrylamide Gel Analysis

Reagents

10× Tris–boric acid–EDTA (TBE) buffer

Tris base, pH 8.3	25 mM
Boric acid	0.5 mM
0.5 M EDTA, pH 8.0	0.5 mM

7% gel

10× TBE	4.0 ml
40% acrylamide/bisacrylamide (38 g/2 g per 100 ml)	7.0 ml
H$_2$O	29 ml
TEMED	24 μl
25% (w/v) ammonium persulfate (APS)	112 μl

Two 20 × 20-cm glass plates are used: one plate is siliconized. The gels are 1.5 mm thick with 10 wells (15 wells maximum). A 7% (w/v) polyacrylamide (38:2, acrylamide : bisacrylamide) gel is prepared in 1.0× TBE buffer. The gel is poured and allowed to polymerize at room temperature and is then equilibrated overnight at 4°. Gels are prerun in 1.0× TBE at 200 V for 1 hr at 4°. The binding assay is loaded onto the gel and electrophoresed at 200 V for 5–7 hr (4°). Subsequently, the siliconized plate is removed from the gel. A piece of Whatman (Clifton, NJ) 3MM paper is placed over the gel surface and the adsorbed gel is gently lifted from the glass plate. The gel is covered with Saran wrap and dried at 50–60° under vacuum for 2 hr. The autoradiograph is developed overnight at −70°. Sequence specificity in the formation of the RNA–protein complex is demonstrated by competition of the shift band by a specific RNA competitor (or anti-ODN) but not by nonspecific, unrelated RNA (or unrelated ODN sequence).

FISH Assay for Disruption of mRNA–RNP Formation and Sorting

In situ hybridization is a well-established technique for the localization of DNA and RNA within tissue sections and cultured cells. With the introduction of synthetic ODN, fluorescent probes, and confocal microscopy, this method provides a sensitive means to quantitatively analyze changes in gene products under different physiologic conditions. We routinely employ biotinylated anti-ODN as a specific probe and Texas Red-Ultra avidin (Leinco Technologies) for the detection of localized mRNAs. We have found that in contrast to cRNA probes, ODN probes offer several advantages, including stability, ease of labeling with nonisotopic reporters, and good cellular penetration. The optimal length of the ODN probe used for FISH varies between 21 and 50 nt.[26] A minimal length of 18 nt ensures specificity in binding. Although the strength of the RNA : DNA hybrid increases with increased length, the chances of nonspecific effects and cost increase. We routinely use anti-ODN probes of 21–24 nt. When combined with immunocytochemistry, the molecular phenotype of the treated cells is determined readily.

FISH Reagents

> 10% paraformaldehyde stock: 10 g paraformaldehyde and 100 ml water
> Heat to 60° for 45–60 min. Add 1 M NaOH dropwise to clear, filter, and store cold. Dilute with an equal volume of 2× PBS before use.
> Phosphate-buffered saline (PBS): 137 mM NaCl, 2.68 M KCl, and 8.10 M Na$_2$HPO$_4$
> Quenching solution: 50 mM NH$_4$Cl in PBS
> Denhardt's solution: 5.0 g Ficoll, 5.0 g bovine serum albumin, and 5.0 g polyvinyl pyrrolidine; add H$_2$O to 500 ml
> Saline, sodium citrate solution (SSC): 60 mM NaCl and 4.0 mM trisodium citrate, pH 7.0
> Pre/hybridization cDNA mix: 50 mM NaH$_2$PO$_4$ 3× SSC 5× Denhardt's solution 50% formamide, 200 μg/ml *Escherichia coli* tRNA, and 1% salmon testis DNA
> Wash buffer: 0.5× SSC and 0.1% SDS
> PBS–saponin–gelatin buffer: 0.25 g gelatin and 0.10 g saponin; make up to 100 ml with PBS
> DABCO mounting media: 10 ml 10× PBS, 90 ml glycerol, and 2.5 g DABCO

---

[26] P. C. Emson, *Trends Neurosci.* **16**, 9 (1993).

Incubation Conditions to Disrupt mRNA-RNP Localization

Hippocampal neurons are grown on coverslips (10^4 neurons, 10^4 glia per coverslip) as described. To inactivate the *cis*-acting mRNA component, sterile, nonderivatized, phosphodiester anti-ODN (1 μM, OPERON) is administered daily for 3–4 days. As controls for sequence specificity, we use sense-ODN and missense-ODN. The expression levels of a given mRNA reflect its rate of synthesis and of degradation. Thus, the efficacy of the anti-ODN is time dependent. In our experience, a significant decrease in dendritic levels of localized ligatin mRNA occurs after 3 days of treatment with TB-RBP Y element anti-ODN. Concurrent with a significant decline in specific signal within dendrites, the level of ligatin mRNA increases significantly within the soma, suggesting that somatodendritic export is altered. Sense-ODN and missense-ODN do not alter the distribution of the localized ligatin mRNA. As an added control, the expression of a mRNA (e.g., neuron-specific enolase) that is restricted to the soma and does not contain the Y element is determined.

Labeling of Anti-ODN Probe

The specific complementary ODN probes used for FISH are tailed with biotinylated-dUTP (Clontech) using a terminal deoxynucleotide transferase reaction. The complementary sequences used by us are directed to the coding region of the mRNA: [ligatin, 5′-GTCTTCTGGGGCTTC-TGAGAG-3′ and neuron-specific enolase 5′-TCTGTATAGACAT-GATGGCTG-3′].

> Terminal deoxynucleotidyltransferase (TdT) tailing reaction (four coverslips): 10 μl ODN (3.0–4.0 μg), 10 μl 5× TdT buffer, 20 μl biotinylated dUTP, 5 μl dCTP, and 2 μl TdT enzyme (20 U/μl); add H_2O to total volume of 50 μl.

The reaction is incubated at 37° for 3 hr. Free biotinylated dUTP is removed by centrifugation on a Select-D G25 spin column (TE midi, 5′-3′ Prime, Inc.).

FISH Analysis

Cultured cells grown on coverslips are fixed in 5% paraformaldehyde in PBS for 30 min. After fixation, the cells are washed three times in 50 mM NH_4Cl followed by PBS (5 min) at 23°. To permeabilize the cells, 150-μl aliquots of 70% ethanol are added (×4 over 5 min). Ethanol–PBS is replaced by 70% ethanol for 20 min. The ethanol is removed and the coverslip is air dried. Subsequently, the cells are rehydrated in PBS for 5 min and then prehybridized in hybridization buffer for 1 hr at 42°. Complementary ODN probes tailed with biotinylated-dUTP are added to the hybridization mix. Cells are "hot-started" for 2 min at 94° and then incubated for 60 min at 37°. To remove the unbound probe, the cells are washed three times (10 min) in wash buffer. The conditions for hybridization and stringency of wash will vary with the base composition, length, and number of mismatches of the ODN probe used. The critical variables are salt concentration, temperature, and formamide concentration; they are determined empirically. We routinely use 50% formamide for hybridization of ODN probes and 0.5× SSC, 0.1% SDS at 37–50° to remove unbound ODN.

Probe Detection

To visualize the bound biotinylated ODN, we use Texas Red-Ultra avidin conjugate (Leinco Technologies). The cells are washed with PBS (5 min) followed by PBS–gelatin–saponin for 60 min (23°) to block nonspecific protein binding. The cells are incubated in Texas Red-Ultra avidin conjugate (10 μl/ml PBS–saponin-gelatin) for 30 min (23°). Cells are then washed (×6, 5 min) with PBS–saponin–gelatin and the coverslips are mounted onto a glass microscope slide with DABCO–PBS mounting medium. Nonspecific labeling is assessed by RNase 1 predigestion and by omitting the specific probe. Cells are scanned with a Zeiss LSM410 (Zeiss Inc., PA) using an argon–krypton laser and a C-Apo 40×/1.2 water-immersion lens. Identical settings are used for anti-ODN and sense-ODN (control)-treated specimens. Quantitation and image analysis are done using a Carl Zeiss laser-scanning microscope system software version 3.84.

Immunocytochemistry

Specific antibodies and immunocytochemistry are used to determine the effect of the ODN treatment. Cells are fixed in 5% paraformaldehyde in PBS for 30 min and permeabilized in PBS–saponin–gelatin for 60 min at 23°. Subsequently, specific primary antibody is added in PBS–saponin–gelatin for 60 min. The immune serum is removed and the cells are washed six times with PBS–saponin–gelatin. The bound primary antibody is detected with biotinylated secondary serum (Vector Labs). Cells are incubated with secondary antibody (60 min, 23°) and then washed six times in PBS–saponin–gelatin to remove unbound secondary antibodies. Specific staining is visualized with Texas Red-Ultra avidin (10 μl/ml, 30 min). Cells are washed six times with PBS and the coverslips are mounted on a microscope slide using DABCO mounting media. Cells are photographed with a Zeiss epifluorescence microscope using a 25× objective (0.8 NA).

[28] Antisense RNA and DNA in *Escherichia coli*

By OLEG MIROCHNITCHENKO and MASAYORI INOUYE

Antisense RNA (asRNA) has been shown to be utilized effectively in *Escherichia coli* as a natural regulatory factor to control a variety of important biological processes. Among them are plasmid replication,[1] IS*10* transposition,[2] osmoregulation of *ompF* gene expression,[3] autoregulation of cAMP-receptor protein synthesis,[4] cell-killing function through activity of *hok* homologs,[5] cell division,[6] and P22 and λ phage development.[7,8] In most cases, asRNA are small untranslatable transcripts that pair with a target RNA sequence and exert a negative control by not allowing the normal interaction of target RNA with other nucleic acids or protein factors or lead to the increased rate of degradation by RNase specific for double-stranded RNA. Some of the asRNA are synthesized in significant quantities and have a high turnover rate, whereas others, which mostly regulate transposon and phage actions, are expressed at low levels but are significantly stable. Members of the asRNA family may be divided into two major groups: asRNAs that are transcribed from the same loci as the target genes and asRNAs whose genes locate at genetic loci different from the target genes acting as *trans*-coded regulators. Antisense RNAs belonging to the first group are completely complementary to their target RNA, whereas asRNAs from the second group usually form only partially complementary hybrids with their target sequences. It has been shown that the mechanism of action of these natural asRNAs is very efficient in regulating the expression and function of specific target genes. Application of a similar approach for the construction of artificially designed asRNAs in *E. coli* has become a powerful tool for investigating specific gene functions and analyzing RNA/RNA and RNA/protein interactions. This technology can also be applied to creating cells resistant to phage infection. The same technique has been employed successfully in eukaryotic cells to inhibit the expression of various genes, including oncogenes and viral genes, and provides exciting implica-

[1] T. Itoh and J. Tomizawa, *Proc. Natl. Acad. Sci. U.S.A.* **77,** 2450 (1980).
[2] R. Simons and N. Kleckner, *Cell* **34,** 683 (1983).
[3] T. Mizuno, M.-Y. Chou, and M. Inouye, *Proc. Natl. Acad. Sci. U.S.A.* **81,** 1966 (1984).
[4] K. Okamoto and M. Freundlich, *Proc. Natl. Acad. Sci. U.S.A.* **83,** 5000 (1986).
[5] L. Poulsen, A. Refin, S. Molin, and P. Anderson, *Mol. Microbiol.* **5,** 1639 (1991).
[6] F. Bouche and J. Bouche, *Mol. Microbiol.* **3,** 991 (1989).
[7] K. Ranade and A. Poteete, *Genes Dev.* **7,** 1498 (1993).
[8] L. Krinke and D. Wulffs, *Genes Dev.* **1,** 1005 (1987).

tions to genetic therapy. It should be noted that significant progress in this direction has been achieved by utilizing antisense oligonucleotides and their derivatives,[9] which will not be discussed in this article.

Retroelements found in gram-negative bacteria allow a novel method of *in vivo* synthesis of specific oligonucleotides, providing an efficient system for the regulation of gene expression in *E. coli*. It is based on the complementarity of single-stranded DNA (asDNA) possessing antisense sequences to specific genes. The principle of the method is to express asDNA as a part of msDNA, which are produced in *E. coli* by bacterial reverse transcriptase.[10] This article describe approaches using both antisense RNA and antisense DNA for the regulation of gene expression in *E. coli*.

Artificial Antisense RNA Genes

General Considerations

Natural *E. coli* asRNA may serve as a model system for the designing of an efficient antisense regulatory system. The following observations should be taken into account for the creation of artificial antisense genes. Most of the natural regulatory asRNAs-targeted expression of specific genes is complementary to the 5′-untranslated regions, including the Shine–Dalgarno sequence and/or coding regions of the individual target mRNAs, resulting in inhibition on the level of translation or the destabilization of the message. Other target sequences of natural asRNAs locate at or near the region where transcription factors interact.

Antisense RNA transcripts that are more stable and that are present in a large copy number usually are more effective in their function. Secondary structures of the natural asRNAs normally contain one or more stems with loops, consisting of five to eight GC-rich bases. Such a secondary structure facilitates the interaction between the loop and the target sequence. Kinetics of the binding reactions of the several natural asRNA systems are well studied and these parameters may be used for designing artificial asRNAs. Computer analysis may be performed to predict efficient annealing between target RNA and asRNA. However, it should be kept in mind that cellular factors could play an important role in the efficiency of the *in vivo* RNA–RNA annealing and stability of the complexes. It is also important to note that a longer asRNA is not necessarily more effective as secondary structures are formed within the complementary region to inhibit efficient

---

[9] C. Cazenave and C. Helene, *in* "Antisense Nucleic Acids and Protens" (J. M. Mol and A. R. van der Krol, eds.), p. 47. Dekker, New York, Basel, 1991.

[10] J. R. Mao, M. Shimada, S. Inouye, and M. Inouye, *J. Biol. Chem.* **270,** 19684 (1995).

pairing with the target RNA. It is also important to avoid any known regulatory sequences, such as AUG and the Shine–Dalgarno sequence within asRNA sequences, which are likely to block the asRNA function by interacting with cellular factors. On the basis of these considerations, one can design a few asRNAs and the best one can be determined empirically.

Construction of Antisense RNA Genes

A variety of commercially available plasmids may be used as expression vectors for asRNA genes. These vectors should contain several key features, along with genes allowing them to be maintained in a particular E. coli strain, desirably in a high copy number. These vectors should have an efficient promoter that can be regulated, convenient restriction sites for the insertion of antisense genes, and a transcription terminator. One very useful strategy is to insert several copies of the asRNA gene or to duplicate the whole expression cassette.

For example, a series of plasmids are designed for asRNA production, which are constructed from PIN-II vector,[11] consisting of the lpp promoter and the lac promoter–operator region (pJDC402, pJDC406, and pJDC408; Fig. 1). These vectors are able to produce asRNA only in the presence of a lac inducer such as isopropyl-β-D-thiogalactopyranoside (IPTG). The expression cassette also contains the lpp terminator. There are several unique restriction sites for cloning the asRNA gene between the promoter sequence and the terminator. In addition, these vectors have restriction sites, which allow the isolation of the whole expression segment, including promoter, asRNA gene, and terminator. Therefore, when this fragment is cloned back into the original construct for asRNA, two copies of the functional asRNA gene can be cloned in one vector. These vectors should be used in an E. coli strain, capable of overproducing the lacI gene product, such as E. coli JA221 (hsdR, leuB6, LacY, thi, recA, ΔtrpE5)/F' (lacI^q, proAB, lacZYA)[11] so that asRNA production can be tightly controlled.

Expression of Antisense RNA

In order to avoid leaky induction of asRNA production in a rich medium, cells carrying a pIN-II vector system should be cultured in M9 medium[12] supplemented with 0.4% glucose, 2 μg/ml thiamin, and 50 μg/ml ampicillin. To induce expression of asRNA, cells are grown to a Klett-Summerson colorimeter reading of 20, at which time IPTG is added to a final concentra-

[11] K. Nakamura and M. Inouye, EMBO J. 1, 771 (1982).
[12] J. Miller, "Experiments in Molecular Genetics." Cold Spring Harbor Press, Cold Spring Habor, NY, 1972.

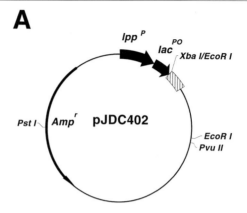

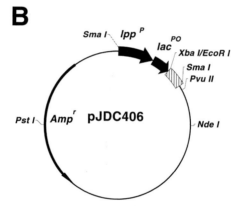

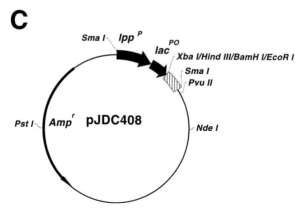

tion of 2 m*M*. At an appropriate time after the addition of IPTG, cell growth is stopped by rapidly chilling the culture on ice. The cells are collected by centrifugation for RNA analysis. To lyse the cell, 1 ml of TRIzol reagent (GIBCO-BRL, Gaithersburg, MD) is added to 1×10^7 cells. After repetitive pipetting, the samples are incubated for 5 min at room temperature. Chloroform (0.2 ml) is added, and after mixing, samples are incubated for an additional 5 min. The samples are then centrifuged at 10,000 g for 15 min at 4°. The aqueous phase is transferred to another tube to which 0.5 ml of 2-propanol is added. The samples are incubated for 10 min at room temperature and centrifuged at 10,000 g for 10 min at 4°. The RNA pellet is washed with 70% ethanol and, after brief drying, redissolved in RNase-free water at 55° for 5 min. The RNA concentration is measured by a spectrophotometer. Quality of the RNA is checked by electrophoresis on a 1.5–2% agarose gel containing 2.2 *M* formaldehyde. Before electrophoresis the RNA sample (5–20 μg) should be incubated for 15 min at 55° in the denaturing solution containing 50% formamide and 1× MOPS running buffer (0.04 *M* MOPS, pH 7.0; 0.01 *M* sodium acetate and 0.001 *M* EDTA) containing 2.2 *M* formaldehyde. A 1/10 volume of loading buffer containing 0.25% (w/v) bromphenol blue and 50% (w/v) glycerol is added to the sample and the mixture is then loaded on the gel. Electrophoresis is carried out at 5 V/cm in 1× MOPS running buffer. After electrophoresis the gel may be stained with ethidium bromide and exposed under UV light to detect RNA bands.

Antisense RNA can be detected by Northern blot analysis. RNA samples from induced and noninduced cultures are purified as described earlier. Duplicate samples are loaded on one side of the gel for subsequent staining. After electrophoresis the gel is washed in 20× SSC solution (3 *M* NaCl, 0.3 *M* sodium citrate, pH 7.0). RNA is transferred from the gel to the GeneScreen membrane (NEN) by capillary blotting or by using the PosiBlot pressure blotter (Stratagene, La Jolla, CA). The membrane is washed with 2× SSC for 5 min and then dried. RNA is fixed to the membrane by baking at 80° for 2 hr. The membrane should be hydrated in 2× SSC before prehybridization. Prehybridization is performed in a solution of 50% for-

---

FIG. 1. Structure of cloning vectors for expression of antisense RNA genes in *E. coli*. (A) pJDC402 was constructed from expression vector pIN-II.[11] It contains the *lac*[po] region downstream of the lipoprotein promoter, thus allowing a high-level inducible expression of an inserted DNA fragment. It contains *Xba*I and *Eco*RI sites for cloning and the ρ-independent transcription termination signal of the *lpp* gene downstream of the cloning sites. (B) pJDC406 was obtained from pJDC402 by deleting a 0.3-kb fragment downstream of the *lpp* terminator, and the whole transcription unit is flanked by *Sma*I sites. (C) pJDC408 was created by inserting a polylinker sequence between the *Xba*I and the *Eco*RI sites in pJDC406.

mamide, 5× SSPE (0.6 M NaCl, 0.04 M NaH$_2$PO$_4$, and 0.005 M EDTA, pH 7.4), 5× Denhardt's solution [0.1% polyvinylpyrrolidone, 0.1% bovine serum albumin (BSA), 0.1% Ficoll 400], 1% SDS, 10% dextran sulfate, 100 μg/ml of carrier DNA at 42° for at least 1 hr. The ^{32}P-labeled sense RNA probe may be prepared by using *in vitro* T7 or SP6 transcription systems. Alternatively, a sense single-stranded DNA probe produced with an M13 system according to Messing[13] can be used. A double-stranded DNA probe could also be used if the asRNA contains sequences that are not present in the target RNA. Hybridization is performed using prehybridization solution containing the probe at 42° for 16–24 hr. The membrane is washed once with 2× SSPE solution for 15 min at room temperature, twice with 2× SSPE, 2% SDS for 30 min at 65°, and once with 0.1× SSPE for 15 min at room temperature.

Analysis of Target mRNA and Protein Expression

Estimation of target mRNA production is essential because it gives information about the effect of asRNA on specific gene expression. For example, on induction by IPTG cells transformed with pJDC402 harboring an asRNA gene against *lpp* mRNA, a significant reduction in the amount of target mRNA is observed,[14] indicating that asRNA is able to influence not only translation of the *lpp* mRNA but also its amount in the cell. The same protocol described for the analysis of the expression of asRNA may be used for the evaluation of the level of expression of the target mRNA. In this case, however, probes used should be complementary only to the *lpp* mRNA.

The next step in the analysis of the effect of asRNA is to evaluate production of the target protein. If the function of the target protein is associated with a certain cellular phenotype, one can test this phenotype, which could be antibiotic resistance, substrate preferences, or morphological changes. To measure the production of target proteins directly, several approaches may be used. If proteins are soluble and easily distinguishable in total cellular extracts from the cells, one-dimensional or two-dimensional gel electrophoresis can be performed. Analysis of the production of the major outer membrane lipoprotein and OmpA in cells transformed with asRNAs against their genes is carried out as follows.[14] *Escherichia coli* JA221 carrying an appropriate recombinant plasmid is grown to a Klett-Summerson colorimeter reading of 30 and then IPTG is added to a concentration of 2 mM. After 1 hr of incubation, 25–50 μCi of [^{35}S] methionine (NEN, 3000 Ci/mmol) per 1 ml of the culture is added. After 1 min the

[13] J. Messing, *Methods Enzymol.* **101**, 20 (1983).
[14] J. Coleman, P. Green, and M. Inouye, *Cell* **37**, 429 (1984).

labeling is terminated by the addition of 1 ml of the cold solution, containing 20 mM sodium phosphate, pH 7.1, 1% formaldehyde, and 1 μg/ml methionine. Cells are then separated by centrifugation, washed, and resuspended in the same buffer. After lysing the cells by ultrasonication, the outer membrane fraction is isolated as described previously[14] and suspended in the sample buffer according to Laemmli.[15] Samples are subjected to SDS–polyacrylamide gel electrophoresis. The effect of asRNA is then determined by comparing the intensities of the bands of interest in the presence and absence of asRNA.

Antisense RNA-Based Immune System against Viral Replication in *E. coli*

Introduction

Antisense RNAs are employed naturally in the developmental control of the prokaryotic bacteriophages, such as P1, P4, P7, P22, and λ.[7,8,16,17] Because general principles of the regulation of these virus genes are similar to those used by endogenous asRNAs in *E. coli*, an antisense approach may be applied for the creation of artificial antivirus systems. They may serve as a model for the designing and testing of most efficient targets and RNA structures for the inhibition of virus reproduction in *E. coli*, as well as a tool to study the mechanisms of virus replication. Phage-resistant bacteria thus engineered may be useful for biotechnological purposes. A general strategy for the construction of antiviral asRNA genes is the same as those described previously. The most efficient antiviral activity in *E. coli* may be obtained with asRNA directed against 5'-untranslated specific viral genes encompassing the Shine–Dalgarno sequence and AUG codon.[18] Phage replication usually proceeds through specific successive stages that require coordinate gene expression. Special attention should be paid in choosing the best gene target for the antisense approach. The use of early regulatory genes, essential for virus replication, is ideal because the amounts of these gene products are relatively small. The target size of the asRNA can be as short as 19 bases, which might broaden the virus specificity.[19] Coexpression of more than one asRNA against different virus genes might

[15] U. Laemmli, *Nature* **227,** 680 (1970).
[16] M. Citron and H. Schuster, *Cell* **62,** 591 (1990).
[17] G. Deho, S. Zangrossi, P. Sabbattini, G. Sironi, and D. Ghisotti, *Mol. Microbiol.* **6,** 3415 (1992).
[18] J. Coleman, A. Hirashima, Y. Inokuchi, P. Green, and M. Inouye, *Nature* **315,** 601 (1985).
[19] A. Hirashima, S. Sawaki, Y. Inokuchi, and M. Inouye, *Proc. Natl. Acad. Sci. U.S.A.* **83,** 7726 (1986).

increase antiviral function. If a sequence that is highly conserved among related viruses is chosen as a target, one can create a strain that has immunity to a group of viruses. The next section describes an example for constructing a strain immune to SP phage.

Inhibition of Reproduction of SP Phage

As a vector for the expression of antivirus asRNA genes, pJDC406 is used (Fig. 1B), which is a derivative of pJDC402 (Fig. 1A) having two *Sma*I sites so that the entire asRNA gene can be excised. This allows one to reinsert the excised gene back into the original plasmid to duplicate the asRNA gene or to combine different asRNA genes in one plasmid. Three genes coding coat protein, replicase, and maturation enzyme are used as targets for asRNA as shown in Fig. 2 and Table I.

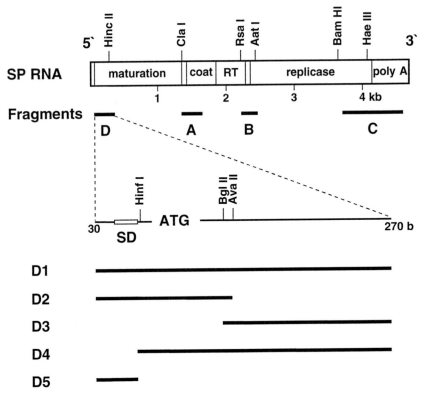

FIG. 2. Genomic structure of SP phage RNA and fragments used to construct asRNA plasmids (A, B, C, D1, D2, D3, D4, and D5). Nucleotide numbers represent those from the 5' end of SP phage RNA.

TABLE I
REGIONS OF SP PHAGE GENOME COVERED BY ANTISENSE SEQUENCES

Plasmid	Target mRNA	Sequences covered	Length of target SP sequence (bp)
A	Coat protein	SD,[a] AUG codon, and 5′-coding region	247
B	Replicase	SD, AUG codon, and 5′-coding region	159
C	3′-terminal region	C-terminal part of replicase 3′-noncoding region and poly(A)tail	518
D1	Maturation protein	SD, AUG codon, and 5′-coding region	240
D2	Maturation protein	SD, AUG codon, and short 5′-coding region	131
D3	Maturation protein	5′-coding region	123
D4	Maturation protein	AUG codon and 5′-coding region	221
D5	Maturation protein	SD and short 5′-noncoding region	19

[a] Shine–Dalgarno sequence.

All SP fragments, listed in Table I, are cloned into the *Xba*I site of pJDC406, which is converted into blunt ends by DNA polymerase I according to Maniatis *et al.*[20] Plasmids are transformed into *E. coli* strain JA221. There are two ways to estimate the effect of asRNA (1) examine the change in the cellular sensitivity of each cell strain to SP phage and (2) estimate the burst size of SP phage for each cell strain. To measure cellular sensitivity to SP phage, *E. coli* clones with asRNA genes are grown at 37° in L broth medium containing 5 mM CaCl$_2$. IPTG is added to a final concentration of 2 mM at a Klett-Summerson reading of 10 to induce asRNA production. At a Klett unit of 50, cells are infected with SP phage (100–200 phages per plate) in 2.5 ml of 0.6% soft agar in L broth medium containing 2 mM IPTG. The number of plaques is determined and the percentage of inhibition may be estimated as the percentage of the total plaque number obtained for control cells carrying only the vector plasmid. The effect of asRNA can be examined by measuring the burst size for each cell strain expressing asRNA, as follows: *E. coli* strains are grown at 37° in L broth medium containing 5 mM CaCl$_2$. To induce asRNA production, IPTG is added as described earlier. SP phage is added at one phage per

[20] T. Maniatis, E. Fritsch, and J. Sambrook, "Molecular Cloning: A Laboratory Manual," Cold Spring Harbor Press, Cold Spring Harbor, NY, 1982.

TABLE II
EFFECT OF asRNA GENES ON PLAQUE FORMATION
OF SP PHAGE

Plasmid[a]	Inhibition (%)
A	69
B	42
C	40
D1	98
D2	98
D3	12
D4	21
D5	94

[a] See Table I.

cell at a Klett unit of about 50. After 5 min, the culture is diluted 10 times in L broth medium, containing anti-SP serum (K-10) and IPTG. Mixtures are incubated at 37° to remove unabsorbed phage particles. At 10, 20, and 30 min after infection, mixtures are diluted to 1 : 10,000. At each point, an aliquot of the cell culture is mixed immediately with the excess of a stationary culture of *E. coli* strain A/λ ($F^+ Su^+ pro^-$), which is used as an indicator. Mixtures are then plated onto L broth agar plates. At 35 min after infection, mixtures are treated with chloroform for cell lysis before an appropriate dilution. The number of plaques on L broth agar plates is measured after a 12- to 16-hr incubation at 37°. To evaluate the specificity of the inhibition one can also examine the effect of the constructs, containing the asRNA coding fragments in a reverse orientation.[21]

Typical results of the inhibition of phage production with asRNA are shown in Table II. Data indicate that the most effective target for the inhibition of SP phage reproduction in *E. coli* is the gene for maturation protein. Strong inhibition was observed when asRNA was complementary to the only 19-base-long target, covering the Shine–Dalgarno sequence. Longer fragments containing regions complementary to AUG and a 5'-coding region exert an even more inhibitory effect on phage reproduction. From these experiments, shorter fragments (D5) possess broad virus specificity, as this construct was able to confer resistance even to Qβ and GA phage.[19] Because similarities at these regions between phage SP and phage Qβ and between phage SP and phage GA are only approximately 50%, the ability of the asRNA expression plasmid with the D5 fragment to inhibit

[21] K. Furuse, A. Hirashima, H. Harigai, A. Ando, K. Watanabe, K. Kurosawa, Y. Inokuchi, and I. Watanabe, *Virology* **97**, 328 (1979).

infection of all three phages may be due to another element within the RNA, which is able to block the replication of phage RNA rather than block the binding of ribosomes to the Shine–Dalgarno regions of the gene for the maturation protein.

Antisense DNA Production by Bacterial Retron

msDNA as a Vector for asDNA Production

Studies of the structure and mechanism of the synthesis of msDNA by bacterial reverse transcriptase lead to the new approach of regulating gene expression, which uses msDNA as a vector for the production of single-stranded DNA. If this asDNA is targeted to the specific mRNA, efficient inhibition of the target gene expression would be expected. This section describes general principles of the construction of asDNA genes and, as an example, the use of this system for the regulation of expression of the *lpp* gene in *E. coli.*

Myxobacteria, some natural isolates of *E. coli,* and a number of other gram-negative bacteria contain a retroelement, called "retron", that is responsible for the production of multicopy single-stranded DNA (msDNA) (for a review, see Inouye and Inouye[22]). msDNA is a DNA–RNA complex consisting of single-stranded DNA, branching out from an internal guanosine residue of an RNA molecule through a 2',5'-phosphodiester linkage[23] (Fig. 3). A retron, usually 1.3–3 kb in length, consists of *msr*, a gene for the RNA region, *msd*, a gene for the DNA region, and the open reading frame for a reverse transcriptase (Fig. 4). The primary transcript contains inverted repeat sequences (a1 and a2). These repeats allow the transcript to form a specific secondary structure. This structure leads to the positioning of the G residue at the end of the stem structure, donating its 2'-OH group for the initiation of the DNA synthesis using the same RNA transcript as a template. The scheme of this pathway is shown in Fig. 4. Structures of *E. coli* retrons are very diverse and share little homology, except for basic features, such as the 2',5'-phosphodiesterase linkage at the internal G residue, a DNA–RNA duplex structure at their 3' ends, and secondary structures in both RNA and DNA strands. Analysis of the requirements for the sysnthesis of msDNA leads to the following conclusions: inverted repeats a1 and a2, the internal G residue for DNA priming, and the secondary structures of RNA corresponding to the *msr* region are absolutely

[22] S. Inouye and M. Inouye, *Virus Genes* **11,** 81 (1996).
[23] B. Lampson, J. Sun, M. Hsu, J. Vallejo-Raminez, S. Inouye, and M. Inouye, *Science* **243,** 1033 (1989).

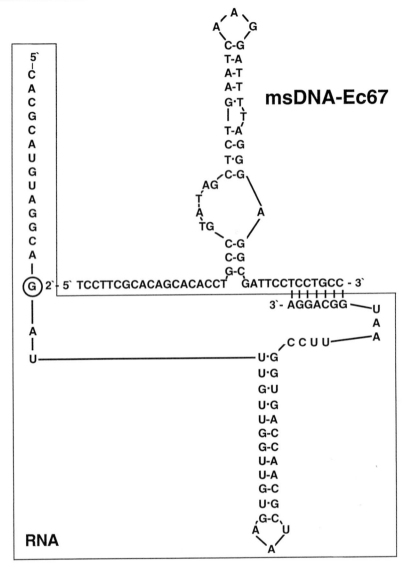

FIG. 3. Proposed secondary structure of msDNA-Ec67.[23] This msDNA consists of a 67-base single-stranded DNA and 58-base RNA molecules. The RNA sequence is boxed. The branching G residue is circled. This mRNA retains all three unique features common to all msDNAs identified so far: a 2′,5′-phosphodiester linkage between the 2′-OH group of an internal guanine residue of the RNA molecule and the 5′ end of the DNA strand, stable secondary structures in both DNA and RNA strands, and a DNA–RNA hybrid at their 3′ ends.

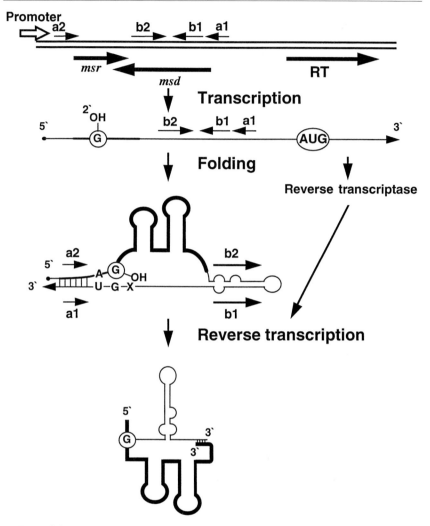

Fig. 4. Schematic diagram of a genetic arrangement of a retron and the pathway of msDNA synthesis. The retron consists of the *msr–msd* region and the gene for RT. Thin arrows with letters indicate the locations of two sets of inverted repeats (a1 and a2; b1 and b2). The primary transcript contains the upstream region of *msr* through the RT gene. When the stable secondary structure is formed from the primary transcript, the branched G residue is placed at the end of the a1 and a2 stem structure. This G residue serves as a primer for DNA synthesis by RT using the same RNA transcript as a template. As the DNA chain is elongated, the template RNA is removed by RNase H. The final product is the branched msDNA.

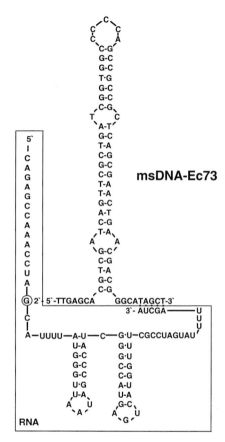

FIG. 5. Structures of the msDNA-Ec73 and its derivatives with antisense sequences used for the regulation of *lpp* expression. (A) msDNA-Ec73 was isolated from a clinical strain of *E. coli* Cl-23. (B and C) msDNA anti-lppN25 and anti-lppN34 containing anti-*lpp* sequences in the loop structure. The anti-*lpp* sequences are underlined. Boxes enclose msdRNA, and the branching G residues are circled. Adapted from Mao *et al.*[10]

necessary for msDNA synthesis by the individual bacterial reverse transcriptases.[24] The upper portions of the stem–loop structure of the *msd* DNA can be deleted from the retron without influence on the production of msDNA.[25] The upper region of msDNA can be replaced with a short asDNA sequence, which is expressed in *E. coli* as part of msDNA. This msDNA vector system has several unique features: it contains its own very

[24] T. Shimamoto, M. Hsu, S. Inouye, and M. Inouye, *J. Biol. Chem.* **268,** 2684 (1993).
[25] M. Shimada, S. Inouye, and M. Inouye, *J. Biol. Chem.* **269,** 14553 (1994).

B

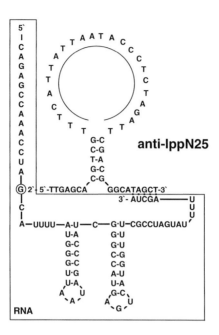

anti-IppN25

RNA

C

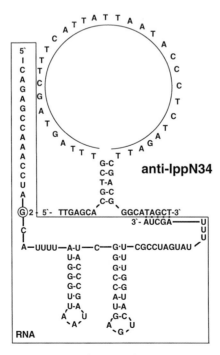

anti-IppN34

RNA

Fig. 5. (*continued*)

efficient replicating mechanism, which can provide up to several thousand copies of asDNA per cell. Because the synthesis of msDNA constitutively occurs inside the cells, it has an advantage over synthetic exogenous oligonucleotides, which should be added continuously to the medium in order to achieve a permanent effect. When asDNA hybridizes with the target mRNA, the resulting DNA/RNA hybrid serves as a substrate for RNase H to increase the asDNA effect.

Construction of Recombinant msDNA

The following procedure is used for obtaining asDNA genes targeted to *lpp* using retron Ec73.[26] PIN-III (lpp^{p-5}) vectors containing anti-*lpp* sequences were constructed as follows. First, to introduce an *Nco*I site in the *msd* region and delete 56 bases in the upper portion of msDNA-Ec73 (Fig. 5A), a two-step polymerase chain reaction (PCR) was performed using a pT7Ec73 *msr-msd*[27] as a template. In the first PCR, two sets of primers were used: 5′ AATCTAGACAGAGCCAAACCTAG 3′ (oligonucleotide 5641) corresponding to bases 10411–10425 and 5′ TACTTGAG-CACCATGGGGCATAGCTAA 3′ (oligonucleotide 5790; the underlined sequence is the *Nco*I site) complementary to bases 10479–10489 and 10546–10556 and containing six bases for the *Nco*I site in the middle and 5′ TCTCTAGATCCTTATGCACCTTGA 3′ (oligonucleotide 5640) complementary to bases 10672–10689 and oligonucleotide 5791, which is the complementary sequence of oligonucleotide 5790. The second PCR was performed using the amplified fragments and oligonucleotides 5641 and 5640 as primers, which created an *Xba*I site at their 5′ and 3′ ends. The amplified fragments were cloned into the *Xba*I site of PIN-III (lpp^{p-5})AI[28] of which the *Eco*RI site was eliminated. Double-stranded nucleotides consisting of anti-*lpp* sequences with *Nco*I sites at the ends were synthesized and cloned into the *Nco*I site of PIN-III (lpp^{p-5})msDNA/*Nco*I, resulting in PIN-III (lpp^{p-5})N25 for msDNA-anti-lppN25 (Fig. 5B) and PIN-III (lpp^{p-5})N34 for msDNA-anti-lppN34 (Fig. 5C), respectively. In these msDNAs, the loop sequences are complementary to the translation initiation region of the mRNA for the major outer membrane lipoprotein, including the Shine–Dalgarno sequence and AUG codon. After confirming the DNA sequence of the constructs, the 0.95-kb *Bam*HI fragment carrying msDNA Ec73 reverse transcriptase (RT-Ec73) from pUC7Xbai73RT[27] was inserted at the *Bam*HI site of these plasmids, and the orientation of RT-Ec73 was determined by restriction digests.

[26] J. Sun, M. Inouye, and S. Inouye, *J. Bacteriol.* **173**, 4171 (1991).
[27] T. Shimamoto, M. Inouye, and S. Inouye, *J. Biol. Chem.* **270**, 581 (1995).
[28] S. Inouye and M. Inouye, *Nucleic Acids Res.* **13**, 3101 (1985).

Analysis of Recombinant msDNA Expression and Its Effect on lpp Production

To isolate recombinant msDNA, *E. coli* strain JA221 carrying plasmids described earlier is grown in M9 medium, and at a Klett unit of 30, msDNA production is induced by the addition of 1 m*M* IPTG. At a Klett unit of 150, cells are collected and total RNA is prepared as described in the first section. RNA samples are treated with RNase A (25 μg/ml) for 10 min at 37° and subjected to 8% polyacrylamide gel electrophoresis together with DNA molecular weight markers. The gel is stained with ethidium bromide and observed under UV light. The RNase-resistant DNA bands are thus observed at expected sizes as shown in Fig. 6. In order to confirm production of msDNA, a reverse transcriptase (RT) extension assay may be used.[28] This method allows detection of msDNA by extension of the DNA strand

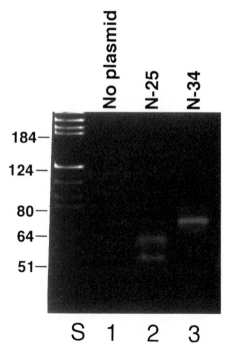

FIG. 6. Production of msDNAs with anti-*lpp* sequences. msDNAs were isolated from cultures of *E. coli* strain JA221 without a plasmid (lanes 1 and 2) and cells harboring pIN-III (lpp^{p-5})N25 (lines 3 and 4) and pIN-III (lpp^{p-5})N34 (lines 5 and 6). Cells growing in M9 medium were induced with 1 m*M* IPTG. Numbers on the left indicate the number of bases of MW markers.[10]

with RT using the RNA portion of msDNA as a template. The DNA extension terminates at the branching guanosine residue. If msDNA is treated by RNase prior to electrophoresis, the shorter band is expected to be seen. The total RNA fraction from 2.5 ml of culture is incubated in 25 μl of reaction mixture, containing 56 mM Tris–HCl (pH 8.3), 75 mM KCl, 10 mM DTT, 3 mM MgCl$_2$, 0.2 mM each dGTP, dATP, and dTTP, 0.2 μM [α-^{32}P] dCTP (NEN, 3000 Ci/mmol), and 100 units of Moloney murine leukemia virus (M-MuLV) reverse transcriptase (Boehringer Mannheim Co.). The mixture is incubated at 37° for 1 hr and then chased with 2 μl of a mixture of dATP, dCTP, dTTP, and dGTP (15 mM each) for 15 min. Samples are then phenol extracted and the products are precipitated with

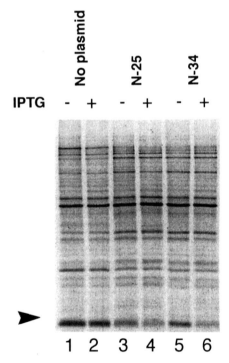

FIG. 7. Inhibition of lipoprotein production by antisense DNA. One milliliter of the same cultures used in Fig. 6 was labeled with 5 μCi of Tran[35] S-label (Amersham) for 10 min at a Klett unit of 150. Membrane fractions were isolated and analyzed by 17% SDS–polyacrylamide gel electrophoresis. JA221 strain alone (lines 1 and 2), JA221 strain harboring pIN-III (lpp[p-5])N25 (lines 3 and 4), and pIN-III (lpp[p-5])N34 (lines 5 and 6). The position of the lipoprotein is indicated with an arrow. Quantitation of lipoprotein using an imaging densitometer Model GS-670 (Bio-Rad) indicated significant inhibition of protein production in the presence of IPTG in both constructs (75% for N25 and 70% for N34).[10]

ethanol. An aliquot of the sample is treated with RNase A (25 μg/ml) and subjected to 6% polyacrylamide gel electrophoresis in the presence of 8% urea. After electrophoresis the gel is dried and exposed to X-ray film.

To detect the effect of antisense msdNA production on *lpp* expression, the outer membrane fraction labeled with [^{35}S]methionine for 10 min is prepared as described previously[14] and analyzed by SDS–gel electrophoresis. Lipoprotein production was inhibited significantly with PIN-III(lpp^{p-5})N25 and PIN-III(lpp^{p-5})N34 after the induction of msdNA by IPTG as shown in Fig. 7.

It has been demonstrated that the *msr* transcript, even if produced independently from the *msd* region, is able to anneal to the *msd* transcript and to initiate cDNA synthesis from the 2'-OH group of the internal branching guanosine residue in the *msr* transcript.[29] If the *msr* RNA designed for a specific bacterial transcriptase containing a complementary sequence to a specific mRNA at the 3' end sequence of the *msr* RNA, cDNA against mRNA is produced, which may stimulate the hydrolysis of RNA by RNase H, thus functioning as an effective asdNA.

Acknowledgments

We thank Dr. Sangita Phadtare for critical reading of the article. This work was supported in part by Grant R01 GM 44012-10 from the National Institutes of Health.

[29] T. Shimamoto, T. Kawanishi, T. Tsuchiya, S. Inouye, and M. Inouye, *J. Bacteriol.* **180,** 2999 (1998).

[29] Utilization of Properties of Natural Catalytic RNA to Design and Synthesize Functional Ribozymes

By Leonidas A. Phylactou, Charlotte Darrah, Louise Everatt, Despina Maniotis, and Michael W. Kilpatrick

The unexpected discovery that the RNA molecule has catalytic properties[1,2] has led to a plethora of interest in the identification and utilization of the variety of catalytic RNA molecules, or ribozymes, that occur in nature. Naturally occurring catalytic RNA motifs can be classified broadly

[1] K. Kruger, P. J. Grabowski, A. J. Zaug, J. Sands, D. E. Gottschling, and T. R. Cech, *Cell* **31,** 147 (1982).
[2] C. Guerrier-Takada, K. Gardiner, T. Marsh, N. Pace, and S. Altman, *Cell* **35,** 849 (1983).

into two groups based on their catalytic activity. The hammerhead, hairpin, and hepatitis delta virus ribozymes can be characterized by their ability to self-cleave a target phospodiester bond, as opposed to the group I and group II intron ribozymes, which can be characterized by their ability to self-splice by a two-step cleavage and ligation mechanism. The former group of small ribozyme motifs is found typically in viral or viroid RNAs, whereas the latter, larger ribozymes are found in bacteria and lower eukaryotes. The detailed analysis of the structure and properties of these ribozymes has identified their potential as molecular tools. In particular, the hammerhead ribozyme, by virtue of its small size and ability to be designed to cleave virtually any target RNA, has been widely touted for the therapy of both genetic and infectious diseases.[3] More recently, the group I intron ribozyme has similarly been proposed as a therapeutic tool by virtue of its potential to repair a defective target RNA.[4] This article outlines the methods by which the hammerhead and group I intron ribozyme motifs can be utilized to design and construct synthetic ribozymes as potential molecular tools.

Ribozyme Design

Hammerhead Ribozyme

The hammerhead ribozyme is a small catalytic RNA molecule whose catalytic activity resides in a core of less than 40 ribonucleotides. This core consists of three base-paired helices connected by two single-stranded regions (Fig. 1). In general, the single-stranded regions that contain the catalytic domain are largely invariant, unlike the stems, which do not contain conserved nucleotides. This lack of conserved sequence allows stems I and III to be designed to bind to any desired target sequence. Thus potential hammerhead ribozymes possess an invariant catalytic domain and flanking sequence (stems I and III) complementary to a target mRNA molecule. Stems I and III will surround the selected cleavage site in the target molecule. The ribozyme is thus designed to cleave *in trans* at the selected target sequence within that mRNA. Although the most commonly found cleavage site in nature is the GUC triplet, mutagenesis studies have revealed that cleavage triplets of the type NUX are tolerated, where N is any nucleotide and X is any nucleotide except G. In-depth analysis of the hammerhead ribozyme cleavage reaction has revealed a hierarchy of preferred cleavage sites, which depend on the relative concentrations of ribozyme and sub-

[3] K. R. Birikh, P. A. Heaton, and F. Eckstein., *Eur. J. Biochem.* **245**, 1, (1997).
[4] L. A. Phylactou, C. Darrah, and M. J. A. Wood., *Nature Genet.* **18**, 378 (1998).

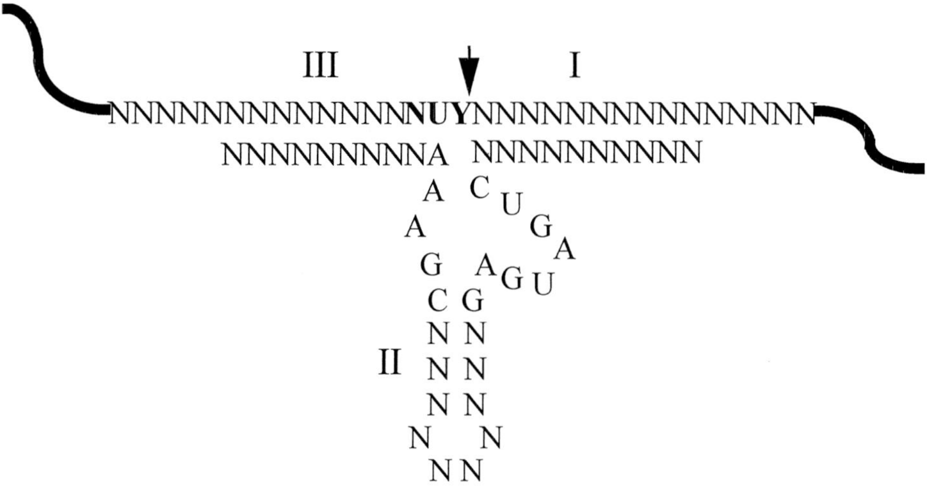

FIG. 1. A *trans*-acting hammerhead ribozyme. The hammerhead ribozyme (bottom strand) is shown in a typical three-stem structure (I, II, and III) bound to its target RNA (top strand). The cleavage site NUY is shown in bold and the actual position of cleavage is denoted by a vertical arrow.

strate. Stems I and III of the ribozyme need to provide sufficient stability of the ribozyme : substrate complex. They need to ensure an adequate association rate of ribozyme and target and that the ribozyme is not displaced from its target before cleavage has occurred. It is important to select ribozyme cleavage sites in the target mRNA that are likely to be accessible to the ribozyme. The very different activities of ribozymes targeted to different sites on an mRNA are at least partly because RNA folds readily into complex secondary structures that can interfere with binding of a ribozyme. It is difficult to select efficient ribozyme cleavage sites on long RNA molecules. Generally, cleavage-susceptible sequences are determined either by trial and error or by predictions of the secondary structure of the target. Programs such as MFold can help to predict the RNA secondary structure and identify potentially accessible cleavage sites.

Group I Intron Ribozyme

The *Tetrahymena* group I intron ribozyme is a larger ribozyme whose catalytic activity resides in a sequence of some 387 nucleotides. As with the hammerhead ribozyme, group I intron ribozyme can be engineered to perform its two-step excision–ligation reaction *in trans*. That is, rather than its self-splicing reaction resulting in the excision of the intron and subsequent ligation of the two flanking exons, a ribozyme can be designed to *trans* splice an exon joined to its 3' end onto a separate 5' exon (Fig. 2). For *trans* splicing to occur, the group I intron ribozyme must possess the 387 nucleotides necessary for catalysis flanked by a target RNA-binding sequence at its 5' end, called the internal guide sequence (IGS), and the 3' exon sequence of the target RNA, containing the sequence necessary to repair the defective target, on its 3' end. The IGS, which will promote binding of the ribozyme to its target via conventional base pairing, allows a great deal of flexibility in target selection, with the only absolute requirement being a single uridine immediately preceding the cleavage site.

In Vitro Synthesis of Ribozymes

Following design, ribozymes can be synthesized to be used in cell culture and animal experimentation. Prior to that, however, ribozymes can be tested in a cell-free environment in the presence of the target RNA. This is an important step in the identification of functional ribozymes as, in a relatively short period of time, the catalytic efficiency of a particular ribozyme can be defined carefully. Ribozymes can generally be synthesized in one of two different ways. They can be directly synthesized chemically or, more commonly, they can be synthesized by *in vitro* transcription off a template containing the ribozyme sequence.

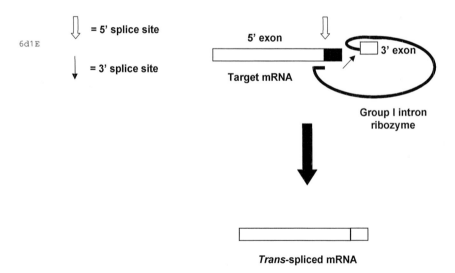

Fig. 2. A group I intron ribozyme designed to repair a defective target RNA. Repair is achieved via a two-step cleavage reaction resulting in the excision of the mutant region (black rectangle) from the target RNA and the ligation of the wild-type sequence (white rectangle).

Ribozyme Synthesis by in Vitro Transcription

In order to synthesize the ribozyme by *in vitro* transcription, it is first necessary to prepare a template containing the desired ribozyme sequence.

Template Production by Cloning. Probably the most popular way of producing ribozymes involves the cloning of the ribozyme sequence, in the form of oligonucleotides, into promoter-containing vectors. Following that, ribozymes can be synthesized by *in vitro* transcription from the linearized constructs. If ribozyme production is desired inside eukaryotic cells, ribozymes can be cloned into the appropriate eukaryotic vectors, which can then be used to transfect target cells (Fig. 3). Because of its large size, the conserved catalytic sequence of the group I intron ribozyme can be transferred from the original plasmid.[5] Both the IGS and the 3' exon can be replaced with sequences specific for the target RNA of interest (Fig. 3).

1. The hammerhead ribozyme sequence can be cloned into vectors, as double-stranded oligonucleotides synthesized on a DNA synthesizer. Because inactive versions of hammerhead ribozymes are used as controls in most cases (e.g., bearing a mutation in the ribozyme catalytic domain), a degenerate base can be incorporated in one of the oligonucleotide strands to enable the identification of both active and inactive clones in a single ligation reaction. Moreover, restriction sites can be added at the 5' end of the oligonucleotides to facilitate easier cloning into the vectors of choice.

Equimolar amounts of oligonucleotides in 10 mM Tris–HCl, 10 mM MgCl$_2$, 10 mM NaCl, and 50 mM dithiothreitol (DTT) at pH 7.9 are incubated at 95° for 5 min, at 65° for 10 min, and cooled gradually to room temperature. Following ethanol precipitation, the annealed oligonucleotides are subjected to restriction digestion to create compatible ends for cloning. Similarly, the vector should be prepared for cloning by digesting it with the appropriate restriction endonucleases. Following incubation, restriction enzymes are removed by phenol/chloroform extraction. A 3–5 M excess of the ribozyme oligonucleotide (100 pmol) over the linearized vector is used for ligation of the annealed ribozyme oligonucleotide into the linearized vector. Positive clones are then identified by dideoxy sequencing. Hammerhead ribozymes are synthesized by *in vitro* transcription after linearization of the ribozyme constructs. Thirty-three nanomoles of template is incubated with all four ribonucleotide triphosphates (ATP, CTP, UTP, and GTP) at a final concentration of 5 mM in the presence of 15 mM MgCl$_2$, 2 mM spermidine, 50 mM Tris–HCl (pH 7.5), 5 mM DTT, 25 units of ribonuclease inhibitor, and 20 units of RNA polymerase in a total volume of 50 μl. All reagents should be added at room temperature to avoid

[5] B. A. Sullenger and T. R. Cech., *Nature* **371**, 619 (1994).

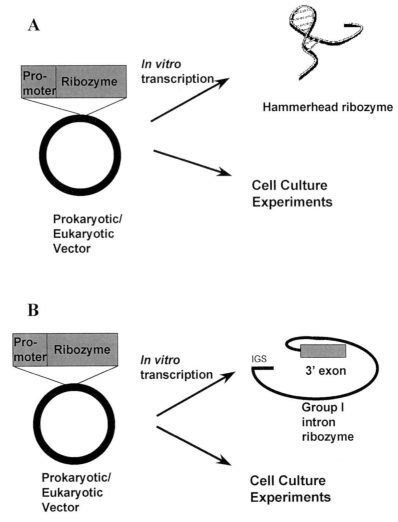

FIG. 3. *In vitro* and *in vivo* synthesis of ribozymes from vectors. Hammerhead (A) and group I intron ribozymes (B) can be cloned into vectors containing prokaryotic promoters and synthesized by *in vitro* transcription. Alternatively, if ribozyme synthesis is desired in cells, ribozyme cloning is carried out into eukaryotic vectors.

precipitation of DNA. The reaction is incubated at 37° for 2–4 hr, stopped with the addition of 4 units of RNase-free DNase I, and further incubated for 15 min at 37°. All enzymes are then removed by phenol/chloroform extraction and the newly synthesized transcript is recovered by ethanol precipitation. The quality and amount of ribozymes can be checked either by reading their absorbance at 260 nm or by denaturing (7 M urea) gel electrophoresis alongside markers of known concentration.

2. Group I intron ribozymes can be constructed by modifying the plasmid containing the catalytic sequence.[5] Multiple cloning sites are present around the IGS and the 3' exons. Thus, oligonucleotides can be synthesized, containing an IGS that will bind to the target RNA. The two complementary oligonucleotides are mixed in 10 mM Tris–HCl, 10 mM MgCl$_2$, 10 mM NaCl, and 50 mM DTT at pH 7.9 and incubated at 95° for 5 min, annealed for 10 min at 65°, and cooled gradually to room temperature. Following ethanol precipitation, the annealed oligonucleotides are subjected to restriction digestion to create compatible ends for cloning. The vector should similarly be prepared for cloning by digestion with the appropriate restriction endonucleases. Following incubation, restriction enzymes are removed by phenol/chloroform extraction. A 3–5 M excess of the ribozyme oligonucleotide (100 pmol) over the linearized vector is used for ligation of the annealed oligonucleotide to the vector. Positive clones are then identified by dideoxy sequencing. Similarly, the 3' exon is cloned in the form of double-stranded DNA. Total RNA is extracted from the cells/tissue expressing the target gene. The RNA is then reverse transcribed with either an oligo(dT) primer or a primer specific for the RNA of interest. Following that, the cDNA is subjected to polymerase chain reaction (PCR) amplification using upstream and downstream primers that will amplify the part of the target cDNA comprising the 3' exon. Restriction sites can be added at the primers' 5' termini to facilitate efficient cloning into the original plasmid containing the ribozyme sequence. A sample of the amplified product is checked by agarose or polyacrylamide gel electrophoresis, and the remaining PCR product is then incubated with restriction enzymes.

Direct Template Production. To avoid the time and expense of cloning the ribozyme sequence adjacent to a prokaryotic promoter, an alternative approach is to produce the hammerhead ribozyme directly off a synthetic DNA template.[6] The template can be either completely double stranded or partially single stranded (Fig. 4). In each case, equimolar amounts of the two complementary strands are annealed by incubating at 94° for 4

[6] L. A. Phylactou, P. Tsipouras, and M. W. Kilpatrick, *Biochem. Biophys. Res. Commun.* **249**, 804 (1998).

Oligo P: 5' AATTTAATACGACTCACTATAG 3'

Oligo R: 3' TTAAATTATGCTGAGTGATATCCCTAGAGGTCGTGACTACTCAGGCACTCCTGCTTTGCGGGAGCT   5'

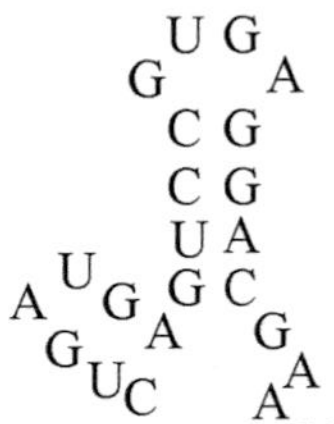

```
                    U  G
                  G     A
                  C  G
                  C  G
                  U  A
                A  U
              G  A  G  C
              G  U  A  G
                 U  C      A  A
Ribozyme Transcript:    5'  GGGAUCUCCAGCA    ACGCCCUCGA   3'
```

FIG. 4. Direct template production for the synthesis of ribozyme by in vitro transcription. A partially single-stranded template for ribozyme production can be prepared by annealing a synthetic oligonucleotide complementary only to the T7 promoter region (Oligo P) to a longer oligonucleotide that also contains the ribozyme sequence (Oligo R). In vitro transcription off such a template using T7 RNA polymerase will produce the desired ribozyme as transcript.

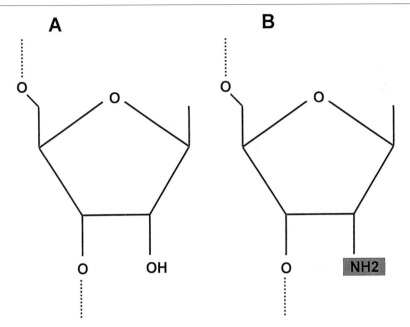

FIG. 5. Chemical synthesis and modification of hammerhead ribozymes. Stable hammerhead ribozymes can be synthesized by a nucleic acid synthesizer by introducing modifications in their 2'-hydroxyl group (A), such as an amino group (B).

min and then cooling slowly to room temperature. The annealed template is then used for *in vitro* ribozyme transcription as described earlier.

Ribozyme Chemical Synthesis

Hammerhead ribozymes can also be produced directly on a nucleic acid synthesizer.[7] An advantage of using this method is the opportunity to introduce modifications to some of the chemical groups on the hammerhead ribozymes, thus protecting them from nuclease attack inside cells. The most important modification is that of the 2'-hydroxyl group of the ribozyme nucleotides, which can be, for example, alkylated or substituted with a fluorine atom or amino group (Fig. 5). The ribozyme can also be protected from exonucleases by changing the 3'- and 5'-terminal phosphates to phosphorothioate internucleotide linkages. The disadvantage of this method of ribozyme synthesis is the high cost needed to produce enough ribozyme for cell culture and animal experimentation. Similarly, it is very difficult to

[7] F. Wincott, A. DiRenzo, C. Shaffer, S. Grimm, D. Tracz, C. Workman, D. Sweedler, C. Gonzalez, S. Scaringe, and N. Usman, *Nucleic Acids Res.* **23**, 2677 (1995).

produce full-length group I intron ribozymes by this approach because of their size.

In Vitro Testing of Ribozymes

In vitro-synthesized ribozymes can then tested for their ability to cleave the target RNA prior to cell culture experiments. This can be done by incubating the ribozyme either with a smaller (minitarget) version of the target mRNA or with total RNA extracts from target-expressing cells.

Testing against Small Target RNA

Target Synthesis by Cloning. A smaller (minitarget) version of the full-length target RNA can be made by *in vitro* transcription of a cloned RT/PCR fragment. Total RNA is extracted from the cells/tissue expressing the target gene. It is then reverse transcribed with either an oligo(dT) primer or a primer specific for the RNA of interest. Following that, the cDNA is subjected to PCR amplification. The upstream and downstream primers are designed to amplify part of the target cDNA, containing the ribozyme cleavage site (Fig. 6). Both primers can have restriction sites at their 5′ ends for more efficient cloning into the vector of choice. A sample from the amplified product is then checked by agarose or polyacrylamide gel electrophoresis. The remainder of the PCR product is then incubated with restriction enzymes to be prepared for cloning. Following successful cloning of the PCR product into the vector of choice, the target construct is linearized and RNA targets are then synthesized by *in vitro* transcription. Our standard protocol uses $[\alpha\text{-}^{32}P]UTP$ to incorporated labeled nucleotide. The method used is similar to that used for ribozyme synthesis. Thirty-three nanomoles of linearized template is incubated with ATP, CTP, and GTP at a final concentration of 0.5 mM, 50 μCi of $[\alpha\text{-}^{32}P]UTP$ (800 Ci/mmol), 15 mM MgCl$_2$, 2 mM spermidine, 50 mM Tris–HCl (pH 7.5), 5 mM DTT, 12.5 units of ribonuclease inhibitor, and 10 units of RNA polymerase in a total volume of 20 μl. All reagents should be added at room temperature to avoid precipitation of DNA. The reaction is incubated at 37° for 60 min, halted with the addition of 4 units of RNase-free DNase I, and incubated further for 15 min at 37°. All enzymes are then removed by phenol/chloroform extraction, and the newly synthesized transcript is recovered by ethanol precipitation in the presence of ammonium acetate. The specific activity of the labeled target is then determined by TCA precipitation followed by liquid scintillation counting or by using a β counter. We usually find that the radioactive ribonucleotide is not limiting during transcription when an

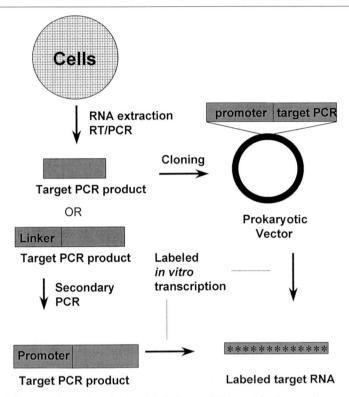

Fɪɢ. 6. Construction and synthesis of labeled target RNA used for *in vitro* ribozyme testing. The template for *in vitro* transcription can be in the form of either a linearized plasmid or a PCR product. The target can be cloned into the vector of choice in the form of a PCR product from total RNA extracted from target-expressing cells. Alternatively, a T7 promoter could be added to the PCR product by a second PCR amplification, thus creating, a template for target synthesis by *in vitro* transcription.

RNA target of less than 400 bases is used. For the synthesis of longer labeled target RNAs, it may be necessary to supplement the reaction with unlabeled UPT in order to synthesize full-length transcripts.

Target Synthesis by Direct Template Production. An alternative way to synthesize the labeled RNA target utilized in our laboratory uses a PCR product as the template for *in vitro* transcription. The PCR product used for transcription is the result of two rounds of amplification reactions (Fig. 6). During the first round, the target cDNA, containing the ribozyme binding site, is amplified; in the second round of amplification, a T7 promoter is added to the 5′ end of the PCR product. Target RNA is then synthesized by *in vitro* transcription using T7 RNA polymerase.[6]

Total RNA is extracted from cells or tissue, expressing the target gene, as described previously. The RNA is then reverse transcribed with either an oligo(dT) primer or a primer specific for the RNA of interest. Following that, the cDNA is subjected to PCR amplification (Fig. 6) using upstream and downstream primers designed to amplify the part of the target cDNA that contains the ribozyme cleavage site. The upstream primer also contains a linker sequence at its 5' end (5'-CTCACTATAGCC), which is used as a recognition signal for the T7 promoter universal upstream primer used in the second PCR amplification. After confirming the efficient synthesis, the first-round PCR product (5 ng) is then used as a template for the second round of amplification. The same downstream primer is used, but as an upstream primer, a universal primer composed of the T7 promoter (5'-AATTTAATACGACTCACTATAG-3') is used. The PCR product is then checked by polyacrylamide gel electrophoresis. Our conditions for both PCR reactions are standard, with an appropriate annealing temperature that depends on the GC content of the bases complementary to the target cDNA sequence.

Synthesis of the target RNA is then carried out by labeled *in vitro* transcription using 33 nmol of template, ATP, CTP, and GTP at a final concentration of 0.5 mM, 50 μCi of [α-^{32}P]UTP (800 Ci/mmol), 15 mM MgCl$_2$, 2 mM spermidine, 50 mM Tris–HCl (pH 7.5), 5 mM DTT, 12.5 units of ribonuclease inhibitor, and 10 units of RNA polymerase in a total volume of 20 μl. All reagents should be added at room temperature to avoid precipitation of DNA. The reaction is incubated at 37° for 60 min, terminated by the addition of 4 units of RNase-free DNase I, and incubated further for 15 min at 37°. All enzymes are then removed by phenol/chloroform extraction, and the newly synthesized transcript is recovered by ethanol precipitation in the presence of ammonium acetate. The specific activity of the labeled target is then determined by TCA precipitation followed by liquid scintillation counting or by using a β counter.

Functional Assay: Hammerhead Ribozymes

Hammerhead ribozymes are incubated with their labeled targets in the presence of magnesium ions at 37° or 50° and the cleavage products are resolved by denaturing gel electrophoresis. Additional care should be taken to prepare ribonuclease-free solutions as contamination will result in degradation of both target and ribozyme. Hammerhead ribozymes can be incubated with the labeled target RNA at different molar ratios in the presence of 50 mM Tris–HCl, pH 7.5, 20 mM MgCl$_2$ in a total volume of 20 μl.[8]

[8] M. W. Kilpatrick, L. A. Phylactou, M. Godfrey, G. Y. Wu, C. H. Wu, and P. Tsipouras, *Hum. Mol. Genet.* **5**, 1939 (1996).

The samples are incubated at either 37° (physiological temperature) or 50°, with cleavage being more efficient at the higher temperature. Incubation times can be adjusted based on ribozyme cleavage efficiency. Following incubation of the ribozyme with its RNA target, the reaction is halted by the addition of 20 mM EDTA, and the labeled cleavage products are separated from the target RNA by denaturing (7 M urea) polyacrylamide gel electrophoresis and are detected by autoradiography (Fig. 7).

Functional Assay: Group I Intron Ribozymes

Group I intron ribozymes are incubated in the presence of magnesium ions and guanosine triphosphate at 37° and *trans*-splicing products are detected by denaturing gel electrophoresis. Following its synthesis, the group I intron ribozyme is preheated in 50 mM HEPES (pH 7), 150 mM NaCl, and 5 mM MgCl$_2$ at 95° for 2 min. In parallel, the labeled target RNA is preheated at 95° for 2 min in the presence of 100 μM GTP. The target is then added to the ribozyme and incubation is continued at 37° in a total volume of 20 μl.[4] Aliquots are removed at a series of time points and reactions are stopped with 20 mM EDTA. Target RNA, cleavage, and

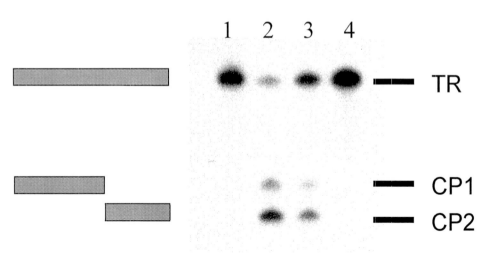

FIG. 7. Example of hammerhead ribozyme-mediated cleavage of a target RNA. Labeled target RNA has been incubated with a hammerhead ribozyme at different time intervals (lanes 2 and 3) or with its catalytically inactive version (lane 4) at 37°. Lane 1 shows incubation of the labeled target in the absence of ribozyme. Labeled target (TR) and cleavage products (CP1 and CP2) have been detected by denaturing polyacrylamide gel electrophoresis.

trans-splicing products are then resolved on a 4% denaturing (7 *M* urea) gel and detected by autoradiography (Fig. 8).

Testing against Total RNA

Alternatively, hammerhead or group I intron ribozymes can be incubated with total RNA extracted from cells or tissue expressing the target gene. The ribozyme of interest is usually incubated with 2 μg of total RNA under the same conditions described earlier: (i) 50 m*M* Tris–HCl, pH 7.5, and 20 m*M* MgCl$_2$ in a total volume of 20 μl for hammerhead ribozymes and (ii) 50 m*M* HEPES, pH 7, 150 m*M* NaCl, 5 m*M* MgCl$_2$, and 100 μ*M* GTP for group I intron ribozymes. The samples are incubated at either 37° or 50° for 1–2 hr. Because the target RNA is not labeled in this case, the effect of the ribozyme can be determined by RT/PCR or RNase protection.

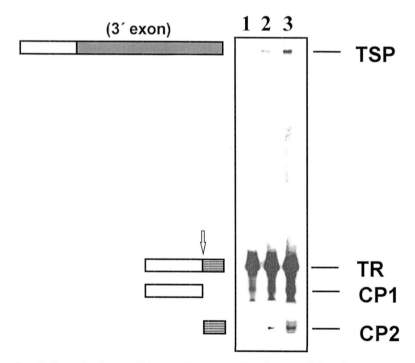

FIG. 8. Example of group I intron ribozyme-mediated *trans* splicing of a target RNA. Labeled target RNA has been incubated with a group I intron ribozyme at different time intervals (lanes 2 and 3) or in the absence of a ribozyme (lane 1) at 37°. Target RNA (TR), *trans*-splicing (TSP), and cleavage (CP) products have been detected by denaturing polyacrylamide gel electrophoresis. The arrow indicates the 5′ splice site.

Detection by RT/PCR. Following incubation with the ribozyme, the total RNA is subjected to cDNA synthesis with either a target-specific downstream primer or an oligo(dT) primer. The cDNA is then used as a template for PCR amplification using target cDNA-specific primers. In the case of hammerhead ribozymes, cDNA-specific primers flanking the ribozyme cleavage site ribozyme are used. However, in the case of group I intron ribozymes, the upstream primer is designed to be specific for the target RNA, whereas the downstream primer should be designed to be specific for the 3' exon that is to be *trans* spliced next to the cleaved target RNA. Detection of *trans*-spliced products is carried out by PCR. In order to determine the effect of the hammerhead or group I intron ribozyme, the PCR amplification should be quantitative, i.e., the amount of the PCR product should reflect the amount of RNA target present.[8] Moreover, coamplification and detection of a housekeeping gene product, such as β-actin are useful for the quantitation of target RNA. The housekeeping cDNA can be synthesized at the same time as the target cDNA. The PCR products are run on a polyacrylamide gel, detected by autoradiography, and quantitated using a phosphorimager.

Detection by RNase Protection. An alternative way to detect and quantitate the amount of target RNA remaining following ribozyme treatment is by using RNase protection. Following ribozyme treatment, a radioactive probe complementary to the target RNA is hybridized with the RNA extracts, and the unhybridized, and therefore unprotected, single-stranded RNAs are then subjected to RNase degradation. The probe sequence can be cloned as cDNA in a vector with a prokaryotic promoter such as T7 or SP6 and synthesized by *in vitro* transcription in the presence of $[\alpha\text{-}^{32}P]UTP$ radiolabeled ribonucleotide. Thirty-three nanomoles of template is mixed with ATP, CTP, and GTP ribonucleotide triphosphates at a final concentration of 0.5 mM, 50 μCi of $[\alpha\text{-}^{32}P]UTP$ (800 Ci/mmol), 15 mM MgCl$_2$, 2 mM spermidine, 50 mM Tris–HCl (pH 7.5), 5 mM DTT, 12.5 units of ribonuclease inhibitor, and 10 units of RNA polymerase in a total volume of 20 μl. All reagents should be added at room temperature to avoid precipitation of DNA. The reaction is incubated at 37° for 60 min, stopped with the addition of 4 units of RNase-free DNase I, and incubated further for 15 min at 37°. All enzymes are then removed by phenol/chloroform extraction, and the newly synthesized transcript is recovered by ethanol precipitation in the presence of ammonium acetate. The specific activity of the labeled target is then determined by TCA precipitation followed by liquid scintillation counting or by using a β counter. During hybridization of the riboprobe with the RNA extracts, the probe must be present in excess over the target RNA in order to achieve quantitative results. After precipitation of the total RNA incubated with the ribozyme, the pellet is redissolved in 20 μl of hybridization solution (80% deionized formamide, 100 mM sodium citrate, pH 6.4, 300 mM sodium acetate, pH 6.4, and 1 mM EDTA). The riboprobe is then added to the solution. The total RNA and riboprobe are denatured for 3 min at 95° and left to anneal between 2 hr and overnight. Any single RNA strands are then digested with RNase A and RNase T1. The reaction is then stopped with EDTA, precipitated with ethanol, and recovered by centrifugation. The protected fragments are then denatured by heating at 95°, separated by denaturing (7 M urea) electrophoresis, detected by autoradiography, and quantitated using a phosphorimager.

Ribozyme Delivery

One of the major current problems in the area of gene therapy is the difficulty involved in delivering and expressing therapeutic genes. The application of ribozyme technology is faced with the same problems. The effective use of ribozymes as therapeutic agents is dependent on efficient delivery of the ribozyme to its target. A number of methods have been and are being developed for the delivery of nucleic acids into cells. Hammerhead ribozymes can be delivered as presynthesized RNA (exogenous delivery) or as DNA constructs encoding the ribozyme sequence for expression within the cell (endogenous delivery). Exogenous ribozyme delivery is essentially a transient approach to inhibition of gene expression due to the short half-life of small RNAs. Endogenous delivery can provide transient or stable expression of the ribozyme, with the latter offering the potential of long-term expression and effect. In addition, the identification of promoter elements that confer cell-type specific expression suggests that it might be possible to design ribozyme constructs whose expression can be regulated in terms of both when and where the ribozyme is expressed. A number of viral vectors have been utilized for ribozyme delivery. Retroviral, adenoviral, and adeno-associated viral vectors have all been successful in expressing a variety of ribozymes. This section describes (i) promoter-containing vectors that can be utilized to produce ribozyme constructs, along with two methods that can be used for the introduction of such constructs into cells, and (ii) delivery of ribozyme by the utilization of adenoviral vectors.

Ribozyme Constructs

Cassettes for cloning and expression of ribozymes need to contain both the ribozyme-coding sequence itself and those sequences necessary for transcription initiation and termination. Obviously, an important part of such an expression cassette is the promoter; in general, Pol II and Pol III promoters have been used for the expression of ribozymes. Pol II promoters have the advantage of allowing tissue-specific expression of the ribozyme, but the disadvantage of generally requiring longer coding sequences, such that the ribozyme tends to be embedded within additional RNA sequences that can interfere with ribozyme structure and function. In contrast, Pol III promoters allow high-level expression of short RNAs; however, Pol III expression is non-tissue specific. Another consideration for ribozyme expression is the desirability of inducible expression. It may well be that for some applications the ability to temporally control expression of the ribozyme is highly desirable. An attractive inducible system that can be utilized for the controlled expression of ribozyme is the tetracycline repressor system developed by Gosen and Bujard.[9] A combination of prokaryotic and viral regulatory elements has produced a mammalian expression system in which expression of the desired transcript can be controlled tightly in both a positive and a negative fashion.

Transfection Systems

A variety of techniques exist for the introduction of nucleic acids into cells. These techniques benefited greatly from the advent of molecular biology, which provided the means to manipulate DNA and prepare quantities of highly purified nucleic acid sequences for introduction into living cells. In addition to viral vector systems, several nonviral transfection methods have been developed. These include electroporation, ballistic transfection, microinjection, receptor-mediated delivery, and artificial utilization of macromolecular complexes such as polycations, polymers, and both cationic and neutral liposomes. Two of these nonviral approaches that have been utilized successfully for the delivery of ribozymes to living cells are receptor-mediated delivery via either the transferrin or the asialoglycoprotein receptor and delivery by artificial liposomes. For delivery via either of these systems, the quality of the nucleic acid used for transfection is critical. Popular methods for purification include column chromatography and cesium chloride gradient.

Transferrin-Polylysine Delivery. Receptor-mediated delivery via the transferrin receptor[10] requires covalent attachment of a polycation such as polylysine to a ligand for a receptor on a cell surface, followed by ionic binding of the nucleic acid to be delivered to the polycation to form a nucleic acid, ligand, polycation complex. During importation of the nucleic acid by receptor-mediated endocytosis, the ligand binds to its receptor,

[9] M. Gosen and H. Bujard, *Proc. Natl. Acad. Sci. U.S.A.* **89,** 5547 (1992).
[10] E. Wagner, M. Zenke, M. Cotton, H. Beug, and M. L. Birnsteil, *Proc. Natl. Acad. Sci. U.S.A.* **87,** 3410 (1990).

and following internalization, the nucleic acid becomes localized to the endosomes and eventually to the lysosomes (Fig. 9). Although the nucleic acid is susceptible to lysosomal nucleases, a significant amount escapes degradation. Indeed, lysosomatropic agents such as chloroquine enhance the expression of genes delivered by receptor-mediated transfection. Additionally, incorporation of viruses such as adenovirus into the transfection complex takes advantage of the ability of the virus to escape lysosomal destruction by disrupting the endosomal membranes and, again, enhancing the expression of the delivered nucleic acid.

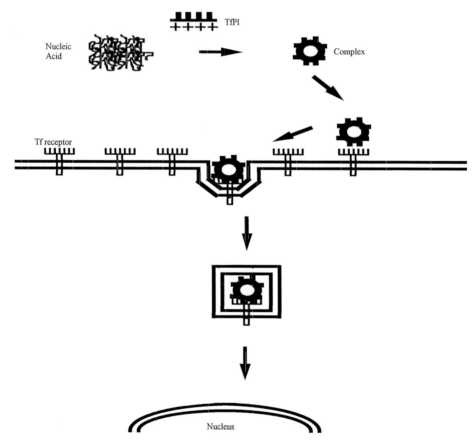

FIG. 9. Receptor-mediated delivery of nucleic acid to cells. Incubation of the nucleic acid to be delivered with the transferrin–polylysine conjugate (TfPl) will produce a nucleic acid–transferrin–polylysine complex. The complex binds to the transferrin receptors on the cell surface and is internalized by receptor-mediated endocytosis.

The ubiquitous expression of the transferrin receptor makes transferrin-fection (receptor-mediated delivery via the transferrin receptor) a popular choice for receptor-mediated transfection. For formation of the nucleic acid : transferrin–polylysine complex, the mass ratio at which the nucleic acid : transferrin–polylysine mixture reaches neutrality is calculated and the appropriate amount of transferrin–polylysine conjugate is incubated with 2–5 mg nucleic acid in 150 mM NaCl in a total volume of 65 μl at room temperature for 30 min. Cells to be transfected are prepared by removal of culture medium, and the nucleic acid : transferrin–polylysine complex is added directly. Serum-free medium is then added to the cells, which are incubated for 4 hr to allow uptake of the nucleic acid. The medium is then replaced with serum-containing medium and incubation is continued for the desired length of time. In general, high levels of expression of transferr-infected nucleic acid have been demonstrated 24–48 hr after transfection.

Liposome-Based Delivery. Since the initial development of synthetic cationic lipids as liposomal vehicles,[11] a variety of liposome formulations have been produced for the transfection of DNA and RNA into mammalian cells. These include the cationic lipid N-[1-(2,3-didoleyloxy)propyl]-n,n,n-trimethylammonium chloride (DOTMA) with the neutral lipid dioleoyl phosphotidylethanolamine (DOPE), the cationic lipid 2,3-dioleyloxy-N-[2(sperminecarboxamido)ethyl]-N,N-dimethyl-1-propanaminium trifluor-oacetate (DOSPA) with DOPE, the cationic lipid N, MI, NII, NIII-Tet-ramethyl-N, NI, NII, NIII-tetrapalmitylspermine (TM-TPS) and DOPE, and the cationic lipid 1,2-dimyristyloxypropyl-3-dimethylhydroxyethylam-monium bromide (DMRIE) with cholesterol. These liposome formulations can be used for the transfection of DNA and RNA and both for the transfection of cells in culture and *in vivo* delivery of nucleic acid.

The optimal ratio of liposome reagent to nucleic acid is generally deter-mined empirically prior to transfection. For transfection, the appropriate amount of liposome reagent is added to 5 μg nucleic acid in 2 ml culture medium. The liposome reagent, nucleic acid mixture is incubated at room temperature for 10–15 min and the culture medium is removed from the cells to be transfected. The liposome–nucleic acid mix is added directly to the cells, which are then incubated for 1 hr. Following this incubation, the cells are overlayed with 4 ml of prewarmed serum-containing medium and the cells are returned to the incubator for the desired length of time. For transient transfection, cells are typically harvested 48 hr after transfection.

Adenoviral Delivery. Adenovirus vectors provide highly efficient gene transfer and expression *in vitro* and *in vivo* and lack their natural pathogenic

[11] P. L. Felgner, T. R. Gadek, M. Holm, R. Roman, H. W. Chan, M. Wenz, J. P. Northrop, G. M. Ringold, and M. Danielsen, *Proc. Natl. Acad. Sci. U.S.A.* **84,** 7413 (1987).

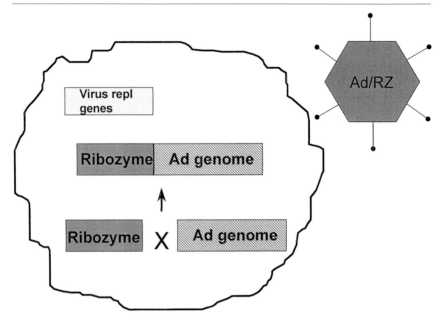

FIG. 10. Construction of ribozyme-expressing adenoviral vectors. The ribozyme sequence is incorporated into the adenoviral genome by homologous recombination after cotransfection inside HEK cells. The resulting recombinant adenovirus, which is replication defective, is grown with the help of replication genes present in HEK cells.

properties because they are deleted for their replication genes. Furthermore, adenoviral vectors can be produced in high titers. Ribozymes can be incorporated into the adenoviral genome and expressed transiently once the virus has reached the nucleus of the host cell. In order to generate a recombinant adenovirus, expressing the ribozyme of choice, the latter has to be introduced into the viral genome by homologous recombination. The virus is then grown inside 293 human embryonic kidney (HEK) cells containing the virus replication genes (Fig. 10). Shuttle vector pJM17 contains the entire genome of Ad5dl309 (adenovirus type 5). The ribozyme sequence can be cloned into another vector under the influence of an appropriate promoter, e.g., the Rous sarcoma virus (RSV) or cytomegalovirus (CMV) promoter and relevant regulatory elements. For more details on the principles of adenoviral vector construction, see Graham and Prevec.[12]

[12] F. L. Graham and L. Prevec, in "Gene Transfer and Expression Protocols" (E. J. Murray, ed.) p. 109. Humana Press, Clifton, NJ.

Five micrograms of each of the two plasmids is used to cotransfect approximately 50% confluent 293 cells in a 25-cm^2 tissue culture flask, using the calcium phosphate transfection method. Transfected cells are left to incubate for 4 hr at 37°. The medium is then replaced [10% fetal calf serum-Dulbecco's modified Eagle's medium (FCS-DMEM)] and cells are incubated further at 37° until a cytopathic effect is seen (usually after 6–10 days). The virus is then released by three rounds of sonication and freeze/thawing (in ethanol/dry ice bath and then at 37°). Cell debris is then collected by centrifugation for 10 min at 3000g, and the prestock of virus is left in the supernatant. Dilutions of the prestock virus, mixed with 10% FCS-DMEM medium, are then used to infect six multiwell plates of 70–80% confluent 293 cells. After an hour the virus is removed by suction and cells are covered with a solution containing 1% low melting point agarose and 1× MEM medium supplemented with antibiotics. Prior to mixing with the MEM and the subsequent addition to cells, the low melting point agarose is melted and then equilibrated at 37° for 1 hr. After overlaying the cells, the medium should solidify when left for 10–15 min in a sterile cabinet. Extra care should be taken when supplementing the poorly attached 293 cells with medium. Cells are then incubated at 37° until infected cells are seen as plaques growing through the agarose medium. Each plaque represents individual recombinant adenoviruses, which under no circumstances should be mixed with each other. Plaques are picked separately and incubated in 293 cell culture medium in sterile tubes. Plaque purification of the recombinant adenovirus is repeated three times to ensure that the virus stock is derived from a single infectious unit. It is then important to determine whether the ribozyme sequence is contained in and expressed from the recombinant viral genome. Therefore, an aliquot of the viral stock can be used to reinfect cells, devoid of the viral replication genes. DNA and total RNA can then be extracted from the infected cells, and Southern blotting and PCR analysis can be performed to detect the presence of the ribozyme sequence in the viral genome. Similarly, RT/PCR analysis, RNase protection, and Northern blotting can be carried out to detect the presence of ribozyme transcripts. Following confirmation that the recombinant virus contains and expresses the ribozyme sequence, a working stock is prepared. To do that, aliquots of the purified stock are used to infect a large number (5–10 175-cm^2 flasks) of 70–80% confluent 293 cells. After the cytopathic effects are seen, infected cells are harvested, and the adenovirus is released by sonication and freeze/thawing as described earlier. The virus is then purified by cesium chloride centrifugation and dialyzed against phosphate-buffered saline (PBS) as described by Graham and Prevec.[12] The purified virus should be aliquoted into tubes and stored frozen at −80°.

Viruses, particularly adenoviruses, are likely to be extremely useful tools for the transport of ribozymes. In addition to being suitable for the administration to cultured cells, they can be utilized for the delivery of ribozyme to whole animals. They can be administered systemically or injected directly into the target organ.

[30] Antisense Oligonucleotides and RNAs as Modulators of pre-mRNA Splicing

By HALINA SIERAKOWSKA,* LINDA GORMAN, SHIN-HONG KANG, and RYSZARD KOLE

Correction of Aberrant Splicing of Human β-Globin pre-mRNA by Antisense Oligonucleotides

Antisense oligonucleotides as sequence-specific downregulators of gene expression (for a review, see Ref. 1) have become used increasingly as sequence-specific research tools[2-4] and as antiviral and anticancer agents in clinical trials and in the clinic[5-8] (see also Crooke[9] and Dean[10] in this volume). We have shown that antisense oligonucleotides can also restore the expression of genes inactivated by specific mutations.[11,12] The restoration of gene expression was accomplished by targeting aberrant splice sites created by mutations that cause genetic diseases such as thalassemia or cystic fibrosis. Blocking of these splice sites with antisense oligonucleotides

---

* On leave of absence from the Institute of Biochemistry and Biophysics, Warsaw, Poland

[1] S. T. Crooke and C. F. Bennett, *Annu. Rev. Pharm. Tox.* **36,** 107 (1996).
[2] G. W. Pasternak and K. M. Standifer, *Trends Pharmacol. Sci.* **16,** 344 (1995).
[3] E. Niggli, B. Schwaller and P. Lipp, *Ann. N.Y. Acad. Sci.* **779,** 93 (1996).
[4] S. Ramchandani, R. A. MacLeod, M. Pinard, E. von Hoffe, and M. Szyf, *Proc. Natl. Acad. Sci. U.S.A.* **94,** 684 (1997).
[5] R. Zhang, J. Yan, H. Shahinian, G. Amin, Z. Lu, T. Liu, M. S. Saag, Z. Jiang, J. Temsamani, and R. R. Martin, *Clin. Pharm. Ther.* **58,** 44 (1995).
[6] M. R. Bishop, P. L. Iversen, E. Bayever, J. G. Shar, T. C. Greiner, B. L. Copple, R. Ruddon, G. Zon, J. Spinolo, M. Arneson, J. O. Armitage, and A. Kessinger, *J. Clin. Oncol.* **14,** 1320 (1995).
[7] S. Agrawal, *Trends Biotech.* **14,** 376 (1996).
[8] J. L. Tonkinson and C. A. Stein, *Cancer Invest.* **14,** 64 (1996).
[9] S. T. Crooke, *Methods Enzymol.* **313** [1] 1999 (this volume).
[10] N. Dean, this volume.
[11] Z. Dominski and R. Kole, *Proc. Natl. Acad. Sci. U.S.A.* **90,** 8673 (1993).
[12] H. Sierakowska, M. J. Sambade, S. Agrawal, and R. Kole, *Proc. Natl. Acad. Sci. U.S.A.* **93,** 12840 (1996).

Thalassemic pre-mRNA   **Thalassemic pre-mRNA + oligo**

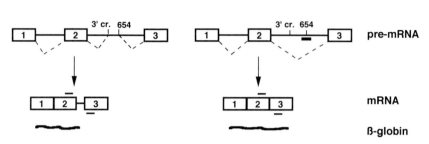

FIG. 1. Splicing of human β-globin IVS2-654 pre-mRNA in the presence of an antisense oligonucleotide. The transcribed pre-mRNA from the human β-globin gene with the IVS2-654 C to T mutation is spliced incorrectly in the absence of the antisense oligonucleotide (left). The oligonucleotide targeted to the aberrant 5' splice sites (right) prevents aberrant splicing and restores correct splicing of the mRNA, which results in transcription of the full-length β-globin polypeptide. Boxes, exons; solid lines, introns; dashed lines indicate both correct and aberrant splicing pathways; the aberrant 5' splice site created by the IVS2-654 mutation and the cryptic 3' splice site (3' cr.) activated upstream are indicated; heavy bar, oligonucleotide antisense to the aberrant IVS2-654 5' splice site; light bars above and below exon sequences indicate primers used in the RT-PCR reaction; heavy line below the mRNA represents the β-globin polypeptide.

prevents aberrant splicing and, by forcing the spliceosome to reselect the original splice sites, restores correct splicing. This results in the generation of correctly spliced and translated mRNA and, therefore, restoration of the activity of the damaged gene. Methods leading to restoration of correct splicing in thalassemic mutants of the human β-globin gene are described in detail as an example of this novel application of the antisense approach.

In the IVS2-654 mutant of the human β-globin gene, a C to T mutation at nucleotide 654 of intron 2 creates an additional aberrant 5' splice site at nucleotide 652 and activates a cryptic 3' splice site at nucleotide 579 of the β-globin pre-mRNA. During pre-mRNA splicing, a fragment of the intron contained between the newly activated splice sites is recognized by the splicing machinery as an exon and is retained in the spliced mRNA. The aberrantly spliced mRNA appears to be relatively unstable and, moreover, due to the presence of a stop codon in the retained intron fragment, leads to translation of a truncated β-globin polypeptide (Fig. 1). This molecular mechanism is responsible for the resultant β-globin deficiency that causes β-thalassemia, an inherited blood disorder in affected individuals.[13] Because the aberrant splicing of IVS2-654 pre-mRNA takes place in the

[13] D. J. Weatherall, in "The Molecular Basis of Blood Diseases" (G. Stamatoyannopoulos et al., eds.), p. 157. Saunders, Philadelphia, 1994.

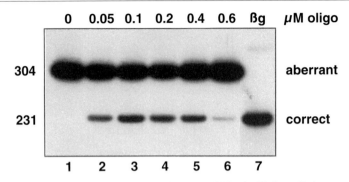

FIG. 2. Correction of splicing of IVS2-654 pre-mRNA in HeLa cells by an antisense oligonucleotide targeted to the aberrant 5′ splice site. Analysis of total RNA by RT-PCR. Lanes 1–6, IVS2-654 HeLa cells treated with increasing concentrations of the oligonucleotide (indicated in micromoles at the top), β g, (lane 7) RNA from human blood. Numbers on the left indicate the size, in base pairs, of the RT-PCR products representing aberrantly (304) and correctly (231) spliced RNAs.

presence of the correct splice sites, one can anticipate that blocking of the aberrant splice sites by antisense oligonucleotides will redirect the splicing machinery from the aberrant to the correct splice sites. Indeed, the outcome of this approach is the restoration of correct splicing of β-globin mRNA and its translation to the β-globin polypeptide (Fig. 1).

In the experiments described in this article, antisense 2′-O-methyl phosphorothioate–oligoribonucleosides were used because their duplexes with pre-mRNA are resistant to RNase H.[14] This property is essential to avoid degradation of the targeted pre-mRNA, which would have led to removal of the splicing substrate.[15] These oligonucleotides are also highly resistant to degradation by other nucleases, resulting in their stability in the cell culture environment[16] and in animal tissues.[17] Furthermore, they form very stable duplexes with RNA with T_m values higher than those of their ribo or deoxyribo analogs.[16]

The IVS2-654 β-globin pre-mRNA was expressed constitutively in a HeLa cell line transfected stably with the β-thalassemic globin gene cloned under the immediate early cytomegalovirus promoter. Cells were treated

[14] C. K. Mirabelli and S. T. Crooke, in "Antisense Research and Applications" (S. T. Crooke and B. Lebleu, eds.), p. 7. CRC Press, Boca Raton, FL, 1993.

[15] P. F. Furdon, Z. Dominski, and R. Kole, Nucleic Acids Res. 17, 9193 (1989).

[16] B. S. Sproat and A. I. Lamond, in "Antisense Research and Applications" (S. T. Crooke and B. Lebleu, eds.), p. 351. CRC Press, Boca Raton, FL, 1993.

[17] R. Zhang, Z. Lu, H. Zhao, X. Zhang, R. B. Diasio, I. Habus, Z. Jiang, R. P. Iyer, D. Yu, and S. Agrawal, Biochem. Pharm. 50, 545 (1995).

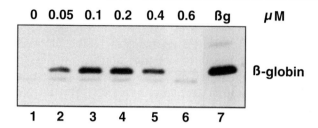

FIG. 3. Restoration of β-globin expression by antisense oligonucleotide in IVS2-654 HeLa cells. Immunoblot of total protein with antihuman hemoglobin antibody. Concentration of the oligonucleotide in micromoles is indicated at the top (lanes 1–6); in lane 7, human globin (Sigma) was used as a marker. After treatment with oligonucleotides, cells were treated with hemin preceding the isolation of proteins. The positions of human β-globin are indicated.

with Lipofectamine[18] in complex with antisense 18-mer phosphorothioate 2'-O-methyl oligoribonucleotides targeted to the aberrant 5' splice site created by the mutation. The total RNA was subsequently isolated and the spliced mRNA was identified by reverse transcription and polymerase chain reaction (RT-PCR) (Fig. 2). RT-PCR was carried out with [α-^{32}P]dATP for 18–20 cycles. Under these conditions the amount of the PCR product was proportional to the amount of input RNA. The relative amounts of PCR products generated from aberrantly and correctly spliced RNAs were also proportional to the input, allowing for quantitative analysis of data. Because the correctly spliced β-globin mRNA generated by the antisense treatment underwent translation, the correction of splicing could be verified by detection in the lysate of oligonucleotide-treated cells of a full-length β-globin polypeptide. This was accomplished by immunoblots using an antihemoglobin antibody[12] (Fig. 3). Note that correction of splicing with antisense oligonucleotides targeted to aberrant splice sites has also been accomplished for similar thalassemic mutations in the β-globin gene, i.e., IVS2-705 and IVS2-745.[19,20]

The fact that the correction of splicing was detected in oligonucleotide-treated cells indicates that several events must have taken place. Clearly, the oligonucleotides had been delivered into the cell and entered the nucleus, the site of splicing. They competed with splicing factors for the aberrant splice site, preventing aberrant splicing and promoting formation of the spliceosome and subsequent splicing at the correct sites. Apart from

[18] P. Hawley-Nelson, V. Ciccarone, G. Gebeyehu, J. Jessee, and P. L. Felgner, Focus 15, 73 (1993).
[19] H. Sierakowska, M. Montague, S. Agrawal, and R. Kole, Nucleotides Nucleosides 16, 1173 (1997).
[20] H. Sierakowska, M. J. Sambade, and R. Kole, RNA, in press.

the potential clinical applications, this system provides an excellent method for testing the efficacy of various antisense oligonucleotides. Correction of splicing may also serve as a measure of oligonucleotide uptake, its hybridization potential, and/or other parameters involved in antisense activity. Because the action of the oligonucleotides generates a new product (spliced mRNA and protein), even minor effects, difficult to discern when antisense oligonucleotides are used as downregulators of gene expression, become readily detectable (see also later).

Procedures

Cell Culture. Culture HeLa cells stably expressing human β-globin thalassemic IVS2-654 pre-mRNA in a monolayer on a 75-cm^2 tissue culture plate in S-MEM to below 50% confluency. The S-MEM medium (Life Technologies) is supplemented with 5% fetal calf, 5% horse sera, 50 μg/ml gentamicin, and 200 μg/ml kanamycin. Twenty-four hours before treatment with oligonucleotides, trypsinize the cells with 1 ml trypsin–EDTA for 3–5 min, suspend them at 10^5 cells/ml medium, and plate in 24-well plates (2 cm^2 well area) at 1 ml per well.

Oligonucleotide Treatment. Dissolve ON-654, a 2'-O-methyl phosphorothioate–oligoribonucleoside (the oligonucleotide used in this study was prepared and purified at Hybridon, Inc., Milford, MA) in water under sterile conditions at 100 μM and store at $-20°$. The oligonucleotide, GCU-AUUACCUUAACCCAG, is antisense to the aberrant 5' splice site in the IVS2-654 pre-mRNA.

Dilute the oligonucleotide to 5 μM or to appropriate lower concentrations in Opti-Mem I (Life Technologies) prewarmed to room temperature. Aliquot 100 μl of oligonucleotide solution into individual Eppendorf tubes. Controls should contain no oligonucleotide.

Suspend 4 μl of Lipofectamine (2 mg/ml, Life Technologies) in 100 μl of Opti-Mem I at room temperature and mix thoroughly by inversion. For multiple samples, increase the reagent volumes as appropriate. Add 100 μl of this suspension to the oligonucleotide solution and mix the contents by pipetting up and down. Incubate the tubes at room temperature for 30 min to form the Lipofectamine–oligonucleotide complex. Subsequently add 800 μl Opti-Mem I at room temperature and mix by inverting five times.

Place the 24-well plate containing IVS2-654 HeLa cells under the hood, aspirate the medium gently, and wash the cells twice for 1 min with 1 ml Opti-Mem I at 37°. This wash removes the serum that interferes with the uptake of the oligonucleotide–Lipofectamine complex. Work with no more than 6 wells per plate to prevent cooling of cells during washes. Transfer 1 ml of the oligonucleotide complex from each Eppendorf tube to the appropriate well and return the plate to the incubator for 10 hr. Remove the transfection medium, add to each well 1 ml of S-MEM culture medium at 37°, and incubate the cells for 24–36 hr in a tissue culture incubator.

Isolation of Total RNA. This procedure follows the recommendations of the manufacturer of TRI reagent (Molecular Research Center, Cincinnati, OH). Aspirate the medium from the wells containing oligonucleotide-treated cells and add 0.8 ml TRI reagent per well at room temperature for 10 min. Pipette the lysate five times up and down, transfer it to Eppendorf tubes, let sit for 3 min, mix by inverting the tubes several times, and touch spin. Proceed directly or store at −80°.

Add 160 μl chloroform per tube, vortex vigorously for 30 sec, and let sit for 5 min. Spin at 14,000g for 20 min at 4°. Transfer 320 μl of the colorless upper aqueous phase to fresh Eppendorf tubes containing 2 μl nuclease-free aqueous glycogen at 20 mg/ml (Boehringer-Mannheim). Vortex the tubes vigorously and touch spin. To each tube add 400 μl 2-propanol, vortex vigorously, invert twice, store on ice for at least 30 min, and centrifuge at 14,000g for 30 min at 4°. Pour off the supernatant and wash the pellet once with 1 ml 75% (v/v) ethanol by vortexing and subsequent centrifugation at 14,000g for 10 min at 4°. Carefully remove as much of the supernatant as possible without disturbing the now fluffy pellet. Dry the pellet, avoiding overdrying, under the hood or in a Speed-Vac. Dissolve the RNA pellet in 60 μl autoclaved distilled water by incubating it for 30–45 min at 45° with intermittent vortexing. The RNA solution can be stored at −20° for at least 1 year.

RNA Analysis by RT-PCR. This procedure follows the recommendation by the manufacturer of the GeneAmp thermostable RNA reverse transcriptase PCR kit (Perkin Elmer-Cetus, Norwalk, CT). All reagents for the reaction, except for the primers and radiolabeled nucleoside triphosphate, are included in the kit. Forward (5′), GGACCCAGAGGTTCTTT-GAGTCC, and reverse (3′), GCACACAGACCAGCACGTTGCCC, primers span positions 21–43 of exon 2 and positions 6–28 of exon 3 of the human β-globin gene, respectively.

Prepare the reverse transcription master mix. For one sample, use 6.4 μl autoclaved deionized water, 2 μl 10X rTth reverse transcription buffer, 2 μl 10 mM MnCl$_2$, 1.6 μl of a mixture of 2.5 mM dGTP, dATP, dTTP, and dCTP, 1 μl (30 pmol) 3′ primer, and 2 μl rTth DNA polymerase. Mix by gentle vortexing. For multiple samples, increase as appropriate.

Add 15 μl of the RT-PCR master mix prewarmed to 37° to each 0.5-ml PCR tube containing 5 μl RNA solution at room temperature and mix the total with the pipette tip. Overlay each sample with two drops of mineral oil and incubate the tubes in DNA thermal cycler (Perkin Elmer) at 70° for 15 min. At the end of this step the samples can be cooled to 4°.

In the meantime, prepare the PCR master mix. For one sample, use 63 μl autoclaved deionized water, 8 μl 10X chelating buffer, 8 μl 10 mM MgCl$_2$, 1 μl (32 pmol) 5' primer, and 2.5 μCi [α-^{32}P]dATP (aqueous solution, 10 μCi/ml, 6000 Ci/mmol, Amersham Life Science). Vortex. For multiple samples, increase as appropriate. Pipette 80 μl of the PCR master mix to each reverse transcription reaction tube, centrifuge at 14,000g for 30 sec, and subject to PCR: 3 min at 95° for 1 cycle followed by 1 min at 95° and 1 min at 65° for 18 cycles.

Analyze the PCR products by electrophoresis on a 1.5 mm × 14 cm × 14 cm nondenatauring polyacrylamide gel. The gel consists of 40 ml 7.5% acrylamide (30:1 acrylamide:bisacrylamide), 0.3 ml 10% ammonium persulfate, and 30 μl TEMED (added last) in 0.5X TEB (0.05 M Tris, 0.04 M boric acid, 0.001 M EDTA). Load 20 μl of the amplified material removed from under the mineral oil and mix with 4 μl 10X loading dye [0.42% bromphenol blue, 0.42% xylene cyanol F.F., and 25% Ficoll (Type 400, Pharmacia, Piscataway, NJ) in water] per lane. The remainder of the samples can be stored at −20° for up to 1 week. Electrophorese at room temperature in 0.5X TEB at constant voltage for a total of 900 Vhr avoiding overheating (e.g., 45 min at 200 V and subsequently for 3 hr at 250 V or overnight at 55 V) until the xylene cyanol dye leaves the gel. Dry the gel in a gel dryer under vacuum at 80° and autoradiograph or analyze in Phosphoimager.

Quantify correctly spliced β-globin mRNA by densitometry of the autoradiograms. Calculate the percentage of correct product relative to the sum of correct and aberrant ones. The results need to be corrected to account for the higher [^{32}P]dAp content of the PCR product derived from aberrantly spliced mRNA than that from correctly spliced mRNA. For IVS2-654 pre-mRNA the correction factor is 1.57.

Separation and Blotting of Proteins. After treatment with oligonucleotides, additional incubation of cells with hemin greatly facilitates the detection of β-globin on immunoblots. Hence, rinse the cells twice with S-MEM at 37° and incubate in S-MEM containing hemin for 4 hr at 37°. To prepare the latter medium, dissolve 3.2 mg hemin (Fluka, Ronkonkoma, NY) in 5 ml 20 mM NaOH and dilute 100-fold with S-MEM (without serum or antibiotics) at 37°.

Wash the hemin-treated cells twice in HBSS, once with PBS, and lyse them for 15 min at room temperature in 75 μl lysis buffer [3% SDS, 63 mM Tris–HCl, pH 6.8, 7% sucrose, 1 mM EDTA, 1 mM phenylmethylsulfonyl fluoride (PMSF), 1 μg/ml pepstatin, 5 μg/ml leupeptin, and aprotinin]. Add the proteolytic inhibitors immediately before use. Scrape the wells with flat-tip sequencing gel-loading pipette tips and transfer the lysate with a 0.5-ml insulin syringe with a 28-gauge 1/2 needle into Eppendorf tubes.

Homogenize the lysate by passing it five times through the syringe. Lysates can be stored at −80°.

Separate proteins by electrophoresis on a 10% polyacrylamide Tricine–SDS gel as follows.[21] The separating gel is 0.75 mm thick, 14 cm wide, and 11 cm high [3.05 ml gel stock, 5 ml gel buffer, 3.54 ml 50% aqueous glycerol, 3.41 ml water; 50 μl 10% ammonium persulfate, 5 μl TEMED. Gel stock: 48% acrylamide, 1.5% bisacrylamide (w/v). Gel buffer: 3 M Tris, 0.3% SDS, adjusted to pH 8.45 with HCl, stored at room temperature]. The stacking gel is 0.75 mm × 14 cm × 2 cm (0.5 ml gel stock, 1.15 ml gel buffer, 4.2 ml water; 50 μl 10% ammonium persulfate and 5 μl TEMED. Ammonium persulfate and TEMED are added immediately before pouring). The cathode buffer consists of 0.1 M Tris, 0.1 M Tricine, 0.1% SDS, pH 8.25. The anode buffer is 0.2 M Tris, adjusted to pH 8.9 with HCl.

To each tube of cell lysate (75 μl) add 4.5 μl 1 M DTT and 6 μl 0.1% aqueous Brilliant blue G (Sigma, St. Louis, MO), heat at 100° for 5 min, and load 40 μl/lane. As markers use globin (Sigma) and SDS–PAGE low range molecular weight protein standards (Bio-Rad, Richmond, CA). To prepare the globin solution, dissolve by heating to 100° for 3 min 1 mg globin in 0.5 ml lysis buffer with 50 μM DTT. Dilute sequentially with heating. Store diluted (40 ng/ml) solution at −20°. Before use, thaw the solution by heating at 45° for 15 min and load 10–30 μl per lane. Electrophorese the gel with anode buffer in the lower chamber and cathode buffer in the upper one for 1 hr at 30 V and subsequently for 5 hr at 150 V until the blue dye reaches 1 cm from gel bottom.

Electroblot the proteins onto 0.2 μm nitrocellulose (Schleicher and Schuel, Keene, NH) at 60 V overnight at room temperature in a Bio-Rad Trans-Blot Cell with a water-cooling coil. If desired, stain the nitrocellulose membrane with 1X Ponceau S [10× stock solution: 2 g Ponceau S (Sigma), 30 g trichloroacetic acid, 30 g sulfosalicylic acid, water to 100 ml] for 1 min at room temperature, wash with several changes of distilled water, and mark the positions of protein standards. Proceed with immunodetection of β-globin or store the air-dried blot wrapped in Saran wrap in the refrigerator.

Immunoblot Detection of β-Globin. Perform all the immunoreactions and rinses with agitation on a clinical rotator at room temperature. Rinse blot in PBST (1% Triton X-100 in PBS) for 15 min. Block for 2 hr with Blotto (5% Carnation nonfat dry milk in PBST. Dissolve the milk in PBS, filter, and add Triton X-100). Incubate with polyclonal affinity-purified chicken antihuman hemoglobin IgG (Accurate Chemicals) diluted 1000-fold with Blotto for 20 min. Wash in Blotto three times for 10 min each

---

[21] H. Schagger and G. von Jagov, *Anal. Biochem.* **166,** 368 (1987).

and with PBST twice for 10 min each. Incubate for 1 hr with the secondary antibody, rabbit antichicken horseradish peroxidase-conjugated IgG (Accurate Chemicals), diluted 2000-fold with Blotto. Wash for 10 min each: twice with Blotto, three times with PBST containing 0.02% Tween 20, and twice with PBST. Proceed with detection of β-globin using the ECL detection system, as recommended by the manufacturer (Amersham).

Correction of Aberrant Splicing of Human β-Globin pre-mRNA by Modified U7 snRNA

The use of antisense oligonucleotides as pharmacological, therapeutic agents is very attractive due to its anticipated sequence specificity and low toxicity. However, especially in treatment of genetic disorders such as thalassemia, a significant drawback of this approach stems from the fact that the oligonucleotides do not remove the offending mutation and therefore the treated patients would require lifelong periodic administrations of these compounds. To circumvent this problem we have developed a novel approach that allows for long-term, possibly permanent, expression of RNA antisense to aberrant thalassemic splice sites in β-globin pre-mRNA. This was accomplished by incorporating the anti-β-globin sequences into the gene for murine U7 small nuclear RNA (snRNA).

U7 snRNA forms a ribonucleoprotein particle (U7 snRNP) that is involved in processing of the 3' end of histone pre-mRNAs.[22–24] In addition to the 62 nucleotide RNA, the particle contains at least two U7-specific proteins and eight so-called Sm proteins, which are also found associated with other snRNAs.[25] The U7 snRNA carries out its function by hybridizing via its 5' end to a sequence within the histone pre-mRNA. This observation indicated that the 5' end of U7 snRNA is available *in vivo* for antisense interactions. Thus it seemed likely that the insertion of appropriate sequences antisense to aberrant splice sites in the β-globin pre-mRNA into the U7 snRNA will change its function and lead to correction of splicing of thalassemic β-globin pre-mRNA.

The modified U7 snRNA was tested on the IVS2-705 mutant of the human β-globin gene (Fig. 4A).[26] This mutation, similar to IVS2-654, creates

[22] G. Galli, H. Hofstetter, H. G. Stunnenberg, and M. L. Birnstiel, *Cell* **34**, 823 (1983).
[23] C. Birchmeier, D. Schümperli, D. Sconzo, and M. L. Birnstiel, *Proc. Natl. Acad. Sci. U.S.A.* **81**, 1057 (1984).
[24] M. L. Birnstiel and F. Schaufele, *in* "Structure and Function of Major and Minor Small Nuclear Ribonucleoprotein Particles" (M. L. Birnstiel, ed.), p. 155. Berlin, 1988.
[25] H. O. Smith, K. Tabiti, G. Schaffner, D. Soldati, U. Albrecht, and M. L. Birnstiel, *Proc. Natl. Acad. Sci. U.S.A.* **88**, 9784 (1991).
[26] L. Gorman, D. Suter, V. Emerick, D. Schumperli, and R. Kole, *Proc. Natl. Acad. Sci. U.S.A.* **95**, 4929 (1998).

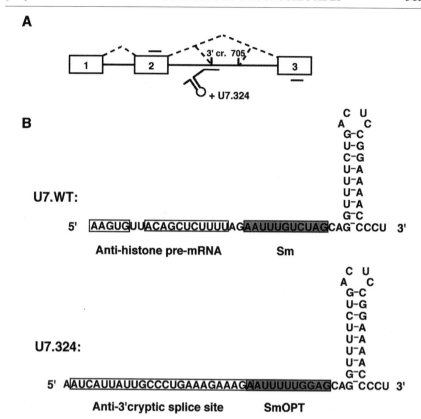

FIG. 4. (A) Use of modified U7 snRNA to correct aberrant splicing. Modified U7 snRNA (U7.324) targeted to the 3' cryptic site activated by a G to T mutation in the IVS2-705 β-globin gene prevents aberrant splicing and restores correct splicing of the pre-mRNA. Boxes, exons; solid lines, introns; dashed lines indicate both correct and aberrant splicing pathways; the aberrant 5' splice site created by the IVS2-705 mutation and the cryptic 3' splice site (3' cr.) activated upstream are indicated; short bars above and below exon sequences indicate primers used in the RT-PCR reaction. (B) Structure of U7 snRNA constructs. Wild-type U7 snRNA contains an antisense sequence to the 3' end of histone pre-mRNA (open boxes), the U7-specific Sm-binding site (gray box), and a stem–loop structure. In the U7.324 construct, the antihistone pre-mRNA sequence has been replaced with a sequence antisense to the β-globin IVS2-705 3' cryptic splice site (open box) and the U7 Sm site has been replaced with the SmOPT sequence (gray box).

an additional 5' splice site and activates a 3' splice site within intron 2 of β-globin pre-mRNA. A 24 nucleotide sequence antisense to the aberrant 3' splice site was inserted into the U7 gene to replace the antihistone sequence of the wild-type U7 snRNA (Fig. 4B). An internal sequence (Sm) was also modified (SmOPT) to improve the stability of the RNA and to

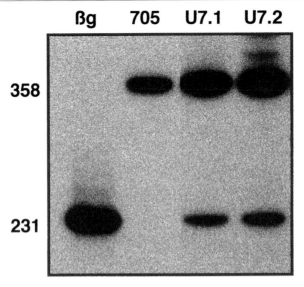

Fig. 5. Correction of aberrant splicing in stable HeLa cell lines expressing U7.324. Total cellular RNA was isolated from HeLa cell lines that stably express IVS2-705 β-globin and U7.324 and analyzed by RT/PCR. Lane 1, β g, RNA from human blood. Lane 2, RNA from IVS-705 HeLa cells. Lanes 3 and 4, stable correction of aberrant splicing of β-globin pre-mRNA in two different clones.

increase its intranuclear concentration.[27] U7.324 snRNA was very effective in the correction of splicing of IVS2-705 pre-mRNA. Importantly, the stable transfection of cells expressing the thalassemic β-globin gene with vectors carrying a modified U7 snRNA gene led to permanent correction of the splicing pattern of the β-globin pre-mRNA (Fig. 5). Consequently, significant amounts of full-length β-globin mRNA accumulated in the cells.

These results suggest a possibility of gene therapy based on the antisense concept. The patients' bone marrow would be transfected *ex vivo* with the antisense U7 vectors and reimplanted. The correction of β-globin pre-mRNA splicing affected by U7.324 snRNA should increase the production of β-globin and reduce the imbalance between the α and the β subunits of hemoglobin, promoting the maturation of erythrocytes and alleviating the symptoms of thalassemia.

Procedures

U7.324 snRNA Construct. The U7 promoter and 3' sequences are included in the construct. The U7-specific Sm-binding site (AAUUUGU-

[27] C. Grimm, B. Stefanovic, and D. Schümperli, *EMBO J.* **12,** 1229 (1993).

CUAG) was replaced with the consensus Sm sequence (AAUUUUUG-GAG).[28] The natural 18 nucleotide sequence complementary to the 3' processing site of histone pre-mRNAs was replaced[29,30] with a 24 nucleotide sequence complementary to the 3' splice site activated by the IVS2-705 mutation (see Fig. 4).

Cell Lines. Culture the HeLa cell line carrying the thalassemic IVS2-705 human β-globin gene[19,20] and the cell lines stably expressing the modified U7 snRNAs (see later) under the same conditions as described in the previous section.

Transient Expression of U7.324 snRNA. The procedure is essentially the same as that described in the first section. Plate HeLa IVS2-705 cells 24 hr before treatment in 24-well plates at 10^5 cells per 2-cm^2 well. Treat the cells for 10 hr with the modified U7.324 plasmid (0.5, 1, 2, and 4 μg/ml) complexed with 4 μl/ml of Lipofectamine in Opti-Mem.

Selection of Cell Lines Stably Expressing U7.324 snRNA. In 24-well plates, cotransfect HeLa IVS2-705 cells with plasmids carrying a hygromycin resistance gene (0.2 μg) and a U7 snRNA expressing plasmid (2 μg) in the presence of Lipofectamine (4 μl/ml) as described in the first section. Twenty-four posttransfection, remove media and wash cells with 1 ml 1× HBSS (Hanks' balanced salt solution). Add 100 μl of 1× trypsin–EDTA (Life Technologies) to each well for approximately 5 min to detach the cells from the plate and supplement the cell suspension with 1.5 ml of serum containing culture medium per well. Mix well by pipetting up and down and plate 0.5 ml of the suspension on each of three 75-cm^2 tissue culture plates containing 10 ml of S-MEM culture medium. Incubate the cells at 37° for 24 hr. Remove the medium and replace it with the same medium containing 250 μg/ml hygromycin. Replace this medium every 3 days. Hygromycin-resistant colonies should appear in 10–14 days.

To isolate the colonies wash the plates once with 5 ml 1× HBSS. Place cloning glass rings (Bellco), with the bottoms coated with vacuum grease, over the colony and push down firmly onto the plate. Add 100 μl of trypsin–EDTA solution to the center of the ring. Incubate at room temperature for 2–3 min. Pipette up and down to detach the cells and transfer to 2-cm^2 wells (24-well plate) containing 1 ml culture medium without hygromycin. Once the cells are ~50% confluent, wash the plates once with 1 ml 1× HBSS and trypsinize with 100 μl of trypsin–EDTA. Add 2 ml of culture medium without hygromycin and pipette up and down to suspend

[28] B. Stefanovic, W. Hackl, R. Luhrmann, and D. Schümperli, *Nucleic Acids Res.* **23,** 3141 (1995).
[29] D. H. Jones and B. H. Howard, *Biotechniques* **10,** 62 (1991).
[30] D. H. Jones and S. C. Winistorfer, *Biotechniques* **12,** 528 (1992).

the cells. Remove 1 ml of cell suspension and transfer to a fresh plate. Once these cells reach ~80% confluence, treat them with TRI reagent and isolate total cellular RNA as described in the first section. Analyze RNA by RT/PCR to detect correction of β-globin splicing affected by stably expressed U7.324 snRNA. Cells remaining on the original plate are cultured for maintenance of the clonal cell line.

RNA Analysis. For analysis of correction of splicing of human β-globin pre-mRNA, carry out RT-PCR exactly as described in the previous section. To detect the expression of U7.324 snRNA, use U7.324 specific forward (GCATAAGCTTAAG<u>CATTATTGCCCTGAA</u>) and reverse (CGTA-GAATTC<u>AGGGGTTTTCCGACCGA</u>) primers; underlined nucleotides overlap with U7 sequences. The RT conditions are the same as described in the first section. The PCR conditions are 3 min at 95° for 1 cycle, 1 min at 95°, 1 min at 55°, and 1 min at 72° for 25 cycles.

Luciferase-Based Assay for Intracellular Activity of Antisense Oligonucleotides

As pointed out in the first section, correction or modulation of splicing pathways provides a positive readout assay for antisense activity. The assay relies on a shift of the splicing machinery from one site to another caused by a block of a splice site with the antisense oligonucleotide. Therefore, the results virtually guarantee that the observed effects are sequence specific and are due to true antisense activity of the compound, which must have migrated into the nucleus, the site of pre-mRNA splicing. The sequence specificity of the effects is also supported by appropriate controls,[12,31] some of which are presented in this section.

We have used RT-PCR as a detection method for the correction of splicing by antisense oligonucleotides. Although the sensitivity of RT-PCR is unmatched, the procedure is relatively laborious and not easily amenable to quantitative analysis. In order to simplify the analysis of correction of splicing, we have devised a system in which the action of antisense oligonucleotides or RNAs results in an upregulation of luciferase activity. This was accomplished by interrupting the coding sequence of the luciferase gene with the IVS2-705 human β-globin intron (Fig. 6).

In cells expressing the modified Luc/705 mRNA stably, the intron is spliced incorrectly, resulting in retention of the intron fragment in the coding sequence of the luciferase mRNA and concomitant inhibition of translation of an active enzyme. Treatment of the cells with antisense oligonucleotides or RNAs targeted to the 5' splice site created by the

[31] S.-H. Kang, M.-J. Cho, and R. Kole, *Biochemistry* **37**, 6235 (1998).

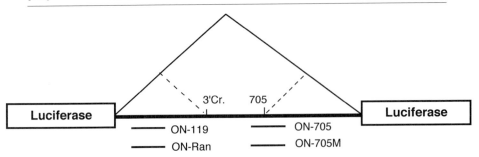

FIG. 6. Splicing of Luc/705 pre-mRNA in HeLa cells. The pre-mRNA is transcribed from a pLuc/705 plasmid in which the coding sequence of luciferase (boxes) is interrupted by the β-globin IVS2-705 intron (solid heavy line). Due to aberrant splicing of the intron (dashed lines), the luciferase is inactive. ON-705 (short bar under the intron) is targeted to the aberrant 5' splice site and restores correct splicing (solid thin line) and luciferase activity. Control oligonucleotides are indicated.

705 mutation resulted in easily detectable, dose-dependent restoration of luciferase activity, which at the optimal concentration of the oligonucleotide (0.4 μM, ON-705, CCUCUUACCUCAGUUACA, Fig. 7) increased approximately 50-fold over the background. The effect was sequence specific because neither an oligonucleotide targeted to an irrelevant site in the intron at nucleotide 119 (ON-119, UGAGACUUCCACACUGAU) nor

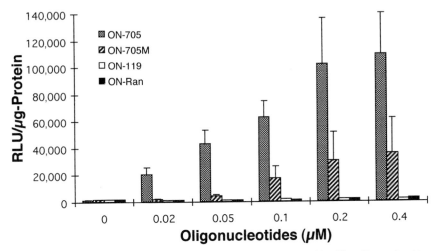

FIG. 7. Restoration of luciferase activity by antisense oligonucleotides. The oligonucleotides used are indicated. Experiments were performed in triplicate, with the line at each bar representing standard deviation. Results are expressed in relative light units (RLU) per microgram of protein.

the one with a random sequence (ON-Ran) exerted any effect on luciferase activity. A single mismatch (ON-705M, CCUCUUACAUCAGUUACA) resulted in an approximately 70% decrease in the antisense effectiveness of ON-705 (Fig. 7).

In HeLa cells transfected with the pLuc/705 construct, RNA expression is controlled by the CMV promoter coupled to the Tet-Off system. In this system, luciferase expression is strictly regulated and is approximately 35-fold higher than in cells transfected with CMV promoter/enhancer-driven luciferase expression vectors.[32] Note also that the luciferase system is approximately 1000-fold more sensitive than, for example, the CAT assay.[33]

The luciferase assay should be very useful in the identification of oligonucleotide carrier molecules and in investigations on the effects of oligonucleotide backbones on their antisense activity. The fact that this assay is incompatible with antisense oligonucleotides that promote cleavage of the target RNA by RNase H does not appear to be a serious limitation. The most promising examples of modified compounds include 2'-O-alkyl-oligoribonucleotides, morpholino oligonucleotides, methyl phosphonates, and the so-called peptide nucleic acids.[34] None of these compounds promote cleavage of the target RNA by RNase H and can therefore be tested in this system.

Procedures

Plasmid and Cell Line Construction. Intron 2 of the IVS2-705 thalassemic mutant of the β-globin gene was inserted between nucleotides 1368 and 1369 of the luciferase cDNA sequence of the pTRE-Luc plasmid (Clontech) by PCR-based *in vivo* recombination.[30] HeLa Tet-Off cells (Clontech) were cotransfected with the resultant recombinant plasmid, pLuc/705, and the hygromycin resistance plasmid using Lipofectamine as a carrier. Stable transfectants were obtained by selection in hygromycin (200 μg/ml) containing Dulbecco's modified Eagle's medium (DMEM) with 10% fetal calf serum as described earlier.

Oligonucleotide Treatment. In a six-well plate treat the cells with a complex of Lipofectamine (4 μl/ml) and oligonucleotides (concentrations as in Fig. 7) in Opti-Mem for 5 hr as described earlier. Replace the medium

[32] D. X. Yin, L. Zhu, and R. T. Schimke, *Anal. Biochem.* **235,** 195 (1996).

[33] J. R. DeWet, K. V. Wood, M. DeLuca, D. R. Helinski, and S. Subramani, *Mol. Cell. Biol.* **7,** 725 (1987).

[34] P. H. Seeberger and M. H. Caruthers, *in* "Applied Antisense Oligonucleotide Technology" (C. A. Stein and A. M. Krieg, eds.), p. 51. Wiley-Liss, New York, 1998.

with DMEM containing 10% fetal calf serum and, after an additional 18 hr, collect the protein for luciferase assay[35] (see later for details).

Luciferase Assay. Remove the culture medium. Rinse the cell monolayer in a six-well plate twice with cold PBS and place the plate on ice. Add 100 μl of cell lysis buffer (0.1 M potassium phosphate, pH 7.8, 1% Triton X-100, 1 mM DTT, 2 mM EDTA) to each well and scrape the cell monolayer with a plastic scraper (Costar). Transfer the cell lysate into Eppendorf tubes and leave it on ice for 15 min. Subsequently centrifuge the lysate at 14,000g for 30 sec at 4°. Collect the cell extract from above the cell debris. At room temperature, mix 5–20 μl of the extract with 100 μl of ATP buffer (30 mM Tricine, 3 mM ATP, 15 mM MgSO$_4$, 10 mM DTT, pH 7.8) in an assay cuvette. Place the cuvette in the luminometer (Analytical Lumines-cence Laboratory) and place up to 5 ml of 1 mM luciferin in the instrument chamber. One hundred microliters of the luciferin solution is added auto-matically into the cuvette. Measure luminescence for 10 sec.

Protein Assay. Measure protein concentration in the cell extract using the bicinchoninic acid assay kit (Sigma). Briefly, prepare an assay solution by adding 1 part of 4% cupric sulfate solution to 50 parts of bicinchoninic acid solution. Mix 5 μl of the cell extract with 200 μl of the assay solution in a well of a 96-well plate. Incubate the plate for 30 min at 37°. Measure absorbance at 595 nm on a plate reader (Bio-Rad Model 3350). Express luciferase activity in relative light units (RLU) per microgram protein.

Addendum

In all the procedures described in this article, the antisense oligonucleo-tides and RNAs were delivered to the cells in complex with Lipofectamine. Work in this laboratory showed that other agents such as DMRIE-C, Cell-fectin (Life Technologies), Superfect, Effectene (Qiagen), ExGEN 500 (Euromedex), and Cytofectin (Sigma) are also useful as carriers. Further-more, in addition to HeLa cells, other cell types, including 3T3[12] and K562 cell lines (Gorman and Kole, unpublished), were subjected successfully to the antisense-mediated modification of splicing pathways.

Acknowledgments

We thank Dr. Sudhir Agrawal from Hybridon Inc. for the oligonucleotides used in this study and Elizabeth Smith for technical assistance. This work was supported in part by grants from Hybridon and National Institutes of Health to R.K.

[35] A. R. Brasier, J. E. Tate, and J. F. Habner, *Biotechniques* **7,** 1116 (1989).

[31] Selective RNA Cleavage by Isolated RNase L Activated with 2-5A Antisense Chimeric Oligonucleotides

By ROBERT H. SILVERMAN, BEIHUA DONG, RATAN K. MAITRA, MARK R. PLAYER, and PAUL F. TORRENCE

Introduction

Cellular ribonucleases often play essential roles in the mechanism of action of antisense oligonucleotides (ODNs) because of their ability to destroy target RNA molecules.[1] RNase L is present in a wide range of cell types in higher vertebrate organisms in either an inactive form or as a highly potent, single-strand-specific endoribonuclease.[2] Activation of RNase L occurs in response to the binding of short, $2',5'$-linked oligoadenylates collectively referred to as 2-5A[3] with the general formula:

$$(p)_x 5'A(2'p5'A)_y; \qquad x = 1 \text{ to } 3, y \geq 2$$

2-5A molecules as small as $pA2'p'5A2'p5'A$ are efficient activators of human RNase L.[4] RNase L has a stringent requirement for both adenylyl residues and $2',5'$-internucleotide linkages for activation and it has high affinity for 2-5A (K_a of 2.5×10^{10} M^{-1}).[4,5] 2-5A induces the formation of RNase L homodimers, both in cell-free systems and in intact mammalian cells, apparently due to conformational changes that unmask protein–protein interaction and catalytic domains.[6–8]

The 2-5A system is part of the antiviral mode of action of interferons. Interferon treatment of mammalian cells results in the induction of a family of 2-5A synthetases that require double-stranded RNA (dsRNA) to catalyze the production of 2-5A from ATP.[3] Virus-infected cells produce and secrete type I interferons that bind to receptors on cells, leading to enhanced

[1] M. D. Matteucci and R. W. Wagner, *Nature* **384,** suppl. (1996).
[2] R. H. Silverman, *in* "Ribonucleases: Structure and Function" (G. D'Alessio and J. F. Riordan, eds.), p. 515. Academic Press, New York, 1997.
[3] I. M. Kerr, and R. E. Brown, *Proc. Natl. Acad. Sci. U.S.A.* **75,** 256 (1978).
[4] M. R. Player and P. F. Torrence, *Pharmacol. Ther.* **78,** 55 (1998).
[5] R. H. Silverman, D. D. Jung, N. L. Nolan-Sorden, C. W. Dieffenbach, V. P. Kedar, and D. N. SenGupta, *J. Biol. Chem.* **263,** 7336 (1988).
[6] B. Dong and R. H. Silverman, *J. Biol. Chem.* **270,** 4133 (1995).
[7] J. L. Cole, S. S. Carroll, E. S. Blue, T. Viscount, and L. C. Kuo, *J. Biol. Chem.* **272,** 19187 (1997).
[8] S. Naik, J. M. Paranjape, and R. H. Silverman, *Nucleic Acid Res.* **26,** 1522 (1998).

levels of 2-5A synthetases and RNase L.[9] Thereafter, viral dsRNA stimulates the production of 2-5A, resulting in RNase L activity and suppression of the viral infection. For instance, the role of the 2-5A system is underscored by the observation that RNase L$^{-/-}$ mice are even more susceptible to encephalomyocarditis virus than wild type mice.[10]

The ability of RNase L to be activated by 2-5A has been exploited for the purpose of degrading RNA molecules of choice.[11–15] 2-5A is modified by attachment through linkers to an antisense cassette.[16] The resulting 2-5A antisense species binds to both an RNA target and to RNase L. Activation of RNase L by the 2-5A moiety causes targeted degradation of the bound RNA molecule. Therefore, 2-5A antisense is a derivative of 2-5A that forces RNase L to selectively cleave RNA targets. The selectivity is due a proximity effect from directing the activated RNase L to the RNA target. Furthermore, addition of the linker/antisense domains to 2-5A suppresses general 2-5A activity.[17] This article describes methods for performing targeted degradation of RNA with purified RNase. Methods for chemical synthesis of 2-5A antisense are not included because they have been described elsewhere.[16]

The 2-5A antisense chimeras in this study have the general formula p5'A2'[p5'A2']$_3$O(CH$_2$)$_4$OpO(CH$_2$)$_4$Op5'(dN)$_m$ and are given as pA$_4$-Bu$_2$-(dN)$_m$, where dN is a deoxyribonucleotide and m is an integer. Unless otherwise indicated, these 2-5A antisense chimeras are 5'-thiophosphorylated. Phosphorothioate substitution in the internucleotide linkages is indicated by a lowercase "s"; e.g., AsC. 5'-Thiophosphorylation is indicated by an uppercase "S". Oligonucleotides are abbreviated as ODN.

[9] G. R. Stark, I. M. Kerr, B. R. G. Williams, R. H. Silverman, and R. D. Schreiber, *Annu. Rev. Biochem.* **67,** 227 (1998).

[10] A. Zhou, J. Paranjape, T. L. Brown, H. Nie, S. Naik, B. Dong, A. Chang, B. Trapp, R. Fairchild, C. Colmenares, and R. H. Silverman, *EMBO J.* **16,** 6355 (1997).

[11] P. F. Torrence, R. K. Maitra, K. Lesiak, A. Zhou, and R. H. Silverman, *Proc. Natl. Acad. Sci. U.S.A.* **90,** 1300 (1993).

[12] A. Maran, R. K. Maitra, A. Kumar, B. Dong, W. Xiao, G. Li, B. R. Williams, P. F. Torrence, R. H. Silverman, *Science* **265,** 789 (1994).

[13] N. M. Cirino, G. Li, W. Xiao, P. F. Torrence, and R. H. Silverman, *Proc. Natl. Acad. Sci. U.S.A.* **94,** 1937 (1997).

[14] M. R. Player, D. L. Barnard, and P. F. Torrence, *Proc. Natl. Acad. Sci. U.S.A.* **95,** 8874 (1998).

[15] S. Kondo, Y. Kondo, G. Li, R. H. Silverman, and J. K. Cowell, *Oncogene* **16,** 3323 (1998).

[16] W. Xiao, M. R. Player, G. Li, W. Zhang, K. Lesiak, and P. F. Torrence, *Nucleic Acid Drug Dev.* **6,** 247 (1996).

[17] R. K. Maitra, G. Li, W. Xiao, B. Dong, P. F. Torrence, and R. H. Silverman, *J. Biol. Chem.* **270,** 15071 (1995).

Methods and Methodology

Expression and Purification of Recombinant Human RNase L Produced in Insect Cells

Although many bacterial, yeast, and insect cell protein expression systems are available, the baculovirus system has given us the best results, perhaps because of the high levels of soluble RNase L obtained. Cloning of the full-length cDNA for human RNase L in the BacPAK6 baculovirus vector (Clontech) has been described previously.[18] Either monolayers or suspension cultures of SF21 insect cells are infected at a multiplicity of infection of 10 plaque-forming units per cell and then incubated for 3 days at 27°. Cells are harvested, washed in phosphate-buffered saline (pH 6.2), frozen as cell pellets on dry ice, and stored at $-70°$. Cells are thawed in four packed cell volumes of buffer A (25 mM Tris–HCl, pH 7.4; 50 mM KCl; 10% glycerol; 1 mM EDTA; 0.1 mM ATP; 5 mM MgCl$_2$; 14 mM 2-mercaptoethanol; and 1 μg/ml of leupeptin). Sonication of the cell suspension is performed six times on ice for 15 sec at 30-sec intervals. After centrifuging at 16,700g at 4° for 30 min, the supernatant is collected and cleared twice more by centrifuging for 10 min.

Purification of RNase L is performed with an FPLC system (Pharmacia, Piscataway, NJ) at 4°. The crude cell extract (about 20 mg of protein in 2 ml of buffer A) is loaded onto a Blue Sepharose CL-6B column (5 × 50 mm, Pharmacia). After washing with 10-column volumes of buffer A at a flow rate of 0.3 ml/min, a linear gradient to 21% buffer B (buffer A supplemented with 1 M KCl) is performed in about 23 min. The ratio of buffer A : buffer B is held constant while the RNase L is eluted (in about 8 ml). The column fractions are monitored for absorbance at 280 nm by a [^{32}P]2-5A filter-binding assay.[18,19] The protein is concentrated and desalted with a Centricon filter unit (Amicon, Danvers, MA). A peak with 2-5A-binding activity, which corresponds to a UV$_{280}$ peak, is identified and pooled (Fig. 1A). RNase L is visualized after SDS–PAGE as an 80-kDa protein (Fig. 2, lane 3). The Blue Sepharose column provides about a sevenfold purification factor (Table I).

The crude preparation of RNase L eluted from the Blue Sepharose column (about 2–2.5 mg of protein in 1 ml of buffer A per preparation) is loaded on a Mono Q (HR 5/5) column (5 × 50 mm, Pharmacia) at a flow rate of 0.4 ml/min. After washing with 10-column volumes of buffer

[18] B. Dong, L. Xu, A. Zhou, B. A. Hassel, X. Lee, P. F. Torrence, and R. H. Silverman, *J. Biol. Chem.* **269,** 14153 (1994).
[19] M. Knight, P. J. Cayley, R. H. Silverman, D. H. Wreschner, C. S. Gilbert, R. E. Brown, and I. M. Kerr, *Nature* **288,** 189 (1980).

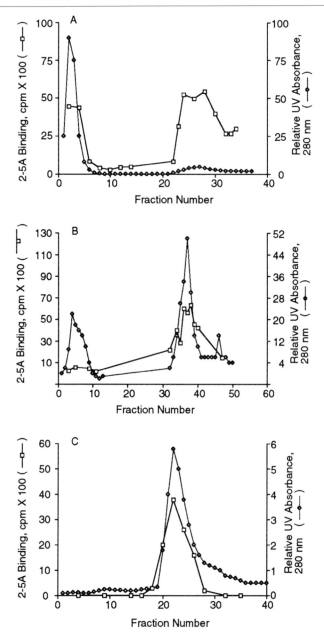

FIG. 1. Purification of human recombinant RNase L produced from a baculovirus vector in insect cells. Sequential purification profiles from FPLC columns of (A) Blue Sepharose CL-6B, (B) Mono Q (HR5/5), and (C) Superose-12 (HR 10/30). 2-5A binding activity was determined on nitrocellulose filters using a radiolabeled $2',5'$-oligoadenylate derivative as described.[18]

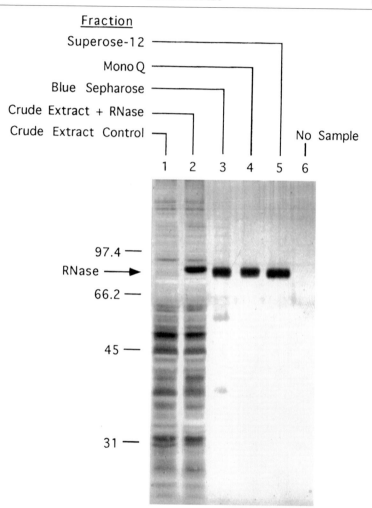

FIG. 2. Silver staining of proteins in different fractions from the purification of RNase L. Crude extracts of insect cells infected with nonrecombinant virus or recombinant virus (4.7 μg of protein each) are shown in lanes 1 and 2, respectively. Lanes 3 to 5 contain about 140 ng of RNase L during different stages of purification, as indicated. Lane 6 is a blank sample to compare background. Reprinted from B. Dong *et al., J. Biol. Chem.* **269** 14153 (1994) with permission.

TABLE I
PURIFICATION OF RNase L[a]

Fraction	Total protein (mg)	Total activity (units[b])	Specific activity (units per μg)	Yield (%)	Purification factor
Crude extract	19.6	66,335	3.4	100	1
Blue Sepharose	2.4	56,507	23.4	85.2	6.9
Mono Q	0.95	40,915	43.3	61.7	12.8
Superose 12	0.81[c]	40,902	50.8	61.7	15.0

[a] Reprinted from B. Dong et al., J. Biol. Chem. **269**, 14153 (1994) with permission.
[b] Units defined in Dong et al. (1994).
[c] In aliquots of about 200 μg per separation.

A, RNase L is eluted in a linear gradient to 40% buffer B in 50 min at 0.4 ml/min. The peak of RNase L (as determined by the 2-5A-binding assay) elutes in about 3 ml of buffer and is observed after 7.5–15 min, corresponding to a KCl concentration of about 120–180 mM (Fig. 1B). An additional purification factor of about two is obtained (Table I).

Between 100 and 200 μg of RNase L per separation, obtained after purification by the previous two steps, in 100 μl of buffer C (buffer A containing 100 mM KCl) is loaded on a Superose-12 (HR 10/30) column (10 × 300 mm) at a flow rate of 0.2 ml/min (Fig. 1C). Fractions containing RNase L (eluted in about 1.4 ml) are analyzed by silver staining of an SDS/10% polyacrylamide gel (Fig. 2). The total purification factor of 15 corresponds to a homogeneous preparation of RNase L (Table I).

Determination of RNase L Activating Ability of Chimeric 2-5A ODNs against Generic Substrate, $C_{11}UUC_7$

Earlier methods of RNase L activity assays have relied on the cleavage of radiolabeled poly(U)[18,20] or on the cleavage of ribosomal RNA (rRNA).[21] The poly(U) degradation approach is somewhat complicated by the appearance of the initial radiolabeled substrate as a ladder when analyzed by gel electrophoresis and the subsequent formation of several different cleavage products.[18] rRNA methodology requires the isolation of rRNA under carefully controlled conditions to provide discrete and sharp bands for gels, which additionally require ethidium bromide visualization with resultant difficulties in quantitation. In addition, the visualized cleavage bands actually contain several different cleavage products.[21]

[20] R. H. Silverman, Anal. Biochem. **144**, 450 (1985).
[21] D. H. Wreschner, T. C. James, R. H. Silverman, and I. M. Kerr, Nucleic Acids Res. **9**, 1571 (1981).

A more recent method employs cleavage of a radiolabeled oligoribonucleotide.[22] Because RNase L cleaves preferentially after UU and UA sequences,[23] Carroll et al.[22] chose the sequence 5'-[^{32}P]pC$_{11}$UUC$_7$ for the kinetic analysis of RNase L. Comparison of the primary cleavage product with cleavage products resulting from RNase PhyM and from base-catalyzed hydrolysis confirmed that 5'-[^{32}P]pC$_{11}$UUp was the primary cleavage product. The addition of RNase L to a reaction mixture containing substrate 5'-[^{32}P]pC$_{11}$UUC$_7$ and (2'-5')pA$_3$ produced an initial lag in product formation. This lag was eliminated by a 30-min preincubation of RNase L with (2'-5')pA$_3$ at 0°. Carroll et al.[22] found the rate of cleavage to be linear between 100 and 600 pM substrate when RNase L was preactivated with 800 nM (2'-5')pA$_3$. The rate of cleavage was half-maximal at 1 nM (2'-5')pA$_3$ and attained maximal velocity at 1 nM (2'-5')pA$_3$ and above. Longer oligoribonucleotides were cleaved more efficiently; for instance, 5'-pC$_5$UUC$_4$ was cleaved 3-fold less efficiently (k_{cat}/K_m) than 5'-[^{32}P]pC$_{11}$UUC$_7$, primarily due to an increase in K$_m$. Furthermore, RNase L catalyzed the cleavage of substrates containing two or three sequential U's with a 20- and 50-fold faster rate than that with which it catalyzed the cleavage of a substrate containing a single U.[22] We have found that the substrate 5'-[^{32}P]pC$_{11}$UUC$_7$ of Carroll et al.[22] provides a very useful approach to measuring the relative activity of 2-5A analogs and 2-5A-antisense chimeras, as it utilizes a readily accessible substrate that can be labeled routinely to high specific activity with ^{32}P and which, on cleavage by RNase L, provides one, or at most two, cleavage products that can be simply quantified. When the RNase L concentration is raised, especially when limiting quantities of activator are present, 5'-[^{32}P]pC$_{11}$UUp becomes the predominant product. Under conditions of increasing activator concentration, the second cleavage product, 5'-[^{32}P]pC$_{11}$Up, is obtained.

The synthetic oligoribonucleotide rC$_{11}$U$_2$C$_7$ can be prepared using protected RNA mononucleotides synthons or obtained by commercial custom synthesis. We have used rC$_{11}$U$_2$C$_7$ prepared by Midland Certified Reagent Co. (Midland, TX) with quite satisfactory results. Usually, the commercial preparation is dialyzed against RNase-free H$_2$O to remove remaining salts and low molecular weight compounds, as polynucleotide kinase can be inhibited by excess sodium or ammonium ions. This ODN is then diluted to approximately 100 A_{260} units/ml in H$_2$O and then divided into 20-μl aliquots that are stored at $-80°$. To label rC$_{11}$U$_2$C$_7$ with ^{32}P, the foregoing

[22] S. S. Carroll, E. Chen, T. Viscount, J. Geib, M. K. Sardana, J. Gehman, and L. C. Kuo, J. Biol. Chem. **271**, 4988 (1996).
[23] D. H. Wreschner, J. W. McCauley, J. J. Skehel, and I. M. Kerr, Nature **289**, 414 (1981).

ODN stock is diluted in water to yield a final concentration of 0.002 $A_{260}/$ 2 μl. To these 2-μl portions of $rC_{11}U_2C_7$, add 5 μl of 5× kinase buffer [350 mM Tris–HCl (pH 7.6), 50 mM $MgCl_2$, 25 mM dithiothreitol], 5 μl (25 μCi) of [γ-^{32}P]ATP (3000 Ci/mmol, New England Nuclear), 1 μl (10 units) of polynucleotide kinase (10,000 units/ml, New England Biolabs, Inc., Beverly, MA), and 12 μl of RNase-free H_2O. Incubate the mixture at 37° for 10 min, add 1 μl of 0.5 M EDTA, and mix. Follow this with 25 μl of 3 M sodium acetate and then 500 μl of ice-cold ethanol and cool at $-80°$ for 30 min. Centrifuge at 15,000 rpm for 10 min, remove the ethanolic supernatant, add 50 μl of RNase-free H_2O to the tube, and store the resulting RNA stock solution at $-80°$. This can be purified additionally on a Chromaspin-10 column (Clontech, Palo Alto, CA). To estimate concentration of the labeled substrate, we have assumed quantitative recovery and an equivalence of 10 A_{260} units to 1 μmol of ODN.

For RNase L-catalyzed cleavage assays, the components are added in the following order: 18 μl cleavage buffer (25 mM Tris–HCl, pH 7.4, 100 mM KCl, 10 mM $MgCl_2$, 100 μM ATP, and 10 mM DTT), 2-5A antisense chimera (2 μl at 10× the required final concentration), and RNase L solution (2 μl) to give a final concentration of 160 nM in enzyme. Concentrations as low as 2 nM can be employed. Cleavage reaction mixtures are kept on ice for 10 min after the addition of RNase L. Finally, the substrate 5'-[^{32}P]p$C_{11}U_2C_7$ (2 μl of a solution to give a final substrate concentration of 10 nM) is added last and the mixture is incubated at 37° for 15 min. RNase-free RNA loading buffer [20 μl of 0.25% (w/v) bromphenol blue, 0.25% xylene cyanol, 1 mM EDTA, 5% (v/v) glycerol] is added, and the samples (7 μl) are applied to a 1-mm 20% PAGE/8 M urea gel, which are then electrophoresed at 350 V for 6 hr at 1°. Gels are then exposed to Kodak (Rochester, NY) X-Omat film, developed, and scanned into a TIFF file using Adobe Photoshop software at 600 dpi. We have used ImageTool software (University of Texas Health Sciences Center, San Antonio, TX) to determine the background subtracted integrated density for an area including each uncleaved substrate (5'-[^{32}P]p$C_{11}U_2C_7$) and cleaved product (5'-[^{32}P]p$C_{11}U_2$). The substrate and product in each lane are measured using the same pixel area in order to correct for differences in amounts of RNA loaded in each lane. The percentage cleavage in each lane may be defined as [background subtracted integrated density of product divided by background subtracted integrated density of uncleaved substrate and cleavage product] × 100. The percentage cleavage is graphed and the concentration necessary to cause 50% cleavage of substrate is determined by extrapolation. If 2-5A is always included as an internal standard, it becomes possible to compare compounds evaluated at different times on different gels.

Figure 3 demonstrates the application of this 5'-[^{32}P]pC$_{11}$U$_2$C$_7$ cleavage assay to determine the ability of two different 2-5A antisense chimeras to activate human recombinant RNase L. NIH 351 has the structure and sequence SpA$_4$-Bu$_2$-AsAsAsAATGGGGCAAAsTsAsA, whereas NIH 489 has the structure and sequence A$_3$-Bu$_2$-AsAsAsAATGGGGCAAAsT-sAsA. NIH 351 is a 2-5A antisense chimera targeted to gene end-intergene-gene start consensus sequences in RSV genomic RNA and has been shown to be a potent inhibitor of respiratory syncytial virus replication in cell culture.[14] NIH 489 is an NIH 351 analog that bears a "disabled" 2-5A moiety, i.e., it contains only the 2-5A core trimer, which itself cannot activate RNase L. Importantly, NIH 498 is at least 100-fold less active than NIH 351 in the inhibition of RSV replication.[14] In line with these considerations, when analyzed for its ability to activate human RNase L, NIH 489 showed no trace of activity even at a concentration of 10^{-6} M, under conditions where NIH 351 affected complete RNA cleavage at 10^{-7} M.

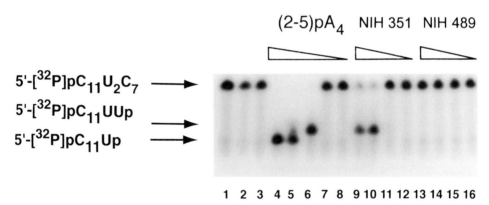

FIG. 3. Ability of 2-5A antisense chimeras to activate purified recombinant human RNase L (11). Two microliters of an appropriate dilution of 2-5A tetramer (p5'A2'p5'A2'-p5'A2'p5'A) or 2-5A antisense chimera (NIH 351 or NIH 489) was added to 14–18 μl of cleavage buffer (25 mM Tris–HCl, pH 7.4, 100 mM KCl, 10 mM MgCl$_2$, 100 M ATP, 10 mM dithiothreitol) together with 2 μl of purified recombinant RNase L (final concentration 160 nM). After 10 min of incubation on ice, the substrate 5'-[^{32}P]rC$_{11}$U$_2$C$_7$ was added (2 μl to give a final concentration of 10 nM). After 15 min at 37°, samples were subjected to 1 mm 20% PAGE–8 M urea gel electrophoresis at 350 V for 6 hr at 1°. Gels were then exposed to film, developed, and scanned (11). Lanes are identified as follows: lanes 1–3, 5'-[^{32}P]rC$_{11}$U$_2$C$_7$ plus RNase L, no 2-5A or 2-5A antisense; lanes 4–8, 2-5A tetramer at 10^{-7}, 10^{-8}, 10^{-9}, 10^{-10}, and 10^{-11} M, respectively; lanes 9–12, NIH 351 at 10^{-6}, 10^{-7}, 10^{-8}, and 10^{-9} M, respectively; and lanes 13–16, NIH 489 at 10^{-6}, 10^{-7}, 10^{-8}, and 10^{-9} M, respectively.

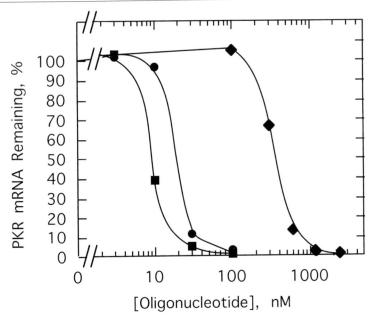

FIG. 4. Cleavage of PKR mRNA (50 nM) by RNase L (12 nM) as a function of p(A2′p5′)$_3$A (■), p(A2′p5′)$_3$A-anti-PKR (●), or p(A2′p5′)$_3$A-sense-PKR (◆) concentration. Incubations were at 37° for 15 min. Loss of intact PKR mRNA was measured by gel electrophoresis and phosphorImage analysis. Reprinted from R. K. Maitra *et al.*, *J. Biol. Chem.* **270**, 15071 (1995) with permission.

Selective Degradation of Target RNA Substrate by RNase L Activated with 2-5A Antisense Chimeras

To determine the ability of 2-5A antisense chimeras to cause targeted RNA degradation, the RNA substrate is produced from a cloned cDNA by *in vitro* transcription and then gel purified.[17,24] As an example, cDNA for the protein kinase PKR is transcribed *in vitro* with T7 RNA polymerase. PKR mRNA is labeled at its 5′ terminus with [γ-^{32}P]ATP to specific activities of 10,000–20,000 cpm/μg of RNA and is purified in 6% polyacrylamide–8 M urea gels followed by elution as described.[24]

2-5A antisense chimeras are mixed with PKR mRNA in buffer containing 25 mM Tris–HCl, pH 7.4, 10 mM magnesium acetate, 8 mM β-

[24] R. H. Silverman, A. Maran, R. K. Maitra, C. Waller, K. Lesiak, G. Li, W. Xiao, and P. F. Torrence, *in* "Antisense Technology: A Practical Approach" (C. Lichtenstein and W. Nellen, eds.), p. 127. Oxford Univ. Press, Oxford, 1997.

TABLE II
CHIMERIC 2-5A ODNs CONTAINING PKR SEQUENCES

ODN	DNA sequence
$p(2'p)_3A$-antiPKR_{19}	5′-GTA CTA CTC CCT GCT TCT G-3′
[M1]$p(2'p)_3A$-antiPKR_{19}	5′-GTA CTA CAC CCT GCT TCT G-3′
[M4]$p(2'p)_3A$-antiPKR_{19}	5′-GTA CTT CAC CCA CCT TCT G-3′
[M10]$p(2'p)_3A$-antiPKR_{19}	5′-CAA GTT CAC GCA CCA ACT G-3′
$p(2'p)_3A$-antiPKR_{15}	5′-CTA CTC CCT GCT TCT-3′
$p(2'p)_3A$-antiPKR_{12}	5′-CTA CTC CCT GCT-3′
$p(2'p)_3A$-antiPKR_9	5′-CTA CTC CCT-3′
$p(2'p)_3A$-antiPKR_6	5′-CTA CTC-3′

mercaptoethanol, and 100 mM KCl on ice for 10–15 min. Subsequently, 5 to 30 ng of RNase L is added to final volumes of 20 μl. Incubations are at 37° and are terminated with the addition of gel sample buffer. Degradation of ^{32}P-labeled RNA is measured by electrophoresis in 6% polyacrylamide–8 M urea gels, autoradiography, and imaging and quantitation is done with a PhosphorImager (Molecular Dynamics, Sunnyvale, CA).

For example, $p(A2'p5')_3A$ had an EC_{50} (the concentration of ODN required to produce a 50% reduction in intact RNA) of about 10 nM (Fig. 4). The chimeric ODN, $p5'(A2'p)_35'A$-antiPKR, $p5'(A2'p)_35'A$-Bu_2-5′-GTA CTA CTC CCT GCT TCT G-3′, containing $p(A2'p5')_3A$ attached to

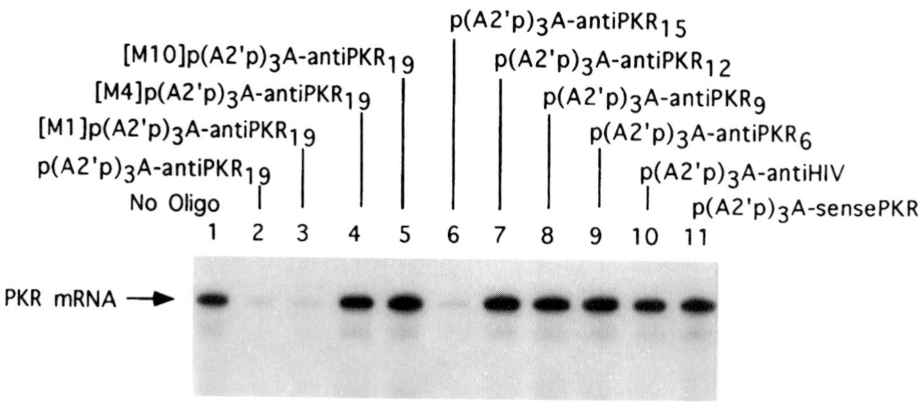

FIG. 5. Effect of ODN length and mismatched bases in 2-5A antisense on cleavage of target RNA by RNase L. Twenty-five nanomolar 5′ ^{32}P-labeled PKR mRNA was incubated with 24 nM purified human RNase L at 37° in the presence of 25 nM 2-5A antisense chimera as indicated. An autoradiogram of a dried gel is shown. Reprinted with permission from W. Xiao *et al., J. Med. Chem.* **40,** 1195 (1997). Copyright 1997 American Chemical Society.

a 19-nucleotide antisense DNA sequence against PKR mRNA (nucleotide numbers $+55$ to $+73$ relative to the start codon) had an EC_{50} of 20 nM (Fig. 4).[17] The reason for the reduced activity of chimeric 2-5A antisense molecules compared to 2-5A per se is not known, but it is believed to be due to steric hinderance in producing the active conformation of RNase L. The inhibition of general 2-5A activity by the linker/DNA domain contributes to the selectivity of the 2-5A antisense method. In contrast, $p5'(A2'p)_35'A$-sensePKR, with a DNA sequence that is complementary to the antisense sequence, possessed an EC_{50} of 400 nM (Fig. 4). These results show that targeted RNA is cleaved at much lower concentrations of ODN when chimeras contain a complementary sequence to the substrate.

To establish that a genuine antisense effect is responsible for the cleavage of mRNA, 2-5A DNA chimeras are produced that contain different numbers of mismatched bases in the DNA cassette. For example, $p(A2'p)3A$ is attached through linkers to the anti-PKR sequence or to the corresponding sequences with 1, 4, or 10 mismatches (Table II).[25] Chimeric ODN with no mismatches or with one mismatch cause the degradation of the target, PKR mRNA. In contrast, chimeras containing 4 or 10 mismatched bases do not cause PKR mRNA degradation (Fig. 5, lanes 2–5).

To determine the optimal antisense length, a panel of 2-5A antisense molecules of different lengths is produced. For instance, 2-5A is attached through linkers to anti-PKR sequences that are 19, 15, 12, 9, and 6 nucleotides in length (Table II and Fig. 5, lanes 2 and 6–9).[25] Chimeras with 19- or 15-nucleotide antisense domains are functional, whereas shorter antisense sequences fail to direct cleavage of PKR mRNA by RNase L. Similarly, 2-5A linked to noncomplementary sequences did not result in cleavage of the target mRNA (Fig. 5, lanes 10 and 11). A close correlation is observed between the activity of chimeric ODNs and their affinity (T_m) for PKR mRNA.[25]

Acknowledgments

Some of these investigations were supported by U.S. Public Health Service Grant 1 PO1 CA 62220 and CA44059 awarded by the Department of Health and Human Services, National Cancer Institute (to R.H.S.), by funds from Atlantic Pharmaceuticals, Inc. (to the Cleveland Clinic Foundation), and by a Cooperative Research and Development Agreement between the National Institute of Health and Atlantic Pharmaceuticals, Inc. (Raleigh, NC) (to P.F.T.).

[25] W. Xiao, G. Li, R. K. Maitra, A. Maran, R. H. Silverman, and P. F. Torrence, *J. Med. Chem.* **40**, 1195 (1997).

Author Index

Buddecke, E., 268–284
Bulinski, J. C., 371(32)
Bunzow, J. R., 56(41)
Buonomo, S. B., 412(47)
Burch, R., 30(138)
Burg, R., 176(6)
Burger, P. C., 371(27)
Burgess, T., 48(20)
Burgess, T. L., 27(122), 279(43)
Burke, J. M., 404(32, 34, 35), 418(80)
Burzynski, J., 205(22)
Busch, H., 87(44)
Busch, R. K., 87(44)
Busen, W., 116(49), 200(19)
Butcher, S. E., 418(80)
Butler, M., 25(113)
Buzby, J. S., 388–397

C

Cabanillas, F., 378(10)
Cadilla, R., 45(190), 156(7)
Cagnon, L., 412(48)
Cahan, M. A., 371(27, 28)
Cai, H., 26(116)
Cai, T., 289(23)
Cai, T.-Q., 269(11), 342(9)
Cairo, M. S., 389(24), 390(28), 397
Calabretta, B., 7(19), 359(10), 389(15)
Calvo, F., 90(49, 50), 95
Cambell, T. B., 405(37)
Campbell, J. M., 21(98), 50(28), 362(14)
Campbell, J. R., 22(102)
Canti, G. F., 270(24)
Cantor, C. R., 170(10), 248(59)
Cantrill, H. L., 32(146)
Cao, L., 269(11), 289(23), 342(9)
Cao, Q., 77(17)
Cao, X., 343(13)
Capaccioli, S., 22(106)
Caparros, T., 268
Capitini, C., 258(8)
Capony, J. P., 205(16)
Carbon, P., 412(46)
Carbonnelle, C., 401(6)
Carcia, R., 267(39)
Carleton, A., 143–156
Carlson, R., 209(29, 30)
Carlsson, S., 268

Carlsson, S. R., 261(27)
Carney, D. H., 371(24)
Carpentier, A., 458(16)
Carroll, S. S., 522(7), 528(22)
Carter, S. G., 45(190)
Caruthers, M. H., 229(19), 236(34), 247(56), 264(34), 272(36), 345(19), 520(34)
Carvalho, D. A., 237(38)
Casaroli-Marano, R. P., 267(39)
Casero, R. A., 354(39)
Casey, S., 162(45)
Casper, M. D., 14(50)
Cassella, A., 32(148)
Castanotto, D., 401–420
Castel, S., 266(38), 268
Castier, Y., 458(16)
Castro, M. M., 39(171)
Catsicas, M., 29(136)
Catsicas, S., 29(136)
Cauchon, G., 22(107), 327(G)
Cayley, P. J., 524(19)
Cazenave, C., 14(57), 116(49), 200(19), 360(11), 420(5), 436(3), 468(9)
Cech, T. R., 402(7, 15–18), 443(2), 447(16), 485(1), 489(5)
Cecil, P. K., 53(34)
Celis, J. E., 261(24)
Ceruzzi, M., 258(14)
Chakel, J. A., 212(32)
Chalfie, M., 181(17)
Challita, P. M, 416(69)
Chan, H., 7(17), 22(104, 105), 49(25), 95(2), 126(56), 258(4), 280(45), 348(25)
Chan, H. W., 348(26), 390(27), 503(11)
Chan, P. P., 74(3)
Chang, A., 523(10)
Chang, L., 390(28)
Charnay, P., 308(22)
Charneau, P., 77(20)
Chassaing, G., 158(22), 355(46, 47)
Chassignol, M., 79(32)
Chatterjee, S., 401(6)
Chaudhary, N., 77(25), 279(38, 39)
Chavany, C., 57(3)
Chemla, E., 458(16)
Chen, C. J., 401(5)
Chen, E., 528(22)
Chen, J., 413(60)
Chen, J. K., 76(10)
Chen, L. L., 355(44, 45)

H

Koster, H., 264(33)
Koszka, C., 363(19, 21)
Kothavale, A., 77(24), 158(24), 178(12)
Koziolkiewicz, M., 289(24)
Krajewski, S., 278(*F*), 281(47)
Krall, W. J., 416(70)
Kramer, T. B., 27(122), 279(43)
Krangel, M. S., 413(55)
Kraszewski, A., 230(21)
Kravtzoff, R., 258(6, 7)
Kreis, T., 363(16)
Krepler, R., 363(21)
Kretzschmar, G., 278(*A*), 279(40, 41)
Krieg, A., 185(20)
Krieg, A. M., 33(154), 160(26), 288(11, 13), 280(22), 520(34)
Krieg, P. A., 440(23)
Krikos, A. J., 410(44)
Krinke, L., 467(8)
Krishnamurthy, R., 44(188)
Kroczek, R. A., 110(39)
Krol, A., 412(46)
Krotz, A., 223(39)
Krug, A., 20(93)
Kruger, K., 443(2), 485(1)
Krzyzanowska, B. K., 228(11, 12)
Kubert, M., 41(176), 258(10), 340(38)
Kuiper, J., 315(32), 342(10)
Kulka, M., 8(22), 342(5)
Kumada, M., 28(127)
Kumar, A., 523(12)
Kumar, S., 165(3)
Kume, A., 232(24)
Kuo, L. C., 522(7), 528(22)
Kuppusamy, P., 354(39)
Kuramoto, E., 32(151)
Kurokawa, K., 28(127)
Kurosawa, K., 476(21)
Kutyavin, I., 164–173
Kwon, Y. K., 461(24)
Kyatkina, N., 94(66)
Kyle, E., 269(8)

L

Labrousse, V., 77(20)
Lacoste, J., 74–95
Lacroix, L., 74–95
Lacy, B. W., 157(15)

Lade, B. N., 447(20)
Laemmli, U. K., 413(59), 473(15)
Lambert, G., 258(5)
Lammineur, C., 79(29)
Lamond, A. I., 14(51), 43(184, 185), 270(26), 508(16)
Lamond, A. T., 317(39)
Lampson, B., 477(23)
Lan, N., 403(21)
Lane, C. L., 232(25)
Lane, D. P., 417(72)
Lanfrancone, L., 417(74)
Langel, U., 157(14)
Langer, R., 28(126)
Langkopf, A., 363(22)
Langner, D., 271(32)
Lanier, M., 350(29)
Lapalainen, K., 262(30)
Larsen, J. H., 162(40)
Larsson, C., 205(18)
Lau, T. T., 442(27)
Laughlin, C., 359(6)
Lavignon, M., 327(Q)
Le Coutre, P., 157(10, 11)
Leal-Pinto, E., 342(11)
Leamon, C. P., 298(5)
Lebleu, B., 4(45), 6(14), 9(28), 14(53), 15(65), 17(77), 18(79), 22(100, 101), 26(117), 27(118), 38(162), 39(169), 158(23), 189–203, 247(57), 327(*A, R*), 343(12), 351(30), 459(20), 508(14, 16)
Leder, P., 181(18), 389(14)
Lederman, S., 269(7)
Lee, C., 209(30)
Lee, K. B., 321
Lee, N. S., 401(6), 410(44)
Lee, R., 151(13), 161(31)
Lee, R. J., 22(107), 327(*G*)
Lee, R. T., 300(9), 321
Lee, S. W., 403(19, 20, 21)
Lee, T. C., 416(68)
Lee, X., 524(18)
Lee, Y. C., 299(8), 300(9), 302(14), 303(18), 313(30)
Leeds, R., 4(8), 23(108a), 35(157, 160, 161), 269(13)
Leeper, N., 180(14)
Lefebvre, I., 459(20)
Lefebvre, J. C., 412(48)
Leffet, L. M., 202(23)

Subject Index

A

ISBN 0-12-182214-1

90038

9 780121 822149